SELECTED CONVERSION FACTORS TO STANDARD INTERNATIONAL UNITS

To Convert From	To	Multiply By
inch	meter (m)	2.54×10^{-2}
inch^2	meter^2 (m^2)	6.452×10^{-4}
inch of mercury (32F)	pascal (Pa)	3.386×10^3
kilocalorie	joule (J)	4.184×10^3
kilogram-force	newton (N)	9.807
kilogram-force-meter	newton-meter (N-m)	9.807
kilogram-force/centimeter^2	pascal (Pa)	9.807×10^4
kilogram-force/meter^2	pascal (Pa)	9.807
kilogram-force/millimeter^2	pascal (Pa)	9.807×10^6
kip (1000 pounds)	newton (N)	4.448×10^3
kip/inch^2 (ksi)	pascal (Pa)	6.895×10^6
kip/inch$^2 \cdot \sqrt{\text{inch}}$ (ksi$\sqrt{\text{in.}}$)	pascal$\cdot \sqrt{m}$ (Pa$\sqrt{m}$)	1.099×10^6
mil	meter (m)	2.54×10^{-5}
millimeter of mercury (mm Hg)	pascal (Pa)	1.333×10^2
poise	pascal-second (Pa-s)	1×10^{-1}
pound-force	newton (N)	4.448
pound-force/inch^2(psi)	pascal (Pa)	6.895×10^3
pound-force/inch$^2 \cdot \sqrt{\text{inch}}$ (psi$\sqrt{\text{in.}}$)	pascal$\cdot \sqrt{m}$ (Pa$\sqrt{m}$)	1.099×10^3
torr [mm Hg, (0 C)]	pascal (Pa)	1.333×10^2

DEFORMATION
AND FRACTURE MECHANICS

DEFORMATION AND FRACTURE MECHANICS OF ENGINEERING MATERIALS

RICHARD W. HERTZBERG
Professor, Metallurgy and Materials Science
and Director,
Mechanical Behavior Laboratory,
Materials Research Center Lehigh University

JOHN WILEY & SONS
New York • Santa Barbara • London • Sydney • Toronto

Library of Congress Cataloging in Publication Data:

Hertzberg, Richard W   1937–
    Deformation and fracture mechanics of engineering materials.

    Bibliography: p.
    Includes index.
    1. Deformation (Mechanics)   2. Fracture of solids.
I. Title.
TA417.6.H46   620.1'123   76-10812
ISBN 0-471-37385-0

Printed in the United States of America

10 9 8 7 6 5 4 3 2

To my wife Linda, and my children Michelle and Jason

PREFACE

This book discusses macroscopic and microscopic aspects of the mechanical behavior of metals, ceramics, and polymers and emphasizes recent developments in materials science and fracture mechanics. The material is suitable for advanced undergraduate courses in metallurgy and materials, mechanical engineering, and civil engineering where a combined materials-fracture mechanics approach is stressed. The book also will be useful to working engineers who want to learn more about mechanical metallurgy and, particularly the fracture-mechanics approach to the fracture of solids. I have assumed that readers have had previous training in strength of materials and basic calculus, and that they have been introduced to metallurgical principles including crystal structure.

My objective is to make the reader aware of several viewpoints held by engineers and materials scientists who are active in the field of mechanical metallurgy: the crystal physics approach and the role of dislocations in controlling mechanical properties; the classical metallurgical approach, which stresses the relationship between microstructure and properties; and the fracture mechanics approach which describes the relationship between material toughness, design stress, and allowable flaw size. The application of each viewpoint in the analysis of certain mechanical responses of solids is illustrated, with the hope that the reader will soon recognize which approach might best explain a given set of data or a particular service failure. I think that unless a proper perspective is gained regarding the limits of applicability of the atomistic, microstructural, and continuum viewpoints the reader will become too involved in the fine points of a concept that may prove to be irrelevant to the problem at hand. Indeed, this is vital to a successful failure analysis.

The book is divided into two sections. Section One is devoted to a study of the deformation of solids. Here, emphasis is placed on the role of microstructure, crystallography, and dislocations in explaining material behavior. Section Two, the larger section, deals with the application of fracture mechanics principles to the subject of the fracture in solids. Although familiarity with some topics discussed in Section One will be useful to the reader, the information is not critical to an understanding of Section Two. Therefore, the reader who wishes to focus on the subject of fracture can proceed from the introductory chapter on tensile behavior of solids (Chapter One) directly to Section Two.

Chapter One examines the different macroscopic mechanical responses of metals, ceramics, and polymers in relation to their respective tensile stress-strain response. Chapters Two to Five constitute a closely related unit that deals with the deformation of crystalline solids. The elements of dislocation theory, discussed in Chapter Two, are source material for the discussion of slip and structure-property relationships in Chapters Three to Five. A detailed treatment of the crystallography of twinning is presented in Chapter Four with an analysis of the cold-worked structure of crystalline solids in terms of mechanical fibering and preferred crystallographic orientations. To acquaint the reader with the multidisciplinary character of time-dependent, high-temperature creep processes in crystalline solids, the topics discussed in Chapter Five include: empirical creep strain relationships with time, temperature, and stress; parametric time-temperature relationships, such as the Larson-Miller parameter used in engineering materials design; and evaluation of creep strain dependence on such material properties as diffusivity, melting point, activation energy, grain size, crystal structure, and elastic modulus. Superplasticity and deformation mechanism maps are also considered. Section One concludes with a discussion of deformation in polymeric materials. Here, again, the mechanical response of these materials is discussed both in terms of their continuum response (as described, for example, with linear viscoelastic relationships and mathematical analogs) and in terms of materials science considerations, involving such topics as the effect of structure on energy damping spectra and the micromechanisms of deformation in amorphous and crystalline polymers.

The subject of fracture is introduced in Section Two by a general overview, ranging from the continuum studies of Leonardo da Vinci in the fifteenth century to current fractographic examinations that employ sophisticated transmission and scanning electron microscopes. The importance of the stress intensity factor and the fracture mechanics approach in analyzing the fracture of solids is developed in Chapter Eight and is compared with the older transition temperature approach to engineering design (Chapter Nine). From this macroscopic viewpoint, the emphasis shifts in Chapters Ten and Eleven to a consideration of the role of microstructural variables in determining material fracture toughness and embrittlement susceptibility. Both environmental embrittlement (such as stress corrosion cracking and liquid metal and hydrogen embrittlement) and intrinsic material embrittlement (such as temper, irradiation, and 300°C embrittlement) are described. The fatigue of solids is discussed at length in Chapters Twelve and Thirteen, and cyclic stress life, cyclic strain life, fatigue crack propagation philosophies, and test data are given. In the final chapter, actual service failures are examined to demonstrate the importance of applying fracture mechanics principles in failure analysis. Several bridge, aircraft, and generator rotor shaft failures are analyzed. In addition, a checklist of information needed to best analyze a service failure is provided for use by the reader. This final chapter can be studied as a unit or as a source for specific case histories that may be considered when a particular point is introduced in an earlier chapter.

A number of scientific colleagues and former students provided valuable assistance in the planning and preparation of the book. Since a complete listing of them would be too lengthy and vulnerable to inadvertent omissions, they cannot be cited individually. I thank those who provided original prints of their previously published photographs that enhance the technical quality of this book. I am grateful to my colleagues at Lehigh University for their many contributions and, most especially, to: P. C. Paris, D. A. Thomas, Y. T. Chou, J. A. Manson, M. R. Notis, N. Zettlemoyer, T. Smith, W. J. Mills, S. Siegler, B. Hayes, W. Walthier, and M. Skibo. The considerable care and exactness shown by Mrs. L. Valkenburg in typing the manuscript is deeply appreciated. I also thank the Alcoa Foundation and the Department of Metallurgy and Materials Science at Lehigh University for their financial support during the preparation of this manuscript.

Finally, this volume, which is the most significant project in my teaching career, could not have been attempted or completed without the understanding and patience of my wife, Linda, and my children, Michelle Ilyce and Jason Lyle. Their sacrifices were great; my gratitude is profound.

Richard W. Hertzberg

Allentown, Pennsylvania
July, 1976

CONTENTS

DEFORMATION
AND FRACTURE MECHANICS

SECTION
ONE
DEFORMATION
OF ENGINEERING
MATERIALS

CHAPTER
ONE
TENSILE RESPONSE
OF MATERIALS

The tensile test is the experimental test method most widely employed to characterize the mechanical properties of materials. From any complete test record, one can obtain important information concerning the material's elastic properties, the character and extent of plastic deformation, yield and tensile strengths, and toughness. That so much information can be obtained from one test justifies its extensive use in engineering materials research. To provide a framework for the varied response to tensile loading in load-bearing materials, several stress-strain plots reflecting different deformation characteristics will be introduced in this chapter.

1.1 DEFINITION OF STRESS AND STRAIN

Before discussing engineering material stress-strain response, it is appropriate to define the terms, stress and strain. This may be done in two generally accepted forms. The first definitions, used extensively in engineering practice, are

$$\sigma_{eng} = \text{engineering stress} = \frac{\text{load}}{\text{initial cross-sectional area}} = \frac{P}{A_0} \qquad (1\text{-}1)$$

$$\epsilon_{eng} = \text{engineering strain} = \frac{\text{change in length}}{\text{initial length}} = \frac{l_f - l_0}{l_0} \qquad (1\text{-}2)$$

where

$$l_f = \text{final gage length}$$

$$l_0 = \text{initial gage length}$$

Alternately, stress and strain may be defined by

$$\sigma_{\text{true}} = \text{true stress} = \frac{\text{load}}{\text{instantaneous cross-sectional area}} = \frac{P}{A_i} \qquad (1\text{-}3)$$

$$\epsilon_{\text{true}} = \text{true strain} = \ln \frac{\text{final length}}{\text{initial length}} = \ln \frac{l_f}{l_0} \qquad (1\text{-}4)$$

The fundamental distinction concerning the definitions for true stress and strain is recognition of the interrelation between gage length and diameter changes associated with plastic deformation. That is, since plastic deformation is a constant-volume process such that

$$A_1 l_1 = A_2 l_2 = \text{constant} \qquad (1\text{-}5)$$

any extension of the original gage length would produce a corresponding contraction of the gage diameter. For example, if a 25-mm (1 in.)*-long sample were to have extended uniformly by 2.5 mm owing to a tensile load P, the real or *true* stress would have to be higher than that computed by the *engineering* stress formulation. Since $l_2 / l_1 = 1.1$, from Eq. 1-5 $A_1 / A_2 = 1.1$, so that $A_2 = A_1 / 1.1$. The *true* stress is then shown to be $\sigma_{\text{true}} = 1.1 P / A_1$ and is larger than the *engineering* value.

The need to define true strain as in Eq. 1-4 stems from the fact that the actual strain at any given time depends on the instantaneous gage length l_i. Consequently, a fixed Δl displacement will result in a decreasing amount of incremental strain, since the gage length at any given time l_i, will increase with each additional Δl increment. Furthermore, it should be possible to define the strain given to a rod by considering the total change in length of the rod as having taken place in either one step or any number of discrete steps. Stated mathematically, $\sum_n \epsilon_n = \epsilon_T$. As a simple example, take the case of a wire drawn in two steps with an intermediate annealing treatment. On the basis of *engineering* strain, the two deformation strains would be $(l_1 - l_0) / l_0$ and $(l_2 - l_1) / l_1$. Adding these two increments does *not* yield a final strain of $(l_2 - l_0) / l_0$. On the other hand, a summation of *true* strains does lead to the correct result. Therefore

$$\ln \frac{l_1}{l_0} + \ln \frac{l_2}{l_1} = \ln \frac{l_2}{l_0} = \epsilon_{\text{true total}}$$

*To convert from inches to millimeters, multiply by 25.4.

EXAMPLE 1.1 A 25-cm(10 in.)*-long rod with a diameter of 0.25 cm is loaded with a 4500-newton (1012 lb)† weight. If the diameter decreases to 0.22 cm, compute the following:

(a) The final length of the rod:

Since $A_1 l_1 = A_2 l_2$ (from Eq. 1-5)

$$l_2 = \frac{A_1}{A_2} l_1 = \frac{\frac{\pi}{4}(0.25)^2}{\frac{\pi}{4}(0.22)^2} (25)$$

$$l_2 = 32.3 \text{ cm}$$

(b) The true stress and true strain at this load:

$$\sigma_{true} = \frac{P}{A_i}$$

$$= \frac{4500}{(\pi/4)(2.2 \times 10^{-3})^2}$$

$$\sigma_{true} = 1185 \text{ MPa } (172{,}000 \text{ psi})^\ddagger$$

$$\epsilon_{true} = \ln \frac{l_f}{l_0}$$

$$= \ln \frac{32.3}{25}$$

$$\epsilon_{true} = 0.256 \text{ or } 25.6\%$$

(c) The engineering stress and strain at this load:

$$\sigma_{eng} = \frac{P}{A_0}$$

$$= \frac{4500}{\frac{\pi}{4}(2.5 \times 10^{-3})^2}$$

$$\sigma_{eng} = 917 \text{ MPa}$$

$$\epsilon_{eng} = \frac{l_f - l_0}{l_0}$$

$$= \frac{32.3 - 25}{25}$$

$$\epsilon_{eng} = 0.292 \text{ or } 29.2\%$$

*To convert from inches to centimeters, multiply by 2.54.
†To convert from pounds to newtons, multiply by 4.448.
‡To convert from psi to pascal, multiply by 6.895×10^3.

The use of true strains offers an additional convenience when considering the constant-volume plastic deformation process in that $\epsilon_x + \epsilon_y + \epsilon_z = 0$. In contrast, we find a less convenient relationship, $(1 + \epsilon_x)(1 + \epsilon_y)(1 + \epsilon_z) = 1$, for the case of engineering strains.

1.2 STRESS-STRAIN CURVES

1.2.1 Elastic Response: Type I

Almost 300 years ago Robert Hooke reported in his classic paper, "Of Spring" the following observations:[1]

"Take a wire string of 20 or 30 or 40 feet long and fasten the upper part ... to a nail, and to the other end fasten a scale to receive the weights. Then with a pair of compasses [measure] the distance [from] the bottom of the scale [to] the ground or floor beneath. Then put ... weights into the ... scale and measure the several stretchings of the said string and set them down. Then compare the several stretchings of the ... string and you will find that they will always bear the same proportions one to the other that the weights do that made them."

This observation may be described mathematically by the equation for an elastic spring:

$$F = kx \qquad (1\text{-}6)$$

where

$$F = \text{applied force}$$

$$x = \text{associated displacement}$$

$$k = \text{proportionality factor often referred to}$$

$$\text{as the spring constant}$$

When the force acts on a cross-sectional area A and the displacement x related to some reference gage length l, Eq. 1-6 may be rewritten as

$$\sigma = E\epsilon \qquad (1\text{-}7)$$

where

$$\sigma = F/A = \text{stress}$$

$$\epsilon = x/l = \text{strain}$$

$$E = \text{proportionality constant (often referred to as}$$

$$\text{Young's modulus or the modulus of elasticity)}$$

Equation 1-7—called Hooke's law—describes a material condition where stresses and strains are proportional to one another, leading to a stress-strain response shown in Fig. 1.1. The modulus of elasticity for several materials is given in Table 1.1. Since E depends on the strength of the interatomic forces that vary with the type of bonding found in a given material, it is relatively insensitive to changes in microstructure. As a result, while heat treatment and minor alloying additions may cause the strength of a steel alloy to change from 210 to 2400 MPa, the modulus of elasticity of both materials remains relatively unchanged—about 200 to 210 GPa.

The modulus of elasticity of metals and ceramics will decrease with increasing temperature. This is related to the fact that the modulus varies inversely with the distance of atom or ion separation raised to the fourth power or more. Since the equilibrium distance of atom separation increases with temperature (i.e., materials expand when heated), the modulus will decrease. The loss of material stiffness with increasing temperature is gradual with only a few percent decrement occurring for a 100°C (180°F) temperature change.

It is found that if the loads are removed from the tensile sample before the point of fracture, the corresponding strain would retrace itself along the same linear plot back to zero. The *reversible* nature of strain in this portion of the σ-ϵ curve is a basic element of elastic strains in any material, whether it is capable of much larger total strain or not. When a material is characterized by such a stress-strain curve and exhibits no plastic deformation, there is great concern for its ability to resist brittle (low energy) premature fracture. This point is treated extensively in Chapters Seven, Eight, and Nine. Typical materials that behave in this manner include glasses, rocks, many ceramics, heavily cross-linked polymers, and some metals at low temperature. Although these materials are not suitable for engineering applications involving tensile loading, they may be used with considerable success in situations involving compression loads for which the material exhibits much greater resistance to fracture. It is not uncommon to

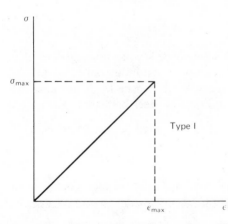

FIGURE 1.1 Type I stress-strain behavior revealing completely elastic material response.

TABLE 1.1a Elastic Properties of Engineering Materials[a]

Material at 20°C	E (GPa)	G (GPa)	v
Metals			
Aluminum	70.3	26.1	0.345
Cadmium	49.9	19.2	0.300
Chromium	279.1	115.4	0.210
Copper	129.8	48.3	0.343
Gold	78.0	27.0	0.44
Iron	211.4	81.6	0.293
Magnesium	44.7	17.3	0.291
Nickel	199.5	76.0	0.312
Niobium	104.9	37.5	0.397
Silver	82.7	30.3	0.367
Tantalum	185.7	69.2	0.342
Titanium	115.7	43.8	0.321
Tungsten	411.0	160.6	0.280
Vanadium	127.6	46.7	0.365
Other Materials			
Aluminum oxide (fully dense)	~415	—	—
Diamond	~965	—	—
Glass (heavy flint)	80.1	31.5	0.27
Nylon 66	1.2–2.9	—	—
Polycarbonate	2.4	—	—
Polyethylene (high density)	0.4–1.3	—	—
Poly(methyl methacrylate)	2.4–3.4	—	—
Polypropylene	1.1–1.6	—	—
Polystyrene	2.7–4.2	—	—
Quartz (fused)	73.1	31.2	0.170
Silicon carbide	~470	—	—
Tungsten carbide	534.4	219.0	0.22

[a]G. W. C. Kaye and T. H. Laby, *Tables of Physical and Chemical Constants*, 14th Ed., Longman Group Limited, London, 1973, p. 31.

find the compressive strength of a brittle solid to be several times greater than the tensile value. Concrete is an excellent example of an industrial material used extensively in compression but not in tension. When tensile loads are unavoidable, the concrete is reinforced by the addition of steel bars that assume the tensile stresses.

1.2.1.1 GENERALIZED HOOKE'S LAW

Hooke's law can be generalized to account for multiaxial loading conditions as well as material anisotropy. Regarding the former, the reader should recall from

TABLE 1.1b Elastic Properties of Engineering Materials[a]

Material at 68°F	E (10^6 psi)	G (10^6 psi)	ν
Metals			
Aluminum	10.2	3.8	0.345
Cadmium	7.2	2.8	0.300
Chromium	40.5	16.7	0.210
Copper	18.8	7.0	0.343
Gold	11.3	3.9	0.44
Iron	30.6	11.8	0.293
Magnesium	6.5	2.5	0.291
Nickel	28.9	11.0	0.312
Niobium	15.2	5.4	0.397
Silver	12.0	4.4	0.367
Tantalum	26.9	10.0	0.342
Titanium	16.8	6.35	0.321
Tungsten	59.6	23.3	0.280
Vanadium	18.5	6.8	0.365
Other Materials			
Aluminum oxide (fully dense)	~60	—	—
Diamond	~140	—	—
Glass (heavy flint)	11.6	4.6	0.27
Nylon 66	0.17	—	—
Polycarbonate	0.35	—	—
Polyethylene (high density)	0.058–0.19	—	—
Poly(methyl methacrylate)	0.35–0.49	—	—
Polypropylene	0.16–0.39	—	—
Polystyrene	0.39–0.61	—	—
Quartz (fused)	10.6	4.5	0.170
Silicon carbide	~68	—	—
Tungsten carbide	77.5	31.8	0.22

[a]G. W. C. Kaye and T. H. Laby, *Tables of Physical and Chemical Constants*, 14th Ed., Longman Group Limited, London, 1973, p. 31.

his studies of the strength of materials that a stress in one direction (say the Y direction) will cause not only a strain in the Y direction but in the X and Z directions as well. Hence

$$\epsilon_{yy} = \frac{\sigma_{yy}}{E} \qquad (1\text{-}8a)$$

$$\epsilon_{xx} = \epsilon_{zz} = -\frac{\nu\sigma_{yy}}{E} \qquad (1.8b)$$

where

$$\sigma_{yy} = \text{stress acting normal to } Y \text{ plane and in } Y \text{ direction}$$

$$\epsilon_{xx}, \epsilon_{yy}, \epsilon_{zz} = \text{corresponding strains in orthogonal directions}$$

$$\nu = \text{Poisson's ratio} \left(= -\frac{\epsilon_{xx}}{\epsilon_{yy}} \right)$$

$$E = \text{modulus of elasticity}$$

From Fig. 1.2, typical normal and shear strain components may be given by

$$\epsilon_{xx} = \frac{\partial u}{\partial x} \tag{1-9a}$$

$$\epsilon_{yy} = \frac{\partial v}{\partial y} \tag{1-9b}$$

$$\gamma_{xy} = \tan\alpha + \tan\beta = \frac{\partial v}{\partial x} + \frac{\partial u}{\partial y} \tag{1-9c}$$

with the other normal and shear strains defined in similar fashion. When

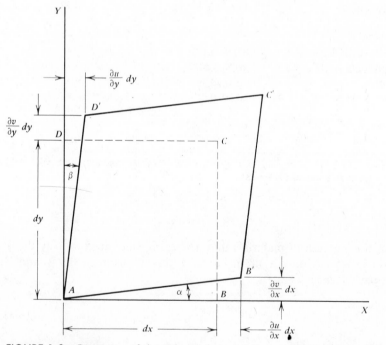

FIGURE 1.2 Distortion of the Z face of a cubical element. The dashed lines indicate the unstrained position of the cube.

multiaxial stresses are applied, the total strain in any given direction is the sum of all strains resulting from each normal and shear stress component. For the case of an isotropic material:

$$\epsilon_{xx} = \frac{\sigma_{xx} - \nu(\sigma_{yy} + \sigma_{zz})}{E}$$

$$\epsilon_{yy} = \frac{\sigma_{yy} - \nu(\sigma_{xx} + \sigma_{zz})}{E}$$

$$\epsilon_{zz} = \frac{\sigma_{zz} - \nu(\sigma_{xx} + \sigma_{yy})}{E}$$

$$\gamma_{xy} = \frac{\tau_{xy}}{G} \qquad (1\text{-}10)$$

$$\gamma_{yz} = \frac{\tau_{yz}}{G}$$

$$\gamma_{xz} = \frac{\tau_{xz}}{G}$$

where

τ_{ij} = stress acting on I plane and in J direction

G = shear modulus

The situation is complicated greatly when the material is anisotropic wherein the elastic constants vary as a function of crystallographic orientation. Since this is the case for practically all crystalline solids, it is important to consider the general loading condition shown in Fig. 1.3. We see that there are three normal

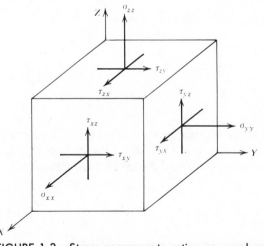

FIGURE 1.3   Stress components acting on a volume element.

and six shear stress components. However, since $\tau_{yx} = \tau_{xy}$, $\tau_{yz} = \tau_{zy}$, and $\tau_{xz} = \tau_{zx}$ (so as to avoid rotation of the cube), only six independent stress components remain that determine the strains of the body. The strains in each direction may be given by

$$
\begin{aligned}
\epsilon_{xx} &= s_{11}\sigma_{xx} + s_{12}\sigma_{yy} + s_{13}\sigma_{zz} + s_{14}\tau_{yz} + s_{15}\tau_{zx} + s_{16}\tau_{xy} \\
\epsilon_{yy} &= s_{21}\sigma_{xx} + s_{22}\sigma_{yy} + s_{23}\sigma_{zz} + s_{24}\tau_{yz} + s_{25}\tau_{zx} + s_{26}\tau_{xy} \\
\epsilon_{zz} &= s_{31}\sigma_{xx} + s_{32}\sigma_{yy} + s_{33}\sigma_{zz} + s_{34}\tau_{yz} + s_{35}\tau_{zx} + s_{36}\tau_{xy} \\
\epsilon_{xy} &= s_{41}\sigma_{xx} + s_{42}\sigma_{yy} + s_{43}\sigma_{zz} + s_{44}\tau_{yz} + s_{45}\tau_{zx} + s_{46}\tau_{xy} \\
\epsilon_{yz} &= s_{51}\sigma_{xx} + s_{52}\sigma_{yy} + s_{53}\sigma_{zz} + s_{54}\tau_{yz} + s_{55}\tau_{zx} + s_{56}\tau_{xy} \\
\epsilon_{zx} &= s_{61}\sigma_{xx} + s_{62}\sigma_{yy} + s_{63}\sigma_{zz} + s_{64}\tau_{yz} + s_{65}\tau_{zx} + s_{66}\tau_{xy}
\end{aligned}
\tag{1-11}
$$

where $s_{ij} =$ elastic compliances. Solving for stresses instead, we have

$$
\begin{aligned}
\sigma_{xx} &= c_{11}\epsilon_{xx} + c_{12}\epsilon_{yy} + c_{13}\epsilon_{zz} + c_{14}\gamma_{yz} + c_{15}\gamma_{zx} + c_{16}\gamma_{xy} \\
\sigma_{yy} &= c_{21}\epsilon_{xx} + c_{22}\epsilon_{yy} + c_{23}\epsilon_{zz} + c_{24}\gamma_{yz} + c_{25}\gamma_{zx} + c_{26}\gamma_{xy} \\
\sigma_{zz} &= c_{31}\epsilon_{xx} + c_{32}\epsilon_{yy} + c_{33}\epsilon_{zz} + c_{34}\gamma_{yz} + c_{35}\gamma_{zx} + c_{36}\gamma_{xy} \\
\tau_{yz} &= c_{41}\epsilon_{xx} + c_{42}\epsilon_{yy} + c_{43}\epsilon_{zz} + c_{44}\gamma_{yz} + c_{45}\gamma_{zx} + c_{46}\gamma_{xy} \\
\tau_{zx} &= c_{51}\epsilon_{xx} + c_{52}\epsilon_{yy} + c_{53}\epsilon_{zz} + c_{54}\gamma_{yz} + c_{55}\gamma_{zx} + c_{56}\gamma_{xy} \\
\tau_{xy} &= c_{61}\epsilon_{xx} + c_{62}\epsilon_{yy} + c_{63}\epsilon_{zz} + c_{64}\gamma_{yz} + c_{65}\gamma_{zx} + c_{66}\gamma_{xy}
\end{aligned}
\tag{1-12}
$$

where $c_{ij} =$ elastic constants.

The reversibility of elastic strains leads to the fact that $s_{ij} = s_{ji}$ and $c_{ij} = c_{ji}$, which reduces the number of independent material constants from 36 to 21. As a result of symmetry considerations, the number of independent constants decreases further, with nine constants required to describe the elastic response of an orthorhombic crystal, five for hexagonal, and only three for cubic crystals. For the latter, the elastic compliance matrix reduces to

$$
s_{ij} =
\begin{bmatrix}
s_{11} & s_{12} & s_{12} & 0 & 0 & 0 \\
s_{12} & s_{11} & s_{12} & 0 & 0 & 0 \\
s_{12} & s_{12} & s_{11} & 0 & 0 & 0 \\
0 & 0 & 0 & s_{44} & 0 & 0 \\
0 & 0 & 0 & 0 & s_{44} & 0 \\
0 & 0 & 0 & 0 & 0 & s_{44}
\end{bmatrix}
\tag{1-13}
$$

It can be shown for the case of cubic crystals that the modulus of elasticity in any given direction may be given by Eq. 1-14 in terms of these three independent elastic constants and the direction cosines of the crystallographic direction under study:

$$
\frac{1}{E} = s_{11} - 2\left[(s_{11} - s_{12}) - \frac{1}{2}s_{44}\right]\left(l_1^2 l_2^2 + l_2^2 l_3^2 + l_1^2 l_3^2\right)
\tag{1-14}
$$

where $l_1, l_2, l_3 =$ direction cosines. Note that the elastic modulus for a given cubic material depends only on the magnitude of the direction cosines, with values for the principal crystallographic directions in the cubic lattice being given in Table 1.2.

TABLE 1.2 Direction Cosines for Principal Directions in Cubic Lattice

Direction	l_1	l_2	l_3
$\langle 100 \rangle$	1	0	0
$\langle 110 \rangle$	$1/\sqrt{2}$	$1/\sqrt{2}$	0
$\langle 111 \rangle$	$1/\sqrt{3}$	$1/\sqrt{3}$	$1/\sqrt{3}$

For example, the modulus in the $\langle 100 \rangle$ direction is given by $1/s_{11}$, since $\sum l_i^2 l_j^2 = 0$. By comparison, $\sum l_i^2 l_j^2 = \frac{1}{3}$ (the maximum value) in the $\langle 111 \rangle$ direction so that $1/E = s_{11} - \frac{2}{3}[(s_{11} - s_{12}) - \frac{1}{2}s_{44}]$. Since the quantity $(s_{11} - s_{12})$ is often greater than $\frac{1}{2}s_{44}$, the modulus of elasticity is usually greatest in the $\langle 111 \rangle$ direction.

The elastic constants for several materials are given in Table 1.3, and their relative elastic anisotropy is tabulated in Table 1.4. Note the large anisotropy exhibited by many of these crystals as compared with the isotropic behavior of tungsten for which $(s_{11} - s_{12}) = \frac{1}{2}s_{44}$. Owing to this equality, the modulus of elasticity in tungsten is independent of the direction cosines (Eq. 1-14).

TABLE 1.3 Stiffness and Compliance Constants for Selected Crystals[a]

Material	$(10^{10}$ Pa$)$			$(10^{-11}$ Pa$^{-1})$		
	c_{11}	c_{12}	c_{44}	s_{11}	s_{12}	s_{44}
Cubic						
Aluminum	10.82	6.13	2.85	1.57	−0.57	3.51
Copper	16.84	12.14	7.54	1.50	−0.63	1.33
Gold	18.60	15.70	4.20	2.33	−1.07	2.38
Iron	23.70	14.10	11.60	0.80	−0.28	0.86
Lithium fluoride	11.2	4.56	6.32	1.16	−0.34	1.58
Magnesium oxide	29.3	9.2	15.5	0.401	−0.096	0.648
Molybdenum[b]	46.0	17.6	11.0	0.28	−0.08	0.91
Nickel	24.65	14.73	12.47	0.73	−0.27	0.80
Sodium chloride[b]	4.87	1.26	1.27	2.29	−0.47	7.85
Spinel (MgAl$_2$O$_4$)	27.9	15.3	15.3	0.585	−0.208	0.654
Titanium carbide[b]	51.3	10.6	17.8	0.21	−0.036	0.561
Tungsten	50.1	19.8	15.14	0.26	−0.07	0.66
Zinc sulfide	10.79	7.22	4.12	2.0	−0.802	2.43

TABLE 1.3 *(Continued)*

	c_{11}	c_{12}	c_{13}	c_{33}	c_{44}	s_{11}	s_{12}	s_{13}	s_{33}	s_{44}
Hexagonal										
Cadmium	12.10	4.81	4.42	5.13	1.85	1.23	−0.15	−0.93	3.55	5.40
Cobalt	30.70	16.50	10.30	35.81	7.53	0.47	−0.23	−0.07	0.32	1.32
Magnesium	5.97	2.62	2.17	6.17	1.64	2.20	−0.79	−0.50	1.97	6.10
Titanium	16.0	9.0	6.6	18.1	4.65	0.97	−0.47	−0.18	0.69	2.15
Zinc	16.10	3.42	5.01	6.10	3.83	0.84	0.05	−0.73	2.84	2.61

[a]Data adapted from H. B. Huntington, *Solid State Physics*, Vol. 7, Academic Press, New York, 1958, p. 213 and K. H. Hellwege, *Elastic, Piezoelectric and Related Constants of Crystals*, Springer-Verlag, Berlin, 1969, p. 3.
[b]Note that $E_{100} > E_{111}$.

TABLE 1.4a   Elastic Anisotropy of Selected Materials

Metal	Relative Degree of Anisotropy $\left[\dfrac{2(s_{11} - s_{12})}{s_{44}} \right]$	E_{111} (GPa)	E_{100} (GPa)	$\left[\dfrac{E_{111}}{E_{100}} \right]$
Aluminum	1.219	76.1	63.7	1.19
Copper	3.203	191.1	66.7	2.87
Gold	2.857	116.7	42.9	2.72
Iron	2.512	272.7	125.0	2.18
Magnesium oxide	1.534	350.1	249.4	1.404
Spinel (MgAl$_2$O$_4$)	2.425	364.5	170.0	2.133
Titanium carbide	0.877	429.2	476.2	0.901
Tungsten	1.000	384.6	384.6	1.00

TABLE 1.4b   Elastic Anisotropy of Selected Materials

Metal	Relative Degree of Anisotropy $\left[\dfrac{2(s_{11} - s_{12})}{s_{44}} \right]$	E_{111} (10^6 psi)	E_{100} (10^6 psi)	$\left[\dfrac{E_{111}}{E_{100}} \right]$
Aluminum	1.219	11.0	9.2	1.19
Copper	3.203	27.7	9.7	2.87
Gold	2.857	16.9	6.2	2.72
Iron	2.512	39.6	18.1	2.18
Magnesium oxide	1.534	50.8	36.2	1.404
Spinel (MgAl$_2$O$_4$)	2.425	52.9	24.8	2.133
Titanium carbide	0.877	62.2	69.1	0.901
Tungsten	1.000	55.8	55.8	1.00

EXAMPLE 1.2 Compute the modulus of elasticity for tungsten and iron in the $\langle 110 \rangle$ direction. From Tables 1.2 and 1.3 we obtain the necessary information regarding elastic compliance values and direction cosines. The modulus of elasticity in the $\langle 110 \rangle$ direction is then determined from Eq. 1-14.

For tungsten

$$\frac{1}{E_{110}} = 0.26 - 2\left\{ [0.26 - (-0.07)] - \frac{1}{2}(0.66) \right\}\left(\frac{1}{4}\right)$$

$$= 0.26 - (0)\left(\frac{1}{4}\right)$$

Therefore,

$$E_{110} = 384.6 \text{ GPa}$$

which is the same value given in Table 1.4 for E_{111} and E_{100}.

For iron

$$\frac{1}{E_{110}} = 0.80 - 2\left\{ [0.80 - (-0.28)] - \frac{1}{2}(0.86) \right\}\left(\frac{1}{4}\right)$$

$$E_{110} = 210.5 \text{ GPa}$$

Note that $E_{111} > E_{110} > E_{100}$ and that E_{110} is in good agreement with the average value of E for a polycrystalline sample (Table 1.1).

For the case of tungsten and any other isotropic material, then,

$$\begin{aligned}
\epsilon_{xx} &= s_{11}\sigma_{xx} + s_{12}\sigma_{yy} + s_{12}\sigma_{zz} \\
\epsilon_{yy} &= s_{12}\sigma_{xx} + s_{11}\sigma_{yy} + s_{12}\sigma_{zz} \\
\epsilon_{zz} &= s_{12}\sigma_{xx} + s_{12}\sigma_{yy} + s_{11}\sigma_{zz} \\
\gamma_{xy} &= s_{44}\tau_{xy} = 2(s_{11} - s_{12})\tau_{xy} \\
\gamma_{xz} &= s_{44}\tau_{xz} = 2(s_{11} - s_{12})\tau_{xz} \\
\gamma_{yz} &= s_{44}\tau_{yz} = 2(s_{11} - s_{12})\tau_{yz}
\end{aligned} \tag{1-15}$$

If we compare Eqs. 1-15 with 1-10, the elastic constants s_{ij} may be described in terms of the familiar strength of materials elastic constants. Therefore

$$s_{11} = \frac{1}{E} \tag{1-16a}$$

$$s_{12} = -\frac{\nu}{E} \tag{1-16b}$$

$$s_{44} = 2(s_{11} - s_{12}) = \frac{1}{G} \tag{1-16c}$$

Finally, from Eq. 1-16,

$$\frac{1}{G} = 2(s_{11} - s_{12}) = 2\left(\frac{1}{E} + \frac{\nu}{E}\right) = \frac{2}{E}(1 + \nu)$$

and

$$G = \frac{E}{2(1+\nu)} \qquad (1\text{-}16\text{d})$$

For the case of hexagonal crystals, the matrix in Eq. 1-11 reduces to

$$\frac{1}{E} = s_{11}\left(1 - l_3^2\right)^2 + s_{33}l_3^4 + (2s_{13} + s_{44})l_3^2\left(1 - l_3^2\right) \qquad (1\text{-}17)$$

where l_1, l_2, l_3 are direction cosines for directions in the hexagonal unit cell. From Eq. 1-17 note that in hexagonal crystals E depends only on the direction cosine l_3, which lies normal to the basal plane. Consequently, the modulus of elasticity in hexagonal crystals is isotropic everywhere in the basal plane.

1.2.1.2 RESILIENCY

The resilience of a material is a measure of the amount of energy that can be absorbed under elastic loading conditions and which is released completely when the loads are removed. From this definition, resilience may be measured from the area under the curve in Fig. 1.1:

$$\text{resilience} = \tfrac{1}{2}\sigma_{max}\epsilon_{max} \qquad (1\text{-}18)$$

where

$$\sigma_{max} = \text{maximum stress for elastic conditions}$$

$$\epsilon_{max} = \text{elastic strain limit}$$

From Eq. 1-7

$$\text{resilience} = \frac{\sigma_{max}^{2}}{2E} \qquad (1\text{-}19)$$

Should an engineering design require a material that allows only for elastic response with large energy absorption (such as in the case of a mechanical spring) the appropriate material to choose would be one possessing a high yield strength but low modulus of elasticity.

1.2.2 Elastic-Homogeneous Plastic Response: Type II

When a material has the capacity for plastic deformation—irreversible flow— the stress-strain curve often assumes the shape of Curve II (Fig. 1.4). Here we see the same elastic region at small strains but now find a smooth parabolic

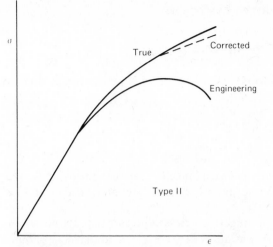

FIGURE 1.4 Type II stress-strain behavior revealing elastic behavior followed by a region of homogeneous plastic deformation. Data plotted on basis of engineering and true stress-strain definitions.

portion of the curve, which is associated with homogeneous plastic deformation processes such as the irreversible movement of dislocations in metals, ceramics, and crystalline polymers and a number of possible deformation mechanisms in amorphorus polymers. That the curve continues to rise to a maximum stress level reflects an increasing resistance on the part of the material to further plastic deformation—a process known as strain hardening. The portion of the true stress-strain curve (from the onset of yielding to the maximum load) may be described empirically by the relationship generally attributed to Hollomon[2]:

$$\sigma = K\epsilon^n \tag{1-20}$$

where

> σ = true stress
> ϵ = true plastic strain
> n = strain hardening coefficient
> K = material constant, defined as the true stress
> at a true strain of 1.0

However, Bülfinger[3] proposed a similar parabolic relationship between stress and strain almost 200 years earlier. The magnitude of the strain hardening coefficient reflects the ability of the material to resist further deformation. In the limit, n may be equal to unity, which represents ideally elastic behavior, or equal to zero, which represents an ideally plastic material. Selected values of strain hardening coefficients for some engineering metal alloys are given in Table 1.5.

TABLE 1.5   Selected Strain Hardening Coefficients

Material	Strain Hardening Coefficient, n
Stainless steel	0.45–0.55
Brass	0.35–0.4
Copper	0.3–0.35
Aluminum	0.15–0.25
Iron	0.05–0.15

(Note that n values are sensitive to thermomechanical treatment; they are generally larger for materials in the annealed condition and smaller in the cold-worked state.)

Such data may be derived by plotting true stress and associated true strain values on log-log paper. If Eq. 1-20 was absolutely correct, a straight line should result with a slope equal to n. However, this is found not always to be the case and reflects the fact that this relationship is only an empirical approximation.[4] (When a nonlinear log-log plot does result for a given material, the strain hardening coefficient is often defined at a particular strain value.) In general, n increases with decreasing strength level and in some alloys with increasing solute content (Fig. 1.5). More will be said about this material property in Chapter Three.

1.2.2.1   STRENGTH LEVELS IN MATERIALS

It has become common practice to define several strength levels that characterize the material's tensile response. The *proportional limit* is that stress level below

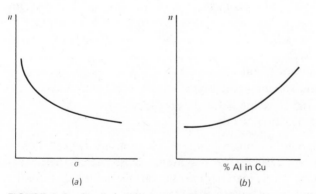

FIGURE 1.5   Strain hardening coefficient is observed to decrease with increasing strength level (*a*), and increase with increasing solute concentration of aluminum in copper (*b*).

which stress is proportional to strain according to Eq. 1-7. The *elastic limit* defines that stress level below which the deformation strains are fully reversible. In most engineering materials, these two quantities are essentially equal. However, it is possible for a metal to exhibit nonlinear but elastic behavior. For example, very high-strength filamentary particles—often called whiskers—can exhibit elastic strains in excess of 2%. In this range of very large elastic strains, the modulus of elasticity reveals its weak dependence on strain—something that is completely obscured when strains are very small. As in the case of Eq. 1-7, Hooke's law, too, is nothing more than an empirical relationship, albeit a good one at small strains. Consequently, the elastic limit is approximately equal to the proportional limit and may be slightly higher in some cases.

A much more important material property is the *yield strength*—a stress level related to the onset of irreversible plastic deformation. This quantity is difficult to define, since the point where plastic flow begins will depend on the sensitivity of the displacement transducer. The more sensitive the gage, the lower the stress level where plastic flow is found. In recent years, special capacitance strain gages have been used to measure strains in the range of 10^{-6}. In fact, a number of studies[5] dealing with the mechanical behavior of materials in the microstrain region have been undertaken as a result of this breakthrough in instrumentation. These investigations have shown, for example, that plastic deformation—the irreversible movement of dislocations—occurs at stress levels many times lower than the conventionally determined engineering yield strength. To arrive at a uniformly accepted method for determination of the yield strength, therefore, a standard test procedure[6] (ASTM Standard E 8-69) has been adopted. The yield strength value is defined in the following manner: (1) Determine the engineering stress-strain curve, such as the one shown in Fig. 1.6; (2) construct a line parallel to the elastic portion of the σ–ϵ curve but offset from the origin by a certain amount (the generally accepted offset is 0.002 or 0.2% strain) and (3) define the yield strength at the intersection of the σ–ϵ curve and the offset line. As cited above, this value is usually defined as the 0.2% offset yield strength. The *ultimate tensile strength* is defined as the maximum load divided by the initial cross-sectional area, while the *true fracture stress* is the load at fracture divided by the final cross-sectional area with correction made for any localized deformation (necking) in the final fracture region (see Section 1.2.2.2). A compilation of tensile properties for a number of engineering materials is given in Table 1.6.

1.2.2.2 PLASTIC INSTABILITY AND NECKING

The true and engineering stress-strain plots from a tensile test reveal basic differences as shown in Fig. 1.4. While the engineering curve reaches a maximum at maximum load and decreases thereafter to fracture, the true curve rises continually to failure. The inflection in the engineering curve is due to the onset of localized plastic flow and the manner in which engineering stress is defined.

TABLE 1.6a Tensile Properties for Selected Engineering Materials[a]

Material	Treatment	Yield Strength (MPa)	Tensile Strength (MPa)	Elongation in 5-cm Gage (%)	Reduction in Area (1.28-cm diameter) (%)
Steel Alloys					
1015	As-rolled	315	420	39	61
1050	"	415	725	20	40
1080	"	585	965	12	17
1340	Q + T (205°C)	1590	1810	11	35
1340	" (425°C)	1150	1260	14	51
1340	" (650°C)	620	800	22	66
4340	" (205°C)	1675	1875	10	38
4340	" (425°C)	1365	1470	10	44
4340	" (650°C)	855	965	19	60
301	Annealed plate	275	725	55	—
304	" "	240	565	60	—
310	" "	310	655	50	—
316	" "	250	565	55	—
403	Annealed Bar	275	515	35	—
410	" "	275	515	35	—
431	" "	655	860	20	—
AFC-77	Variable	560–1605	835–2140	10–26	32–74
PH 15-7 Mo	"	380–1450	895–1515	2–35	—
Titanium Alloys					
Ti-5 Al-2.5 Sn	Annealed	805	860	16	40
Ti-8 Al-1 Mo-1 V	Duplex annealed	950	1000	15	28
Ti-6 Al-4 V	Annealed	925	995	14	30
Ti-13 V-11 Cr-3 Al	Solution + age	1205	1275	8	—
Magnesium Alloys					
AZ31B	Annealed	103–125	220	9–12	—
AZ80A	Extruded bar	185–195	290–295	4–9	—
ZK60A	Artificially aged	215–260	295–315	4–6	—
Aluminum Alloys					
2219	-T31, -T351	250	360	17	—
2024	-T3	345	485	18	—
2024	-T6, -T651	395	475	10	—
2014	-T6, -T651	415	485	13	—
6061	-T4, -T451	145	240	23	—
7049	-T73	475	530	11	—
7075	-T6	505	570	11	—
7075	-T73	415	505	11	—
7178	-T6	540	605	11	—
Plastics					
ABS	Medium impact	—	46	6–14	—
Acetal	Homopolymer	—	69	25–75	—
Poly(tetra-fluorethylene)	—	—	14–48	100–450	—

TABLE 1.6a (Continued)

Material	Treatment	Yield Strength (MPa)	Tensile Strength (MPa)	Elongation in 5-cm Gage (%)	Reduction in Area (1.28-cm diameter) (%)
Poly(vinylidene fluoride)	—	—	35–48	100–300	—
Nylon 66	—	—	59–83	60–300	—
Polycarbonate	—	—	55–69	130	—
Polyethylene	Low density	—	7–21	50–800	—
Polystyrene	—	—	41–54	1.5–2.4	—
Polysulfone	—	69	—	50–100	—

[a]Databook 1974, Metal Progress, mid-June issue, 1974.

TABLE 1.6b Tensile Properties for Selected Engineering Materials[a]

Material	Treatment	Yield Strength (ksi)	Tensile Strength (ksi)	Elongation in 2-in. Gage (%)	Reduction in Area (0.505-in. diameter) (%)
Steel Alloys					
1015	As-rolled	46	61	39	61
1050	"	60	105	20	40
1080	"	85	140	12	17
1340	Q + T (400°F)	230	260	11	35
1340	" (800°F)	167	183	14	51
1340	" (1200°F)	90	116	22	66
4340	" (400°F)	243	272	10	38
4340	" (800°F)	198	213	10	44
4340	" (1200°F)	124	140	19	60
301	Annealed plate	40	105	55	—
304	" "	35	82	60	—
310	" "	45	95	50	—
316	" "	36	82	55	—
403	Annealed bar	40	75	35	—
410	" "	40	75	35	—
431	" "	95	125	20	—
AFC-77	Variable	81–233	121–310	10–26	32–74
PH 15-7 Mo	"	55–210	130–220	2–35	—
Titanium Alloys					
Ti-5 Al-2.5 Sn	Annealed	117	125	16	40
Ti-8 Al-1 Mo-1 V	Duplex annealed	138	145	15	28
Ti-6 Al-4 V	Annealed	134	144	14	30
Ti-13 V-11 Cr-3 Al	Solution + age	175	185	8	—

TABLE 1.6b (*Continued*)

Material	Treatment	Yield Strength (ksi)	Tensile Strength (ksi)	Elongation in 2-in. Gage (%)	Reduction in Area (0.505-in. diameter) (%)
Magnesium Alloys					
AZ31B	Annealed	15–18	32	9–12	—
AZ80A	Extruded bar	27–28	42–43	4–9	—
ZK60A	Artificially aged	31–38	43–46	4–6	—
Aluminum Alloys					
2219	-T31,-T351	36	52	17	—
2024	-T3	50	70	18	—
2024	-T6,-T651	57	69	10	—
2014	-T6,-T651	60	70	13	—
6061	-T4,-T451	21	35	23	—
7049	-T73	69	77	11	—
7075	-T6	73	83	11	—
7075	-T73	60	73	11	—
7178	-T6	78	88	11	—
Plastics					
ABS	Medium impact	—	6.8	6–14	—
Acetal	Homopolymer	—	10	25–75	—
Poly(tetra-fluorethylene)	—	—	2–7	100–450	—
Poly(vinylidene fluoride)	—	—	5.1–7	100–300	—
Nylon 66	—	—	8.6–12	60–300	—
Polycarbonate	—	—	8–10	130	—
Polyethylene	Low density	—	1–3	50–800	—
Polystyrene	—	—	6–9	1.5–2.4	—
Polysulfone	—	10	—	50–100	—

[a]*Databook 1974, Metal Progress*, mid-June issue, 1974.

To understand this, consider for a moment the following sequence of events. When the stress reaches a critical level, plastic deformation will occur at the weakest part of the test sample, somewhere along the gage length. This local extension under tensile loading will cause a simultaneous area constriction so that the true local stress is higher at this location than anywhere else along the gage length. Consequently, it would be expected that all additional deformation would concentrate in this most highly stressed region. Such is the case in an ideally plastic material. For all other materials, however, this localized plastic deformation strain hardens the material, thereby making it more resistant to further damage. At this point, the applied stress must be increased to produce additional plastic deformation at the second weakest position along the gage length. Here again the material strain hardens and the process continues. On a

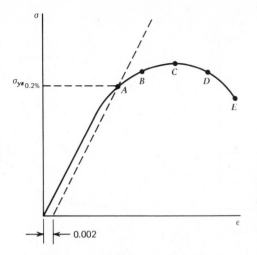

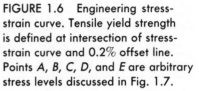

FIGURE 1.6   Engineering stress-strain curve. Tensile yield strength is defined at intersection of stress-strain curve and 0.2% offset line. Points *A*, *B*, *C*, *D*, and *E* are arbitrary stress levels discussed in Fig. 1.7.

macroscopic scale, the gage length extends uniformly in concert with a uniform reduction in cross-sectional area. (Recall that plastic deformation is a constant-volume process.) With increasing load, a point is reached where the strain hardening capacity of the material is exhausted and the nth local area contraction is no longer balanced by a corresponding increase in material strength. At this maximum load, further plastic deformation is localized in the necked region, since the stress increases continually with areal contraction even though the applied load is decreasing as a result of elastic unloading in the test bar outside the necked area. Eventually the neck will fail. Since engineering stress is based on A_0, the decreasing load on the sample after the neck has formed will result in the computation of a decreasing stress. By comparison, the decreasing load value is more than offset by the decrease in instantaneous cross-sectional area such that the true stress continues to rise to failure even after the onset of necking.

1.2.2.3   STRAIN DISTRIBUTION IN TENSILE SPECIMEN

The total strain distribution along the specimen gage length is shown schematically in Fig. 1.7 for various stress levels as indicated on the engineering stress-strain curve (Fig. 1.6). Owing to the variation of elongation along the gage length of the tensile specimen, researchers occasionally report both the total strain, $(l_f - l_0)/l_0$ or $\ln(l_f/l_0)$, and the uniform strain, which is related to the elongation just prior to local necking (Curve *C* in Fig. 1.7). It should be emphasized that the total strain reported for a given test result will depend on the gage length of the test bar. From Fig. 1.7, it is clear that as the gage length decreases, the elongation involved in the necking process becomes increasingly more dominant. Consequently, total strain values will increase the shorter the

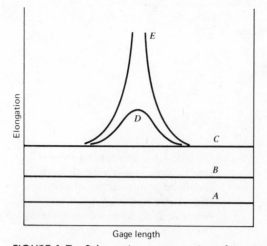

FIGURE 1.7 Schematic representation of specimen elongation along the gage length. Uniform extension occurs up to the onset of necking (*C*). Additional displacements are localized in necked region (*D* and *E*). Stress levels *A, B, C, D,* and *E* are identified in Fig. 1.6.

TABLE 1.7 Round Tension Test Specimen Dimensions[6]

Diameter (D)		Gage Length (L)	
mm	(in.)	mm	(in.)
12.5	(0.5)	50	(2.0)
8.75	(0.345)	35	(1.375)
6.25	(0.25)	25	(1.0)
4.0	(0.16)	16	(0.63)
2.50	(0.1)	10	(0.394)

gage length. For this reason both specimen size and total strain data should be reported. ASTM has standardized specimen dimensions to minimize variability in test data resulting from such geometrical considerations. As noted in Table 1.7, the gage length to diameter ratio is standardized to a value of about 4.

1.2.2.4 EXTENT OF UNIFORM STRAIN

From the standpoint of material usage in an engineering component, it is desirable to maximize the extent of uniform elongation prior to the onset of localized necking. It may be shown that the amount of uniform strain is related to the magnitude of the strain hardening exponent.

$$P = \sigma A$$

$$dP = \sigma dA + A d\sigma \tag{1-21}$$

Recalling that necking occurs at maximum load

$$dP = 0$$

so that

$$\frac{d\sigma}{\sigma} = -\frac{dA}{A}$$

From Eq. 1-5

$$A dl + l dA = 0$$

$$-\frac{dA}{A} = \frac{dl}{l}$$

Since $dl/l \equiv d\epsilon$, we find

$$\sigma = \frac{d\sigma}{d\epsilon} \tag{1-22}$$

By using the Hollomon relation (Eq. 1-20)

$$K\epsilon^{n} = Kn\epsilon^{n-1}$$

Therefore

$$n = \epsilon \tag{1-23}$$

The true plastic strain at necking instability is numerically equal to the strain hardening coefficient.

In addition to the necking strains, a triaxial stress state exists in the vicinity of the neck (Fig. 1.8). The radial (σ_r) and transverse (σ_t) stresses that are induced are developed as a result of a Poisson effect. In effect, the more highly stressed material within the neck wishes to pull in to accommodate the large local extensions (Fig. 1.7). Since the material immediately adjacent to the necked area experiences a much lower stress level, these regions will resist such contractions by exerting induced tensile stresses that act to retard deepening of the neck. Consequently, the triaxial stress field acts to plastically constrain the material from deforming in the reduced area. To provide for such plastic flow, the axial stress must be increased. The stress values recorded on the true stress-strain curve (Fig. 1.4) after the onset of necking reflect the higher axial stresses

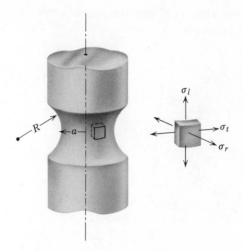

FIGURE 1.8 Triaxial tension stress distribution within necked region which acts to constrain additional deformation in the neck.

necessitated by the triaxial stress condition. In terms of the radius of curvature of the neck contour R and the radius of the minimum cross-sectional area a, Bridgman was able to correct the applied axial stress (σ_{app}) to determine the true stress (σ_{true}) that would be necessary to deform the material were it not for the presence of the neck. The corrected true stress-strain curve shown in Fig. 1.4 may be determined from the Bridgman[7] relation

$$\frac{\sigma_{true}}{\sigma_{app}} = \frac{1}{(1+2R/a)\left[\ln(1+a/2R)\right]} \qquad (1\text{-}24)$$

It is seen from this formula that the stress necessary to produce a given level of plastic deformation will increase with increasing notch root acuity for a given notch depth.

At some critical point, the triaxial tensile stress condition within the necked region causes small particles within the microstructure to either fracture or separate from the matrix. The resulting microvoids then undergo a period of growth and eventual coalescence, producing an internal, disc-shaped crack oriented normal to the applied stress axis. Final fracture then occurs by a shearing-off process along a conical surface oriented 45 degrees to the stress axis. This entire process produces the classical cup-cone fracture surface appearance shown in Fig. 1.9. Sometimes the circular region in the middle of the sample (called the fibrous zone) is generated entirely by slow, stable crack growth, while the smooth shear walls are formed at final failure. Usually the fibrous zone contains a series of circumferential ridges reflecting slight undulations in the stable crack propagation direction. (Another example of these ridges is shown in Fig. 10.35.) However, test conditions can be altered to suppress the extent of the slow, stable crack growth region; instead, the crack continues to

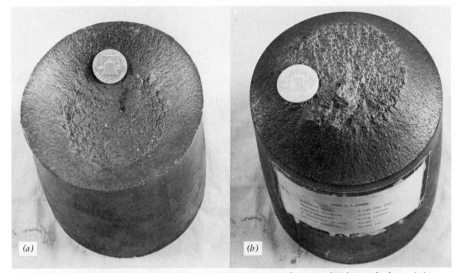

FIGURE 1.9 Typical cup-cone fracture appearance of unnotched tensile bar. (*a*) cup portion, (*b*) cone portion. (Courtesy of Richard Sopko, Lehigh University.)

grow on the same plane but in unstable fashion at a much faster rate. This new region, defined as the radial zone, contains radial markings (Fig. 1.10) often associated with the fracture of oriented inclusions in test bars prepared from rod stock. (More will be said of this fracture detail in Chapter Fourteen.) The relative amount of fibrous, radial, and shear lip fracture zones has been found to depend on the strength of the material and the test temperature[8] (Fig. 1.11). Since the internal fracture process depends on plastic constraint resulting from the tensile triaxiality within the neck, the crack nucleation process could be suppressed if hydrostatic pressure were to be introduced. Indeed, Bridgman[9] demonstrated that when a sufficiently large hydrostatic pressure is applied, necking can proceed uninterrupted almost to where the sample draws down to a point (Fig. 1.12).

1.2.2.5 TOUGHNESS

Another important material characteristic is its resistance to fracture (measured in units of energy). We may define a brittle material as one absorbing little energy while a tough material would require a large expenditure of energy in the fracture process. For an unnotched tensile bar, the energy to break may be estimated from the area under the stress-strain curve.

$$\text{energy} = \int_0^{\epsilon_f} \sigma d\epsilon \qquad (1\text{-}25)$$

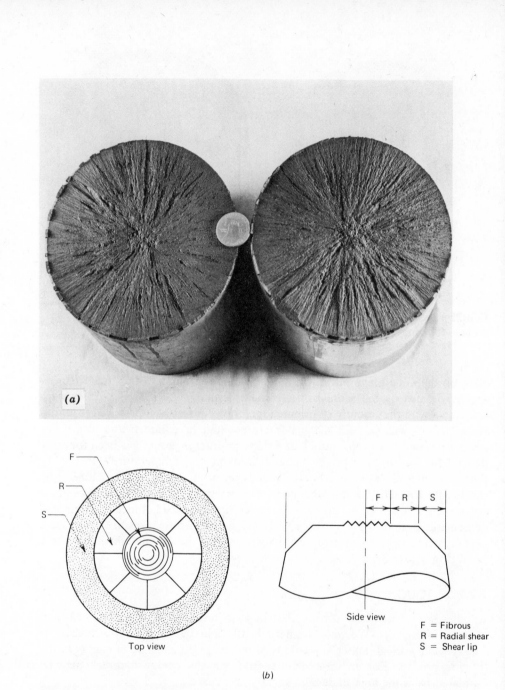

(a)

(b)

Top view

Side view

F = Fibrous
R = Radial shear
S = Shear lip

FIGURE 1.10 Extent of fibrous radial and shear lip zones, (*a*) macrofractograph (courtesy of Richard Sopko, Lehigh University); (*b*) schematic showing zone location (after Larsen et al.,[8] reprinted by permission of the American Society for Metals).

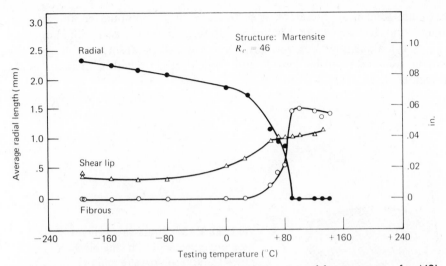

FIGURE 1.11 Effect of test temperature on relative size of fracture zones for AISI 4340 steel heat treated to R_c 46. (After Larson et al.,[8] reprinted by permission of the American Society for Metals.)

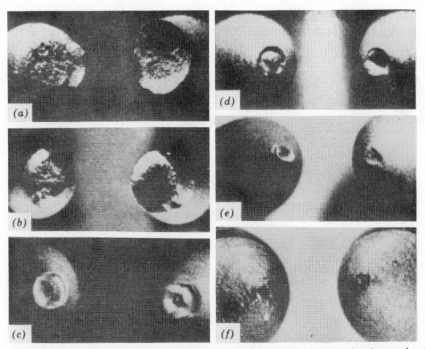

FIGURE 1.12 Effect of increasing hydrostatic pressure in suppressing internal void formation within necked region. (a) Atmospheric pressure, 10×; (b) 235 MPa hydrostatic pressure, 10×; (c) 1000 MPa, 12×; (d) 1290 MPa, 12×; (e) 1850 MPa, 12×; and (f) 2680 MPa, 18×. (After Bridgman,[9] reprinted by permission of the American Society for Metals.)

Maximum toughness, therefore, is achieved with an optimum combination of strength and ductility; neither high strength (e.g., glass) nor exceptional ductility (e.g., taffy) alone provides for large fracture energy absorption (Fig. 1.13). Material toughness will be considered in much greater detail in Chapters Seven through Eleven.

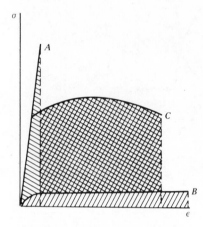

FIGURE 1.13 Stress-strain curves for strong material with little plastic flow capacity *A*, low strength but high ductility material *B*, and material with optimum combination of strength and ductility for maximum toughness *C*.

1.2.3 Elastic-Heterogeneous Plastic Response: Type III

Occasionally, a test specimen will produce a stress-strain curve that exhibits a series of serrations that are superimposed on the parabolic portion of Fig. 1.4 after the normal range of elastic response. Such behavior, as shown in Fig. 1.14, reflects nonuniform or heterogeneous deformation within the material. While serrated stress-strain response is usually not experienced in materials testing, it is

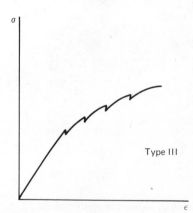

FIGURE 1.14 Type III stress-strain behavior reflecting elastic behavior followed by heterogeneous plastic flow. The latter can be caused by twin controlled deformation or solute atom-dislocation interactions.

known to occur under at least two different conditions. When face-centered cubic metals are tested at low temperatures and under high strain rates, they tend to deform plastically by a twinning process rather than by slipping along glide planes. (The nature of plastic deformation in metals is discussed in greater detail in Chapters Two through Four.) When this happens, extension of the gage length proceeds in discrete bursts associated with twin band nucleation and growth. Whenever the instantaneous strain rate in the specimen exceeds the rate of motion of the test machine crosshead, a load drop will occur. A similar stress strain response is found in other metals that deform by twinning.

Serrated stress-strain curves are encountered also in body-centered cubic iron alloys containing carbon in solid solution and in dilute solid solutions of aluminum. It has been argued that the Portevin-Le Chatelier effect (serrated σ–ϵ curve) is due to solute atom or vacancy interactions with lattice dislocations.[10] When a sufficiently large stress is applied, dislocations can break free from solute clusters and cause a load drop. If the solute atoms can diffuse quickly enough to retrap these dislocations, then more load must be applied once again to continue the deformation process.

1.2.4 Elastic-Heterogeneous Plastic-Homogeneous Plastic Response: Type IV

In many body-centered cubic iron based alloys and some nonferrous alloys, a relatively narrow region of heterogeneous plastic deformation (with a range of approximately 1 to 3% strain) separates the elastic region from the homogeneous plastic flow portion of the stress-strain curve (Fig. 1.15). This segment of the curve is caused by a dislocation—solute atom interaction related to that mentioned in the previous section. After being loaded elastically to A, defined as the upper yield point, the material is observed to develop a local deformation band (Fig. 1.16); the sudden onset of plastic deformation associated with this Lüder

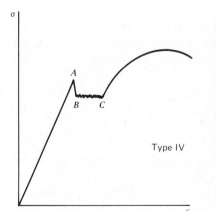

FIGURE 1.15 Type IV stress-strain behavior exhibiting a narrow heterogeneous deformation region between initial elastic and final homogeneous flow regions. Onset of local yielding occurs at upper yield point A with corresponding load drop to B defined as the lower yield point. After passage of Lüder band throughout the gage section, homogeneous deformation commences at C.

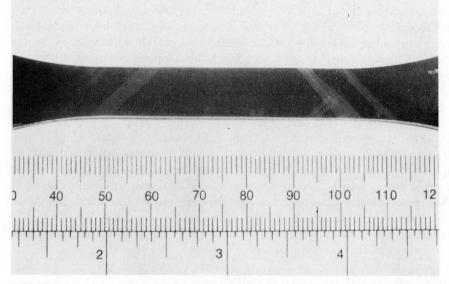

FIGURE 1.16 Concentrated deformation (Lüder) bands formed in plain carbon steel test sample. Band will grow across the gage section before homogeneous deformation develops at point *C* from Fig. 1.15.

band is responsible for the initial load drop to *B*, defined as the lower yield point. Since the upper yield point is very sensitive to minor stress concentrations, alignment of the specimen in the test grips, and other related factors, measured values reflect considerable scatter. For this reason, the yield strength of materials exhibiting Type IV behavior (Fig. 1.15) is usually reported as the lower yield point value. The remainder of the heterogeneous segment of deformation is consumed in the passage of the Lüder band across the entire gage section. (Occasionally more than one band may propagate simultaneously during this period.) When deformation has spread to all parts of the gage length, the material then continues to deform in a homogeneous manner, similar to that described for Type II behavior. Yield points are found also in ionic and covalent materials and are discussed more fully in Section 3.4.

1.2.5 Elastic-Heterogeneous Plastic-Homogeneous Plastic Response: Type V

Type V behavior may be found in the deformation of some crystalline polymers. While possessing an upper yield point and subsequent load drop (Fig. 1.17) similar to that observed in Type IV response, the final deformation stage is decidedly different. Meinel and Peterlin[11] have rationalized the shape of the

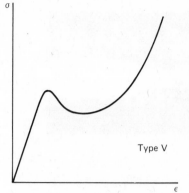

FIGURE 1.17 Type V stress-strain behavior usually found in crystalline polymers. Minimum in curve reflects competition between breakdown of initial structure and its subsequent reorganization into a highly oriented, strong material.

Type V curve as reflecting the competition of two events. Initially, yielding occurs along with a breakdown of the original crystalline structure within the polymer; this produces the initial load drop and a general whitening of the sample in the region of greatest deformation. If failure does not occur soon after the necking process begins, continuing strains will lead to a reorganization of the broken-down structure into new, highly oriented, and strong units. As more and more of this new structure is produced, the polymer offers more resistance to deformation—it strain hardens—and the stress-strain curve begins to rise once again. Strain hardening continues by molecular alignment (sometimes called cold drawing) up to the fracture stress. As the cold-drawn regions become more highly oriented, the milky white regions associated with initial crystallite breakdown gradually become clear once again (Fig. 1.18). Deformation of crystalline polymers is discussed further in Chapter Six.

The tensile response of elastomers bears strong resemblance to Type V behavior, with the exception of the load drop at intermediate strains (the slope of the stress-strain curve is always positive for these materials). Elastomer or rubber elasticity is distinguished by two basic characteristics: very large elastic strains (often in excess of 100%) and elastic moduli that increase with increasing temperature. The latter response is opposite to that found in other materials (including rigid polymers). Rubber elasticity is related primarily to the straightening of amorphous polymer chains from their curled positions into partially extended conformations. As a result, the extent of elastic deformation is great, while the elastic moduli are very low because of the small contribution of actual polymer chain stretching. That is, a curled chain of length l is extended so that its end-to-end length approaches l with little additional chain lengthening attributed to the more difficult covalent bond extension mode. The straightening of the chains is responsible for the rapidly increasing apparent hardening of the material at large strains (Fig. 1.17). When the applied loads are relaxed, the chains return to a curled position, indicating the latter conformation to be preferred.

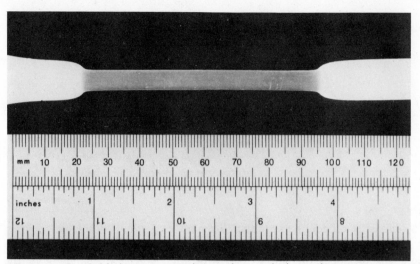

FIGURE 1.18 Cold drawing in polypropylene, which produces greater optical transparency in gage section as a result of enhanced molecular alignment.

By simple application of the first and second laws of thermodynamics, it is possible to demonstrate that the elastic modulus of rubber should increase with increasing temperature. From the first law of thermodynamics

$$dU = dQ + dW \tag{1-26}$$

where

dU = change in internal energy
dQ = change in heat absorbed or released
dW = work done on the system

For a reversible process, the second law of thermodynamics gives

$$dQ = TdS \tag{1-27}$$

where

T = temperature
dS = change in entropy

If an elastomeric rod of length l is extended by an amount dl owing to a tensile force F, the work dW done on the rod is Fdl. Combining Eqs. 1-26 and 1-27 with the expression for dW we have

$$dU = TdS + Fdl \tag{1-28}$$

At constant temperature

$$F = \left(\frac{\partial U}{\partial l} \right)_T - T \left(\frac{\partial S}{\partial l} \right)_T \qquad (1\text{-}29)$$

where

$\left(\dfrac{\partial U}{\partial l} \right)_T$ = related to the strain energy associated with the application of a load

$\left(\dfrac{\partial S}{\partial l} \right)_T$ = related to the change in entropy or order of the rod as it is stretched

Since the chains prefer a random curled configuration, their initial degree of order is low and their entropy high. (Because of the very high degree of order of atoms in metals and ceramics, their entropy term by comparison is negligible.) However, when a tensile load is applied, the entropy decreases as the chains become straightened and aligned. As a consequence, $(\partial S/\partial l)_T$ is negative. The force required to extend the elastomer rod, therefore, increases with increasing temperature.

1.3 TEMPERATURE AND STRAIN RATE EFFECTS ON TENSILE BEHAVIOR

Brief mention was made in Section 1.2.2.4 of a temperature-induced transition in macroscopic fracture surface appearance. Since this transition most often parallels important changes in the strength and ductility of the material, some additional discussion is indicated. It is known that the general flow curve for a given material will decrease with increasing temperature T and decreasing strain rate $\dot{\epsilon}$ (Fig. 1.19). The magnitude of these changes vary with the material; body-centered cubic metals (e.g., iron, chromium, molybdenum, and tungsten), and ceramic materials are much more sensitive to T and $\dot{\epsilon}$ than face-centered cubic metals (e.g., aluminum, copper, gold, and nickel) with polymeric solids being especially sensitive. Over the years, a number of investigators have sought to define the overall response of a material in terms of some generalized equation of state:

$$\sigma = f(\epsilon, \dot{\epsilon}, T) \qquad (1\text{-}30)$$

(Various aspects of Eq. 1-30 will be discussed in subsequent chapters.) For example, on the basis of simple rate theory, Bennett and Sinclair[12] proposed that

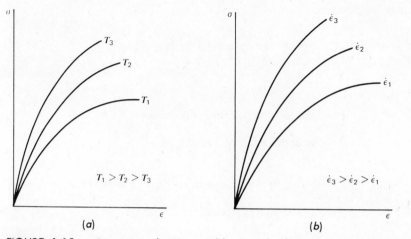

FIGURE 1.19   Diagrams showing yield strength change as a function of (a) temperature and (b) strain rate.

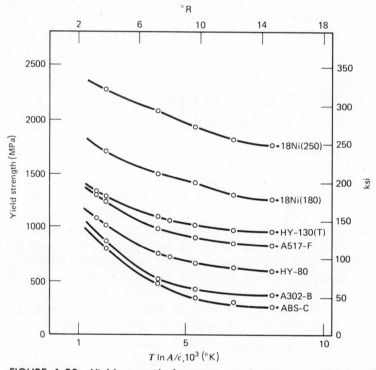

FIGURE 1.20   Yield strength for seven steels in terms of the Bennett-Sinclair parameter, $T \ln A/\dot{\epsilon}$.[13] (Reprinted with permission from A. K. Shoemaker and S. I. Rolfe, *Engineering Fracture Mechanics*, **2**, p. 319, © 1971, Pergamon Press).

the yield strength of iron and other body-centered cubic transition metals be described in terms of a rate-temperature parameter, $T\ln(A/\dot{\epsilon})$, where A is a frequency factor with an approximate value of $10^8/\sec$ for these materials. As seen in Fig. 1.20, the parameter provides good correlation for the case of seven steels. While the lower strength steels reveal a somewhat larger yield strength sensitivity to T and $\dot{\epsilon}$ than do the stronger alloys at low $T\ln(A/\dot{\epsilon})$ levels, the seven curves are remarkably similar, reflecting comparable *absolute* changes in yield strength with temperature and strain rate variations. It is important to recognize, however, that the *relative* change in yield strength with $T\ln(A/\dot{\epsilon})$ is much greater in the lower strength alloys.

REFERENCES

1. R. Hooke, "Of Spring" 1678, as discussed in S. P. Timoshenko, *History of the Strength of Materials*, McGraw-Hill, New York, 1953, p. 18.
2. J. H. Hollomon, *Trans.*, AIME, **162**, 1945, p. 268.
3. G. B. Bülfinger, *Comm. Acad. Petrop.*, **4**, 1735, p. 164.
4. A. W. Bowen and P. G. Partridge, *J. Phys. D: Appl. Phys.*, **7**, 1974, p. 969.
5. *Microplasticity*, C. J. McMahon, Jr., ed. (Advances in Materials Research, Vol. 2), Interscience, New York, 1968.
6. 1971 Annual Book of ASTM Standards, ASTM, Standard E8-69, p. 205.
7. P. W. Bridgman, *Trans.*, ASM, **32**, 1944, p. 553.
8. F. R. Larson and F. L. Carr, *Trans.*, ASM, **55**, 1962, p. 599.
9. P. W. Bridgman, *Fracturing of Metals*, ASM, Metals Park, Ohio, 1948, p. 246.
10. A. H. Cottrell, *Vacancies and Other Point Defects in Metals and Alloys*, Institute of Metals, London, 1958, p. 1.
11. G. Meinel and A. Peterlin, *J. Polym. Sci.*, Part A-2, **9**, 1971, p. 67.
12. P. E. Bennett and G. M. Sinclair, *Trans.*, ASME, *J. Basic Eng.*, **88**, 1966, p. 518.
13. A. K. Shoemaker and S. T. Rolfe, *Eng. Fract. Mech.*, **2**, 1971, p. 319.

PROBLEMS

1.1 A rod is found to creep at a fixed rate over a period of 10,000 hr. when loaded with a 1000-newton weight. If the initial diameter of the rod is 10 mm and its initial length 200 mm, it is found that the steady state creep rate is 10^{-5} hr^{-1}. Calculate the following:
(a) The final length of the rod after 100 hr., 10,000 hr.
(b) The engineering and true strains after these two time periods.
(c) The engineering and true stress after these two time periods.

1.2 Calculate the energy absorbed in fracturing a tensile specimen when the material obeys the Hollomon parabolic stress-strain relationship.

1.3 A 200-mm-long rod with a diameter of 2.5 mm is loaded with a 2000-

newton weight. If the diameter decreases to 2.2 mm, compute the following:

(a) The final length of the rod.

(b) The *true* stress and *true* strain at this load.

(c) The *engineering* stress and strain at this load.

1.4 Calculate the elastic moduli for sodium chloride and nickel in the $\langle 100 \rangle$ and $\langle 111 \rangle$ directions. Compare the anisotropy in these two materials.

1.5 Calculate the elastic modulus for gold in the $\langle 110 \rangle$ direction and compare your results with the E_{100} and E_{111} values reported in Table 1.4 and the polycrystalline isotropic value given in Table 1.1.

1.6 Define true strain in terms of engineering strain.

CHAPTER
TWO
ELEMENTS OF DISLOCATION THEORY

2.1 STRENGTH OF A PERFECT CRYSTAL

In Chapter One, the elastic limit of a given material was defined as that stress level above which strains are irreversible. The objective of this chapter is to consider the manner by which permanent deformations are generated and to estimate the magnitude of stresses necessary for such movement. To begin, consider atom movements along a particular crystallographic plane leading to the displacement of the upper half of a cube relative to the bottom (Fig. 2.1). If atom A is to move to position B, atom B to position C, etc., a *simultaneous* translation of all atoms on the slip plane must occur. Since this would involve simultaneous rupture of all the interatomic bonds acting across the slip plane (e.g., bonds $A-A'$, $B-B'$, $C-C'$, etc.), the necessary stress would have to be large. From Fig. 2.2a we see that the equilibrium atom positions within the crystalline lattice are located at P and R, with a separation of b units. Midway between P and R the energy is a maximum at Q, which represents a metastable equilibrium position. The exact shape of the energy curve shown in Fig. 2.2a depends on the nature of the interatomic bonds. Since this is not known precisely, a sinusoidal wave form is assumed for simplicity in this analysis. Locating the atoms on the slip plane anywhere other than at an equilibrium position, such as P and R, requires a force defined by the slope of the energy curve (Fig. 2.2b) at that position where $F \equiv -dE/dx$. For example, to place an atom between P and Q (that is, $0 < x < b/2$), a force acting to the right is required to counteract the

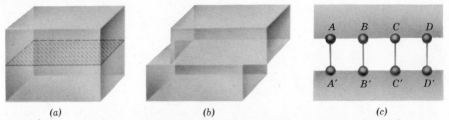

(a)   *(b)*   *(c)*

FIGURE 2.1   Movement of solid cube along a particular slip plane. (*a*) Undeformed cube with anticipated slip plane; (*b*) slipped cube revealing relative translation of part of cube; (*c*) atom position showing bonds across slip plane (*A–A′*, *B–B′*, *C–C′*, *D–D′*).

tendency for the atom to move back to the equilibrium site at P. Note that between P and Q, the slope of the energy curve is everywhere positive so that the force curve is positive also from $0 \leqslant x \leqslant b/2$. When $b/2 < x < b$, the atom wants to slide into its new equilibrium position at R (that is, the energy decreases continually from Q to R). To prevent this, a force acting to the left is needed to keep the atom stationary at some location between Q and R. This force is in an opposite direction to that needed between P and Q and is, therefore, negative. Note that in the region between Q and R, the slope of the

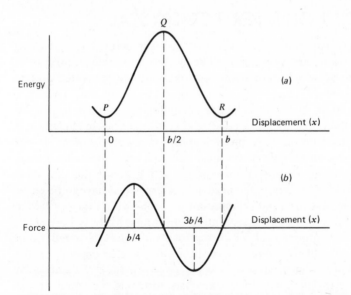

FIGURE 2.2   Diagram showing periodic nature of lattice. (*a*) Variation of energy with atom position in lattice. Preferred atom sites are at P and R, associated with minimum energy; (*b*) variation in force acting on atoms throughout lattice. Force is zero at equilibrium site positions and maximum at $b/4$, $3b/4$, $5b/4$,...,$(2n-1)b/4$.

energy curve is negative. Therefore, the corresponding portion of the force curve in this same region must also be negative.

From the above discussion it is clear that the shear stress necessary to move the atoms on the slip plane varies periodically from zero at P, Q, and R to a maximum value at $b/4$ and $3b/4$. Therefore, the shear stress may be expressed in the following form (from an analysis due to Frenkel[1]), based on an assumed sinusoidal variation in energy throughout the lattice:

$$\tau = \tau_m \sin \frac{2\pi x}{b} \tag{2-1}$$

where

$\tau = $ applied shear stress

$\tau_m = $ maximum theoretical strength of crystal

$x = $ distance atoms are moved

$b = $ distance between equilibrium positions

Plastic flow (that is, irreversible deformation) will then occur when the upper part of the cube (Fig. 2.1) is translated a distance greater than $b/4$ because of an applied shear stress τ_m.

For elastic strains, the shear stress may also be defined by Hooke's law

$$\tau = G\gamma \tag{2-2}$$

The shear strain may be approximated for small values by

$$\gamma \approx \frac{x}{a} \tag{2-3}$$

where $a = $ distance between slip planes. Combining Eqs. 2-1, 2-2, and 2-3 with $\sin 2\pi x/b$ approximated by $2\pi x/b$ for small strains

$$G\frac{x}{a} \approx \tau_m \frac{2\pi x}{b} \tag{2-4}$$

Upon rearranging

$$\tau_m \approx \frac{Gb}{2\pi a} \tag{2-5}$$

For most crystals b is of the same order as a, so Eq. 2-5 may be rewritten in the form

$$\tau_m \approx \frac{G}{2\pi} \tag{2-6}$$

TABLE 2.1   Theoretical and Experimental Yield Strengths in Various Materials[2]

Material	$G/2\pi$		Experimental Yield Strength		
	GPa	10^6 psi	MPa	psi	Error
Silver	12.6	1.83	0.37	55	$\sim 3 \times 10^4$
Aluminum	11.3	1.64	0.78	115	$\sim 1 \times 10^4$
Copper	19.6	2.84	0.49	70	$\sim 4 \times 10^4$
Nickel	32	4.64	3.2–7.35	465–1065	$\sim 1 \times 10^4$
Iron	33.9	4.92	27.5	3,990	$\sim 1 \times 10^3$
Molybdenum	54.1	7.85	71.6	10,385	$\sim 8 \times 10^2$
Niobium	16.6	2.41	33.3	4,830	$\sim 5 \times 10^2$
Cadmium	9.9	1.44	0.57	85	$\sim 2 \times 10^4$
Magnesium (basal slip)	7	1.02	0.39	55	$\sim 2 \times 10^4$
Magnesium (prism slip)	7	1.02	39.2	5,685	$\sim 2 \times 10^4$
Titanium (prism slip)	16.9	2.45	13.7	1,985	$\sim 1 \times 10^3$
Beryllium (basal slip)	49.3	7.15	1.37	200	$\sim 4 \times 10^4$
Beryllium (prism slip)	49.3	7.15	52	7,540	$\sim 1 \times 10^3$

Because of the approximations made in this analysis, especially with regard to the form of the energy-displacement curve, the magnitude of the theoretical shear strength τ_m from Eq. 2-6 is of an approximate nature. More realistic estimates place τ_m in the range of $G/30$. Nevertheless, it is instructive to compare theoretical strength values calculated with Eq. 2-6 with experimentally determined shear strengths for single crystals of various materials. From Table 2.1, it is immediately obvious that very large discrepancies exist between theoretical and experimental values for all materials tabulated. Without question, the lack of precision regarding computations based on Eq. 2-6 is *not* responsible for these large errors. Rather, the discrepancies must be accounted for in a different manner.

2.2   THE NEED FOR LATTICE IMPERFECTIONS—DISLOCATIONS

In 1934 Taylor, Orowan, and Polanyi postulated independently the existence of a lattice defect that would allow the cube in Fig. 2.1 to slip at much lower stress levels.[3,4] By introducing an extra half plane of atoms into the lattice (Fig. 2.3),

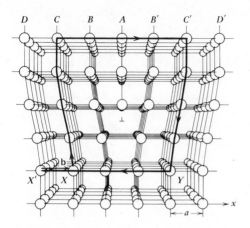

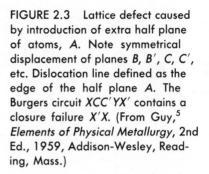

FIGURE 2.3 Lattice defect caused by introduction of extra half plane of atoms, *A*. Note symmetrical displacement of planes *B*, *B'*, *C*, *C'*, etc. Dislocation line defined as the edge of the half plane *A*. The Burgers circuit *XCC'YX* contains a closure failure *X'X*. (From Guy,[5] *Elements of Physical Metallurgy*, 2nd Ed., 1959, Addison-Wesley, Reading, Mass.)

they were able to show that atom bond breakage on the slip plane could be restricted to the immediate vicinity of the bottom edge of the half plane (called the dislocation line). As the dislocation line moves through the crystal, bond breakage across the slip plane occurs *consecutively* rather than *simultaneously* as was necessary in the perfect lattice (Fig. 2.1). The consecutive nature of bond breakage is shown in Fig. 2.4 where the extra half plane is shown at different locations during its movement through the crystal. The end result of the movement of this half plane is the same as shown in Fig. 2.1—the upper half of the cube has been translated relative to the bottom half by an amount equal to the distance between equilibrium atomic positions **b**. The major difference, however, is the fact that it takes much less energy to break one bond at a time than all the bonds at once. This concept is analogous to moving a large floor rug across the room. If you have ever tried to grab the edge of the rug and pull it to a new position, you know that it is nearly impossible to move a rug in this manner. In this case, the "theoretical shear stress" necessary to move the rug is strongly dependent on the frictional forces between the rug and the floor. If you persisted in your task you probably discovered that the rug could be moved quite easily in several stages by first creating a series of buckles at the edge of the rug and then propagating them, one at a time, across the rug by shuffling your feet behind each buckle. In this way you were able to move the rug by increments equal to the size of the buckle. Since the only part of the rug to move at any given time was the buckled segment, there was no need to overcome the frictional forces acting on the whole rug. Since the lattice dislocation is a similar work-saving "device," one may reconcile the large errors between theoretical and experimental yield strengths (Table 2.1) by assuming the presence of dislocations in the crystals that were examined.

Before we begin to deal with the specific character of dislocations, it is natural to wonder whether the analysis leading to Eq. 2-6 is correct after all. What is needed, of course, are test data for crystals possessing *no* dislocations. Fortu-

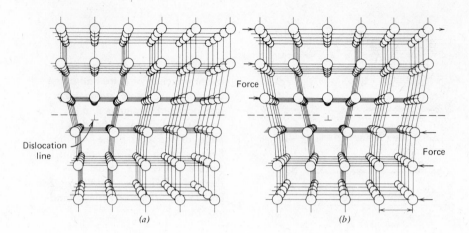

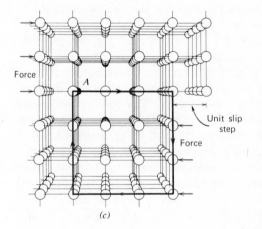

FIGURE 2.4 Successive positions of dislocation as it moves through crystal. Note that final offset of crystal resulting from passage of dislocation is the same as simultaneous movement of the entire crystal. Note the perfect Burgers circuit in (c). (From Guy,[5] *Elements of Physical Metallurgy*, 2nd Edition, 1959, Addison-Wesley, Reading, Mass.)

TABLE 2.2 Theoretical and Experimental Strengths of Dislocation-Free Crystal (Whiskers)[6]

Material	Theoretical Strength $(G/2\pi)$		Experimental Strength		
	GPa	10^6 psi	GPa	10^6 psi	Error
Copper	19.1	2.77	3.0	0.44	~6
Nickel	33.4	4.84	3.9	0.57	~8.5
Iron	31.8	4.61	13	1.89	~2.5
B_4C	71.6	10.4	6.7	0.98	~10.5
SiC	132.1	19.2	11	1.60	~12
Al_2O_3	65.3	9.47	19	2.76	~3.5
C	156.0	22.6	21	3.05	~7

nately, such perfect crystals—called whiskers—have been prepared in the laboratory. The strengths of these extraordinary crystals, shown in Table 2.2, are seen to be in close agreement with theoretical maximum values computed from Eq. 2-6. On this basis, the Frenkel analysis is verified.

2.3   LATTICE RESISTANCE TO DISLOCATION MOVEMENT—THE PEIERLS STRESS

From Fig. 2.3 it is clear that the insertion of the extra half plane of atoms has perturbed the lattice and caused atoms to be pushed aside laterally, particularly in the upper half of the crystal. For example, atoms along planes B and C are displaced to the left while atoms in planes B' and C' are displaced to the right. Since the forces acting on these groups of atoms are equal and of opposite sign (that is, pairing atoms in plane B with those in B' and those in C with atoms in plane C'), movement of the extra half plane A either to the left or right would be met by self-balancing forces on the other atoms within the distorted region. On this basis, the force necessary to move a dislocation would be zero. However, Cottrell[4] pointed out that although the above situation should prevail when the dislocation occurs in a symmetrical position (such as the one shown for plane A in Fig. 2.3), it would not hold true when the dislocation passes through nonsymmetrical positions. Consequently, some force is necessary to move the dislocation through the lattice. An important characteristic of this force (called the Peierls-Nabarro or Peierls force) is that its magnitude varies periodically as the dislocation moves through the lattice.

It is known that the magnitude of the Peierls force depends to a large extent on: (1) the width of the dislocation, W, which represents a measure of the distance over which the lattice is distorted because of the presence of the dislocation (Fig. 2.5) and (2) the distance between similar planes a. The Peierls stress has been shown to depend on W and b in the form

$$\tau_{p-n} \propto G e^{-2\pi W/b} \tag{2-7}$$

where $W = a/1 - \nu$.

From Eq. 2-7, the Peierls stress for a given plane is seen to decrease with increasing distance between like planes. Since the distance between planes varies inversely with their atomic density, slip is preferred on closely packed planes. In addition, the Peierls stress depends on the dislocation width, which is dependent on atomic structure and the nature of the atomic bonding forces. For example, when the bonding forces are spherical in distribution and act along the line of centers between atoms, the dislocation width is large. Since this type of bonding is found in close-packed structures, it is seen that the Peierls stress in face-centered cubic and close-packed hexagonal crystals is low. By contrast, when bonding forces are highly directional (as in the case of covalent, ionic, and

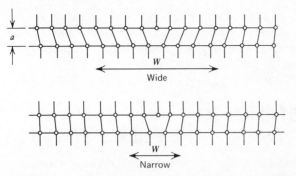

FIGURE 2.5 Characteristic width of an edge dislocation which affects the Peierls-Nabarro stress.[7] (Reproduced by courtesy of the Council of the Institution of Mechanical Engineers from *The Properties of Materials at High Rates of Strain* by A. H. Cottrell.)

body-centered cubic crystals), the dislocation width is narrow and the Peierls stress correspondingly large. Although several attempts have been made to compute precisely the magnitude of the Peierls stress in a given lattice, considerable difficulties arise because the exact shape of the force-displacement curve is unknown. Foreman, Jaswon, and Wood[8] showed that when the amplitude of an assumed sinusoidal force-displacement law was reduced by half, the width of the dislocation increased four-fold. This, in turn, had the effect of reducing the computed Peierls stress value by more than *six orders of magnitude*. Until the force-displacement relationship between atoms can be defined more precisely, the magnitude of the Peierls stress in crystals can be described only in qualitative terms.

2.3.1 Peierls Stress Temperature Sensitivity

One such qualitative characteristic of the Peierls stress relates to the temperature sensitivity of the yield strength. Since the Peierls stress depends on the short-range stress field of the dislocation core, it is sensitive to the thermal energy in the lattice and, hence, to the test temperature. At low temperatures, where thermal enhancement of dislocation motion is limited, the Peierls stress is large. In crystals that have wide dislocations, however, the increase in Peierls stress with decreasing temperature is insignificant, since the Peierls stress is negligible to begin with. The situation is quite different in crystals that contain narrow dislocations. Although the Peierls stress may be small at elevated temperatures, it rises rapidly with decreasing temperature and represents a large component of the yield strength in the low-temperature regime. The yield strength temperature sensitivity of several engineering materials is shown in Table 2.3. The large Peierls stress in ceramic materials is partly responsible for their limited ductility

TABLE 2.3 Relationship Between Dislocation Width and Yield Strength Temperature Sensitivity

Material	Crystal Type	Dislocation Width	Peierls Stress	Yield Strength Temperature Sensitivity
Metal	FCC	wide	very small	negligible
Metal	BCC	narrow	moderate	strong
Ceramic	ionic	narrow	large	strong
Ceramic	covalent	very narrow	very large	strong

at low and moderate temperatures. However, the Peierls stress decreases rapidly with increasing temperature, thereby enhancing plastic deformation processes in these materials at high temperatures.

2.3.2 Effect of Dislocation Orientation on Peierls Stress

The Peierls stress as described above represents an upper bound to the stress necessary to move a dislocation through a crystal. In fact, dislocations will seldom lie completely along directions of lowest energy, or energy valleys, within the lattice. Rather, the dislocation line will contain bends or kinks that lie across energy peaks at some angle (Fig. 2.6a). The angle θ that the kink makes with the rest of the dislocation line, as well as its length l, is a direct consequence of the balance between two competing factors. On one hand, the dislocation will prefer to lie along the energy valleys such that the kink length is minimized and the kink angle maximized [90 degrees (Fig. 2.6b)]. (It should be noted that to create such a sharp kink angle will increase the energy of the dislocation, since the energy of any curved segment of a dislocation line increases with decreasing radius of curvature.) On the other hand, the dislocation line tries to be as short as possible to minimize its self-energy and in the limit would prefer a straight-line configuration (Fig. 2.6c; see also Section 2.6). The degree to which the kinked dislocation line approaches either extreme will depend strongly on ΔE, the amplitude of the periodic energy change along the crystal. When this amplitude is large, the dislocation line will prefer to lie along energy troughs such that short sharp kinks will be formed. Alternately, when ΔE is small, long undulating kinks (more like gradual bends) will be observed.

The relative ease of movement of both the dislocation line segments lying along energy troughs and the kink sections, respectively, is now considered. Since the kinked sections are located across higher energy portions of the crystal, they can move more easily than the line segments along the energy troughs, which must overcome the maximum energy barrier if they are to move. Upon application of a shear stress, the kinked segments shown in Fig. 2.6a move to the left or right (depending on the sense of the stress), which in effect allows the entire dislocation line to move in a perpendicular direction from one energy

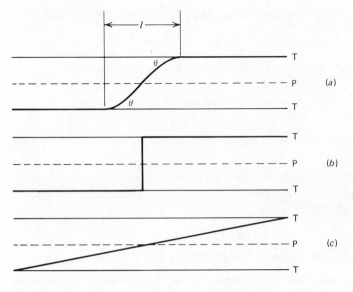

T = Trough
P = Peak

FIGURE 2.6 Position of dislocation line containing kinks with respect to energy troughs within lattice. (a) Typical configuration showing kink of length *l* with angle θ between kink segment and segment lying along energy trough; (b) sharp kink formed when magnitude of energy fluctuation in lattice is large. In this case, $l \to 0$ and $\theta \to 90°$; (c) broad kink formed when energy fluctuation in lattice is small. Here, $l \to \infty$ and θ becomes very small.

trough to the adjacent one. The lateral movement of such a kink may be likened to the motion of a whip that has been snapped. Consequently, the introduction of kinks into dislocations eases their movement through the lattice.

It may be concluded, then, that the lattice resistance to the movement of a dislocation depends on both the magnitude of the Peierls stress and the orientation of the dislocation line within the periodically varying energy field in the lattice. Since both factors will depend on ΔE, which depends on the force-displacement relationship between atoms, the importance of the latter relationship is emphasized. Unfortunately, current lack of specific knowledge concerning the force law severely hampers quantitative treatments of dislocation-lattice interactions.

2.4 CHARACTERISTICS OF DISLOCATIONS

The reader should understand certain fundamental characteristics of dislocations. A dislocation is a lattice *line* defect that defines the boundary between slipped and unslipped portions of the crystal. Two basically different dislocations can be identified. The *edge* dislocation is defined by the edge of the extra half plane of atoms shown in Fig. 2.3. Note how this extra half plane is wedged into the top half of the crystal. As a result, the upper part of the crystal is compressed on either side of the half plane, while the region below the dislocation experiences considerable dilatation. By convention, the bottom edge of the half plane shown in Fig. 2.3 is defined as a *positive* edge dislocation. Had the extra half plane been introduced into the lower half of the crystal, the regions of localized compression and dilatation would be reversed and the dislocation line defined as a *negative* edge dislocation. Clearly, if a crystal contained both positive and negative edge dislocations lying on the same plane, their combination would result in mutual annihilation and the elimination of two high-energy regions of lattice distortion.

The movement of an edge dislocation and its role in the plastic deformation process may be understood more clearly by considering its Burgers circuit. (The Burgers circuit is a series of atom to atom steps along lattice vectors that generate a closed loop about any location in the lattice.) In a perfect lattice (Fig. 2.4c), the Burgers circuit beginning at A and progressing an equal and opposite number of lattice vectors in the horizontal and vertical directions will return to its starting position. When this occurs, the lattice contained within the circuit is considered perfect. When an edge dislocation is present in the lattice, the circuit does not close (Fig. 2.3). The vector needed to close the Burgers circuit (X'X) is called the Burgers vector **b** of the dislocation and represents both the magnitude and direction of slip of the dislocation.

Another important feature of **b** is its orientation relative to the dislocation line. For the edge dislocation, **b** is oriented normal to the line defect. Ordinarily, plastic flow via edge dislocation movement is restricted to that one plane defined by the dislocation line and its Burgers vector. Such *conservative* motion will occur with the edge dislocation moving in the same direction as **b** (i.e., the direction of slip). From Section 2.3, the planes on which dislocations move are usually those of greatest separation and atomic density.

It is possible for an edge dislocation to undertake *nonconservative* motion; that is, movement out of its normal glide plane. This can occur by removal of a row of atoms, such as by the diffusion of lattice vacancies to the bottom of the extra half plane (Fig. 2.7). In this manner, the dislocation *climbs* from one plane to another where conservative glide may occur once again. Since vacancy diffusion is a thermally activated process, dislocation climb becomes an important process only at elevated temperatures above about one-half the melting point of the material. This mechanism will be discussed again in Chapter Five.

The other line defect, called the *screw* dislocation is defined by the line AB in Fig. 2.8, the latter being generated by displacement of one part of the crystal relative to the other. The Burgers circuit about the screw dislocation assumes the shape of a helix, very much like a spiral staircase, wherein a 360-degree rotation

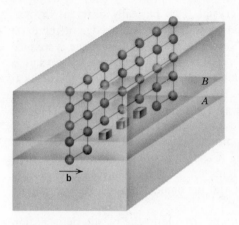

FIGURE 2.7 Dislocation climb involving vacancy (□) diffusion to edge dislocation allowing its movement to climb from plane *A* to plane *B*.

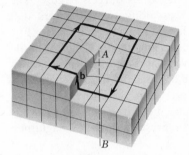

FIGURE 2.8 Screw dislocation *AB* resulting from displacement of one part of crystal relative to the other. Note *AB* is parallel to **b**.

produces a translation equal to one lattice vector in a direction parallel to the dislocation line *AB*. A right-handed screw dislocation is defined when a clockwise, 360-degree rotation causes the helix to advance one lattice vector. The same advance resulting from a 360-degree counterclockwise rotation is a left-handed screw dislocation. Thus, the screw dislocation and its Burgers vector are mutually parallel, unlike the orthogonal relationship found for the edge dislocation. Note that while the slip direction is again parallel to **b** as was found for the edge dislocation, the direction of movement of the screw dislocation is perpendicular to **b**. To better visualize this fact, take a piece of paper and tear it part way across its width. Note that the movement of your hands (the shear direction parallel to the Burgers vector) is perpendicular to the movement of the terminal point (the screw dislocation) of the tear.

Unlike the edge dislocation, a unique slip plane cannot be identified for the screw dislocation. Rather, an infinite number of potential slip planes may be defined, since the dislocation line and Burgers vector are parallel to one another. In fact, the movement of a screw dislocation is confined to those sets of planes that possess a low Peierls-Nabarro stress. Even so, the screw dislocation possesses greater mobility than the edge dislocation in moving through the lattice.

The movement of the screw dislocation from one slip plane to another takes place by a process known as *cross-slip* and may be understood by examining Fig. 2.9. At the onset of plastic deformation, the screw dislocation XY is seen to be moving on plane A (Fig. 2.9a). If continued movement on this plane is impeded by some obstacle, such as a precipitate particle, the screw dislocation can cross over to another equivalent plane, such as B, and continue its movement (Fig. 2.9b). Since the Burgers vector is unchanged, slip continues to occur in the same direction, though on a different plane. Movement of the screw dislocation may continue on plane B or return to plane A by a second cross-slip process (Fig. 2.9c). A summary of the basic differences between edge and screw dislocations is presented in Table 2.4.

Since many dislocations in a crystalline solid are curved like the one shown in Fig. 2.10a, they take on aspects of both edge and screw dislocations. With **b** the

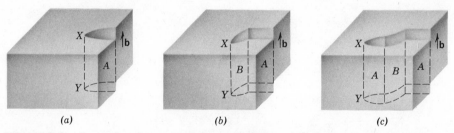

(a)   (b)   (c)

FIGURE 2.9   Cross-slip of a screw dislocation XY from (a) plane A to (b) plane B to (c) plane A. Slip always occurs in direction of Burgers vector **b**.

TABLE 2.4   Characteristics of Dislocations

Dislocation characteristic	Type of Dislocation	
	Edge	Screw
Slip Direction	Parallel to **b**	Parallel to **b**
Relationship between dislocation line and **b**	Perpendicular	Parallel
Direction of dislocation line movement relative to **b**	Parallel	Perpendicular
Process by which dislocations may leave the glide plane	Nonconservative climb	Cross-slip

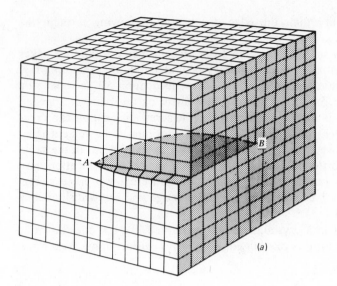

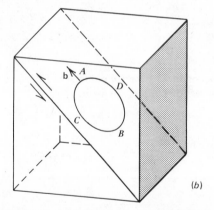

FIGURE 2.10 Curved dislocations containing edge and screw components. (a) Dislocation AB is pure screw at A and pure edge at B; (b) dislocation loop that grows out radially with shear stress applied parallel to **b**.

same along the entire length of the dislocation, the dislocation is seen to be pure screw at A and pure edge at B. The reader should verify this by constructing a Burgers circuit around the dislocation at A and B. It follows that at all points between A and B the dislocation possesses both edge and screw components. For this reason, AB is called a mixed dislocation. Another example of a mixed dislocation is the dislocation loop. From Fig. 2.10b the loop is seen to be pure positive edge at A and pure negative edge at B while being pure right-handed screw at D and pure left-handed screw at C. Everywhere else the loop contains both edge and screw components. When a shear stress is applied parallel to $\mathbf{b}$, we see from Table 2.4 that the loop will expand radially.

Dislocations can terminate at a free surface or at a grain boundary but never within the crystal. Consequently, dislocations either must form closed loops or networks with branches that terminate at the surface (Fig. 2.11). A basic characteristic of a network junction point or node involving at least three dislocation branches is that the sum of the Burgers vectors is zero.

$$\mathbf{b}_1 + \mathbf{b}_2 + \mathbf{b}_3 = 0 \qquad (2\text{-}8)$$

Furthermore, when these dislocations are of the same sense

$$\mathbf{b}_1 = \mathbf{b}_2 + \mathbf{b}_3 \qquad (2\text{-}9)$$

and the Burgers vector of one dislocation is equal to the sum of the other two Burgers vectors. This holds when two dislocations combine to form a third or when one dislocation dissociates into two additional dislocations.

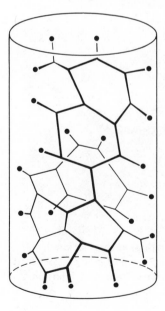

FIGURE 2.11 Diagram illustrating network arrangement of dislocations in crystal. Dislocations can terminate only at a node, in a loop, or at a grain boundary or free surface.[7] (Reproduced by courtesy of the Council of the Institution of Mechanical Engineers from *The Properties of Materials at High Rates of Strain* by A. H. Cottrell.)

2.5 OBSERVATION OF DISLOCATIONS

Much effort has been devoted to the direct examination of dislocations within the crystal. One successful technique involves chemical or electrolytic etching of polished free surfaces. By carefully controlling the strength of the etchant, the high-energy dislocation cores exposed at the surface are attacked preferentially with respect to other regions on the polished surface. The result is the formation of numerous etch pits, each corresponding to one dislocation (Fig. 2.12). This technique has been used to study the effect of applied stress on dislocation velocity. Notable experiments by Johnston and Gilman[9] have identified the stress-induced, time-dependent change in dislocation position by repeated etching, with each new etch pit representing the new location of the moving dislocation.

The most widely used technique for the study of dislocations involves their direct examination in the transmission electron microscope. Since electrons have little penetrating power, the specimens used for such studies are very thin films

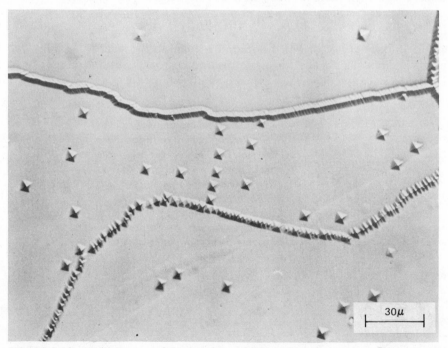

FIGURE 2.12 Etch pits on polished surface of lithium fluoride, each associated with an individual dislocation. The etch pit lineage indicates alignment of many dislocations in the form of low-angle boundaries (see Section 2.6). (From Gilman and Johnston,[23] reprinted by permission of General Electric Co.)

—only about 0.1 to 0.2 μm thick. Since dislocations are lattice defects, their presence perturbs the path of the diffracted electron beam relative to its path in a perfect crystal. As a result, various images are produced, depending on the prevailing diffraction conditions. Often, dislocations appear as single dark lines like those shown in planar array in Fig. 2.13a. Each dislocation lies along a particular plane and extends from the top to the bottom of the foil (Fig. 2.13b).

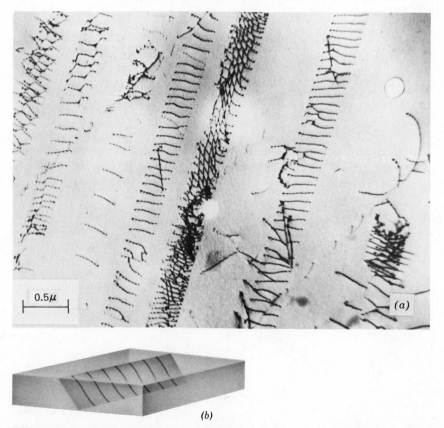

FIGURE 2.13   Observation of individual dislocations in thin foil. (a) Planar arrays of dislocations in 18 Cr-8 Ni stainless steels (from Michalak,[10] *Metals Handbook* Vol. 8, copyright American Society for Metals, 1973 used with permission); (b) diagram showing position of dislocations on the glide plane in the foil (after Hull[11]).

As a result, the viewer sees only the projected length of the dislocation line, with the actual length being dependent on the foil thickness and angle of the plane containing the dislocations. Much progress was made about 15 years ago toward the understanding of electron diffraction images and the identification of numerous dislocation configurations. Some important publications in this area are cited at the end of the chapter.[11-17]

2.6 ELASTIC PROPERTIES OF DISLOCATIONS

As might be expected, there is an elastic stress field associated with the distorted lattice surrounding a dislocation. It is easy to describe the stresses developed around a screw dislocation (Fig. 2.14a). By rolling the cylindrical element out flat (Fig. 2.14b), the shear strain $\gamma_{\theta z}$ is seen in polar coordinates to be

$$\gamma_{\theta z} = \frac{b}{2\pi r} \tag{2-10}$$

From Hooke's law, the corresponding stress is

$$\tau_{\theta z} = G\gamma_{\theta z} = \tau_{z\theta} = \frac{Gb}{2\pi r} \tag{2-11}$$

Since displacements are generated only in the z direction, the other stress components are zero. Equation 2-11 shows that the stress $\tau_{\theta z}$ becomes infinitely large as r approaches zero. Since this is unreasonable, there exists a limiting

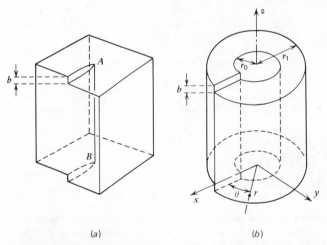

(a)   (b)

FIGURE 2.14 Elastic distortions surrounding screw dislocation. (From Hull,[11] reprinted with permission from Hull, *Introduction to Dislocations*, Pergamon Press, 1965.)

distance r_0 from the dislocation center (estimated to be 5 to 10 Å) within which Eq. 2-11 is no longer applicable. For the rectangular coordinates shown in Fig. 2.14b, the shear stresses surrounding the screw dislocation can also be given by

$$\tau_{xz} = \tau_{zx} = -\frac{Gb}{2\pi} \frac{y}{x^2+y^2}$$

$$\tau_{yz} = \tau_{zy} = \frac{Gb}{2\pi} \frac{x}{x^2+y^2}$$

$$(2\text{-}12)$$

Again all other stresses are zero.

The stress field surrounding an edge dislocation is more complicated, since both hydrostatic and shear stress components are present. In rectangular coordinates these stresses are given by[4]

$$\sigma_{xx} = -\frac{Gby}{2\pi(1-\nu)} \frac{(3x^2+y^2)}{(x^2+y^2)^2}$$

$$\sigma_{yy} = +\frac{Gby}{2\pi(1-\nu)} \frac{(x^2-y^2)}{(x^2+y^2)^2}$$

$$\tau_{xy} = \tau_{yx} = \frac{Gbx}{2\pi(1-\nu)} \frac{(x^2-y^2)}{(x^2+y^2)^2}$$

$$(2\text{-}13)$$

$$\sigma_{zz} = \nu(\sigma_{xx}+\sigma_{yy})$$

$$\tau_{xz} = \tau_{zx} = \tau_{yz} = \tau_{zy} = 0$$

where ν = Poisson's ratio. Comparing Eq. 2-13 with Fig. 2.15, we found a region of pure compression directly above the edge dislocation ($X=0$) and pure tension below the bottom edge of the extra half plane. Along the slip plane ($Y=0$) the stress is pure shear. For all other positions surrounding the dislocation, the stress field is found to contain compressive and/or tensile components as well as a shear component.

The elastic strain energy is another elastic property of a dislocation. For the simple case of the screw dislocation, this quantity may be given by

$$E_{\text{screw}} = \frac{1}{2} \int_{r_0}^{r_1} \tau_{\theta z} b \, dr$$

$$(2\text{-}14)$$

Note that the energy is defined for the region outside the core of the dislocation r_0 to the outer boundaries of the crystal r_1 (see Fig. 2.14). Combining Eqs. 2-11

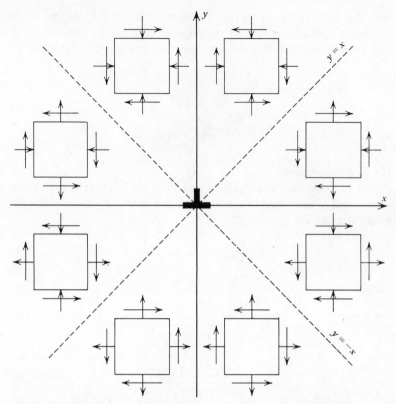

FIGURE 2.15 Elastic stress field surrounding an edge dislocation. (From Read,[3] *Dislocations in Crystals.* Copyright 1953, McGraw-Hill Book Co. Used with permission of McGraw-Hill Book Company.)

and 2-14, we get

$$E_{\text{screw}} = \frac{1}{2} \int_{r_0}^{r_1} \frac{Gb^2}{2\pi} \frac{dr}{r} = \frac{Gb^2}{4\pi} \ln \frac{r_1}{r_0} \qquad (2\text{-}15)$$

The elastic energy of the edge dislocation is slightly larger and given by

$$E_{\text{edge}} = \frac{Gb^2}{4\pi(1-\nu)} \ln \frac{r_1}{r_0} \qquad (2\text{-}16)$$

Since a general dislocation contains both edge and screw components, its energy is intermediate to the limiting values given by Eqs. 2-15 and 2-16. For purposes of our discussion it is sufficient to note that

$$E = \alpha Gb^2 \qquad (2\text{-}17)$$

where

> E = energy of any dislocation

> α = geometrical factor with α taken between 0.5 and 1.0

A particularly important consequence of Eqs. 2-15 to 2-17 is that slip will usually occur in close-packed directions so as to minimize the Burgers vector of the dislocation. The preferred slip directions in major crystal types are given in Chapter Three. Equations 2-15 to 2-17 also allow one to determine whether or not a particular dislocation reaction will occur. From Eq. 2-9, such a reaction will be favored when

$$b_1{}^2 > b_2{}^2 + b_3{}^2$$

(neglecting possible anisotropy effects associated with G).

Two other elastic properties of a dislocation are its line tension and the force needed to move the dislocation through the lattice. The line tension T is described in terms of its energy per unit length and is given by

$$T \propto Gb^2 \tag{2-18}$$

The line tension acts to straighten a dislocation line to minimize its length, thereby lowering the overall energy of the crystal (see Fig. 2.6c). Consequently, it is necessary to apply a stress τ in order that a dislocation line remains curved. This stress is shown to increase with increasing line tension T and decreasing radius of curvature R where

$$\tau \propto \frac{T}{bR} \tag{2-19}$$

Combining Eqs. 2-18 and 2-19 we find that

$$\tau \propto \frac{Gb}{R} \tag{2-20}$$

This relationship will be referred to in Section 2.8.

Finally, the force acting on a dislocation is found to depend on the intrinsic resistance to dislocation movement through the lattice, the Peierls-Nabarro stress (Section 2.3), and interactions with other dislocations. As shown by Read[3] for the case of parallel dislocations, screw dislocations will always repel one another when the Burgers vectors of both dislocations are of the same sign; they will always be attracted to one another when the sign of the Burgers vectors are opposite. In either case, the magnitude of the force is inversely proportional to the distance between the two dislocations. The force between two edge dislocations is complicated by a reversal in sign when the horizontal distance between two dislocations becomes less than the vertical distance between the two parallel

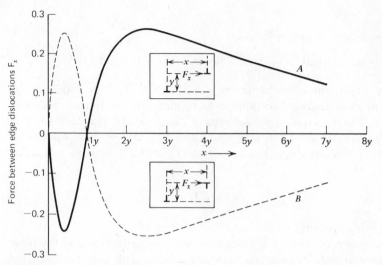

FIGURE 2.16 Force between parallel edge dislocations. Curve *A* corresponds to dislocations of the same sign. Force reversal when $X < Y$ causes like dislocations to become aligned as in a tilt boundary. Curve *B* corresponds to dislocations of opposite

sign. Unit of force F_x is $\dfrac{Gb^2}{2\pi(1-\nu)}\ \dfrac{x(x^2-y^2)}{(x^2+y^2)^2}$.[7] (From A. H. Cottrell, *The Properties of*

Materials at High Rates of Strain, Institute of Mechanical Engineering, London, 1957.)

slip planes (Fig. 2.16). Consequently, like edge dislocations are attracted to one another when $x < y$. As a result, like edge dislocations can form stable arrays of dislocations located vertically above one another in the form of simple tilt boundaries (Fig. 2.12).

2.7 DISLOCATION-DISLOCATION INTERACTIONS

Since dislocations of the same sign will repel one another and not coalesce, they will tend to pile up (each with a unit Burgers vector) against a barrier on the slip plane (Fig. 2.17). As one might expect, a large stress concentration is developed at the leading edge of the pileup, which can lead to premature fracture in certain materials (see Chapter Ten for further discussion of this point).

Equally important are the intersections of dislocations moving on different slip planes. Two different edge dislocation intersections are shown in Fig. 2.18. In the first case, where the Burgers vectors of the two dislocations are at right angles, the intersection of dislocation AB leads to a simple lengthening of dislocation XY. On the other hand, dislocation XY with a Burgers vector $\mathbf{b}_1$ cuts dislocation AB, producing a jog PP' that has a length equal to that of $\mathbf{b}_1$. The

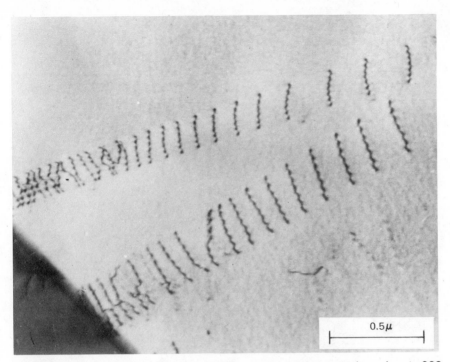

FIGURE 2.17   Dislocation pileups on two systems against a grain boundary in 309 stainless steel ($\gamma = 35$ mJ/m^2). (Courtesy of Anthony Thompson, Rockwell International.)

Burgers vector of PP', however, is $\mathbf{b}_2$—the same as dislocation AB to which it belongs. Since $\mathbf{b}_2$ and PP' are normal to one another, PP' is of the edge type. It may be shown that this jog does not impede the movement of the dislocation AB. When the Burgers vectors of the edge dislocations are parallel, the jogs produced are different in character. As shown in Fig. 2.18b, dislocation XY with its Burgers vector $\mathbf{b}_1$ produces a jog PP' in dislocation AB. Since the Burgers vector $\mathbf{b}_2$ in dislocation AB is parallel to the jog PP', the jog is of the screw type. Similarly, the jog QQ' in dislocation XY is found also to be of the screw type. The screw jogs PP' and QQ' have greater mobility than the edge dislocations to which they belong. Consequently, their presence does not impede the overall motion of the dislocation. In summary, jogs generated in edge dislocations will not affect the movement of the dislocation.

The same cannot be said for intersections involving screw dislocations. As illustrated in Fig. 2.19a the intersection of an edge and screw dislocation will produce a jog PP' in the edge dislocation AB and another jog QQ' in the screw dislocation XY. Since each jog assumes the same Burgers vector as its dislocation, it may be seen that PP' and QQ' are both edge jogs. From the above

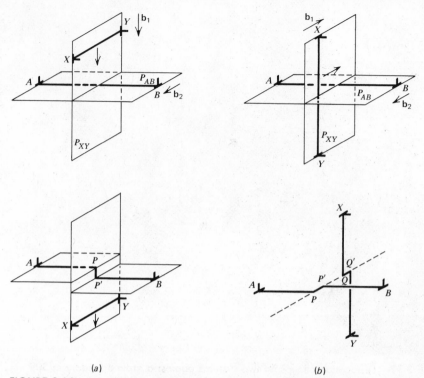

(a)   (b)

FIGURE 2.18   Intersection of two edge dislocations. (a) Burgers vectors are at right angles and produce an edge jog PP' in dislocation AB (from Read,[3] Dislocations in Crystals. Copyright 1953, McGraw-Hill Book Co. Used with permission of McGraw-Hill Book Company); (b) Burgers vectors are parallel and produce two screw jogs PP' and QQ'. (From Hull,[11] reprinted with permission from Hull, Introduction to Dislocations, Pergamon Press, 1965.)

discussion, PP' will not impede the motion of dislocation AB, whereas QQ' will restrict the movement of the screw dislocation XY. The same can be said for the edge type jogs PP' and QQ' found in the screw dislocations AB and XY, respectively, shown in Fig. 2.19b. The restriction placed on the mobility of the screw dislocations is caused by the inability of the edge jog to move on any plane other than that defined by the jog QQ' and $\mathbf{b}_2$ (i.e., the plane $QQ'YZ$; see Fig. 2.20). Consequently, when a shear stress is applied parallel to $\mathbf{b}_2$, the screw segments XQ and $Q'Y$ will produce displacements parallel to $\mathbf{b}_2$, while the screw dislocation lines move to DE and FG, respectively. The only way that the edge jog QQ' can follow along plane $EFQ'Q$ is by nonconservative motion involving vacancy-assisted dislocation climb. As shown schematically in Fig. 2.21 for the case of small jogs with heights of one or two atom spacings, the screw dislocation first bows out under application of a shear stress and then moves

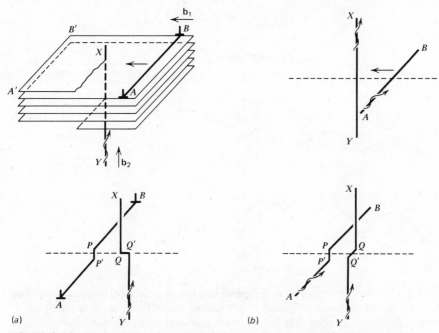

FIGURE 2.19   Intersection of screw dislocation *XY* with (*a*) edge dislocation *AB* to form two edge jogs *PP'* and *QQ'* (from Read,[3] *Dislocations in Crystals.* Copyright 1953, McGraw-Hill Book Co. Used with permission of McGraw-Hill Book Company); (*b*) another screw dislocation *AB* which forms two edge jogs *PP'* and *QQ'* (from Hull,[11] reprinted with permission from Hull, *Introduction to Dislocations*, Pergamon Press, 1965.) Edge jogs in screw dislocations impede their motion.

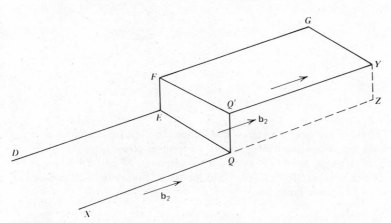

FIGURE 2.20   Screw dislocation *XY* containing an edge jog *QQ'* which can move conservatively on plane *QQ'YZ* but nonconservatively on plane *EFQ'Q* when screw components *XQ* and *Q'Y* move to *DE* and *FG*, respectively.

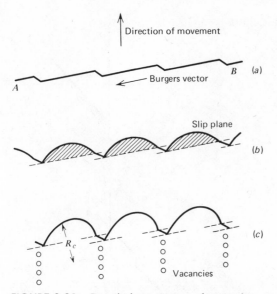

FIGURE 2.21 Detailed movement of jogged screw dislocation. (a) Jogged dislocation under zero stress; (b) applied shear stress causes screw component to bow out between edge jogs; (c) edge jogs follow screw segments by nonconservative climb, leaving behind a trail of vacancies. (From Hull,[11] reprinted with permission from Hull, *Introduction to Dislocations*, Pergamon Press, 1965.)

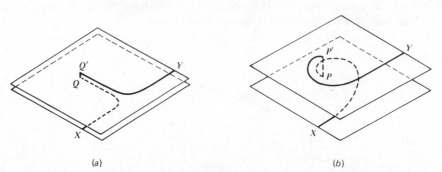

FIGURE 2.22 Effect of jog height on screw dislocation mobility. (a) Intermediate jog height QQ' causes long-edge segments (dipoles) to form as screw segments glide through crystal; (b) large jog height PP' allows screw segments XP and YP' to move independently of one another. (From Gilman and Johnston,[18] reprinted with permission of the authors and Academic Press, Inc.)

farther only by dragging along the edge jogs which leave behind a trail of vacancies. When the jog height is greater as a result of multiple dislocation-dislocation intersections (e.g., about 50 to 100Å in silicon iron), too many vacancies would be required for climb of the jog. As a result, long-edge dislocation segments (called dipoles) are left behind as the screw segments of the dislocation advance through the crystal (Fig. 2.22a). When the jog height is even larger (e.g., greater than 200 Å in silicon iron), the screw segments XP and $P'Y$ move independently of one another (Fig. 2.22b). Examples of the three height categories of edge jogs in screw dislocations are shown in Fig. 2.23 for the case of silicon iron.

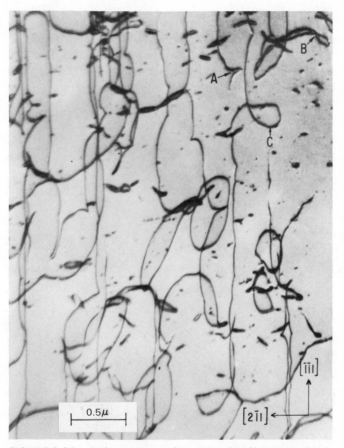

FIGURE 2.23　Dislocations in silicon-iron thin film. Note dipole trails at A, pinched off dipoles at B, and independent dislocation movement at the large jog at C. (From Low and Turkalo,[19] reprinted with permission from Low, *Acta Met.*, **10** (1962), Pergamon Press.)

2.8 DISLOCATION MULTIPLICATION

Since slip offsets are clearly visible in a light microscope (e.g., see Figs. 3.1, 3.5, 3.7), they must be in the range of $1 \mu m$ in height. Since the typical Burgers vector for a dislocation is of the order of 2 to 3×10^{-8} cm, there is a requirement for approximately 10^4 dislocations on each slip plane to create the slip step. That so many dislocations of the same sign should lie on the same plane *before* the crystal is stressed is highly unlikely. A possible alternative explanation is that additional dislocations must have been generated during deformation. This view is supported by the observations of electron microscopists who have found the dislocation densities in thin metal films to increase from 10^4–10^5 to 10^{11}–10^{12} dislocations/cm^2 as one proceeds from the annealed to heavily cold-worked state.

A widely accepted mechanism for dislocation generation is based on the Frank-Read source. In this model, a segment of a dislocation line is considered to be pinned either by foreign atoms or particles, or by interaction with other dislocations (Fig. 2.24). If a shear stress is applied to the crystal, the segment $A–B$ will bow out with a radius given by

$$\tau \propto \frac{Gb}{R} \tag{2-20}$$

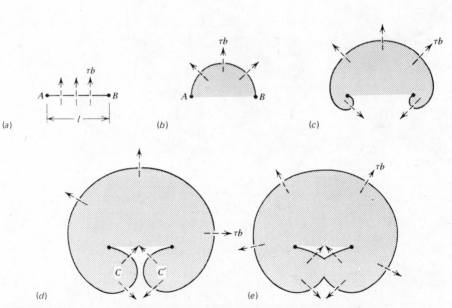

FIGURE 2.24 Frank-Read source for dislocation multiplication. Slipped area is shaded. Loop instability point is reached when shear stress $\tau \approx Gb/l$. (From Read,[3] *Dislocations in Crystals*. Copyright 1953, McGraw-Hill Book Co. Used with permission of McGraw-Hill Book Company.)

Dislocation bowing will increase with increasing applied stress while the radius of curvature decreases to the point where R equals half the pinned segment length, l (Fig. 2.24b). At this point, the loop becomes unstable and begins to bend around itself (Fig. 2.24c). The stress necessary to produce this instability is given by

$$\tau \approx \frac{Gb}{l} \qquad (2\text{-}21)$$

where $l =$ distance between pinning points. Finally, the loop pinches off at C and C', since these two regions correspond to two screw dislocations of opposite sign (Fig. 2.24d). After this has occurred, the loop and cusp ACB straighten, leaving the same segment AB as before but with an additional loop containing the same Burgers vector as the original segment (Fig. 2.24e). Upon further application of the stress, the segment AB can bow again to form a second loop while the initial loop moves out radially. With continued application of the shear stress, this source can generate an unlimited number of dislocations. In reality, however, the source is eventually shut down by back stresses produced by the pileup of dislocation loops against unyielding obstacles. The photograph shown in Fig. 2.25 represents a classic illustration of a Frank-Read source in a silicon crystal.

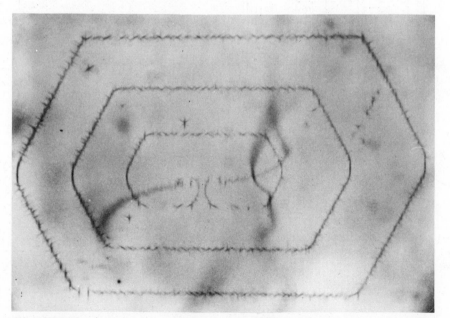

FIGURE 2.25 Photomicrograph showing Frank-Read source in silicon crystal. (From Dash,[20] reprinted with permission of General Electric Co.)

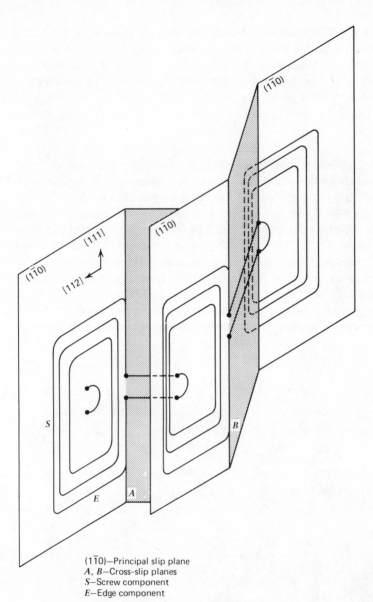

(1$\bar{1}$0)—Principal slip plane
A, B—Cross-slip planes
S—Screw component
E—Edge component

FIGURE 2.26   Dislocation multiplication by double cross-slip mechanism. (From Low and Guard,[22] reprinted with permission from Low, *Acta Met.*, **7** (1959), Pergamon Press.)

Another closely related dislocation generation mechanism has been suggested by Koehler[21] as modified by Low and Guard.[22] The basic feature of this model is that through the process of cross-slip, a screw dislocation can generate additional Frank-Read sources. From Fig. 2.26, we see that a screw dislocation segment has cross-slipped twice to resume movement on a plane parallel to the initial slip plane. Note the additional dislocation loops that may be generated by this process.

REFERENCES

1. J. Frenkel, *Zeit. Phys.*, **37**, 1926, p. 572.
2. W. J. McG. Tegart, *Elements of Mechanical Metallurgy*, Macmillan, New York, 1966.
3. W. T. Read, Jr., *Dislocations in Crystals*, McGraw-Hill, New York, 1953.
4. A. H. Cottrell, *Dislocations and Plastic Flow in Crystals*, Clarendon Press, Oxford, England, 1953.
5. A. G. Guy, *Elements of Physical Metallurgy*, 2nd Ed., Addison-Wesley, Reading, Mass., 1959.
6. W. H. Sutton, B. W. Rosen, and D. G. Flom, *SPE J.*, **72**, 1964, p. 1203.
7. A. H. Cottrell, *The Properties of Materials at High Rates of Strain*, Institute of Mechanical Engineering, London, 1957.
8. A. J. Forman, M. A. Jaswon, and J. K. Wood, *Proc. Phys. Soc.*, **A64**, 1951, p. 156.
9. W. G. Johnston and J. J. Gilman, *J. Appl. Phys.*, **30**, 1959, p. 129.
10. J. T. Michalak, *Metals Handbook*, Vol. 8, ASM, Metals Park, Ohio, 1973, p. 218.
11. D. Hull, *Introduction to Dislocations*, Pergamon Press, Oxford, England, 1965.
12. P. B. Hirsch, *J. Inst. Met.*, **87**, 1959, p. 406.
13. P. B. Hirsch, *Metall. Rev.*, **4**, 1959, p. 101.
14. A. Howie, *Metall. Rev.*, **6**, 1961, p. 467.
15. P. Kelly and J. Nutting, *J. Inst. Met.*, **87**, 1959, p. 385.
16. P. B. Hirsch, *Prog. Met. Phys.*, **6**, 1956, p. 236.
17. J. B. Newkirk and J. H. Wernick, eds., *Direct Observations of Imperfections in Crystals*, Interscience, New York, 1962.
18. J. J. Gilman and W. G. Johnston, *Solid State Phys.*, **13**, 1962, p. 147.
19. J. R. Low and A. M. Turkalo, *Acta Met.*, **10**, 1962, p. 215.
20. W. C. Dash, *Dislocations and Mechanical Properties of Crystals*, J. C. Fisher, ed., John Wiley, New York, 1957.
21. J. S. Koehler, *Phys. Rev.*, **86**, 1952, p. 52.
22. J. R. Low and R. W. Guard, *Acta Met.*, **7**, 1959, p. 171.
23. J. J. Gilman and W. G. Johnston, *Dislocations and Mechanical Properties of Crystals*, J. C. Fisher, W. G. Johnston, R. Thomson, and T. Vreeland, Jr., eds., John Wiley, New York, 1957, p. 116.

PROBLEMS

2.1 Demonstrate mathematically that dislocations at the head of a pileup will not combine to form a super dislocation with a Burgers vector of nb where $n=2$, $3, 4 \ldots n$.

2.2 The following face-centered cubic dislocation reaction will occur:

$$\frac{a}{2}[110] \rightarrow \frac{a}{6}[21\bar{1}] + \frac{a}{6}[121]$$

(a) Prove that the reaction will occur.
(b) What kind of dislocations are the $(a/6)\langle 121 \rangle$?
(c) What kind of crystal imperfection results from this dislocation reaction?
(d) What determines the distance of separation of the $(a/6)[21\bar{1}]$ and the $(a/6)[121]$ dislocations?

2.3 Distinguish between climb and cross-slip and discuss the role of stacking fault energy with regard to the latter.

2.4 Discuss the nature of the Peierls stress with regard to a dislocation and describe the role of the Peierls stress in determining the preferred slip plane in a crystal and the yield-strength temperature dependence of the crystal.

2.5 Why do dislocation loops tend to be circular? Why then are they angular for silicon as shown in Fig. 2.25?

CHAPTER
THREE
SLIP IN
CRYSTALLINE
SOLIDS

We saw in the previous chapter that plastic deformation occurs primarily by sliding along certain planes with one part of a crystal moving relative to another. This block-like nature of slip produces crystal offsets (called slip steps) in amounts given by multiples of the unit dislocation displacement vector **b** (see Fig. 3.1). To minimize the Peierls stress, slip occurs predominantly on crystallographic planes of maximum atomic density. In addition, slip will occur in the close-packed direction, which represents the shortest distance between two equilibrium atom positions and, hence, the lowest energy direction.

3.1   CRYSTALLOGRAPHY OF SLIP

As shown in Fig. 3.2, the dominant slip systems (combinations of slip planes and directions) vary with the material's crystal lattice, since the respective atomic density of planes and directions are different. For the case of face-centered cubic (FCC) crystals, slip occurs most often on {111} octahedral planes and in ⟨110⟩ directions which are parallel to cube face diagonals. In all, there are 12 such slip systems (four {111} planes and three ⟨110⟩ slip directions for each {111} plane). Other FCC slip systems have been found but will not be considered here since they are activated only by unusual test conditions.

In body-centered cubic (BCC) crystals, slip occurs in the ⟨111⟩ cube diagonal direction and on {110} dodecahedral planes. In addition, slip may occur on

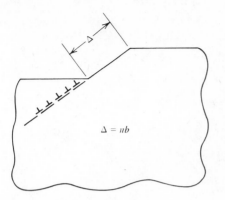

$\Delta = nb$

FIGURE 3.1 Diagram showing slip offset due to *n* dislocations leaving the crystal.

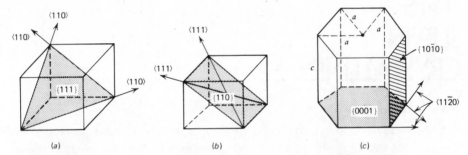

⟨110⟩

⟨110⟩

⟨110⟩

{111}

⟨111⟩

⟨111⟩

⟨110⟩

{110}

{1010}

a

a

a

c

{0001}

⟨1120⟩

(a)   (b)   (c)

FIGURE 3.2 Diagram showing predominant slip systems in (*a*) FCC; (*b*) BCC; and (*c*) HCP crystals.

{112} and {123} planes as well. A total of 48 possible slip systems can be identified, based on combinations of these three slip planes and the common ⟨111⟩ slip direction. The four-fold greater number of slip systems in BCC as compared to FCC crystals does not mean that the former lattice provides more ductility; in fact, the reverse is true because FCC crystals have a much lower Peierls-Nabarro stress and contain more mobile dislocations.

Prediction of the preferred slip systems in hexagonal close-packed (HCP) materials is not an easy task. On the basis of the Peierls stress argument wherein the most densely packed planes of greatest separation would be the preferred slip planes, one would expect the active slip planes in hexagonal crystals to vary with the c/a ratio (see Fig. 3.2c). That is, if the c/a ratio is less than that for ideal packing (1.633) then the prism planes become atomically more dense relative to the basal plane. For this case, {1010} prism slip would be preferred in the ⟨1120⟩ close-packed direction (Fig. 3.2). This has been found true for the case of zirconium and titanium but not for cobalt, magnesium, or beryllium (Table 3.1). On the other hand, when $c/a > 1.633$, the basal plane should be the preferred slip plane as shown for the case of zinc and cadmium. The exceptions mentioned above have been carefully studied. It has been suggested[1] that the true atomic density of certain nonbasal planes are greater than that originally computed. Careful examination of atomic positions within the hexagonal lattice reveals that certain planes, such as the $\{10\bar{1}0\}$, $\{10\bar{1}1\}$ and $\{10\bar{1}2\}$ planes, are not equally spaced but rather may be considered to represent corrugated or puckered planes (Fig. 3.3). To illustrate this point, we see that while the average distance between $\{10\bar{1}0\}$ planes is $a\sqrt{3}/2$, the distances between adjacent $\{10\bar{1}0\}$ planes can be either $a\sqrt{3}/6$ or $a\sqrt{3}/3$, respectively. If one considers the two planes spaced $a\sqrt{3}/6$ apart as being one corrugated plane, then the distance between corrugated planes remains $a\sqrt{3}/2$ while the effective atomic density is doubled. On the basis of this argument, the c/a ratio where a transition from basal to prism slip would be expected is changed from 1.633 to 1.732 (i.e., $\sqrt{3}$). This is found where the ratio of basal to prism planar atomic density or interplanar spacing is unity. For the case of planar atomic densities $\rho_{\{10\bar{1}0\}}/\rho_{\{0001\}}=(2/ac)/(2/a^2\sqrt{3})=c/a\sqrt{3}$ while the ratio of interplanar spacings may be given by $d_{\{10\bar{1}0\}}/d_{\{0001\}}=(a\sqrt{3}/2)/(c/2)=a\sqrt{3}/c$. From Table 3.1 we see that the observed predominant slip plane in cobalt, magnesium, and beryllium still does not conform to the plane predicted on the basis of atomic density considerations. Consequently, researchers have sought with little success other explanations to account for observed behavior of these three metals. For example, it was suggested that preferential segregation of solute atoms might cause the prism plane in beryllium to be strengthened relative to the basal plane. However, Partridge[1] summarized the findings of others and showed that while

TABLE 3.1   Predicted and Observed Dominant Slip Planes in Hexagonal Crystals[1]

Metal	c/a Ratio	Ratio of Projected Planar Atomic Densities $\left(\dfrac{\rho_{\{10\bar{1}0\}}}{\rho_{\{0001\}}}=\dfrac{2/ac}{2/a^2\sqrt{3}}\right)$	Predicted Slip Plane	Observed Slip Plane
Be	1.568	1.10	$\{10\bar{1}0\}$	$\{0001\}$
Ti	1.587	1.09	$\{10\bar{1}0\}$	$\{10\bar{1}0\}$
Zr	1.593	1.09	$\{10\bar{1}0\}$	$\{10\bar{1}0\}$
Mg	1.623	1.07	$\{10\bar{1}0\}$	$\{0001\}$
Co	1.623	1.06	$\{10\bar{1}0\}$	$\{0001\}$
—	1.732	1.00	$\{10\bar{1}0\}$ or $\{0001\}$	—
Zn	1.856	0.94	$\{0001\}$	$\{0001\}$
Cd	1.886	0.92	$\{0001\}$	$\{0001\}$

both prism and basal slip became easier with increased purity of the beryllium crystal, the *relative* propensity for basal slip versus prism slip actually increased. Some correlation between predicted and observed slip planes has been found for the case of lithium additions to magnesium crystals. For this alloy, the c/a ratio decreased with increasing lithium content, suggesting prism slip to be more likely. In fact, prism slip was found. This confirmation was reinforced by the observation that the addition of lithium also served to strengthen the basal plane while weakening the prism plane.

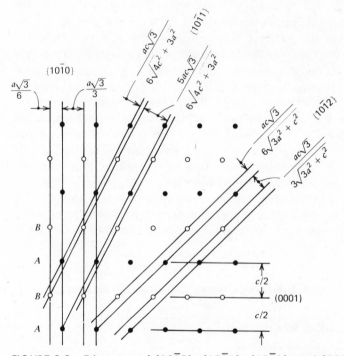

FIGURE 3.3   Edge view of $\{10\bar{1}0\}$, $\{10\bar{1}1\}$, $\{10\bar{1}2\}$, and $\{0001\}$ planes revealing interplanar spacing and atom positions. (After Partridge,[1] reprinted with permission of The Metals Society.)

Thus far, my discussion has focused on the importance of the relative atomic density of crystallographic planes and directions in deciding whether a particular plane and direction combination could serve as a potential slip system. Certain slip-system combinations of ceramic crystals seem reasonable on the basis of density considerations, but are negated by the effects of strong directional bonding in covalent crystals or electrostatic interactions in ionic crystals. Since such atomic movements would be energetically unfavorable, the number of potential slip systems in these materials is restricted as is their overall ductility (except at relatively high test temperatures). A compilation of reported slip systems in selected ceramic crystals is given in Table 3.2.

TABLE 3.2   Observed Slip Systems in Selected Ceramics[39]

Material	Structure Type	Preferred Slip Sytem
C, Ge, Si	Diamond cubic	$\{111\}\langle\bar{1}10\rangle$
NaCl, LiF, MgO	Rock salt	$\{110\}\langle1\bar{1}0\rangle$
CsCl	Cesium chloride	$\{110\}\langle001\rangle$
CaF_2, UO_2, ThO_2	Fluorite	$\{001\}\langle1\bar{1}0\rangle$
TiO_2	Rutile	$\{101\}\langle10\bar{1}\rangle$
$MgAl_2O_4$	Spinel	$\{111\}\langle\bar{1}10\rangle$
Al_2O_3	Hexagonal	$\{0001\}\langle11\bar{2}0\rangle$

The ductility of a material depends also on its ability to withstand a general homogeneous strain involving an arbitrary shape change of the crystal. Von Mises[2] showed this to be possible when five independent slip systems are activated. Allowing one slip system to account for each of the six independent components of strain (Fig. 1.3), a total of six such systems would seem to be indicated; however, plastic deformation is a constant-volume process where $\epsilon_{xx} + \epsilon_{yy} + \epsilon_{zz} = 0$, thereby reducing to five the number of independent slip systems. An independent slip system is defined as one producing a crystal shape change that cannot be reproduced by any combination of other slip systems. On this basis, Taylor[3] showed that for the 12 possible $\{111\}\langle110\rangle$ slip systems in FCC crystals, only five are independent. Furthermore, Taylor found there to be 384 different combinations of five slip systems that could produce a given strain, the activated combination being the one for which the sum of the glide shears is a minimum. Likewise, Groves and Kelly[4] found 384 combinations of five sets of $\{110\}\langle111\rangle$ slip systems to account for slip in BCC metals. Since slip in BCC can occur also on $\{112\}\langle111\rangle$ and $\{123\}\langle111\rangle$ systems, the total number of combinations of five independent slip systems becomes incredibly large. Chin and coworkers[5,6] have applied computer techniques to identify the preferred slip-system combinations for the case of BCC metals.

Difficulties arise when one seeks five independent slip systems in the hexagonal materials. Of the three possible $\{0001\}\langle11\bar{2}0\rangle$ slip systems, only two are independent.[4] Similarly, only two independent $\{10\bar{1}0\}\langle11\bar{2}0\rangle$ slip systems can be identified from the three possible prism slip systems. Although four independent pyramidal $\{10\bar{1}1\}\langle11\bar{2}0\rangle$ slip systems may be identified from a total of six such systems, the resulting deformations can be produced by simultaneous operation of the two independent basal and prism slip systems, respectively. Consequently, a fifth independent slip system is still needed. Besides some deformation twinning (see Chapter Four), additional nonbasal slip with a c axis Burgers vector component is necessary to explain the observed ductility in hexagonal engineering alloys.[7]

3.2   PARTIAL DISLOCATIONS

As we noted in Chapter Two, the strain energy associated with a dislocation is proportional to Gb^2 where b is the dislocation Burgers vector. It is then possible to have the dislocation $\mathbf{b}_1$ dissociate into two dislocations $\mathbf{b}_2$ and $\mathbf{b}_3$ such that the vector sum is the same, $\mathbf{b}_1 = \mathbf{b}_2 + \mathbf{b}_3$ (Fig. 3.4a). The likelihood that dislocation $\mathbf{b}_1$ will do this depends on the difference in strain energy between $\mathbf{b}_1$ and the pair of dislocations $\mathbf{b}_2$ and $\mathbf{b}_3$, often referred to as Shockley partial dislocations. From Eq. 2-17, the dissociation will occur when

$$Gb_1{}^2 > Gb_2{}^2 + Gb_3{}^2$$

Using the FCC lattice as a model, it can be shown that the whole dislocation $\mathbf{b}_1$, oriented in the $\langle 110 \rangle$ close-packed direction, can dissociate into two dislocations of type $\langle 112 \rangle$. For example

$$\frac{a}{2}[\bar{1}01] \rightarrow \frac{a}{6}[\bar{2}11] + \frac{a}{6}[\bar{1}\bar{1}2]$$

From Eq. 2-17

$$\frac{a^2}{4}(1+1) > \frac{a^2}{36}[4+1+1] + \frac{a^2}{36}[1+1+4]$$

$$\frac{a^2}{2} > \frac{a^2}{3}$$

(For simplicity, any anisotropy in elastic shear modulus has been ignored.) Therefore, the dislocation reaction will proceed in the direction indicated. Because of the reduction in strain energy and the fact that the partials have similar vector components, these partials will tend to repel one another and move apart. The extent of separation will depend on the nature of the change in stacking sequence that occurs between $\mathbf{b}_2$ and $\mathbf{b}_3$. Movement of these partial dislocations produces a change in the stacking sequence from the FCC type—$ABCABCABC$—to include a local perturbation involving the formation of a layer of HCP material—$ABCBCABC$. Examples of stacking faults are shown in Fig. 3.4b. For an FCC crystal, the layer of HCP material that is introduced will elevate the total energy of the system. Therefore, the equilibrium distance of

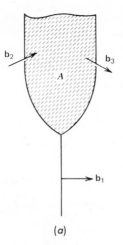

(a)

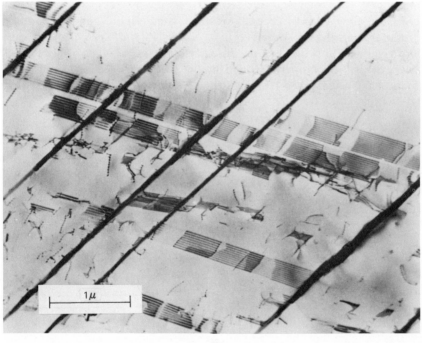

(b)

FIGURE 3.4 Dislocation dissociation of b_1 into b_2 and b_3 Shockley partial dislocations. (a) Shockley partials b_2 and b_3 surrounding stacking fault region A. (b) Long stacking fault ribbons (bands of closely spaced lines) in low SFE 18Cr-8Ni stainless steel. Faults are bounded at ends by partial dislocations. Thin black bands are mechanical twins. (After Michalak,[37] reprinted by permission from *Metals Handbook*, Vol. 8, copyright American Society for Metals, 1973.)

77

separation of two partials reflects a balance of the net repulsive force between the two partial dislocations containing Burgers vector components of the same sign and the energy of the associated stacking fault. According to Cottrell,[8] this separation distance varies inversely with the stacking fault energy (SFE) and may be given by

$$d = \frac{G(\mathbf{b}_2 \mathbf{b}_3)}{2\pi\gamma}$$

(3-1)

where

$$d = \text{partial dislocation separation}$$

$$\mathbf{b}_2, \mathbf{b}_3 = \text{partial dislocation Burgers vectors}$$

$$G = \text{shear modulus}$$

$$\gamma = \text{stacking fault energy}$$

The SFE of alloy crystals depends on their composition, and comparative values for pure metals also differ. Typical values for different elements and alloys are given in Table 3.3. For the case of copper-based alloys, Thornton et al.[9] showed SFE to be strongly affected by the material's electron/atom ratio. They found that when $e/a > 1.1$, the stacking fault energy usually decreased below 20 mJ/m².

TABLE 3.3   Selected Stacking Fault
Energies for FCC Metals

Metal	Stacking Fault Energy (mJ/m² = ergs/cm²)
Brass	< 10
Stainless steel	< 10
Ag	~25
Au	~75
Cu	~90
Ni	~200
Al	~250

3.2.1   Movement of Partial Dislocations

The movement of the two Shockley partial dislocations is restricted to the plane of the fault, since movement of either partial on a different plane would involve energetically unfavorable atomic movements. Therefore, cross-slip of an extended screw dislocation around obstacles is not permitted without thermally activated processes and, as a consequence, the slip offsets seen on a polished surface will be straight (Fig. 3.5). Such is the case for a material of low stacking fault energy and widely separated partial dislocations. This type of dislocation movement is called *planar glide*. By the application of a suitably large stress, however, it is possible to squeeze the partial dislocations together against a barrier to form a whole dislocation. If this recombined dislocation is of the screw type, it may cross-slip (Fig. 3.6). As you might imagine, the stress necessary to recombine the partial dislocations will depend on the equilibrium distance of separation of the partials, which in turn depends on the magnitude of the stacking fault energy (Eq. 3-1). For materials with low stacking fault energy, partial dislocation separation is large (on the order of 10 to 20 b) and the force necessary for recombination is large. Conversely, little stress is necessary to recombine partial dislocations in a high stacking fault energy material where partial dislocation separation is small (on the order of 1 b or less). When cross-slip is easy, slip offsets on a polished surface take on a wavy pattern (Fig. 3.7), and this deformation is called *wavy glide*.

One major implication of the dependence of cross-slip on stacking fault energy is the dominant role the latter plays in determining the strain hardening characteristics of a material. When the stacking fault energy is low, cross-slip is restricted so that barriers to dislocation movement remain effective to higher stress levels than in material of higher stacking fault energy. That is to say, the low stacking fault energy material strain hardens to a greater extent. It is then

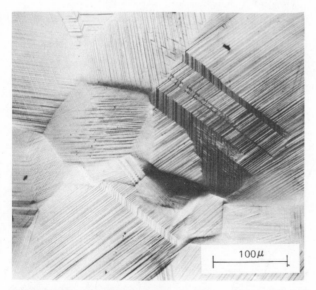

FIGURE 3.5 Photomicrograph revealing planar glide indicative of plastic flow in low stacking fault energy materials.

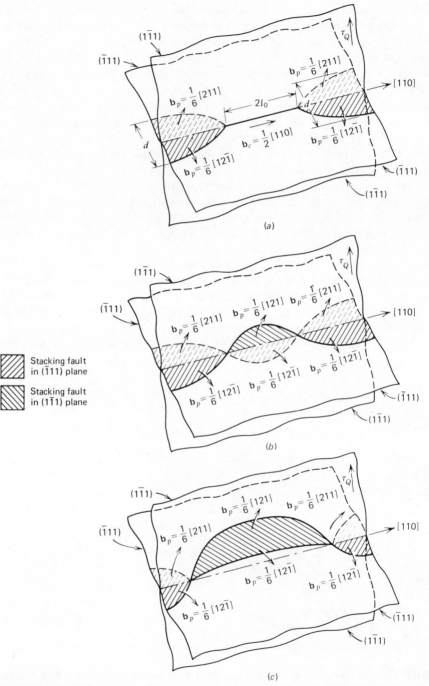

FIGURE 3.6 Three stages of cross-slip of an extended screw dislocation from the $(\bar{1}11)$ to $(1\bar{1}1)$ plane. (a) Partial dislocations have been recombined over length $2l_0$; (b) combined segment now moves and dissociates in $(1\bar{1}1)$ cross-slip plane; (c) further movement of extended partials on $(1\bar{1}1)$ cross-slip plane. (After Seeger,[29] reprinted by permission of General Electric Company.)

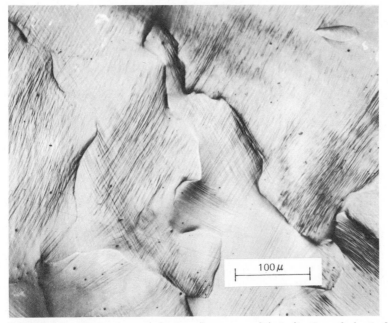

FIGURE 3.7 Photomicrograph revealing wavy glide indicative of plastic flow in high stacking fault energy materials.

TABLE 3.4 Slip Character and Strain Hardening Coefficients for Several Metals

Metal	Stacking Fault Energy (mJ/m^2)	Strain Hardening Coefficient	Slip Character
Stainless steel	< 10	~0.45	Planar
Cu	~90	~0.3	Planar/wavy
Al	~250	~0.15	Wavy

possible to relate strain hardening coefficients (Table 1.5) with stacking fault energy values (Table 3.3) as shown in Table 3.4. Note that the strain hardening coefficient increases with decreasing stacking fault energy while the slip character changes from a wavy to planar mode.

3.3 GEOMETRY OF SLIP

It has been shown that the onset of plastic deformation in a single crystal takes place when the shear stress acting on the incipient slip plane and in the slip direction reaches a critical value. From Fig. 3.8, we see that the cross-sectional area of the slip plane is given by

$$A_{\text{slip plane}} = \frac{A_0}{\cos \phi} \tag{3-2}$$

where

$A_0 =$ cross-sectional area of single crystal rod

$\phi =$ angle between the rod axis and the normal to the slip plane

Furthermore, the load on this plane resolved in the slip direction is given by

$$P_{\text{resolved}} = P \cos \lambda \tag{3-3}$$

where

$P =$ axial load

$\lambda =$ angle between load axis and slip direction

By combining Eq. 3-2 and 3-3 the resolved shear stress acting on the slip system

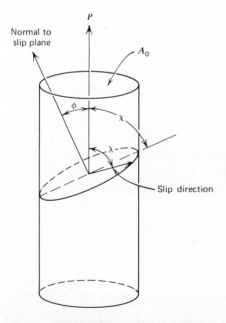

FIGURE 3.8   Diagram showing orientation of slip plane and slip direction in crystal relative to the loading axis.

is

$$\tau_{RSS} = \frac{P}{A} \cos\phi \cos\lambda \qquad (3\text{-}4)$$

where $\cos\phi\cos\lambda$ represents an orientation factor (often referred to as the Schmid factor). Plastic deformation will occur when the resolved shear stress τ_{RSS} reaches a critical value τ_{CRSS}, which represents the yield strength of the single crystal. From Eq. 3-4, we see that yielding will occur on the slip system possessing the greatest Schmid factor. Consequently, if only a few systems are available, such as in the case of basal slip in zinc and cadmium, the necessary load for yielding can vary dramatically with the relative orientation of the slip system (i.e., the Schmid factor).[10,11] For example, the axial stress necessary for yielding anthracene crystals varies dramatically with crystal orientation (Fig. 3.9), while the critical resolved shear stress is unchanged.[12] (Note that the curve drawn in Fig. 3.9b was computed from Eq. 3-4 using a value of 137 kPa.)

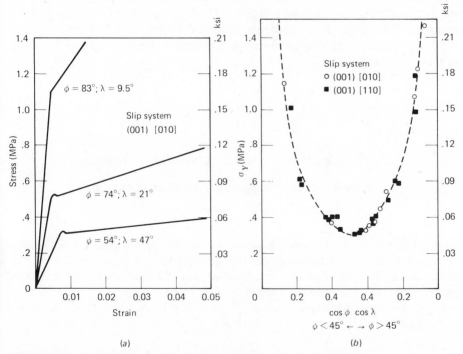

FIGURE 3.9   Yield behavior of anthracene single crystals. (a) Axial stress-strain curves for crystals possessing different orientations relative to the loading axis; (b) axial stress for many crystals plotted versus respective Schmid factors. Dotted curve represents relationship given by Eq. 3-4 where $\tau_{CRSS} = 137$ kPa. (After Robinson and Scott,[12] reprinted with permission from Robinson, *Acta Met.*, 15 (1967), Pergamon Press.)

Furthermore, the stress normal to the slip plane

$$\sigma_n = \frac{P}{A} \cos^2 \phi \qquad (3\text{-}5)$$

can vary considerably without affecting the onset of yielding. For example, Andrade and Roscoe[11] found for cadmium that τ_{CRSS} varied within 2% for all crystal orientations examined, while the normal stress σ_n experienced a 20-fold change.

Let us now consider what happens to the single crystal once it begins to yield. From Fig. 3.10 we see that slip on planes oriented χ degrees away from the tensile axis can occur in two ways. First, the planes can simply slide over one another without changing their relative orientation to the load axis. This would be analogous to offsetting groups of playing cards on a table. Since such lateral movement of the crystal planes in a tensile bar would be forbidden by lateral constraints imposed by the specimen grips, the slip planes are forced to rotate with $\chi_i < \chi_0$. X-ray diffraction studies have shown that crystal planes undergo pure rotation in the middle of the gage length but experience simultaneous rotation and bending near the end grips. If we focus attention on the simpler mid-region of the sample, it can be shown that the reorientation of the slip plane varies directly with the change in length of the specimen gage length according to the relationship[13]

$$\frac{L_i}{L_0} = \frac{\sin \chi_0}{\sin \chi_i} \qquad (3\text{-}6)$$

where

L_0, L_i = gage length before and after plastic flow, respectively

χ_0, χ_i = angle between slip plane and stress axis before and

after plastic flow, respectively

(Note that $\chi + \phi = 90$ degrees. However $\lambda + \phi = 90$ degrees *only* when the two vectors are coplanar.) By analogy, the deformation-induced rotation of slip planes is similar to the rotation of individual venetian blind slats—the more you pull on the cord, the more the individual slats deflect.

From the work of Schmid and Boas,[13] when $\chi_0 = \lambda_0$, the shear strain γ, after a given amount of extension, is found to be

$$\gamma = \frac{1}{\sin \chi_0} \left\{ \left[\left(\frac{L_i}{L_0} \right)^2 - \sin^2 \lambda_0 \right]^{\frac{1}{2}} - \cos \lambda_0 \right\} \qquad (3\text{-}7)$$

Note that γ is determined by the initial orientation of the glide elements and by

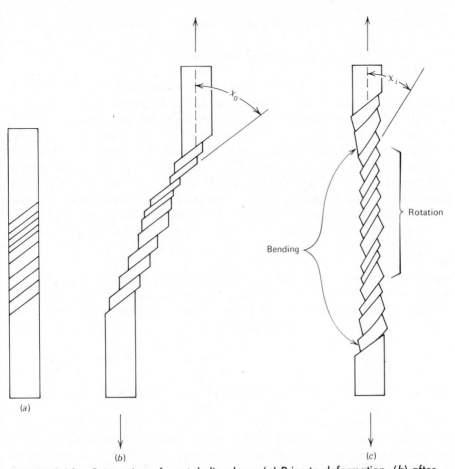

FIGURE 3.10 Orientation of crystal slip plane. (*a*) Prior to deformation; (*b*) after deformation without grip constraint where crystal segments move relative to one another but with no slip plane rotation; (*c*) after deformation with grip constraint revealing slip plane rotation in gage section (note $\chi_i < \chi_0$).

the amount of extension. Furthermore, Eq. 3-7 is valid when only one slip system is active, since multiple slip involves an undefined amount of crystal rotation from each system. The resolved shear stress is given by

$$\tau = \frac{P}{A} \sin\chi_0 \left[1 - \frac{\sin^2\lambda_0}{(L_i/L_0)^2} \right]^{\frac{1}{2}} \tag{3-8}$$

For a detailed discussion of other relationships involving the shear stresses and strains in single crystals, see Schmid and Boas.[13]

It is instructive to trace the path of rotation of the slip plane. This is accomplished most readily with the aid of a stereographic projection.* For the purpose of this discussion, some basic understanding of this portrayal method is desirable. For the crystal block shown in Fig. 3.11a, imagine that a normal to each plane is extended to intersect an imaginary reference sphere that surrounds the block. Now place a sheet of paper (called the projection plane) tangent to the sphere. Next, take a position at the other end of the sphere diameter, which is oriented normal to the projection plane. From this position (called the point of projection) draw lines through the points on the reference sphere and continue on to the projection plane (Fig. 3.11b). The points on the projection plane then reflect the relative position of various planes (or plane normals) with planar angle relationships faithfully reproduced. For convenience, standard stereographic projections are used to portray the relative positions of major planes, such as those shown in Fig. 3.12. Since a cubic crystal is highly symmetrical, the relative orientation of a crystal can be given with respect to any triangle within the stereographic projection. As a result, attention is usually focused on the central section of the projection. In Fig. 3.13, for example, we see the axis of a rod in terms of its angular relationship with the (001), (011), and ($\bar{1}$11) planes, respectively. That is, P is the normal to the plane lying perpendicular to the rod axis. When this rod is stressed to τ_{CRSS}, the crystal will yield on that slip system possessing the greatest Schmid factor and begin to rotate. For all orientations within triangle I (sometimes referred to as the standard triangle), the (111) [$\bar{1}$01] slip system possesses the greatest Schmid factor and will be the first to operate.[14,15] The rotation occurs along a great circle (corresponding to the trace of a plane on the reference sphere that passes through the center of the sphere) of the stereographic projection and toward the [$\bar{1}$01] slip direction. For simplicity, it is easier to consider rotation of the stress axis relative to the crystal than vice versa, so that P is seen to move toward the [$\bar{1}$01] pole. As the crystal rotates λ will decrease while ϕ increases. In situations where $\lambda_0 > 45° > \phi_0$, rotation of the crystal will bring about an increase in the Schmid factor, since both λ_i and ϕ_i would approach 45 degrees. As a result, yielding can continue at a lower load and the crystal is said to have undergone *geometrical softening*. Conversely, when $\phi_0 > 45° > \lambda_0$, crystal rotation will bring about a reduction in the Schmid factor, thereby increasing the load necessary for further deformation on the initial slip system. Bear in mind that this *geometrical hardening* is distinct from strain hardening which involves dislocation-dislocation interactions. Geometrical hardening continues as the crystal axis moves toward the [001]–[$\bar{1}$11] tie line. As soon as the relative crystal orientation crosses over into the adjacent triangle II, the Schmid factor on the primary slip system becomes less than that associated with the ($\bar{1}$11)[011] system, the latter being the slip system that would have operated had the crystal been oriented initially within triangle II. This

*See B. D. Cullity, *Elements of X-ray Diffraction*, Addison-Wesley Publishing Co., Reading, Mass., 1956, for a treatment of stereographic projection.

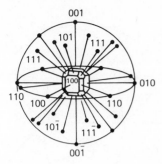

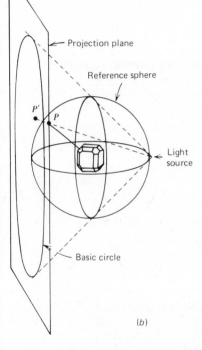

(a)

(b)

FIGURE 3.11 Geometric construc-
tions to develop a stereographic
projection. (*a*) Intersection of plane
normals on reference sphere. (After
C. W. Bunn, *Chemical Crystallography*, Oxford-Clarendon Press, 1946,
p. 30). (*b*) Projection of poles on the
reference sphere to the projection
plane. (After N. H. Polakowski and
E. J. Ripling, *Strength and Structure
of Engineering Materials* © 1966,
p. 83. Reprinted by permission of
Prentice-Hall, Inc., Englewood Cliffs,
N.J.)

newly activated slip system (the conjugate system) now causes the crystal to
rotate along a different great circle toward the [011] direction of the conjugate
slip system. Shortly, however, this movement returns the axis of the crystal to
within the bounds of triangle I where primary slip resumes. The ultimate effect
of this jockeying back and forth between primary and conjugate slip systems is
the movement of the crystal axis along the [001]–[$\bar{1}$11] tie line to a location
where further crystal rotations in either slip direction occur along the same great
circle. This point is reached when the load axis is parallel to the [$\bar{1}$12] direction.

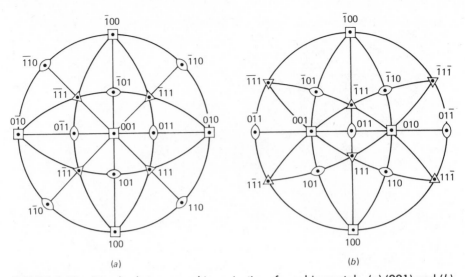

FIGURE 3.12 Standard stereographic projections for cubic crystals. (a) (001) and (b) (011) projections.

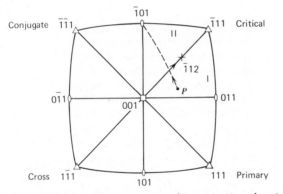

FIGURE 3.13 (001) stereographic projection showing lattice rotation for FCC crystals during tensile elongation.

The above analysis reflects the classic geometrical arguments proposed originally by Taylor and Elam.[14,15] In reality, some alloy crystals exhibit "overshooting," wherein the primary slip system continues to operate well into triangle II even though the Schmid factor of the conjugate system is greater. Similarly, the conjugate system, once activated, may continue to operate into triangle I (Fig. 3.14). Koehler[16] proposed that overshooting was caused by weakening of the primary system by passage of dislocations that destroyed precipitates and other solute atom clusters. Consequently, he argued that slip would be easier if continued on the softened primary plane. Alternately, Piercy et al.[17] argued that overshooting resulted from a "latent hardening" process involving increased resistance to conjugate slip movement resulting from the dislocation debris found on the already activated primary system. That is, for slip to occur on the conjugate system, dislocations on this plane would have to cut across many dislocations lying on the primary plane. By comparison, then, Koehler[16] argued that overshooting resulted from a relative weakening of the primary plane while Piercy et al.[17] argued that the conjugate plane was strengthened relative to the primary plane by a latent hardening mechanism. By careful experimentation, the latent hardening theory was proven correct.

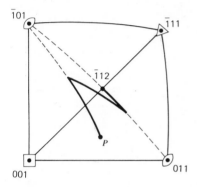

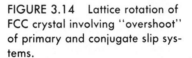

FIGURE 3.14 Lattice rotation of FCC crystal involving "overshoot" of primary and conjugate slip systems.

From Fig. 3.13, two other slip systems can be identified. These are denoted as the cross-slip system $(1\bar{1}1)$ $[\bar{1}01]$ and the critical slip system $(\bar{1}11)$ $[0\bar{1}1]$. The critical system is not encountered very often; the cross-slip system is the system involving the movement of screw dislocations that have cross-slipped out of the primary slip plane. Note that the slip direction is the same in this case.

3.4 YIELD BEHAVIOR

The critical resolved shear strengths for selected materials were given in Table 2.1. The magnitude of these values depends on several factors, including the intrinsic lattice resistance to dislocation motion (the Peierls stress) and the presence of solute atoms and other dislocations that obstruct the movement of glide dislocations. Yielding can occur gradually, similar to that described by Type II stress-strain polycrystalline behavior (see Chapter One), or discontinuously as in the case of Type IV behavior. As mentioned in Chapter One, the yield point can be produced by impurity atom pinning of dislocations. According to current theory,[18,19] after some dislocations tear away from their impurity atom clusters, they multiply rapidly by a multiple cross-slip mechanism

(Section 2.8). As a result, the number of mobile dislocations increases sharply, yielding becomes easier, and the load necessary for continued deformation decreases abruptly. Although this explanation may be appropriate for iron single crystals containing small solute additions of interstitial carbon and nitrogen, it does not explain similar yield point behavior in other materials such as silicon, germanium, and lithium fluoride. Johnston[20] and Hahn[21] have proposed that yield point behavior in these crystals is related to an initially low mobile dislocation density and a low dislocation velocity stress sensitivity. Regarding the latter, studies by Stein and Low,[22] Gilman and Johnston,[23,24] and others demonstrated that the dislocation velocity v depends on the resolved shear stress as given by Eq. 3-9

$$v = \left(\frac{\tau}{D} \right)^{m}$$

(3-9)

where

$$v = \text{dislocation velocity}$$

$$\tau = \text{applied resolved shear stress}$$

$$D, m = \text{material properties}$$

Defining the plastic strain rate by

$$\dot{\epsilon}_p \propto Nbv$$

(3-10)

where

$\dot{\epsilon}_p = \text{plastic strain rate}$

$N = \text{number of dislocations per unit area free to move about and multiply}$

$b = \text{Burgers vector}$

$v = \text{dislocation velocity}$

Johnston[20] argued that when the initial mobile dislocation density in these materials is low, the plastic strain rate would be less than the rate of movement of the test machine crosshead and little overall plastic deformation would be detected. At higher stress levels, the dislocations would be moving at a higher velocity and also begin to multiply rapidly such that the total plastic strain rate would then exceed the rate of crosshead movement. To balance the two rates, the dislocation velocity would have to decrease. From Eq. 3-9, this may be accomplished by a drop in stress, the magnitude of which would depend on the stress sensitivity parameter m. If m were very small (less than 20 as in the case of covalent- and ionic-bonded materials as well as in some BCC metals), then a large drop in load would be required to reduce the dislocation velocity by the necessary amount. If m were large (greater than 100 to 200 as found for FCC metal crystals), only a small load drop would be required to effect a substantial change in dislocation velocity. The severity of the yield drop is depicted in Fig. 3.15a for a range of dislocation velocity stress sensitivity values. Note the magnitude of the yield drop increasing with decreasing m. If there are many free dislocations present at the outset of the test, they may multiply more gradually at lower stress levels, precluding the occurrence of a sudden avalanche of dislocation generation at higher stress levels. The corresponding decrease in magnitude of the yield drop with increasing initial mobile dislocation density is shown in Fig. 3.15b and 3.15c. From the above discussion, a yield point is pronounced in crystals that (1) contain few mobile dislocations at the beginning of the test, (2) have the potential for rapid dislocation multiplication with increasing plastic strain, and (3) exhibit relatively low dislocation velocity stress sensitivity. Since many ionic- and covalent-bonded crystals possess these characteristics,[25] yield points are predicted and found experimentally in these materials.

For the case of carbon and nitrogen locked dislocations in iron, dislocation mobility is essentially zero prior to the upper yield point where dislocations are finally able to tear away from interstitial atmospheres. It is theorized that the unpinning of some dislocations, their rapid multiplication, and weak velocity stress sensitivity (i.e., low m value) all contribute to the development of a yield point in engineering iron alloys. By contrast, most FCC metals have an initially high mobile dislocation density and a very high dislocation velocity stress sensitivity, thereby making a yield drop an unlikely event in most of these materials.

3.5 GENERALIZED SINGLE-CRYSTAL STRESS-STRAIN RESPONSE

Let us now consider the overall stress-strain response of single crystals. From Fig. 3.16, the resolved shear stress-shear strain curve is seen to contain several distinct regions: an initial region of elastic response where the resolved shear stress is less than τ_{CRSS}; Stage I (region of easy glide); Stage II (region of linear hardening); and Stage III (region of dynamic recovery or parabolic hardening). The latter three regions involve different aspects of the plastic deformation process for a given crystal. It is known that the extent of Stages I, II and III depends on such factors as the test temperature, crystal purity, initial dislocation density, and initial crystal orientation.[26,27] It should be noted that Stage III closely resembles the stress-strain response of the polycrystal form of the same material.

A number of theories have been proposed to explain the strain hardening process in crystals, including the reason for the dramatic changes in strain

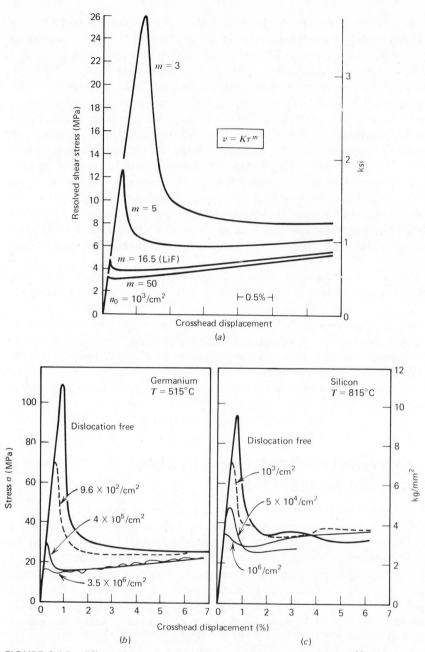

FIGURE 3.15 Effect of (a) stress sensitivity m in LiF (From Johnston,[20] reprinted with permission of American Institute of Physics.) and (b, c) initial mobile dislocation density n_0 in Ge and Si on severity of yield drop (From Patel and Chaudhuri,[38] reprinted with permission of American Institute of Physics.)

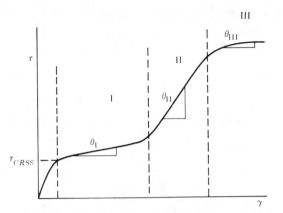

FIGURE 3.16 Shear stress-strain curve for single crystal revealing elastic behavior when $\tau < \tau_{CRSS}$ and Stage I, II, III plastic response when $\tau > \tau_{CRSS}$. θ_I, θ_{II}, θ_{III} measure the strain hardening rate in each region.

hardening rate associated with the three stages of plastic deformation. An extensive literature[28] has developed regarding these theories, all of which have focused on some of the dislocation interaction mechanisms described in the previous chapter. Seeger[29] and Friedel,[30] for example, argued that rapid strain hardening in Stage II resulted from extensive formation of dislocation pileups at strong obstacles such as Cottrell-Lomer locks. Mott[31] proposed that heavily jogged dislocations produced by dislocation-dislocation interactions (see Section 2.7) would be more resistant to movement, thereby enhancing the hardening rate. Unfortunately, a certain degree of confusion has arisen in this field because of the varying importance of certain dislocation interactions in different alloy crystals. One wonders then why the three distinct stages of deformation are so reproducible from one material to another and why the work hardening coefficient θ_{II} associated with Stage II deformation is almost universally constant at $G/300$. For these reasons, the "mesh length" theory of strain hardening proposed by Kuhlmann-Wilsdorf[32,33] is appealing pedagogically, since it does not depend on any specific dislocation model which might be appropriate for one material but not for another . Her theory may be summarized as follows: In Stage I a heterogeneous distribution of low-density dislocations exists in the crystal. Since these dislocations can move along their slip planes with little interference from other dislocations, the strain hardening rate θ_I is low. The easy glide region (Stage I) is considered to end when a fairly uniform dislocation distribution of moderate density is developed but not necessarily in lockstep with the onset of conjugate slip where a marked increase in dislocation-dislocation interactions would be expected. At this point Kuhlmann-Wilsdorf theorizes the existence of a quasi-uniform dislocation array with clusters of dislocations surrounding cells of relatively low dislocation density. The stress necessary for further plastic deformation is then seen to depend on the mean free dislocation length $\bar{l}$ (Fig. 3.17) in a manner similar to that necessary for the activation of a Frank-Read source where

$$\tau \propto \frac{Gb}{\bar{l}} \qquad (3\text{-}11)$$

Since the dislocation density is proportional to $1/\bar{l}^2$, Eq. 3-11 may be written in the form

$$\Delta\tau \propto Gb\sqrt{\rho} \qquad (3\text{-}12)$$

where

$\rho =$ dislocation density

$\Delta\tau =$ incremental shear stress necessary to overcome dislocation barriers

This relationship has been verified experimentally for an impressive number of materials[34] and represents a necessary requirement for any strain hardening theory. With increasing plastic deformation, ρ increases resulting in a decrease in the mean free dislocation length $\bar{l}$. From Eqs. 3-11 and 3-12, the stress necessary for further deformation then increases. Kuhlmann-Wilsdorf suggests[32] that there is a continued reduction in cell size and an associated increase in flow stress throughout the linear hardening region. In other words, the *character* of the dislocation distribution remains unchanged, only the *scale* of the distribution changes (Fig. 3.17). With further deformation, the number of free dislocations within the cell interior decreases to the point where glide dislocations can move relatively unimpeded from one cell wall to another. Since the formation of new cell walls (and hence a reduction in $\bar{l}$) are believed to depend on such interactions, a point would be reached where the cell size $\bar{l}$ would stabilize or at best decrease slowly with further deformation. According to Kuhlmann-Wilsdorf,[33] this condition signals the onset of Stage III and a lower strain hardening rate, since $\bar{l}$ would not decrease further.

The role of stacking fault energy is considered important with regard to the onset of Stage III. Seeger[29] has argued that Stage III begins when dislocations can cross-slip around their barriers, a view initially supported by Kuhlmann-Wilsdorf. From Seeger's point of view, Stage III would occur sooner for high stacking fault energy materials since cross-slip would be activated at a lower stress. Conversely, a low stacking fault energy material, such as brass, would require a larger stress necessary to force the widely separated partial dislocations to recombine and hence cross-slip. More recently, Kuhlmann-Wilsdorf[33] suggested that the mesh length theory could also explain the sensitivity of τ_{III} to stacking fault energy by proposing that enhanced cross-slip associated with a

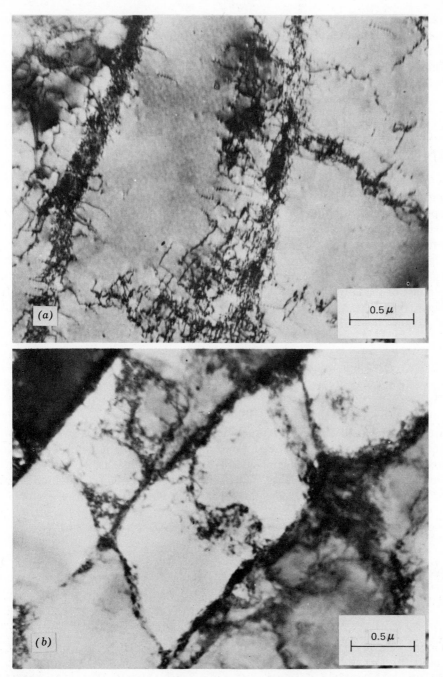

FIGURE 3.17   Cell structure formation in iron deformed at room temperature to (*a*) 9% strain and (*b*) 20% strain. Note the decreasing cell size with increasing plastic strain. (From Michalak,[37] reprinted by permission from *Metals Handbook* Vol. 8, copyright American Society for Metals, 1973.)

high value of stacking fault energy would accelerate the dislocation rearrangement process. Consequently, l would become stabilized at a lower stress level. Setting aside for the moment the question of the correctness of the Seeger versus Kuhlmann-Wilsdorf interpretations, it is sufficient for us to note that both theories account for the inverse dependence of τ_{III} on stacking fault energy.

3.6 RELATION BETWEEN SINGLE-CRYSTAL AND POLYCRYSTALLINE STRESS-STRAIN CURVES

Before concluding the discussion of single-crystal stress-strain curves, it is appropriate to consider whether one can relate qualitative and quantitative aspects of the stress-strain response of single-crystal and polycrystalline specimens of the same material. For one thing, the early stages of single-crystal deformation would not be expected in a polycrystalline sample because of the large number of slip systems that would operate (especially near grain boundary regions) and interact with one another. Consequently, the tensile stress-strain response of the polycrystalline sample is found to be similar only to the Stage III single-crystal shear stress-strain plot. A number of attempts have been made to relate these two stress-strain curves. From Eq. 3-4

$$\sigma = \frac{P}{A} = \tau \frac{1}{\cos\phi\cos\lambda} = \tau M \tag{3-13}$$

where $M = 1/(\cos\phi\cos\lambda)$

Assuming the individual grains in a polycrystalline aggregate to be randomly oriented, M would vary with each grain such that some average orientation factor $\overline{M}$ would have to be defined. Since there are 384 combinations of the five necessary slip systems to accomplish an arbitrary shape change, $\overline{M}$ is not easy to compute. From Section 3.1, Taylor[3] determined the preferred combination to be the one for which the sum of the glide shears was minimized. As a result it may be shown[25] that

$$\epsilon = \frac{\gamma}{\overline{M}} \tag{3-14}$$

By combining Eq. 3-13 and 3-14 it is seen that

$$\frac{d\sigma}{d\epsilon} = \overline{M}^2 \frac{d\tau}{d\gamma} \tag{3-15}$$

For the case of $\{111\}\langle110\rangle$ slip in FCC metals and $\{110\}\langle111\rangle$ slip in BCC metals, Taylor[3] and Groves and Kelly[4] showed $\overline{M}$ equal to 3.07. More recently, Chin et al.[5,6] analyzed the more difficult case of $\{110\}\langle111\rangle + \{112\}\langle111\rangle + \{123\}\langle111\rangle$ slip in BCC crystals and found $\overline{M} = 2.75$. In either case, one can see from Eq. 3-15 that the strain hardening rate of a polycrystalline material is many times greater than its single-crystal counterpart.

The presence of grain boundaries has an additional effect on the deformation behavior of a material by serving as an effective barrier to the movement of glide dislocations. From the work of Petch[35] and Hall,[36] the yield strength of a polycrystalline material could be given by

$$\sigma_{ys} = \sigma_i + k_y d^{-\frac{1}{2}} \tag{3-16}$$

where

σ_{ys} = yield strength of polycrystalline sample

σ_i = overall resistance of lattice to dislocation movement

k_y = "locking parameter," which measures relative hardening

      contribution of grain boundaries

d = grain size

It is readily seen that grain refinement techniques (e.g., normalizing alloy steels) provide additional barriers to dislocation movement and enhance the yield strength. As will be shown in Chapter Ten, improved toughness also results from grain refinement.

Conrad[19] has demonstrated clearly that σ_i may be separated into two components: σ_{ST}, which is not temperature sensitive but structure sensitive where dislocation-dislocation, dislocation-precipitate, and dislocation-solute atom interactions are important and σ_T, which is strongly temperature sensitive and related to the Peierls stress. The yield strength of a material may then be given by

$$\sigma_{ys} = \sigma_T + \sigma_{ST} + k_y d^{-\frac{1}{2}} \tag{3-17}$$

Short-range order	Long-range order	Very long-range
Peierls stress	dislocation stress	structural size
effects ($<10\,\text{Å}$)	field effects	effects ($>10^4\,\text{Å}$)
	($100-1000\,\text{Å}$)	

Note that the overall yield strength of a material depends on both short- and long-range stress field interactions with moving dislocations.

REFERENCES

1. P. G. Partridge, *Met. Rev.*, **12**, 1967, p. 118.
2. R. von Mises, *Z. Ang. Math. Mech.*, **8**, 1928, p. 161.
3. G. I. Taylor, *J. Inst. Met.*, **62**, 1938, p. 307.
4. G. W. Groves and A. Kelly, *Phil. Mag.*, **8**, 1963, p. 877.
5. G. Y. Chin and W. L. Mammel, *Trans. Met. Soc.*, AIME, **239**, 1967, p. 1400.
6. G. Y. Chin, W. L. Mammel, and M. T. Dolan, *Trans. Met. Soc.*, AIME, **245**, 1969, p. 383.
7. W. J. McG. Tegart, *Phil. Mag.*, **9**, 1964, p. 339.
8. A. H. Cottrell, *Dislocations and Plastic Flow in Crystals*, Clarendon Press, Oxford, England, 1953.
9. P. R. Thornton, T. E. Mitchell, and P. B. Hirsch, *Phil. Mag.*, **7**, 1962, p. 1349.
10. D. C. Jillson, *Trans. Met. Soc.*, AIME, **188**, 1950, p. 1129.
11. E. N. Andrade and R. Roscoe, *Proc. Roy. Soc.*, **49**, 1937, p. 166.
12. P. M. Robinson and H. G. Scott, *Acta Met.*, **15**, 1967, p. 1581.
13. E. Schmid and W. Boas, *Plasticity of Crystals*, Hughes, London, 1950, p. 55.
14. G. I. Taylor and C. F. Elam, *Proc. Roy. Soc.*, **A108**, 1925, p. 28.
15. C. F. Elam, *Distortion of Metal Crystals*, Oxford University Press, London, 1935.
16. J. S. Koehler, *Acta Met.*, **1**, 1953, p. 508.
17. G. R. Piercy, R. W. Cahn, and A. H. Cottrell, *Acta Met.*, **3**, 1955, p. 331.
18. A. H. Cottrell, *Report of the Bristol Conference on Strength of Solids*, Physical Society, London, 1948, p. 30.
19. H. Conrad, *JISI*, **198**, 1961, p. 364.
20. W. G. Johnston, *J. Appl. Phys.*, **33**, 1962, p. 2716.
21. G. T. Hahn, *Acta Met.*, **10**, 1962, p. 727.
22. D. F. Stein and J. R. Low, Jr., *J. Appl. Phys.*, **31**, 1960, p. 362.
23. J. J. Gilman and W. G. Johnston, *J. Appl. Phys.*, **31**, 1960, p. 687.
24. W. G. Johnston and J. J. Gilman, *J. Appl. Phys.*, **30**, 1959, p. 129.
25. W. J. McG. Tegart, *Elements of Mechanical Metallurgy*, Macmillan, New York, 1966.
26. J. Garstone and R. W. K. Honeycombe, *Dislocations and Mechanical Properties of Crystals*, J. Fisher, ed., John Wiley, New York, 1957, p. 391.
27. L. M. Clarebrough and M. E. Hargreaves, *Progress in Materials Science*, vol. 8, Pergamon Press, London, 1959, p. 1.
28. J. P. Hirth and J. Weertman, eds., *Work Hardening*, Gordon and Breach, New York, 1968.
29. A. Seeger, *Dislocations and Mechnical Properties of Crystals*, J. C. Fisher, ed., John Wiley, New York, 1957, p. 243.
30. J. Friedel, *Phil. Mag.*, **46**, 1955, p. 1169.
31. N. F. Mott, *Trans. Met. Soc.*, AIME, **218**, 1960, p. 962.
32. D. Kuhlmann-Wilsdorf, *Trans. Met. Soc.*, AIME, **224**, 1962, p. 1047.
33. D. Kuhlmann-Wilsdorf, *Work Hardening*, J. P. Hirth and J. Weertman, eds., Gordon and Breach, New York, 1968, p. 97.
34. H. M. Otte and J. J. Hren, *Exp. Mech.*, **6**, 1966, p. 177.
35. N. J. Petch, *JISI*, **173**, 1953, p. 25.
36. E. O. Hall, *Proc. Phys. Soc.*, **B64**, 1951, p. 747.

37. J. T. Michalak, *Metals Handbook*, Vol. 8, ASM, Metals Park, Ohio, 1973, p. 218.
38. J. R. Patel and A. R. Chaudhuri, *J. Appl. Phys.*, **34**, 1963, p. 2788.
39. A. Kelly and G. W. Groves, *Crystallography and Crystal Defects*, Addison-Wesley, Reading, Mass. 1970, p. 175.

PROBLEMS

3.1 From the work of D. C. Jillson, *Trans.*, AIME **188**, 1950, p. 1129, the following data were taken relating to the deformation of zinc single crystals.

ϕ	λ	F(newtons)
83.5	18	203.1
70.5	29	77.1
60	30.5	51.7
50	40	45.1
29	62.5	54.9
13	78	109.0
4	86	318.5

The crystals have a normal cross-sectional area of 122×10^{-6} m^2.
ϕ = angle between loading axis and normal to slip plane
λ = angle between loading axis and slip direction
F = force acting on crystal when yielding begins
(a) Name the slip system for this material.
(b) Calculate the resolved shear τ_{RSS} and normal σ_n stresses acting on the slip plane when yielding begins.
(c) From your calculations, does τ_{RSS} or σ_n control yielding?
(d) Plot on graph paper the Schmid factor versus the normal stress P/A acting on the rod.

3.2 A bar of BCC iron containing carbon and nitrogen solute additions is stressed to the point where general yielding occurs and the test interrupted prior to necking. Describe the stress-strain response phenomenologically and in terms of dislocation dynamics. Now immediately reload the test bar into the plastic strain range and describe the new stress-strain response. Explain any different response in terms of dislocation dynamics.

3.3 What would have been the stress-strain response of the reloaded bar had the second loading taken place several weeks later or had the bar been given a moderate temperature ageing treatment prior to being reloaded.

3.4 Prove to yourself that the c/a ratio for ideal packing in the case of a hexagonal closed-packed structure is 1.633.

3.5 What crystallographic factors usually determine the planes and directions on which slip occurs? How does one of these factors determine the temperature sensitivity of the yield strength? (Refer to Chapter Two.)

3.6 Calculate the Schmid factor for an FCC single-crystal rod oriented with the $\langle 100 \rangle$ direction parallel to the loading axis.

3.7 Sketch a single-crystal shear stress-strain curve. Label the regions.
(a) Will aluminum (SFE$\approx$300 mJ/m^2) or stainless steel (SFE$\approx$10 mJ/m^2) end Stage II and begin Stage III sooner?
(b) What is the difference in slip character between aluminum and stainless steel?

CHAPTER FOUR
DEFORMATION TWINNING AND STRUCTURE OF PLASTICALLY DEFORMED METALS

4.1 DEFORMATION TWINNING

As was noted in the previous chapter, the simultaneous operation of at least five independent slip systems is required to maintain continuity at grain boundaries in a polycrystalline solid. Failure to do so will lead to premature fracture. If a crystal possesses an insufficient number of independent slip systems, twin modes may be activated in some metals to provide the additional deformation mechanisms necessary to bring about an arbitrary shape change.

4.1.1 Comparison of Slip and Twinning Deformations

The most obvious difference between a slipped versus a twinned crystal is the shape change resulting from these deformations. Whereas slip involves a simple translation across a slip plane such that one rigid portion of the solid moves relative to the other, the twinned body undergoes a shape change (Fig. 4.1).

According to Bilby and Crocker,[1] "A deformation twin is a region of a crystalline body which had undergone a homogeneous shape deformation in such a way that the resulting product structure is identical with that of the parent, but oriented differently." As pointed out in Chapter Two, dislocation movement associated with slip will take place in multiples of the unit displacement—the dislocation Burgers vector. By contrast, the shape change found in

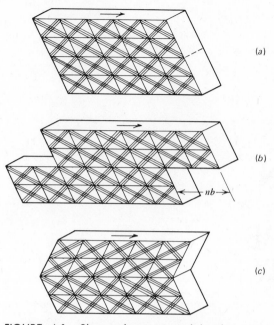

FIGURE 4.1 Shape change in solid cube caused by plastic deformation. (a) Undistorted cube; (b) slipped cube with offsets nb; (c) Twinned cube revealing reorientation within twin. Displacements are proportional to distance from twin plane.

the twinned solid results from atom movements taking place on all planes in fractional amounts within the twin. In fact, we see from Fig. 4.1c that the displacement in any plane within the twin is directly proportional to its distance from the twin-matrix boundary. Examining more closely these twinning displacements in a simple cubic lattice, it is seen that the twinning process has effected a rotation of the lattice such that the atom positions in the twin represent a mirror image of those in the untwinned material (Fig. 4.2). By contrast, slip occurs by translations along widely spaced planes in whole multiples of the displacement vector, so that the relative orientation of different regions in the slipped cube remains unchanged.

The differences associated with these deformation mechanisms are revealed when one examines the deformed surface of a prepolished sample (Fig. 4.3). Offsets due to slip are revealed as straight or wavy lines (depending on the stacking fault energy of the material) with no change in contrast noted on either side of the slip offset. Twin bands do exhibit a change in contrast, since the associated lattice reorientation within the twin causes the incident light to be reflected away from the objective lens of the microscope. After repolishing and etching the sample, only twin band markings persist, since they were associated with a reorientation of the lattice (Figs. 4.3b and 4.3d).

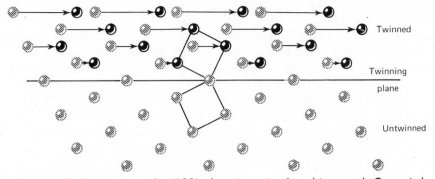

FIGURE 4.2 Twinning on the (120) plane in a simple cubic crystal. Grey circles represent original atom positions. Black circles are final atom positions.

FIGURE 4.3 Surface markings resulting from plastic deformation. (a) Prepolished and deformed zinc revealing slip lines (upper left to lower right markings) and twin bands (large horizontal band); (b) Same as (a) but repolished and etched to show only twin bands; (c) Prepolished and deformed brass revealing straight slip lines (reflecting low stacking fault energy) and preexisting annealing twins; (d) Same as (c) but repolished and etched to show only annealing twins.

4.1.2 Geometry of Twin Formation[2]

Consider the growth of a twin over the upper half of a crystalline unit sphere. Any point on the sphere will be translated from coordinates X, Y, Z to X', Y', Z', where $X = X'$, $Z = Z'$ and $Y' = Y + SZ$ (Fig. 4.4). Since S represents the magnitude of the shear strain, we see that the shear displacement on any plane is directly proportional to the distance from the twinning plane (called the composition plane). Therefore, the equation for the distorted sphere is given by

$$X'^2 + Y'^2 + Z'^2 = 1 = X^2 + Y^2 + 2SZY + S^2Z^2 + Z^2 = 1 \qquad (4\text{-}1)$$

or

$$X^2 + Y^2 + 2SZY + Z^2(S^2 + 1) = 1 \qquad (4\text{-}2)$$

which defines a quadric surface. Specifically, the distorted sphere forms an ellipsoid whose major axis is inclined to η_1 by an angle ϕ. It is clear from Fig. 4.4 that most planes contained within the sphere are either foreshortened or extended. For example, consider the movement of points A and B, which are translated by the twinning deformation to A' and B', respectively. If AO and BO

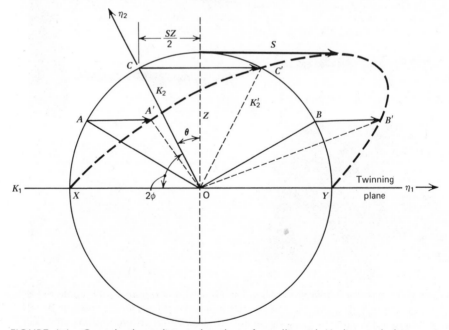

FIGURE 4.4 Crystal sphere distorted to that of an ellipsoid. Undistorted planes are K_1 and K_2, separated by angle 2ϕ. Note foreshortening of plane OA after twinning while plane OB is extended.

represent the traces of two different planes, it is clear that AO has been foreshortened ($A'O$) while BO has been stretched ($B'O$). Only two planes remain undistorted after the twin shear has been completed. The first is the composition plane, designated as K_1; the direction of the shear is given by η_1. The second undistorted plane is the one shown in profile by the line OC. (Note that $OC = OC'$.) This plane is designated as the K_2 plane, where η_2 is defined by the line of intersection of the K_2 plane and the plane of shear (the plane of this page). The final position of this second undistorted plane is designated as the K_2' plane. Therefore, all planes located between X and C will be compressed, while all planes located between C and Y will be extended. Typical values for K_1, K_2, η_1, and η_2 are shown in Table 4.1 and discussed in the following sections. By definition,[5] when K_1 and η_2 are rational and K_2 and η_1 are not, we speak of this twin as being of the *first kind*. The orientation change resulting from this twin can be accounted for by reflection in the K_1 plane or by a 180-degree rotation about the normal to K_1. When K_2 and η_1 are rational but K_1 and η_2 are not, the twin is of the *second kind*. The twin orientation in this case may be achieved either by a 180-degree rotation about η_1 or by reflection in the plane normal to η_1. When all twin elements are rational, the twin is designated as *compound*. This occurs often in crystals possessing high symmetry (such as most metals), where the reflection and rotation operations are equivalent.

The magnitude of the shear strain S in the unit sphere is given by the angle 2ϕ

TABLE 4.1   Observed Twin Elements in Metals[3,4]

Metal	Crystal Structure	c/a Ratio	K_1	K_2	η_1	η_2	S	$(l'/l)_{max}$
Al, Cu, Au, Ni, Ag, γ-Fe	FCC		$\{111\}$	$\{11\bar{1}\}$	$\langle 11\bar{2} \rangle$	$\langle 112 \rangle$	0.707	41.4%
αFe	BCC		$\{112\}$	$\{\bar{1}\bar{1}2\}$	$\langle \bar{1}\bar{1}1 \rangle$	$\langle 111 \rangle$	0.707	41.4
Cd	HCP	1.886	$\{10\bar{1}2\}$	$\{\bar{1}012\}$	$\langle 10\bar{1}\bar{1} \rangle$	$\langle 10\bar{1}1 \rangle$	0.17	8.9
Zn	HCP	1.856	$\{10\bar{1}2\}$	$\{\bar{1}012\}$	$\langle 10\bar{1}\bar{1} \rangle$	$\langle 10\bar{1}1 \rangle$	0.139	7.2
Mg	HCP	1.624	$\{10\bar{1}2\}$	$\{\bar{1}012\}$	$\langle 10\bar{1}\bar{1} \rangle$	$\langle 10\bar{1}1 \rangle$	0.131	6.8
			$\{11\bar{2}1\}$	$\{0001\}$	$\langle 11\bar{2}\bar{6} \rangle$	$\langle 11\bar{2}0 \rangle$	0.64	37.0
Zr	HCP	1.589	$\{10\bar{1}2\}$	$\{\bar{1}012\}$	$\langle 10\bar{1}\bar{1} \rangle$	$\langle 10\bar{1}1 \rangle$	0.167	8.7
			$\{11\bar{2}1\}$	$\{0001\}$	$\langle 11\bar{2}\bar{6} \rangle$	$\langle 11\bar{2}0 \rangle$	0.63	36.3
			$\{11\bar{2}2\}$	$\{11\bar{2}\bar{4}\}$	$\langle 11\bar{2}\bar{3} \rangle$	$\langle 22\bar{4}3 \rangle$	0.225	11.9
Ti	HCP	1.587	$\{10\bar{1}2\}$	$\{\bar{1}012\}$	$\langle 10\bar{1}\bar{1} \rangle$	$\langle 10\bar{1}1 \rangle$	0.167	8.7
			$\{11\bar{2}1\}$	$\{0001\}$	$\langle 11\bar{2}\bar{6} \rangle$	$\langle 11\bar{2}0 \rangle$	0.638	36.9
			$\{11\bar{2}2\}$	$\{11\bar{2}\bar{4}\}$	$\langle 11\bar{2}\bar{3} \rangle$	$\langle 22\bar{4}3 \rangle$	0.255	11.9
Be	HCP	1.568	$\{10\bar{1}2\}$	$\{\bar{1}012\}$	$\langle 10\bar{1}\bar{1} \rangle$	$\langle 10\bar{1}1 \rangle$	0.199	10.4

between the two undistorted planes K_1 and K_2. From Fig. 4.4

$$\tan \theta = SZ/2/Z = \frac{S}{2} \qquad (4\text{-}3)$$

Since

$$\theta + 2\phi = 90°$$

$$\cot 2\phi = \frac{S}{2} \qquad (4\text{-}4)$$

4.1.3 Elongation Potential of Twin Deformation

Hall[2] has shown that the total deformation strain to be expected from a completely twinned crystal may be given by

$$\frac{l'}{l} = \left[1 + S \tan \chi \right]^{\frac{1}{2}} \qquad (4\text{-}5)$$

where

$$l, l' = \text{initial and final length, respectively}$$

$$\tan \chi = \frac{S \pm \sqrt{S^2 + 4}}{2}$$

From Eq. 4-5, the maximum potential elongation of the metals shown in Table 4.1 is quite small, particularly in HCP crystals, which undergo $\{10\bar{1}2\}$ type twinning. Although the twinning reaction contributes little to the total elongation of the sample, the rotation of the crystal within the twin serves mainly to reorient the slip planes so that they might experience a higher resolved shear stress and thereby contribute more deformation by slip processes.

4.1.4 Twin Shape

From the above geometrical analysis, one would assume twinned regions to be bounded by two parallel composition planes representing the two twin-matrix coherent interfaces. In practice, twins are found often to be lens-shaped, so that the interface must consist of both coherent and noncoherent segments. These noncoherent portions of the interface can be described in terms of particular dislocation arrays (Fig. 4.5). Mahajan and Williams[6] have reviewed the literature and found that twin formation has been rationalized both in terms of heterogeneous nucleation at some dislocation arrangement or by homogeneous nucleation in a region of high stress concentration. It is worth noting that dislocations

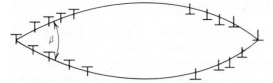

FIGURE 4.5 Schematic of lens-shaped twin with dislocations to accommodate noncoherent twin-matrix interface regions. Lens angle β increases with decreasing twin shear and increasing ability of matrix to accommodate the twin strain concentration.

are also needed to account for the requirement of a much lower stress to move a twin boundary than the theoretically expected value.

Cahn[5] postulated that the lens angle β should increase with decreasing shear strain. Since the magnitude of β controls the permissible thickness of the lens, Cahn's postulate correctly predicts the empirical fact that twin thickness increases with decreasing shear strain. More recently, Friedel[7] also concluded that the optimum lens thickness to length ratio should increase with decreasing shear strain. Since twin formation involves discontinuous deformations, some type of lattice accommodation is necessary along the perimeter of the twin lens. When the parent lattice possesses limited ductility, the lens angle β is kept small and the strain discontinuity accommodated by crack formation.[8] At the other extreme, lattice plane bending and/or slip may be introduced to "smear out" the strain discontinuity resulting from the twin. If the crystal is able to slip readily, the lens angle β can increase, thereby enabling the twin to thicken. Therefore, we find that in ductile crystals, the thickness of deformation twins increases with decreasing twin shear strain (Fig. 4.6).

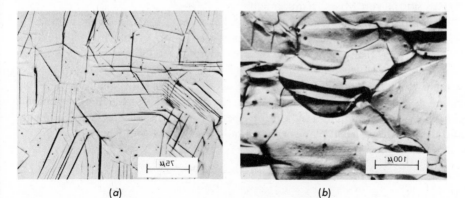

(a)   (b)

FIGURE 4.6 Prepolished and subsequently deformed surfaces revealing shape of twins. (a) Narrow deformation twins in α-Fe ($S=0.707$); (b) broad deformation twins in Mg ($S=0.131$). (After Eckelmeyer and Hertzberg,[9] copyright American Society for Metals, 1970.)

4.1.5 Stress Requirements for Twinning

It is now generally known that the twin initiation stress is much greater than the stress needed to propagate a preexistent twin. Similarly, the *nucleation* of a twin is associated with a sudden load drop (responsible for serrated stress-strain curves as shown in Fig. 1.14), while the *growth* of a twin exhibits smoother loading behavior. Both observations point to the need for a large stress concentration to nucleate the twin, although the stress concentration apparently is not needed for twin growth. The prerequisite of a stress concentration to initiate plastic deformation via a twinning mode raises serious doubts regarding the existence of a critical resolved shear stress (CRSS) for twinning, as a counterpart to the well-documented CRSS for slip. Furthermore, the large degree of scatter in reported "CRSS for twinning" in various materials raises additional doubts as to its existence.

4.1.6 Twinning in HCP Crystals

Among the three major unit cells found in metals and their alloys, twinning is most prevalent in HCP materials. Over a broad temperature range, twinning and slip are highly competitive deformation processes. We saw from Section 3.1 that regardless of the c/a ratio (that is, whether basal or prism slip was preferred), an insufficient number of independent slip systems can operate to satisfy the von Mises requirement.[10] Since alloys of magnesium, titanium, and zinc are known to possess reasonable ductility, some other deformation mechanisms must be operative. While combinations of basal, prism, and pyramidal slip do not provide the necessary five independent slip systems necessary for an arbitrary shape change in a polycrystalline material, deformation twinning often is necessary to satisfy von Mises' requirement.

Twinning in HCP metals and alloys has been observed on a number of different planes (Table 4.1). One twin mode common to many HCP metals is that involving $\{10\bar{1}2\}$ planes. One of three possible sets of these planes is shown in Fig. 4.7a. Activation of one particular set will depend on the respective Schmid factors. Naturally, twinning will occur on the $\{10\bar{1}2\}$ plane and $\langle\bar{1}011\rangle$ direction that experiences the highest resolved shear stress. For additional clarification, the angular relationships between the undistorted $\{10\bar{1}2\}$ planes and the prism and basal planes are shown in Fig. 4.7b. From Fig. 4.7b

$$\tan\theta = \frac{c}{\sqrt{3}\,a} \tag{4-6}$$

Since $2\phi + 2\theta = 180°$ and $\tan 2\phi = 2/S$

$$\tan 2\phi = 2/S = \tan(180-2\theta) \tag{4-7}$$

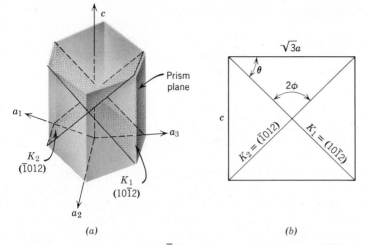

FIGURE 4.7 One set of $\{10\bar{1}2\}$ type K_1 and K_2 planes in HCP crystal. (a) Inclined view of prism, basal, and two undistorted planes; (b) important planes viewed on edge along a_2 direction.

With trigonometric identities it may be shown that

$$S = \frac{\tan^2\theta - 1}{\tan\theta} \tag{4-8}$$

Combining Eqs. 4-6 and 4-8 and rearranging we find

$$S = \left[(c/a)^2 - 3\right]\frac{\sqrt{3}\,a}{3c} \tag{4-9}$$

From Eq. 4-9, it is seen that the sense of the twin deformation is opposite for HCP metals exhibiting c/a ratios $\gtrless \sqrt{3}$. When $c/a = \sqrt{3}$, the analysis predicts that $S = 0$ and that twinning would not occur by the $\{10\bar{1}2\}$ mode. Stoloff and Gensamer[11] have verified this in a magnesium crystal alloyed with cadmium to produce a c/a ratio of $\sqrt{3}$. The reversal in sense of the twin deformation is seen when the response of beryllium ($c/a = 1.568$) and zinc ($c/a = 1.856$) are compared using strain ellipsoid diagrams. The relevant interplanar angles in each metal are determined by

$$\cos\theta = \frac{h_1h_2 + k_1k_2 + \frac{1}{2}(h_1k_2 + h_2k_1) + \frac{3a^2}{4c^2}l_1l_2}{\left\{\left[h_1^2 + k_1^2 + h_1k_1 + \frac{3a^2}{4c^2}l_1^2\right]\left[h_2^2 + k_2^2 + h_2k_2 + \frac{3a^2}{4c^2}l_2^2\right]\right\}^{\frac{1}{2}}} \tag{4-10}$$

and are given in Table 4.2.

TABLE 4.2   Interplanar Angles in Beryllium and Zinc

	$\{10\bar{1}2\}-\{0001\}$	$\{10\bar{1}2\}-\{\bar{1}012\}$	$\{10\bar{1}2\}-\{10\bar{1}0\}$
Beryllium	42°10′	84°20′	47°50′
Zinc	46°59′	86°02′	43°01′

For the case of beryllium, the basal plane bisects the acute angle separating the $\{10\bar{1}2\}$ planes, and the prism plane bisects its supplement. In addition, the prism plane may be positioned simply by the fact that it must lie 90 degrees away from the basal plane. From Fig. 4.8, we see that the twinning process in beryllium involves compression of the basal plane and tension of the prism plane. Consequently, if a single crystal were oriented with the basal plane parallel to the loading direction, the crystal would twin if the loads were compressive but not if the loads were tensile. The crystal would be able to twin in tension only if the basal plane were oriented perpendicular to the loading axis.

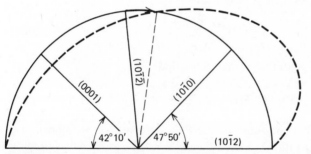

FIGURE 4.8   Strain ellipsoid for beryllium revealing twin related foreshortening of basal plane and extension of prism plane. Twinning by $\{10\bar{1}2\}$ mode will occur when compression is applied parallel to basal plane or tension applied parallel to prism plane.

The situation is completely opposite for zinc. Here, because the prism plane bisects the acute angle between K_1 and K_2, zinc will twin when the applied stress causes compression of the prism or extension of the basal plane (Fig. 4.9). The response of any HCP metal that twins by the $\{10\bar{1}2\}$ mode is summarized in Fig. 4.10. When $c/a < \sqrt{3}$ twinning will occur if compressive loads are applied parallel to the basal plane or tensile loads applied parallel to the prism planes. The opposite is true for the case of $c/a > \sqrt{3}$ where twinning occurs when the prism plane is compressed or the basal plane extended.

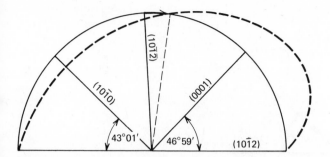

FIGURE 4.9   Strain ellipsoid for zinc revealing twin related foreshortening of prism plane and extension of basal plane. Twinning by $\{10\bar{1}2\}$ mode will occur when compression is applied parallel to prism plane or tension applied parallel to basal plane.

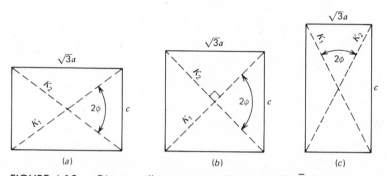

FIGURE 4.10   Diagram illustrating conditions for $\{10\bar{1}2\}$ twinning in hexagonal crystals. (a) When $c/a < \sqrt{3}$, twinning results when the basal plane is compressed or the $\{10\bar{1}0\}$ planes stretched; (b) when $c/a = \sqrt{3}$, no twinning by $\{10\bar{1}2\}$ mode occurs; (c) when $c/a > \sqrt{3}$ twinning occurs when the prism planes are compressed or the basal planes stretched.

The other twin modes shown in Table 4.1 operate under certain conditions. For example, the necessary circumstances for twinning in zirconium on the three most important modes are shown in Fig. 4.11. Note that twinning will occur by the $\{10\bar{1}2\}$ and $\{11\bar{2}1\}$ modes when tensile stresses are applied parallel to the prism plane, while compressive loads are necessary to activate $\{11\bar{2}2\}$ twinning. Although more total elongation is possible from twin modes involving large shear strains (Eq. 4-5), they are generally not preferred, since the strain energy of the twin increases with S^2. Therefore, if the resolved shear stress for given K_1 and η_1 twin elements is sufficient, twinning will occur via the mode possessing the lowest shear strain. However, there are other factors associated with twinning in noncubic crystals that qualify the above statement to some degree (see Section 4.1.6.1).

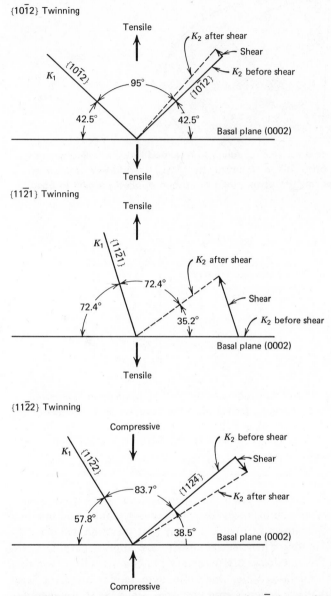

FIGURE 4.11 Twinning modes in zirconium. $\{10\bar{1}2\}$ and $\{11\bar{2}1\}$ modes activated by tension parallel to prism plane. $\{11\bar{2}2\}$ mode activated by tension parallel to basal plane. (After Reed-Hill,[3] reprinted with permission from Gordon & Breach Science Publishers.)

As might be expected, there is not only competition between twin modes but also between slip and twinning as the dominant deformation mechanism under specific test conditions. Reed-Hill[3] examined the likelihood of either prism slip or {10$\bar{1}$2} twinning in zirconium and found these mechanisms to be complementary (Fig. 4.12).

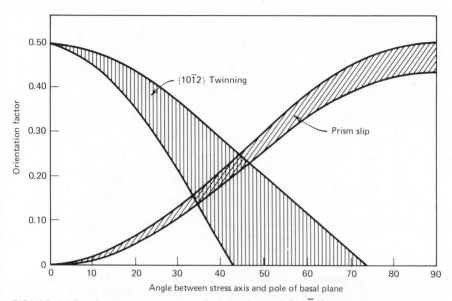

FIGURE 4.12   Competitive aspects of prism slip and {10$\bar{1}$2} twinning in zirconium. (After Reed-Hill,[3] reprinted with permission from Gordon & Breach Science Publishers.)

4.1.6.1   ATOM "SHUFFLES" TO RESTORE THE HCP LATTICE

From the geometrical considerations of twinning, it was shown that the lattice must be restored after twinning. This will occur when all the atoms in the lattice are of the same kind and located at lattice points. In HCP crystals and other noncubic materials, where the atoms do not assume positions coincident with lattice points, homogeneous shear of atoms in amounts proportional to the distance from the composition plane (Fig. 4.2) occurs only on alternate planes. The atoms on the intervening planes are forced to "shuffle" to their new positions with the minimum expenditure of energy. Note from Fig. 4.13 that the movement of these "shuffling" atoms involves components perpendicular to the composition plane. It is not clear how the twin shear, parallel to the η_1 direction, can produce such movement.

It does seem reasonable that twin modes with lower shuffle requirements are more highly preferred. In fact, Laves[12] found empirically that twinning would

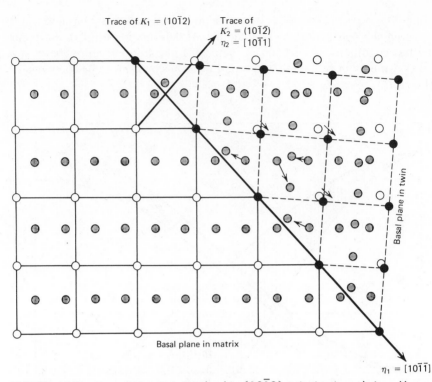

FIGURE 4.13 Atom movements involved in $\{10\bar{1}2\}$ twinning in cadmium. Homogeneous shear occurs on every other plane, while other atoms must "shuffle" to twinned positions. Shuffles often involve displacement components perpendicular to composition plane. (After Hall,[2] reproduced from *Twinning and Diffusionless Transformations in Metals*, published by Newnes-Butterworths 1954.)

occur if the arithmetic mean of the shuffle vectors was less than 20% of the mean interatomic distance. More recently, Bilby and Crocker[1] established a set of general criteria for the incidence of a particular twinning mode that stipulate that the twin shear should be small, the shuffle mechanism simple, the shuffle magnitude small and its direction parallel rather than perpendicular to η_1.

It may be concluded, then, that while the geometrical analysis describes the twin process in terms of the *initial* and *final* atom positions, it does not explain the complex atom movements that occur often *during* the process.

4.1.7 Twinning in BCC Crystals

Twinning in BCC materials has been examined most closely for the case of ferritic steels, because of their engineering significance. Twin formation in steels (called Neumann bands) occurs most readily under high strain rate and/or low-temperature test conditions.

The twin plane is found to be of type {112}, with the shearing direction parallel to [$\bar{1}\bar{1}1$]. What is intriguing again is the fact that twinning will depend on the direction of shear; twinning will occur in the [$\bar{1}\bar{1}1$] direction but not in the opposite [11$\bar{1}$] direction.[13] The reason for this circumstance is made clear by consideration of the atom movements when viewed normal to the (110) plane (Fig. 4.14a). Each atom is designated as to its lattice position (corner or body-centered) and its location relative to the plane of the page. Note that the displacement distance YY' is $\sqrt{3}\,a_0/6$, but is $\sqrt{3}\,a_0/3$ for the distance ZY' where Y' is the mirror image atom position after twinning is complete. Since XY' is twice the interplanar distance between {112} planes,

$$XY' = 2d_{\{112\}} = \frac{2a}{\sqrt{h^2 + k^2 + l^2}} = \frac{2a}{\sqrt{6}} \tag{4-11}$$

Since

$$XZ = a$$

$$ZY' = \sqrt{XZ^2 - XY'^2}$$

$$= \sqrt{a^2 - \frac{2a^2}{3}}$$

$$ZY' = \frac{\sqrt{3}\,a}{3} \tag{4-12}$$

Since $ZY = \sqrt{3}\,a/2$, it may be seen that

$$YY' = \frac{\sqrt{3}\,a}{2} - \frac{\sqrt{3}\,a}{3} = \frac{\sqrt{3}\,a}{6} \tag{4-13}$$

Therefore, the twinning process would rather proceed from Y to Y' than from Z to Y'.

The same conclusion may be reached if one were to examine these same atom movements with the plane of the page serving as the (112) composition plane (Fig. 4.14b). In addition to ZY' being twice as great as YY', atom Z would have to surmount the saddle between the adjoining body-centered atoms, whereas atom Y would not have to overcome this difficulty in its movement to Y'.

As stated in the previous section, restoration of the crystal after twinning is simplified when the atoms are located at lattice sites and are of the same kind. When the atoms are ordered, complications arise. For example, an ordered BCC

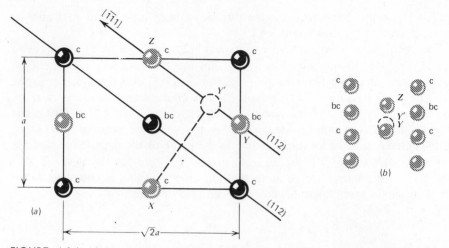

FIGURE 4.14   Atom movements associated with twinning in α-iron. (*a*) (110) projection showed twinning on (112) plane in [$\overline{1}1$1] direction. Black circles represent atoms lying in plane of paper. Grey circles represent atoms above plane of paper; (*b*) (112) projection showing ease of movement in [$\overline{1}1$1] direction as compared with [11$\overline{1}$] direction.

$L2_0$ superlattice will twin on the (112) plane but in the [11$\overline{1}$] direction, even though the [$\overline{1}1$1] direction involves half the shear strain. This is because the [11$\overline{1}$] shear will restore the ordered BCC lattice, while the [$\overline{1}1$1] shear produces a tetragonal structure.[14,15] Since a twin process must restore the lattice in the twin to its original type, the latter shear direction is not permissible.

4.1.8   Twinning in FCC Crystals

Deformation twinning is found least frequently in FCC metals except under cryogenic temperature conditions, extremely high strain rates, and in certain alloys. Since the twin elements, $\{111\}$, $\{11\overline{1}\}$, $\langle11\overline{2}\rangle$, and $\langle112\rangle$ produce a large twin strain (0.707), it would appear that slip processes are more highly favored (i.e., partial dislocation motion along close-packed planes) than twin-related movements. By comparison, it should be pointed out that while *deformation* twinning is found only under extreme conditions, some FCC alloys may exhibit many *annealing* twins. These twins result from accidents associated with the growth of recrystallized grains from previously deformed material possessing a high density of stacking faults. As a grain grows with a particular packing sequence, for example, *ABCABC*, the stacking pattern will be altered as a result of the presence of the stacking fault, and the packing sequence might become ABCABCBACBA. The *C* plane in the middle of this planar grouping serves as the composition plane, with atoms on one side being reflected on the other side. Note that the twin boundary itself consists of a layer of hexagonal packing— that is, *BCB*. Since the incidence of these growth accidents depends strongly on the number of stacking faults present in the lattice, the density of annealing twins found in an FCC crystal will vary inversely with the stacking fault energy of the material. Brass, for example, reveals many annealing twins while aluminum shows none. Since the stacking fault probability depends also on the extent of deformation, the number of annealing twins found in a given material should increase with increasing prior cold work. As such, the number of annealing twins found in a recrystallized material provides a clue as to the deformation history of the material.

4.2 STRUCTURE OF PLASTICALLY DEFORMED METALS

4.2.1 Mechanical Fibering

Having examined the two major mechanisms by which crystalline solids may plastically deform (slip and twinning), attention is now given to the overall consequence of such irreversible atom movements. Let us begin this discussion by presuming that we have taken a cube of equiaxed polycrystalline material and changed its shape by some mechanical process such as rolling, drawing, or swaging (a combination of drawing and twisting). By the principle of similitude, the conversion of the cube into a thin plate should be reflected by a change in the size and shape of the grains within the solid. That is, the equiaxed grains should be flattened and spread out, as shown in Fig. 4.15. The alignment of the grain structure in the direction of mechanical working is known as *mechanical fibering* and is exhibited most dramatically in forged products such as the one shown in Fig. 4.16. Here the grains have been molded to parallel the contour of the forging dies. Engineers have found that the fracture resistance of a forged component can be enhanced considerably when the *forging flow lines* are oriented parallel to major stress trajectories and normal to the path of a potential crack. As such, forged parts are considered to be superior to comparable castings because of the benefits derived from the deformation-induced microstructural anisotropy. Japanese armor makers took advantage of this fact in their manufacture of the Samuri sword.[16] They heated a billet of iron to elevated temperature where it was folded back upon itself by repeated blows of forging hammers, and then placed back in the furnace until ready for another forging and folding operation. This was done 10 to 20 times, resulting in a sword blade containing 2^{10} to 2^{20} layers of the original billet. After masterful decoration and a special heat treatment, this aesthetically appealing and structurally sound weapon was ready for its deadly purpose. Of course, when a forged

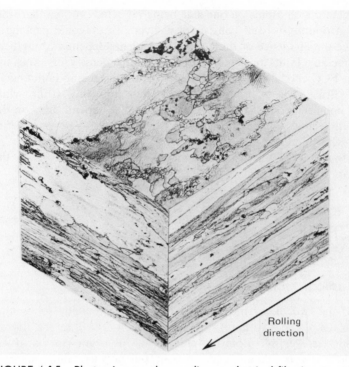

Rolling
direction

FIGURE 4.15 Photomicrograph revealing mechanical fibering associated with rolling of 7075-T651 aluminum plate. (Courtesy of J. Staley, Alcoa Aluminum Company.)

product is used improperly, the flow lines act as readily available paths for easy crack propagation. (This situation is considered at greater length in Chapter Ten.) The reader should note that mechanical fibering involves not only alignment of grains but also alignment of inclusions. For example, in standard steel-making practice, the hot rolling temperature for billet breakdown exceeds the softening point for manganese sulfide inclusions commonly found in most steels. Consequently, these inclusions are strung out in the rolling direction and flattened in the rolling plane, as shown in Fig. 4.17. The deleterious nature of these aligned inclusions relative to the fracture properties of steel plate is discussed further in Chapter Ten, along with recently developed procedures aimed at inclusion shape control.

A drawing operation will convert our reference cube into a long, thin wire or rod. Again, by similitude, the grains are found to be sausage-shaped and elongated in the drawing direction. In a transverse section normal to the rod axis, the grains should appear equiaxed while being highly elongated parallel to the rod axis. For very large draw ratios in BCC metals, however, the grains take on a ribbon-like appearance, because of the nature of the deformation process in

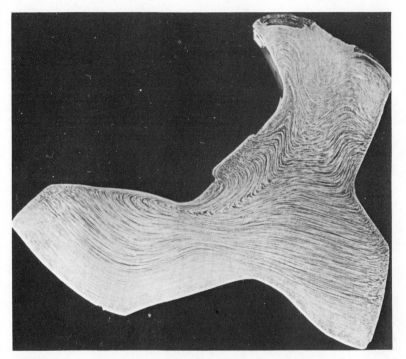

FIGURE 4.16   Flow lines readily visible in forged component. (Courtesy of G. Vander Voort, Bethlehem Steel Company.)

the BCC lattice.[18] When a metal is swaged, the elongated grain structure along the rod axis is maintained (Fig. 4.18a), and the transverse section reveals a beautiful spiral nebula pattern, reflecting the twisting action of the rotating dies during the swaging process[19] (Fig. 4.18b).

In discussing the deformation structure of metals, it is important to keep in mind the temperature of the operation. It is known that the highly oriented grain structure in a wrought product, which has a very high dislocation density (10^{11} to 10^{13} dislocations/cm^2), remains stable only when the combination of stored strain energy (related to the dislocation substructure) and thermal energy (determined by the deformation temperature) is below a certain level. If not, the microstructure becomes unstable and new strain-free equiaxed grains are formed by combined recovery, recrystallization, and grain growth processes. These new grains will have a much lower dislocation density (in the range of 10^4 to 10^6 dislocations/cm^2). When mechanical deformation at a given temperature causes the microstructure to recrystallize spontaneously, the material is said to have been *hot worked*. If the microstructure were stable at that temperature, the metal experienced *cold working*. The temperature at which metals undergo hot working

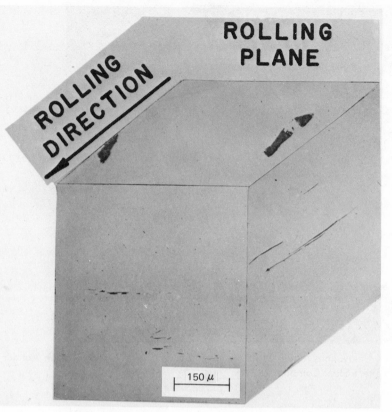

FIGURE 4.17 Alignment of manganese sulfide inclusions on rolling plane in hot-rolled steel plate. (After Heiser and Hertzberg,[17] reprinted with permission of the American Society of Mechanical Engineers.)

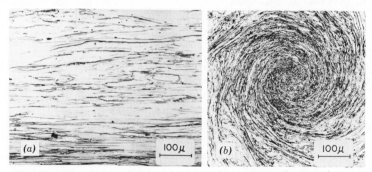

FIGURE 4.18 Longitudinal and transverse sections of swaged tungsten wire reduced by 87%. (After Peck and Thomas,[19] reprinted with permission of the Metallurgical Society of AIME.)

varies widely from one alloy to another but is found generally to occur at about one-third the absolute melting temperature. Accordingly, lead is hot worked at room temperature, while tungsten may be cold worked at 1500°C.

4.2.2 Crystallographic Textures (Preferred Orientations)

Recalling from Section 3.3 that single-crystal deformation involves lattice rotation, it should not be surprising to find individual grains in a polycrystalline aggregate undergoing similar reorientation. As might be expected, lattice reorientation in a given grain is impeded by constraints introduced by contiguous grains, thereby making the development of crystallographic textures in polycrystalline aggregates a complex process. In addition, the preferred orientation is found to depend on a number of additional variables, such as the composition and crystal structure of the metal and the nature, extent, and temperature of the plastic deformation process.[20] As a result, the texture developed by a metal usually is not complete, but instead may be described by the strength of one orientation component relative to another.

Crystallographic textures are portrayed frequently by the pole figure, which is essentially a stereographic projection showing the distribution of *one* particular set of $\{hkl\}$ poles in orientation space. That is, X-ray diffractometer conditions are fixed for a particular diffraction angle and X-ray wavelength so that the distribution of one set of $\{hkl\}$ poles in the polycrystalline sample can be monitored. To illustrate, consider the single-crystal orientation responsible for the (100) stereographic projection shown in Fig. 3.12a. The (100) pole figure for this crystal would reveal (100) diffraction spots at the north, south, east, and west poles and at the center of the projection (Fig. 4.19a). No information concerning the location of $\{110\}$, $\{111\}$, or $\{hkl\}$ poles is collected, since diffraction conditions for these planes are not met. Their location would have to be surmised based on the position of the (100) poles and the known angular relationship between the $\{100\}$ and $\{hkl\}$ poles. It is possible, of course, to change diffraction conditions to "see" the location of these other poles but then the $\{100\}$ poles would "disappear" from the $\{hkl\}$ pole figure. Figure 4.19b shows the same crystal as in Fig. 4.19a, but with its orientation portrayed by a (110) pole figure. It is important to appreciate that although these two pole figures look different, they convey the same information—the orientation of the crystal. By analogy, different $\{hkl\}$ pole figures represent different languages by which the same thought (the preferred orientation) is conveyed.

When wires or rods are produced, such as by drawing or swaging, a uniaxial preferred orientation may develop in the drawing direction, with other crystallographic poles distributed symmetrically about the wire axis. For a [100] crystallographic wire texture, such as heavily deformed silver wire, the texture is given by Fig. 4.19c as portrayed by a (100) pole figure. Note the presence of $\{100\}$ poles at the north (and south) pole of the projection corresponding to the

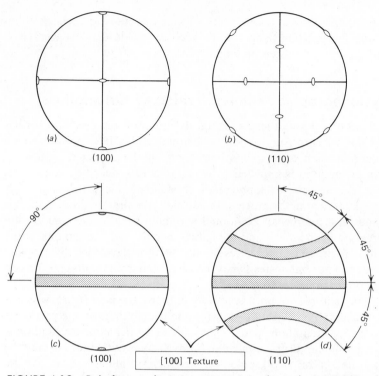

FIGURE 4.19 Pole figures depicting orientation of metals. (*a*) (100) pole figure for crystal orientation shown in Fig. 3.12*a*; (*b*) (110) pole figure for same orientation; (*c*) (100) pole figure for [100] wire texture; (*d*) (110) pole figure for [100] wire texture. Note rotational symmetry in (*c*) and (*d*).

drawing direction and the smearing out of the other {100} poles across the equator, the latter reflecting the rotational symmetry found in wire textures. The same texture is shown in Fig. 4.19*d* via a (110) pole figure. Here the rotational symmetry of the wire texture is again evident while the [100] wire texture must be inferred from the relative position of the {110} poles. Naturally, a (111) pole figure would present yet another interpretation of the same [100] wire texture. (The reader is advised to sketch the (111) pole figure for the [100] wire texture for his or her edification.) Typical wire textures for a number of FCC metals and alloys are given in Fig. 4.20, where the variation of texture with stacking fault energy (SFE) is shown clearly.[21] The explanation for the SFE dependence of texture transition and for the reversal in texture at very low stacking fault energies has been the subject of considerable debate. Cross-slip,[22,23,24] mechanical twinning,[25,26] overshooting,[27,28] and extensive movement of Shockley partial dislocation[29] mechanisms have been proposed as possible contributing factors toward development of both wire and sheet textures.

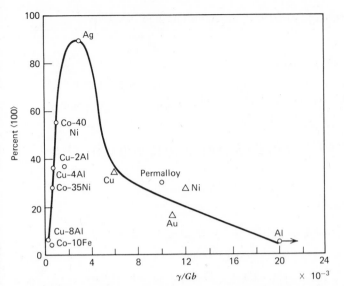

FIGURE 4.20 Relationship between strength of [100] component in FCC wire texture and stacking fault energy parameter, γ/Gb. (After English and Chin,[21] reprinted with permission from Chin, *Acta Met.*, **13** (1965), Pergamon Press.)

For the case of BCC metals, the wire texture is uncomplicated and found to be [110]. In HCP metals, texture is found to vary with the c/a ratio. When $c/a < 1.633$, a [$10\bar{1}0$] texture may be developed with the basal plane lying parallel to the rod axis.[20] By contrast, texture development is more complex when $c/a > 1.633$.

Although wire textures may be defined by one component—the direction parallel to the wire axis—sheet textures are given by both the crystallographic plane oriented parallel to the rolling plane and the crystallographic direction found parallel to the rolling direction. Hence, rolling textures are described by the notation (hkl) [uvw], where (hkl) corresponds to the plane parallel to the rolling plane and [uvw] to the direction parallel to the rolling direction. Rolling textures are very complex, with several different components often existing simultaneously (Table 4.3). For this reason, more than one pole figure is desirable to properly identify the texture of a given sheet. In addition, the distribution of poles is described with the use of contour lines, each representing the density of a pole relative to that found in a randomly oriented sample. For example, the (111) and (200) pole figures for heavily worked silver are shown in Fig. 4.21. (Note the spread in texture from the ideal orientation.) Although no poles lie at the north pole (parallel to the rolling direction) or at the center of the projection (normal to the rolling plane), either pole figure provides the necessary information to deduce the sheet texture to be (110) [$\bar{1}12$]. This texture is often

TABLE 4.3 Typical Rolling Textures in Selected Engineering Alloys[20]

FCC	
Brass, silver, stainless steel	$(110)[\bar{1}12] + (110)[001]$
Copper, nickel, aluminum	$(123)[\bar{4}12] + (146)[\,\overline{211}\,] + (112)[11\bar{1}]$

BCC	
Iron, tungsten, molybdenum, tantalum, niobium	$(001)[\bar{1}10]$ to $(111)[\bar{1}10] + (112)[\bar{1}10]$ to $(111)[\bar{2}11]$

HCP	
Magnesium, cobalt $(c/a \approx 1.633)$	$(0001)[2\bar{1}\bar{1}0]$
Zinc, cadmium $c/a > 1.633$	(0001) plane tilted ± 20–$25°$ from rolling plane about a $[10\bar{1}0]$ transverse direction axis
Titanium, zirconium, beryllium $c/a < 1.633$	(0001) plane tilted ± 30–$40°$ from rolling plane about a $[10\bar{1}0]$ rolling direction axis

referred to as the *brass* or *silver* texture, typical of FCC materials that possess low stacking fault energy. In copper, nickel, and aluminum, which possess intermediate and high SFE, respectively, the major textural components are (123) $[\bar{4}12]$, (146) $[\bar{2}11]$, and (112) $[\bar{1}11]$. This more complicated preferred orientation is called the *copper* texture. Since the SFE for an alloy depends on solute content (i.e., changes in the electron to atom ratio), the texture of a metal can change from copper to brass type with increasing alloy additions. As seen in Fig. 4.22, increasing the zinc content in copper can effect a transition in rolling texture from *copper* to *brass* type.[31] (Compare the similarity between Figs. 4.21*a* and 4.22*c*.) Some researchers have argued that the importance of SFE in controlling the type of deformation texture is related to the relative ease by which cross-slip occurs, the brass texture being generated when cross-slip is more difficult. Others have suggested that the brass texture develops when mechanical twinning or deformation faulting is relatively easy. For this reason, the preferred orientation should also be sensitive to the temperature of deformation, since cross-slip, mechanical twinning, and faulting are thermally dependent processes. In studying the rolling texture in high-purity silver, Hu and Cline[30] found that cold rolling at 0°C produced a typical $\{110\}\langle 211 \rangle$ *brass* or *silver* texture. However, when the silver was rolled at 200°C, near $\{123\}\langle 412 \rangle$ and $\{146\}\langle 211 \rangle$ components were observed, reflecting a *copper* type texture. A similar *brass* to *copper* type texture transition was found when 18-8 stainless steel was rolled at 200°C and 800°C, respectively.[32] Conversely, a reverse *copper* to

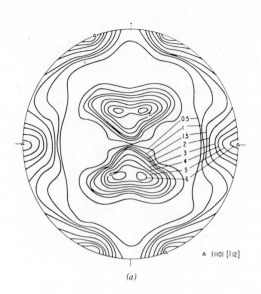

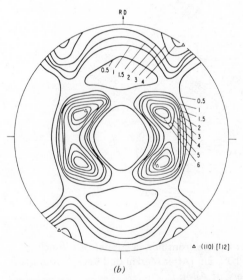

FIGURE 4.21   Rolling texture of $99.999^+\%$ pure silver, rolled 91% at 0°C. (a) (111) pole figure; (b) (200) pole figure. (After Hu and Cline,[30] reprinted with permission from Hu and the American Institute of Physics.)

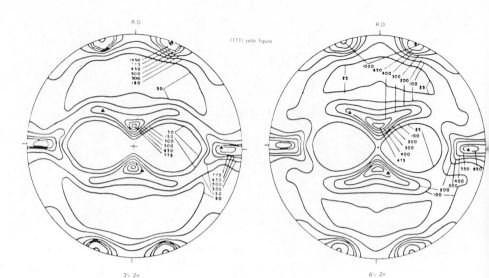

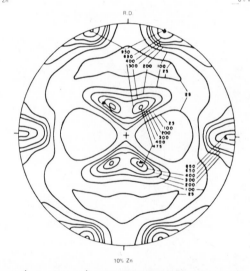

FIGURE 4.22   Copper to brass type sheet texture transition in copper as a function of zinc content. (a) 3% Zn; (b) 6% Zn; (c) 10% Zn. (After Merlini and Beck,[31] reprinted with permission of the Metallurgical Society of AIME.)

brass texture transition was realized for copper when the rolling temperature was reduced from ambient to $-196°C$.[33] Finally, by combining the effects of alloy content and deformation temperature on SFE, Smallman and Green[22] demonstrated for the silver-aluminum alloy, that the *brass* to *copper* rolling texture transition temperature increased with decreasing initial stacking fault energy.

REFERENCES

1. B. A. Bilby and A. G. Crocker, *Proc. Roy. Soc.*, **A288**, 1965, p. 240.
2. E. O. Hall, *Twinning and Diffusionless Transformations in Metals*, Butterworth, London, 1954.
3. R. E. Reed-Hill, *Deformation Twinning*, Gordon and Breach, New York, 1964, p. 295.
4. P. G. Partridge, *Met. Rev.*, **12**, 1967, p. 169.
5. R. W. Cahn, *Adv. Phys.*, **3**, 1954, p. 363.
6. S. Mahajan and D. F. Williams, *Int. Metall. Rev.*, **18**, 1973, p. 43.
7. J. Friedel, *Dislocations*, Pergamon Press, Oxford, 1964.
8. D. Hull, *Fracture of Solids*, Interscience, New York, 1963, p. 417.
9. K. E. Eckelmeyer and R. W. Hertzberg, *Met. Trans.*, **1**, 1970, p. 3411.
10. R. Von Mises, *Z. Ang. Math. Mech.*, **8**, 1928, p. 161.
11. N. S. Stoloff and M. Gensamer, *Trans. Met. Soc.*, AIME, **227**, 1963, p. 70.
12. F. Laves, *Naturwiss*, **30**, 1952, p. 546.
13. R. Clark and G. B. Craig, *Prog. Met. Phys.*, **3**, 1952, p. 115.
14. R. W. Cahn and J. A. Coll, *Acta Met.*, **9**, 1961, p. 138.
15. V. S. Arunachalam and C. M. Sargent, *Scripta Met.*, **5**, 1971, p. 949.
16. C. S. Smith, *A History of Metallography*, University of Chicago Press, Chicago, 1960, p. 40.
17. F. A. Heiser and R. W. Hertzberg, *J. Basic Neg.*, **93**, 1971, p. 71.
18. J. T. Michalak, *Metals Handbook*, Vol. 8, ASM, Metals Park, Ohio, 1973, p. 220.
19. J. F. Peck and D. A. Thomas, *Trans. Met. Soc.*, AIME, **221**, 1961, p. 1240.
20. H. Hu, *Texture*, **1**(4), 1974, p. 233.
21. A. T. English and G. Y. Chin, *Acta Met.*, **13**, 1965, p. 1013.
22. R. E. Smallman and D. Green, *Acta Met.*, **12**, 1964, p. 145.
23. I. L. Dillamore and W. T. Roberts, *Acta Met.*, **12**, 1964, p. 281.
24. N. Brown, *Trans. Met. Soc.*, AIME, **221**, 1961, p. 236.
25. J. S. Kallend and G. J. Davies, *Texture*, **1**, 1972, p. 51.
26. G. Wassermann, *Z. Met.*, **54**, 1963, p. 61.
27. E. A. Calnan, *Acta Met.*, **2**, 1954, p. 865.
28. J. F. W. Bishop, *J. Mech. Phys. Sol.*, **3**, 1954, p. 130.
29. H. Hu, R. S. Cline and S. R. Goodman, *Recrystallization Grain Growth and Textures*, ASM, Metals Park, Ohio, 1965, p. 295.
30. H. Hu and R. S. Cline, *J. Appl. Phys.*, **32**, 1961, p. 760.
31. A. Merlini and P. A. Beck, *Trans. Met. Soc.*, AIME, **203**, 1955, p. 385.
32. S. R. Goodman and H. Hu, *Trans. Met. Soc.*, AIME, **230**, 1964, p. 1413.
33. H. Hu and S. R. Goodman, *Trans. Met. Soc.*, AIME, **227**, 1963, p. 627.

PROBLEMS

4.1 Draw the strain ellipsoid for magnesium ($c/a = 1.623$), locate the major planes and describe those conditions that would produce twinning in the metal for the case where K_1 and K_2 are of the $\{10\bar{1}2\}$ type. Confirm the shear strain value given in Table 4.1.

4.2 Draw the (100) pole figure for the following textures:
(a) (100) [001] rolling texture
(b) [110] wire texture

4.3 Draw the (111) pole figure for the [100] wire texture in silver and for the [110] wire texture in iron wires.

4.4 From Figure 4.20, we see that copper exhibits a mixed [100]+[111] wire texture. Draw both (100) and (111) pole figures to portray this duplex texture.

4.5 The tensile strength for cold-rolled magnesium alloy AZ31B plate is approximately 160 MPa for specimens tested either parallel or perpendicular to the rolling direction. When similarly oriented specimens are compressed, the yield strength is only 90 MPa. Why? (*Hint*: Consider the possible deformation mechanisms available in the magnesium alloy and any crystallographic texture that might exist in the wrought plate.)

4.6 For the same magnesium alloy discussed in Problem 3, what would happen to a test bar loading in simple bending?

4.7 An HCP alloy, known as Hertzalloy 200, has a c/a ratio of 1.800.
(a) What is the most probable slip system?
(b) For each of the following diagrams, determine whether slip will occur and whether twinning will occur (consider only $\{10\bar{1}2\}$ twinning).

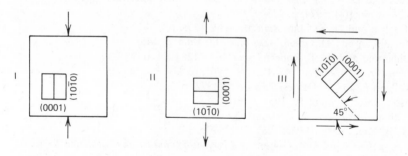

4.8 Three different investigators are each given a cube-shaped single crystal of an HCP metal. It is not known whether the cubes are of the same material. The investigators are to cut out tensile and compression specimens and measure the loads necessary for yielding. In addition, they are to observe the nature of deformation in each case. A summary of their results is given in the accompanying table.

From these results, two investigators concluded that three different materials were involved in the test program. The third investigator claimed that the materials were the same.
(a) Who was correct?
(b) Could both conclusions be correct?

(c) If the materials were the same, what additional information would have to be known and how could it be used to support the claim of similar materials?

	Investigator I	Investigator II	Investigator III
Tension Test	Low loads for yielding. Only slip observed.	Low loads for yielding. Little slip and much twinning.	High loads for yielding. Little slip and no twinning.
Compression Test	Low loads for yielding. Only slip mechanism.	High loads for yielding. Little slip and no twinning.	Low loads for yielding. Little slip and much twinning.

CHAPTER
FIVE
HIGH-TEMPERATURE DEFORMATION RESPONSE OF CRYSTALLINE SOLIDS

Brief mention was made in Section 1.3 of the effect of strain rate and temperature on the mechanical response of engineering materials. It was shown that tensile strength increased with increasing strain rate and decreasing temperature. In Section 2.3.1, the temperature sensitivity of strength in crystalline solids was related to the role played by the Peierls-Nabarro stress in resisting dislocation movement through a given lattice. In addition, the potential importance of temperature in controlling crystalline deformation through thermally activated edge dislocation climb was mentioned in Section 2.4. In the next two chapters, temperature and strain rate effects on mechanical properties are explored more extensively. Particular attention is given to the time- and temperature-dependent deformation characteristics of solids.

5.1 CREEP OF SOLIDS: AN OVERVIEW

For the most part, our discussions of deformation in solids thus far have been limited to the instantaneous deformation response—elastic or plastic—to the application of a load. A time-dependent change in the observed strain adds a new dimension to the problem. As shown in Fig. 5.1, after a load has been applied, the strain increases with time until failure finally occurs. For convenience, researchers have subdivided the creep curve into three regimes, based on the similar response of many materials. After the initial instantaneous strain

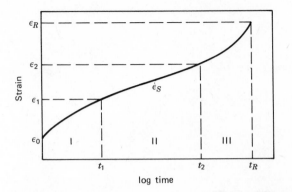

FIGURE 5.1 Typical creep curve showing three regions in strain-time space.

ϵ_0, materials often undergo a period of transient response where the strain rate $d\epsilon/dt$ decreases with time to a minimum steady-state value that persists for a substantial portion of the material's life. Appropriately, these two regions are referred to in the literature as the transient or primary creep stage and the steady-state creep stage, respectively. Final failure with a rupture life t_R then comes soon after the creep rate increases during the third, or tertiary, stage of creep.

It is generally believed that the varying creep response of a material (Fig. 5.1) reflects a continually changing interaction between strain hardening and softening (recovery) processes, which strongly affect the overall strain rate of the material at a given temperature and stress. Strain hardening at elevated temperatures is believed to involve subgrain formation associated with the rearrangement of dislocations,[1] while thermally activated cross-slip and edge dislocation climb represent two dominant recovery processes (see Chapter Three). It is logical to conclude that the decrease in strain rate in Stage I must be related to substructure changes that increase overall resistance to dislocation motion. Correspondingly, the constant strain rate in Stage II would indicate a stable substructure and a dynamic balance between hardening and softening processes. Indeed, Barrett et al.[2] verified that the substructure in Fe-3 Si was invariant during Stage II. At high stress and/or temperature levels, the balance between hardening and softening processes is lost, and the accelerating creep rate in the tertiary stage is dominated by a number of weakening metallurgical instabilities. Among these microstructural changes are localized necking, corrosion, intercrystalline fracture, microvoid formation, precipitation of brittle second phase particles, and resolution of second phases that originally contributed toward strengthening of the alloy. In addition, the strain hardened grains may recrystallize, thereby further destroying the balance between material hardening and softening processes.

The engineering creep strain curve shown schematically in Fig. 5.1 reflects the material response under constant tensile loading conditions and represents a convenient method by which most elevated temperature tests are conducted. However, from Eq. 1-3, the true stress increases with increasing tensile strain. As a result, a comparable true creep strain-time curve should differ significantly were the test to be conducted under constant *stress* rather than constant *load* conditions (Fig. 5.2). This is especially true for Stage II and III behavior. As a general rule, data being generated for engineering purposes are obtained from constant load tests, while more fundamental studies involving the formulation of mathematical creep theories should involve constant stress testing. In the latter instance, the load on the sample is lowered progressively with decreasing specimen cross-sectional area. This is done either manually or by the incorporation of automatic load-shedding devices in the creep stand load train.

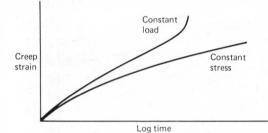

FIGURE 5.2   Creep curves produced under constant load and constant stress conditions.

The creep response of materials depends on a large number of material and external variables. Certain material factors are considered in more detail later in this chapter. For the present, attention will be given to the two dominant external variables—stress and test temperature—and how they affect the shape of the creep-time curve. Certainly, environment represents another external variable because of the importance of corrosion and oxidation in the fracture process. Unfortunately, consideration of this variable is not within the scope of this book.

The effect of temperature and stress on the minimum creep rate and rupture life are the two most commonly reported data for a creep or creep rupture test, although different material parameters are sometimes reported.[3,4] The rupture life at a given temperature and stress is obtained when it is necessary to evaluate the response of a material for use in a short-life situation, such as for a rocket engine nozzle ($t_R \approx 100$ sec) or a turbine blade in a military aircraft engine ($t_R \approx 100$ hr). In such short-life situations, the dominant question is whether the component will or will not fail, rather than by how much it will deform. As a result, the details of the creep-time curve are not of central importance to the engineering problem. For this reason, *creep rupture* tests usually provide only one datum—the rupture life t_R. Rupture life information is sometimes used in the design of engineering components that will have a service life up to 10^5 hours. An example of such data is given in Fig. 5.3 for the high-temperature, iron based alloy, S590. As expected, the rupture life t_R is seen to decrease with increasing test temperature and stress. When preparing this plot, Grant and Bucklin[5] chose to separate the data for a given temperature into several discrete regimes. This was done to emphasize the presence of several metallurgical instabilities that they identified metallographically and which they believed to be responsible for the change in slope of the $\log \sigma$-$\log t_R$ curve.

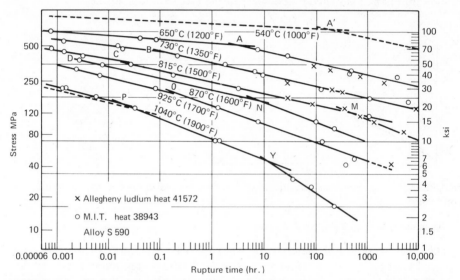

FIGURE 5.3 Stress-rupture life plot at several test temperatures for iron based alloy S-590. (From N. J. Grant and A. G. Bucklin,[5] copyright American Society for Metals, 1950.)

For long-life material applications, such as in a nuclear power plant designed to operate for several decades, component failure obviously is out of the question. However, it is equally important that the component not creep excessively. For long-life applications, the minimum creep rate represents the key material response for a given stress and test temperature. To obtain this information, *creep* tests are performed into Stage II, where the steady-state creep rate $\dot{\epsilon}_s$ can be determined with precision. Therefore, the *creep* test focuses on the early deformation stages of creep and is seldom carried to the point of fracture. As one might expect, the accuracy of $\dot{\epsilon}_s$ increases with the length of time the specimen experiences Stage II deformation. Consequently, $\dot{\epsilon}_s$ values obtained during instrumented creep rupture tests are not very accurate because of the inherently short time associates with the creep rupture test. The magnitude of $\dot{\epsilon}_s$ often depends strongly on stress. As a result, steady-state creep rate data are usually plotted against applied stress, as shown in Fig. 5.4. The significance of the $\dot{\epsilon}_s$ differences between $\alpha-$ and $\gamma-$ iron[6] at the allotropic transformation temperature is discussed in Section 5.2.

Since the creep and creep rupture tests are similar (though defined over different stress and temperature regimes), it would seem reasonable to assume the existence of certain relationships among various components of the creep curve (Fig. 5.1). In his text, Garofalo[3] summarized a number of log-log relation-

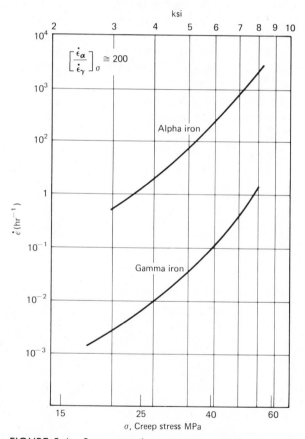

FIGURE 5.4   Stress-steady state creep rate for α- and γ-iron at 910°C. (From O. D. Sherby and J. L. Lytton,[6] reprinted with permission of the American Institute of Mining, Metallurgical and Petroleum Engineers, 1956.)

ships between t_R and other quantities, such as $t_2\text{-}t_1$, t_2, and the steady-state creep rate $\dot{\epsilon}_s$. Regarding the latter, Monkman and Grant[7] identified an empirical relationship between t_R and $\dot{\epsilon}_s$ with the form

$$\log t_R + m \log \dot{\epsilon}_s = B \qquad (5\text{-}1)$$

where

$$t_R = \text{rupture life}$$

$$\dot{\epsilon}_s = \text{steady-state creep rate}$$

$$m, B = \text{constants}$$

For a number of aluminum, copper, titanium, iron, and nickel base alloys, Monkman and Grant found $0.77 < m < 0.93$ and $0.48 < B < 1.3$. To a first approximation, then, the rupture life was found to be inversely proportional to $\dot{\epsilon}_s$. This would allow t_R to be estimated as soon as $\dot{\epsilon}_s$ was determined. Of course, the magnitude of t_R could be estimated from Eq. 5-1 only after the validity of the relationship for the material in question was established and the two constants identified.

A number of other empirical relationships have been proposed to relate the primary creep strain to time at stress and temperature. Garofalo[3] summarized the work of others and showed that for low temperatures $(0.05 < T_h{}^* < 0.3)$ and small strains, a number of materials exhibit *logarithmic creep*:

$$\epsilon_t \propto \ln t \qquad (5\text{-}2)$$

where

$$\epsilon_t = \text{true strain}$$

$$t = \text{creep time}$$

In the range $0.2 < T_h < 0.7$ another relationship has been employed with the form

$$\epsilon_t = \epsilon_{0_t} + \beta t^m \qquad (5\text{-}3)$$

where

$$\epsilon_{0_t} = \text{instantaneous true strain accompanying application}$$

$$\text{of the load}$$

$$\beta, m = \text{time-independent constants}$$

*T_h represents the homologous temperature—the ratio of ambient to melting point temperatures (absolute units).

Creep response in materials according to Eq. 5-3 is often referred to in the literature as *parabolic creep* or *β flow*. Since $0 < m < 1$ in transient creep, both Eqs. 5-2 and 5-3 reflect a decreasing strain rate with time. The strain rate $\dot{\epsilon}$ can be derived from Eqs. 5-2 and 5-3 with the form

$$\dot{\epsilon} \propto t^{-n} \qquad (5\text{-}4)$$

as suggested by Cottrell,[8] where

$$\dot{\epsilon} = \text{strain rate}$$

$$t = \text{time}$$

$$n = \text{constant.}$$

It is generally found that n decreases with increasing stress and temperature. At low temperatures when $n = 1$, Eq. 5-4 describes logarithmic creep (see Eq. 5-2). In the parabolic creep regime at higher temperatures, $m = 1 - n$. To provide a transition from Stage I to Stage II creep, another term $\dot{\epsilon}_s t$, has to be added to Eq. 5-3 to account for the steady-state creep rate in Stage II. Hence

$$\epsilon_t = \epsilon_{0_t} + \beta t^m + \dot{\epsilon}_s t \qquad (5\text{-}5)$$

where

$$\dot{\epsilon}_s = \text{steady-state creep rate in Stage II reflecting}$$

a balance between strain hardening and recovery processes.

When $m = \frac{1}{3}$, Eq. 5-5 reduces to the relationship originally proposed by Andrade[9] in 1910.

5.2 TEMPERATURE—STRESS—STRAIN RATE RELATIONSHIPS

Since the creep life and total elongation of a material depends strongly on the magnitude of the steady-state creep rate $\dot{\epsilon}_s$ (Eqs. 5-1 and 5-5), much effort has been given to the identification of those variables that strongly affect $\dot{\epsilon}_s$. As mentioned in Section 5.1, the external variables, temperature and stress, exert a strong influence along with a number of material variables. Hence the steady-state creep rate may be given by

$$\dot{\epsilon}_s = f(T, \sigma, \epsilon, m_1, m_2) \qquad (5\text{-}6)$$

where

T = absolute temperature

σ = applied tensile stress

ϵ = creep strain

m_1 = various intrinsic lattice properties, such as

the elastic modulus G and the crystal structure

m_2 = various metallurgical factors, such as grain and

subgrain size, stacking fault energy, and thermo-

mechanical history

It is important to recognize that m_2 also depends on T, σ, and ϵ. For example, subgrain diameter decreases markedly with increasing stress. Consequently, there exists a subtle but important problem of separating the effect of the major test variables on the structure from the deformation process itself that controls the creep rate. Dorn, Sherby, and coworkers[10-14] suggested that where $T_h > 0.5$ for the steady-state condition, the structure could be defined by relating the creep strain to a parameter θ

$$\epsilon = f(\theta) \tag{5-7}$$

where

$\theta = te^{-\Delta H/RT}$ described as the temperature-compensated

time parameter

t = time

ΔH = activation energy for the rate controlling process

T = absolute temperature

R = gas constant

The activation energy ΔH, shown schematically in Fig. 5.5 represents the energy barrier to be overcome so that an atom might move from A to the lower energy location at B. Upon differentiating Eq. 5-7 with respect to time, one finds

$$Z = f(\epsilon) = \dot{\epsilon}e^{\Delta H/RT} \tag{5-8}$$

which describes the strain rate-temperature relationship for a given stable structure and applied stress. When the rate process is given by the minimum

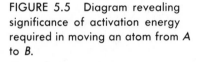

Energy

A

ΔH

B

Distance

FIGURE 5.5 Diagram revealing significance of activation energy required in moving an atom from A to B.

creep rate $\dot{\epsilon}_s$ and its logarithm plotted against $1/T$, a series of parallel straight lines for different stress levels is predicted from Eq. 5-8 (Fig. 5.6). The slope of these lines, $\Delta H/2.3R$, then defines the activation energy for the controlling creep process. The fact that the isostress lines were straight in Fig. 5.6 suggests that only one process had controlled creep in the TiO_2 single crystals throughout the stress and temperature range examined. Were different mechanisms to control the creep rate at different temperatures, the $\log \dot{\epsilon}_s$ vs. $1/T$ plots would be nonlinear. When multiple creep mechanisms are present and act in a concurrent and dependent manner, the slowest mechanism would control $\dot{\epsilon}_s$. The overall strain rate would take the form

$$\frac{1}{\dot{\epsilon}_T} = \frac{1}{\dot{\epsilon}_1} + \frac{1}{\dot{\epsilon}_2} + \frac{1}{\dot{\epsilon}_3} + \cdots + \frac{1}{\dot{\epsilon}_n} \qquad (5\text{-}9)$$

where

$$\dot{\epsilon}_T = \text{overall creep rate}$$

$$\dot{\epsilon}_{1,2,3,\ldots,n} = \text{creep rates associated with } n \text{ mechanisms}$$

For the simple case where only two mechanisms act interdependently

$$\dot{\epsilon}_T = \frac{\dot{\epsilon}_1 \dot{\epsilon}_2}{\dot{\epsilon}_1 + \dot{\epsilon}_2} \qquad (5\text{-}10)$$

Conversely, if the n mechanisms were to act independently of one another, the fastest one would control. For this case, $\dot{\epsilon}_T$ would be given by

$$\dot{\epsilon}_T = \dot{\epsilon}_1 + \dot{\epsilon}_2 + \dot{\epsilon}_3 + \cdots + \dot{\epsilon}_n \qquad (5\text{-}11)$$

To determine the activation energy for creep over a small temperature interval, where the controlling mechanism would not be expected to vary, researchers often make use of the temperature differential creep test method.

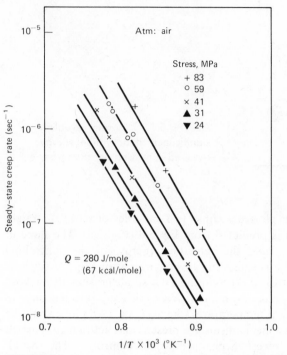

FIGURE 5.6 Log steady-state creep rate versus reciprocal of absolute temperature for rutile (TiO$_2$) at various stress levels. (From W. M. Hirthe and J. O. Brittain,[15] reprinted with permission from the American Ceramic Society, copyright 1963.)

After a given amount of strain at temperature T_1, the temperature is changed abruptly to T_2, which may be slightly above or below T_1. The difference in the steady-state creep rate associated with T_1 and T_2 is then recorded (Fig. 5.7). If the stress is held constant and the assumption made that the small change in temperature does not change the alloy structure, then Z is assumed constant. From Eq. 5-8 the activation energy for creep may then be calculated by

$$\Delta H_C = \frac{R \ln \dot{\epsilon}_1 / \dot{\epsilon}_2}{(1/T_2 - 1/T_1)} \tag{5-12}$$

where

$$\Delta H_C = \text{activation energy for creep}$$

$$\dot{\epsilon}_1, \dot{\epsilon}_2 = \text{creep rates at } T_1 \text{ and } T_2, \text{ respectively.}$$

This value of ΔH_C should correspond to the activation energy determined by a data analysis like that shown in Fig. 5.6, as long as the same mechanism controls

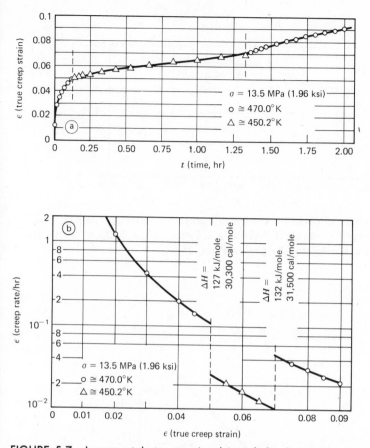

FIGURE 5.7 Incremental step test involving slight change in test temperature to produce change in steady-state creep rate in aluminum. (From J. E. Dorn, *Creep and Recovery*, reprinted with permission from American Society for Metals, copyright 1957.)

the creep process over the expanded temperature range in the latter instance. As shown in Fig. 5.8, this is not always the case. The activation energy for creep in aluminum is seen to increase with increasing temperature up to $T_h \approx 0.5$, whereupon ΔH_C remains constant up to the melting point. Similar results have been found in other metals.[16] It would appear that different processes were rate controlling over the test temperature range.[14] Furthermore, it should be recognized that ΔH_C may represent some average activation energy reflecting the integrated effect of several mechanisms operating simultaneously and interdependently (see Section 5.3).

Dorn,[13] Garofalo,[3] and Weertman[17] have compiled a considerable body of

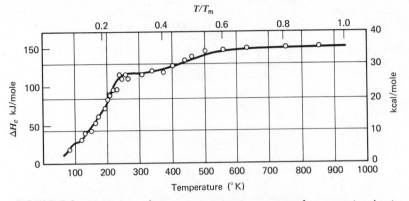

FIGURE 5.8   Variation of apparent activation energy for creep in aluminum as a function of temperature. (From O. D. Sherby, J. L. Lytton, and J. E. Dorn,[14] reprinted with permission from Sherby and Pergamon Press, 1957.)

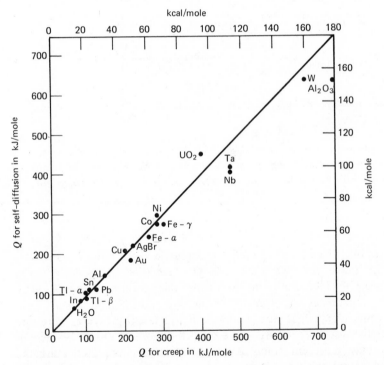

FIGURE 5.9   Correlation between activation energy for self-diffusion and creep in numerous metals and ceramics. (From J. Weertman,[17] reprinted with permission from American Society for Metals, copyright 1968.)

data to demonstrate that at $T_h \geqslant 0.5$, ΔH_C is most often equal in magnitude to ΔH_{SD}, the activation energy for self-diffusion (Fig. 5.9); this fact strongly suggests the latter to be the creep rate controlling process in this temperature regime. While the approximate equality between ΔH_C and ΔH_{SD} seems to hold for many metals and ceramics at temperatures equal to and greater than half the melting point, some exceptions do exist, particularly for the case of intermetallic and nonmetallic compounds. It is found that small departures from stoichiometry of these compounds have a pronounced effect on ΔH_C, which in turn affects the creep rate. For example, a reduction in oxygen content in rutile from TiO_2 to $TiO_{1.99}$ causes a reduction in ΔH_C from about 280 to 120 kJ/mole (67–29 kcal/mole)* with an associated 100-fold increase in $\dot{\epsilon}_s$.[15] For the more general case, however, the creep process is found to be controlled by the diffusivity of the material

$$D = D_0 e^{-\Delta H_{SD}/RT} \qquad (5\text{-}13)$$

where

$$D = \text{diffusivity, cm}^2/\text{sec}$$

$$D_0 = \text{diffusivity constant} \approx 1 \text{cm}^2/\text{sec}$$

$$\Delta H_{SD} = \text{activation energy, J/mole}$$

$$R = \text{gas constant, J/}^\circ\text{K}$$

$$T = \text{absolute temperature, }^\circ\text{K}$$

Combining Eqs. 5-8 and 5-13, the steady-state creep rate at different temperatures can be normalized with respect to D to produce a single curve as shown in Fig. 5-10. This is an extremely important finding since it allows one to conveniently portray a great deal of data for a given material. For example, we see from a re-examination of Fig. 5.4, that at the allotropic transformation temperature, the creep rate in γ-iron (FCC lattice) is found to be approximately 200 times slower than that experienced by α-iron (BCC lattice).[6] This substantial difference is traced directly to the 350-fold lower diffusivity in the close-packed FCC lattice in γ-iron. Similar findings were reviewed by Sherby and Burke[18] for the allotropic transformation from HCP to BCC in thallium. Therefore, it is appropriate to briefly consider those factors that strongly influence the magnitude of D. Sherby and Simnad[19] reported an empirical correlation showing D to be a function of the type of lattice, the valence, and the absolute melting point of the material.

*To convert from kcal to kJ, multiply by 4.19.

$$D = D_0 e^{-(K_0 + V)T_m/T} \tag{5-14}$$

$K_0 =$ dependent on the crystal structure and equal to 14 for BCC lattice, 17 for FCC and HCP lattices, and 21 for diamond-cubic lattice

$V =$ valence of the material

$T_m =$ absolute melting temperature

The constants, K_0, are estimates associated with an assumed diffusivity constant $\approx 1 \ \mathrm{cm}^2/\mathrm{sec}$.

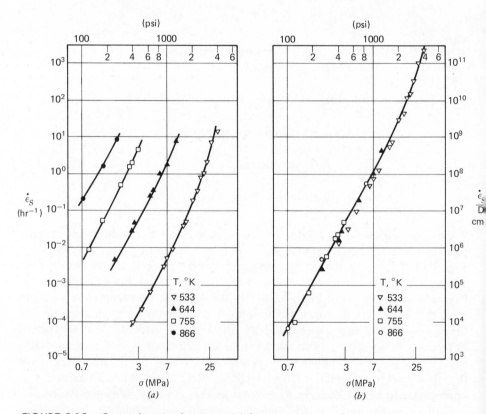

FIGURE 5.10 Creep data in aluminum. (a) Stress versus steady-state creep rate $\dot{\epsilon}_s$ at various test temperatures; (b) data normalized by plotting stress versus $\dot{\epsilon}_s$ divided by the diffusion coefficient. (From O. D. Sherby and P. M. Burke,[18] reprinted with permission from Sherby and Pergamon Press, 1968.)

Combining Eqs. 5-13 and 5-14

$$\Delta H_{SD} = RT_m (K_0 + V) \tag{5-15}$$

it is seen that the activation energy for self-diffusion increases (corresponding to a reduction in D) with increasing melting point, valence, packing density, and degree of covalency. Consequently, although refractory metals with high melting points, such as tungsten, molybdenum, and chromium, would seem to hold promise as candidates for high-temperature service, their performance in high-temperature applications is adversely affected by their open BCC lattice, which enhances diffusion rates. From Eq. 5-15, ceramics are identified as the best high-temperature materials because of their high melting point and the covalent bonding that often exists.

It is important to recognize that creep rates for all materials cannot be normalized on the basis of D alone because other test variables affect the creep process in different materials. A semi-empirical relationship with the form

$$\frac{\dot{\epsilon}_s kT}{DGb} = A \left(\frac{\sigma}{G} \right)^n \tag{5-16}$$

has been proposed[1] to account for other factors where

$$\dot{\epsilon}_s = \text{steady-state creep rate}$$

$$k = \text{Boltzman's constant}$$

$$T = \text{absolute temperature}$$

$$D = \text{diffusivity}$$

$$G = \text{shear modulus}$$

$$b = \text{Burgers vector}$$

$$\sigma = \text{applied stress}$$

$$A, n = \text{material constants}$$

Here again we see that creep is assumed to be diffusion controlled. Even after normalizing creep data with Eq. 5-16, a three-decade scatter band still exists for the various metals shown in Fig. 5.11. While some of this difference might be attributable to actual test scatter or relatively imprecise high-temperature measurements of D and G, other as yet unaccounted for variables most likely will account for the remaining inexactness. For example, there appears to be a trend toward higher creep rates in FCC metals and alloys possessing high stacking fault energy (SFE). Whether the SFE variable should be incorporated into either A or n is the subject of current discussion.[20–22] The role of substructure on A and n must also be identified more precisely.

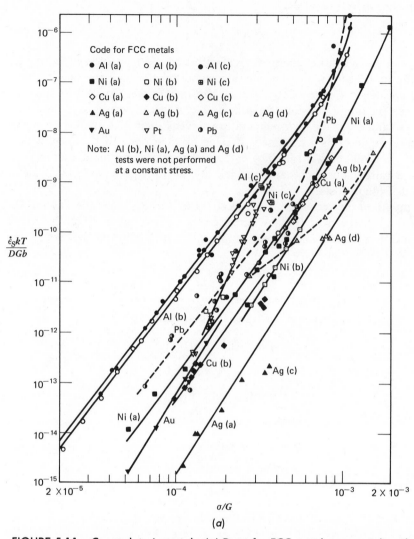

FIGURE 5.11 Creep data in metals. (*a*) Data for FCC metals—materials with high stacking fault energy tend to have higher steady-state creep rates. (*b*) Data for BCC metals. (From A. K. Mukherjee, J. E. Bird, and J. E. Dorn,[1] copyright American Society for Metals, 1969.)

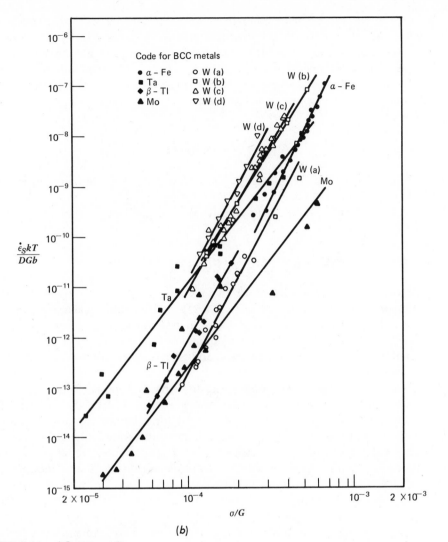

$\dfrac{\dot{\epsilon}_S kT}{DGb}$

Code for BCC metals

- ● α – Fe
- ■ Ta
- ◆ β – Tl
- ▲ Mo
- ○ W (a)
- □ W (b)
- △ W (c)
- ▽ W (d)

σ/G

(b)

FIGURE 5.11   (*Continued*)

One important factor in Eq. 5-16 is the stress dependency of the steady-state creep rate. It is now generally recognized that $\dot{\epsilon}_s$ varies directly with σ at low stresses and temperatures near the melting point. At intermediate to high stresses and at temperatures above 0.5 T_m, $\dot{\epsilon}_s \propto \sigma^5$ (so-called power law creep) while at very high stress levels $\dot{\epsilon} \propto e^{\alpha\sigma}$. Garofalo[23] showed that power law and exponential creep represented limiting cases for a more general empirical relationship

$$\dot{\epsilon}_s \propto (\sinh \alpha\sigma)^n \tag{5-17}$$

When $\alpha\sigma < 0.8$ Eq. 5-17 reduces to power law creep but approximates exponential creep when $\alpha\sigma > 1.2$. An explanation for the changing stress dependence of $\dot{\epsilon}_s$ in several operative deformation mechanisms is discussed in the next section.

5.3 DEFORMATION MECHANISMS

At low temperatures relative to the melting point of crystalline solids, the dominant deformation mechanisms are slip and twinning (Chapters Three and Four). However, at intermediate and high temperatures, other mechanisms become increasingly important and dominate material response under certain conditions. It is with regard to these additional deformation modes that attention will now be focused.

Over the years a number of theories have been proposed to account for the creep data trends discussed in the previous sections. In fact, the empirical form of Eq. 5-16 takes account of mathematical formulations for several proposed creep mechanisms. At low stresses and high temperatures, where the creep rate varies with applied stress, Nabarro[24] and Herring[25] theorized that the creep process was controlled by stress-directed atomic diffusion. Such *diffusional creep* is believed to involve the migration of vacancies along a gradient from grain boundaries experiencing tensile stresses to boundaries undergoing compression (Fig. 5.12); simultaneously atoms would be moving in the opposite direction, leading to elongation of the grains and the test bar. This gradient is produced by a stress-induced decrease in energy to create vacancies when tensile stresses are present and a corresponding energy increase for vacancy formation along compressed grain boundaries. Nabarro-Herring creep can be described by Eq. 5-16 when $A \approx 7 \, (b/d^2)$ ($d =$ grain diameter) and $n = 1$, such that[21]

$$\dot{\epsilon}_s \approx \frac{7\sigma D_v b^3}{kTd^2} \tag{5-18}$$

where

$D_v =$ volume diffusivity through the grain interior

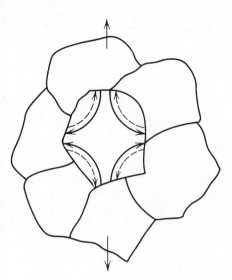

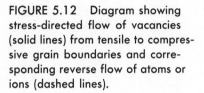

FIGURE 5.12 Diagram showing stress-directed flow of vacancies (solid lines) from tensile to compressive grain boundaries and corresponding reverse flow of atoms or ions (dashed lines).

As one might expect, $\dot{\epsilon}_s$ is seen to increase with increasing number of grain boundaries (i.e., smaller grain size). A closely related *diffusional creep* process described by Coble[26] involves atomic or ionic diffusion along grain boundaries. Setting $A \approx 50(b/d)^3$ and $n = 1$, Eq. 5-16 reduces to the Coble relationship

$$\dot{\epsilon}_s \approx \frac{50\sigma D_{gb}b^4}{kTd^3} \tag{5-19}$$

(Note that Coble creep is even more sensitive to grain size than is Nabarro-Herring creep.) In complex alloys and compounds there is a problem in deciding which particular atom or ion species controls the diffusional process and along what path such diffusion takes place. This is usually determined from similitude arguments. That is, if ΔH_C is numerically equal to ΔH_{SD} for element A along a particular diffusion path, then it is presumed that the self-diffusion of element A had controlled the creep process.

At intermediate to high stress levels and test temperatures above 0.5 T_m, creep deformation is believed to be controlled by diffusion controlled movement of dislocations. Several of these theories have been evaluated by Mukherjee et al.,[1] with the Weertman[17,27] model being found to suffer from the least number of handicaps and found capable of predicting best the experimental creep results described in Section 5.2. Weertman proposed that creep in the above-mentioned stress and temperature regime was controlled by edge dislocation climb away from dislocation barriers. Again using Eq. 5-16 as the basis for comparison, Bird et al.[21] showed that when A is constant and $n \approx 5$, *dislocation creep* involving the climb of edge dislocations could be estimated by

$$\dot{\epsilon}_s \approx \frac{ADGb}{kT}\left(\frac{\sigma}{G}\right)^5 \qquad (5\text{-}20)$$

The actual Weertman relationship expresses the shear strain rate $\dot{\gamma}_s$ in terms of the shear stress τ by

$$\dot{\gamma}_s \propto \tau^2 \sinh \tau^{2.5} \qquad (5\text{-}21)$$

As such, the transition from power law to exponential creep mentioned earlier is readily predicted from Eq. 5-21. Weertman[27] theorized that the onset of exponential creep ($\dot{\epsilon}_s \propto e^{\alpha\sigma}$) at high stress levels was related to accelerated diffusion, because of an excess vacancy concentration brought about by dislocation-dislocation interactions.

Another high-temperature deformation mechanism involves grain boundary sliding. The problem in dealing with grain boundary sliding, however, is that it does not represent an independent deformation mechanism; it must be accommodated by other deformation modes. For example, consider the shear-induced displacement of the two grains in Fig. 5.13a. At sufficiently high temperatures, the local grain boundary stress fields can cause diffusion of atoms from the compression region BC to the tensile region AB by either a Nabarro-Herring or Coble process. As might be expected, the rate of sliding should depend strongly on the shape of the boundary. Raj and Ashby[28] demonstrated that $\dot{\epsilon}_s$ increased rapidly as the ratio of perturbation period λ to perturbation

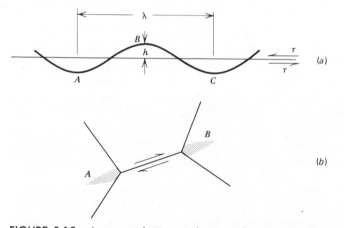

FIGURE 5.13 Accommodation mechanisms for grain boundary sliding. (*a*) Shear along boundary accommodated by diffusional flow of vacancies from region *AB* to *BC*; (*b*) grain boundary sliding accommodated by dislocation climb within contiguous grains *A* and *B*.

height h increased. Furthermore, when λ is small and the temperature relatively low, diffusion is found to be controlled by a grain boundary path. On the other hand, when λ is large and the temperature relatively high, volume diffusion controls the grain boundary sliding process.[28] Consequently, grain boundary sliding may be accommodated by diffusional flow, which is found to depend on both the temperature and the grain boundary morphology. For this case, the sliding rate would be directly proportional to stress (see Eqs. 5-18 and 5-19). By examining this problem from a different perspective, one finds that Nabarro-Herring and Coble creep models are themselves dependent on grain boundary sliding! From Fig. 5.14, note that the stress directed diffusion of atoms from compression to tension grain boundaries causes the grain boundaries to separate from one another (Fig. 5.14b). Grain boundary sliding is needed, therefore, to maintain grain contiguity during diffusional flow processes (Fig. 5.14c).[28-30] On the basis of this finding, Raj and Ashby concluded that Nabarro-Herring and Coble diffusional creep mechanisms were "identical with grain boundary sliding with diffusional accommodation."[28]

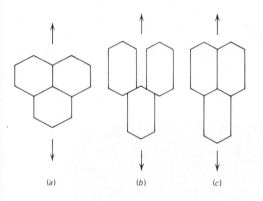

(a) (b) (c)

FIGURE 5.14 Stress-induced diffusional flow elongates grains and could lead to grain separation (b) but is accommodated by grain boundary sliding which brings grains together (c).

For the internal boundary shown in Fig. 5.13b, grain boundary sliding could be accommodated by dislocation creep within grains A and B. Matlock and Nix[31] examined this condition for several metals and found that the grain boundary sliding strain rate contribution was proportional to σ^{n-1}, where n is the exponent associated with the dislocation creep mechanism ($\approx$4–5). Unfortunately this stress sensitivity does not agree with any presently known theoretical predictions.

It is apparent from the above discussion that these high-temperature deformation mechanisms all depend on atom or ion diffusion but differ in their sensitivity to other variables such as G, d, and σ. As such, a particular strengthening mechanism may strengthen a material *only* with regard to a particular deformation mechanism but not another. For example, an increase in alloy grain size will suppress Nabarro-Herring and Coble creep along with grain boundary sliding, but will not substantially change the dislocation climb process.[1] As a result, the rate controlling creep deformation process would shift from one mechanism to another. Consequently, marked improvement in alloy performance requires simultaneous suppression of several deformation mechanisms. This point is considered further in Section 5.5.

5.4 SUPERPLASTICITY

As we have just seen, fine-grained structures are to be avoided in high-temperature, load bearing components since this would bring about an increase in creep strains resulting from Nabarro-Herring, Coble, and grain boundary sliding creep mechanisms. In fact, recent metallurgical developments[33] reveal improved creep response in alloys possessing either no grain boundaries (i.e., single-crystal alloys) or highly elongated boundaries (produced by unidirectional solidification) oriented parallel to the major stress axis. Where the opposite of creep resistance (i.e., easy flow) is required, such as in hot forming processes, fine-grained structures are preferred. Some such materials are known to possess *superplastic* behavior[34] with total strains in excess of 1000% (Fig. 5.15). These large strains, generated at low stress levels, drastically improve the formability of certain alloys.

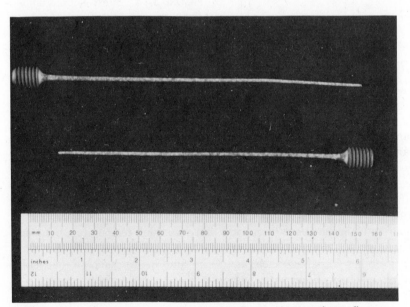

FIGURE 5.15 Tensile specimen having experienced superplastic flow.

Expressing the flow stress-strain rate relationship (Eq. 1-30) in the form

$$\sigma = \frac{F}{A} = K\dot{\epsilon}^m \qquad (5\text{-}22)$$

where

$F =$ applied force

$A =$ cross-sectional area

$K =$ constant

$$\dot{\epsilon} = \frac{1}{l}\frac{dl}{dt} = -\frac{1}{A}\frac{dA}{dt}$$

$m =$ strain rate sensitivity factor

superplasticity is found when m is large[34-36] and approaches unity. Figures 5.16 and 5.17 show the normalized stress-strain rate relationship for loading in the superplastic region. After substituting for $\dot{\epsilon}$ and rearranging, the change in

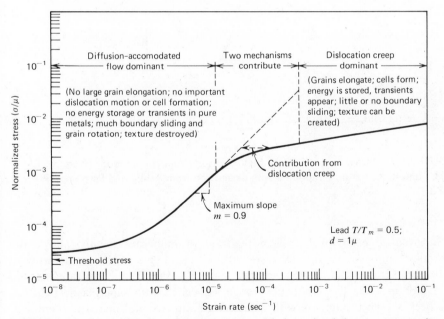

FIGURE 5.16 Normalized stress versus strain rate plot in lead showing intermediate region associated with superplastic behavior. (From M. F. Ashby and R. A. Verall,[37] reprinted with permission from Ashby and Pergamon Press, 1973.)

cross-sectional area with time, dA/dt, is given by

$$\frac{-dA}{dt} = \frac{F^{(1/m)}}{K} \left[\frac{1}{A^{(1-m/m)}} \right]$$ (5-23)

In the limit, as the rate sensitivity factor m approaches unity, note that dA/dt depends only on the applied force and is independent of any irregularities in the specimen cross-sectional area, such as incipient necks and machine tool marks which are maintained but not worsened. That is, the sample undergoes extensive deformation without pronounced necking.

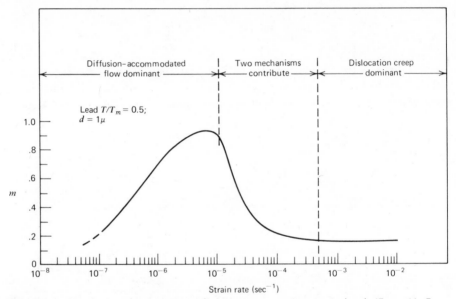

FIGURE 5.17 Strain rate sensitivity factor versus strain rate in lead. (From M. F. Ashby and R. A. Verall,[37] reprinted with permission from Ashby and Pergamon Press, 1973.)

Superplastic behavior has been reported in numerous metals, alloys and ceramics[34] and associated in all cases with: (1) A fine grain size (on the order of 1–10 μm), (2) deformation temperature $>0.5\ T_m$, and (3) a strain rate sensitivity factor $m > 0.3$. The strain rate range associated with superplastic behavior has been shown to increase with decreasing grain size and increasing temperature, as shown schematically in Fig. 5.18. There has been considerable debate, however, regarding the mechanisms responsible for the superplastic process. Avery and Backofen[36] originally proposed that a combination of deformation mechanisms involving Nabarro-Herring diffusional flow at low stress levels and dislocation climb at higher stresses were rate controlling. The applicability of the Nabarro-Herring creep model in the low stress regime has been questioned, based on experimental findings and theoretical considerations. First, it is generally found that m is of the order 0.5 rather than unity, the latter associated with Nabarro-Herring creep. Furthermore, Nabarro-Herring creep would lead to the formation of elongated grains proportional in length to the entire sample. To the contrary, equiaxed grain structures are preserved during superplastic flow. More recent theories have focused with greater success on grain boundary sliding arguments, with diffusion controlled accommodation[37-39] as the operative deformation mechanism associated with superplasticity at low stress levels.

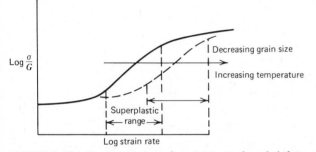

FIGURE 5.18 Temperature and grain size induced shift in strain rate range associated with superplastic behavior.

As mentioned above, the formability of a material is enhanced greatly when in the superplastic state, while forming stresses are reduced substantially. To this end, grain refinement is highly desirable. The alert reader will immediately recognize, however, that once an alloy is rendered superplastic through a grain refinement treatment, it no longer possesses the optimum grain size for high-temperature load applications. To resolve this dichotomy, researchers are currently seeking to develop duplex heat treatments to optimize both hot-forming and load-bearing properties of an alloy.[40] For example, a nickel base superalloy to be used in a gas turbine engine may first receive a grain refining heat treatment to provide superplastic response during a forging operation. Once the alloy has been formed into the desired component, it is given another heat treatment to coarsen the grains so as to suppress Nabarro-Herring, Coble, and grain boundary sliding creep processes during high-temperature service conditions. Other commercial applications of superplasticity are described by Hubert and Kay.[41]

5.5 DEFORMATION MECHANISM MAPS

It is important for the materials scientist and the practicing engineer to identify the deformation mechanisms that dominate a material's performance under a particular set of boundary conditions. This can be accomplished by solving the various constitutive equations for each deformation mechanism (e.g., Eqs. 5-16 to 5-20) and recognizing their respective interdependence or independence (Eqs. 5-9 and 5-11). Solutions to these equations reveal over which range of test variables a particular mechanism is rate controlling. Ashby and co-workers[42-44] have displayed such results pictorially in the form of maps in stress-temperature space based on the original suggestion by Weertman.[17] Typical deformation mechanism maps for pure silver and germanium are shown in Fig. 5.19, where most of the high-temperature deformation mechanisms discussed in Section 5.3 (as well as pure glide) are shown. Each mechanism is rate controlling within its stress-temperature boundaries. Consistent with the previous discussion, dislocation creep is seen to dominate the creep process in both materials at relatively high stresses and homologous temperatures above 0.5. For the FCC metal, diffusional creep by either Nabarro-Herring or Coble mechanisms dominates at high temperatures but lower stress levels. The virtual absence of these two diffusional flow mechanisms in covalently bonded diamond-cubic germanium is traced to its larger activation energy for self-diffusion and associated lower diffusivity. The boundaries separating each deformation field are defined by equating the appropriate constitutive equations (Eqs. 5-16 to 5-20) and solving for stress as a function of temperature. This amounts to the boundary lines representing combinations of stress and temperature, wherein the respective strain rates from the two deformation mechanisms are equal. Triple points in the deformation map would occur when a particular stress and temperature would produce equal strain rates from three mechanisms.

The maps shown in Fig. 5.19 do not portray a grain boundary sliding region, since uncertainties exist regarding the appropriate constitutive equation for this mechanism (see discussion, Section 5.3). Recent studies[44] have shown, however, that the dislocation creep field can be subdivided with a grain boundary sliding contribution existing at lower stress levels associated with lower creep strain rates. Regarding the latter point, it is desirable to portray on the deformation map the strain rate associated with a particular stress-temperature condition, regardless of the rate controlling mechanism. This may be accomplished by plotting on the diagram contours of iso-strain rate lines calculated from the

---

FIGURE 5.19 Deformation mechanism map for (a) pure silver and (b) germanium, showing stress-temperature space where different deformation mechanisms are rate controlling. Grain size in both materials is $32\,\mu$m. Elastic boundaries determined at a strain rate of 10^{-8}/sec. (From M. F. Ashby,[42] reprinted with permission from Ashby and Pergamon Press, 1972.)

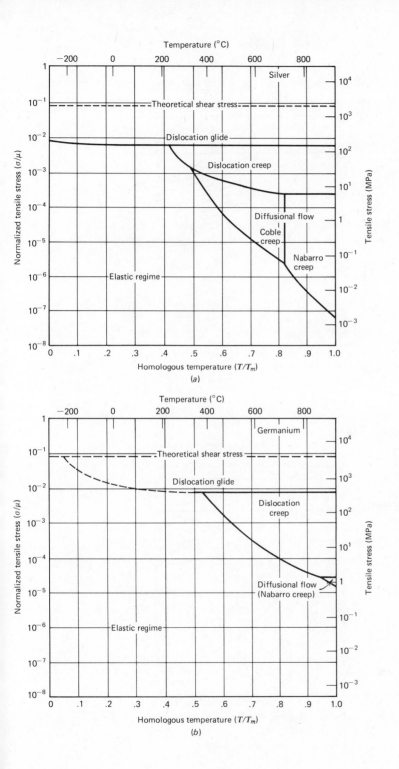

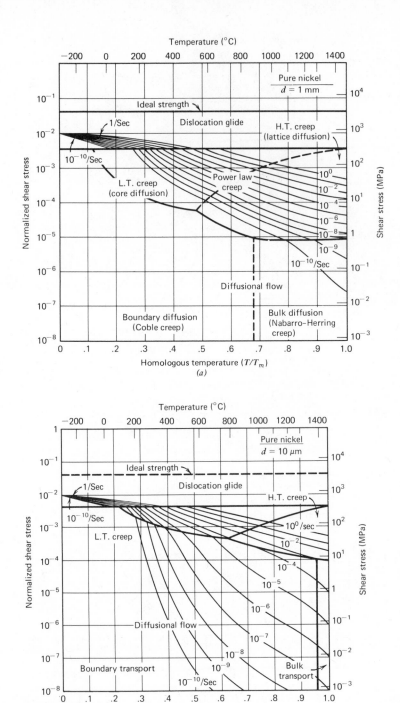

(a)

(b)

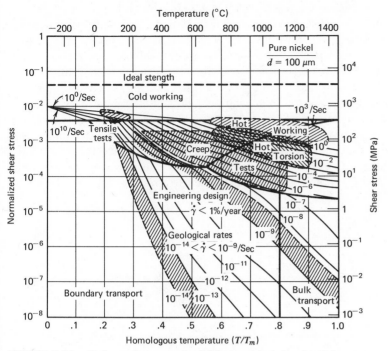

FIGURE 5.21   Deformation map for 100 μm nickel showing laboratory test regimes relative to deformation fields experienced by the material. (From M. F. Ashby,[43] reprinted with permission of the Institute of Metals.)

constitutive equations. Examples of such modified maps are given in Fig. 5.20 for pure nickel prepared with two different grain sizes. These maps allow one to pick any two of the three major variables—stress, strain rate, and temperature—which then identifies the third variable as well as the dominant deformation mechanism. This is particularly useful in identifying the location of testing domains (such as creep and tensile tests) relative to the stress-temperature-strain rate domains experienced by the material (e.g., hot working, hot torsion, and geological processes) (Fig. 5.21). Note that in most instances, the laboratory test domains do not conform to the material's application experience. Certainly a better correspondence would be more desirable.

FIGURE 5.20   Deformation map for (a) 1 mm and (b) 10 μm grain size nickel. Iso-strain rate lines superimposed on map. Dislocation climb region divided into low-temperature (core diffusion) and high-temperature (volume diffusion) regions. Note lower strain rates in more coarsely grained material. (M. F. Ashby,[43] reprinted with permission of the Institute of Metals.)

There are two additional points to be made regarding Fig. 5.20. First, the dislocation climb field has been divided into low- and high-temperature segments, corresponding to dislocation climb controlled by dislocation core and lattice diffusion, respectively. Furthermore, since Coble creep involves grain boundary diffusion, three diffusion paths are represented on these maps. Second, a large change in grain size in pure nickel drastically shifts the iso-strain rate contours and displaces the deformation field boundaries. For example, at $T_h = 0.5$ and a strain rate of 10^{-9}/sec, a 100-fold decrease in grain size causes the creep rate controlling process to shift from low temperature dislocation creep to Coble creep. Furthermore, the stress necessary to produce this strain rate decreases by almost three orders of magnitude! Both the expansion of the Coble creep regime and the much lower stress needed to produce a given strain rate

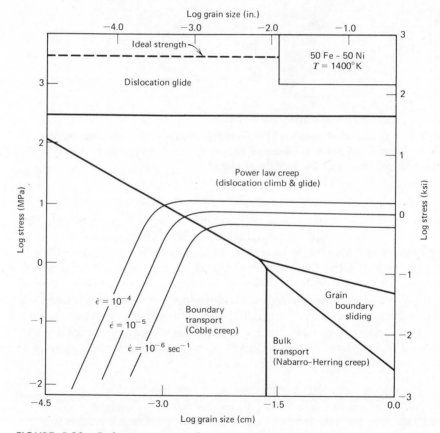

FIGURE 5.22 Deformation map for 50Fe-50Ni in stress-grain size space at a temperature of 1400°K. Note inclusion of grain boundary sliding field. (Courtesy of Michael R. Notis, Lehigh University.)

reflect the strong inverse dependence of grain size on the rate of this mechanism (Eq. 5-19). The Nabarro-Herring creep domain also expands for the same reason (Eq. 5-18). Since grain size effects on deformation maps are large, some researchers[29,45] have further modified the maps to include grain size as one of the dominant variables along with stress and iso-strain rate contour lines. These diagrams, such as the one shown in Fig. 5.22, portray the deformation field boundaries at a fixed temperature, where the grain size dependence of each deformation mechanism is clearly indicated. (Note the lack of grain size dependence in the dislocation creep region.)

Figure 5-23 provides one final map comparison by showing the effect of nickel base superalloy (MAR-M200) multiple strengthening mechanisms in shrinking the dislocation climb domain relative to that associated with pure nickel. In addition, the creep strain rates in the stress-temperature region associated with gas turbine material applications are reduced substantially. By combining alloying additions *and* grain coarsening, the iso-strain rate contours are further displaced, thereby providing additional creep resistance to the material.[33] In summary, it must be recognized that displacement of a particular boundary resulting from some specific strengthening mechanism does not in itself eliminate an engineering design problem. It may simply shift the rate controlling deformation process to another mechanism. The materials designer then must suppress the strain rate of the new rate controlling process with a different flow attenuation mechanism. As such, the multiple strengthening mechanisms built into high-temperature alloys are designed to counteract simultaneously a number of deformation mechanisms much in the same manner as an all-purpose antibiotic attacks a number of bacterial infections that may assault living organisms.

5.6 PARAMETRIC RELATIONSHIPS— EXTRAPOLATION PROCEDURES FOR CREEP RUPTURE DATA

It goes without saying that an engineering alloy will not be used for a given elevated temperature application without first obtaining a profile of the material's response under these test conditions. Although this presents no difficulty in short-life situations, such as for the rocket engine nozzle or military gas turbine blade, the problem becomes monumental when data are to be collected for prolonged elevated temperature exposures, such as encountered in a nuclear power plant. If the component in question is to withstand 30 or 40 years of uninterrupted service should there not be data available to properly design the part? Were this to be done, however, final design decisions concerning material selection would have to wait until all creep tests were concluded. Not only

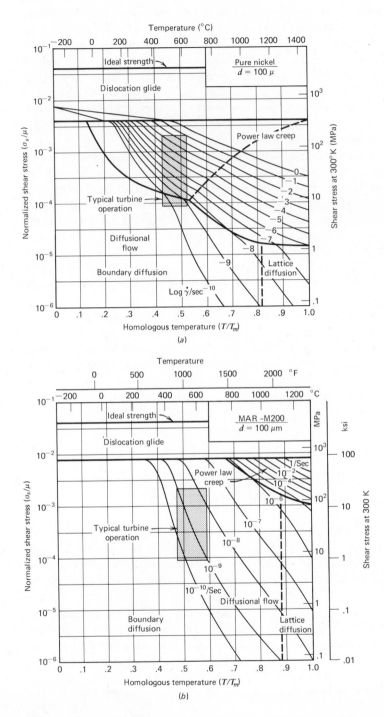

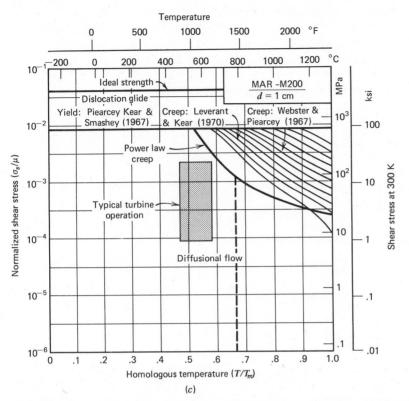

FIGURE 5.23 Deformation maps for (a) nickel (100 μm), (b) MAR-M200 nickel base alloy (100 μm), and (c) MAR-M200 (1 cm). Creep rate is suppressed by multiple strengthening mechanisms and grain coarsening. (From M. F. Ashby,[43] reprinted with permission from the Institute of Metals.)

would the laboratory costs of such a test program be prohibitively expensive, but all plant construction would have to cease and the economies of the world would stagnate. In addition, while such tests were being conducted, superior alloys most probably would have been developed to replace those originally selected. Assuming that some of these new alloys were to replace the older alloys in the component manufacture, a new series of long-time tests would have to be initiated. Obviously, nothing would ever be built!

The practical alternative, therefore, is to perform certain creep and/or creep rupture tests covering a convenient range of stress and temperature and then to *extrapolate* the data to the time-temperature-stress regime of interest. A considerable body of literature has been developed that examines parametric relationships (of which there are over 30) intended to allow one to extrapolate experimental data beyond the limits of convenient laboratory practice. A textbook[4] on the subject has even been written. Although it is beyond the scope of this book to consider many of these relationships to any great length, it is appropriate to consider two of the more widely accepted parameters.

The Larson-Miller parameter is, perhaps, most widely used. Larson and Miller[46] correctly surmised creep to be thermally activated with the creep rate described by an Arrhenius-type expression of the form

$$r = Ae^{-\Delta H/RT} \qquad (5\text{-}24)$$

where

$$r = \text{creep process rate}$$

$$\Delta H = \text{activation energy for the creep process}$$

$$T = \text{absolute temperature}$$

$$R = \text{gas constant}$$

$$A = \text{constant}$$

Eq. 5-24 also can be written as

$$\ln r = \ln A - \frac{\Delta H}{RT} \qquad (5\text{-}25)$$

After rearranging and multiplying by T, Eq. 5-25 becomes

$$\Delta H / R = T(\ln A - \ln r) \qquad (5\text{-}26)$$

Since $r \propto (1/t)$ (also suggested by Eq. 5-1), Eq. 5-24 can be written as

$$\frac{1}{t} = A'e^{-\Delta H/RT} \qquad (5\text{-}27)$$

Therefore,

$$-\ln t = \ln A' - \frac{\Delta H}{RT} \qquad (5\text{-}28)$$

and after rearranging Eq. 5-28, multiplying by T, and converting $\ln t$ to $\log t$

$$\Delta H / R = T(C + \log t) \qquad (5\text{-}29)$$

which represents the most widely used form of the Larson-Miller relationship.

Assuming ΔH to be independent of applied stress and temperature (not always true as demonstrated earlier) the material is thought to exhibit a particular Larson-Miller parameter $[T(C+\log t)]$ for a given applied stress. That is to say, the rupture life of a sample at a given stress level will vary with test temperature in such a way that the Larson-Miller parameter $T(C+\log t)$ remains unchanged. For example, if the test temperature for a particular material with $C=20$ were increased from 800°C to 1000°C, the rupture life would decrease from an arbitrary value of 100 hours at 800°C to 0.035 hours at 1000°C. The value of this parametric relationship is shown by examining the creep rupture data in Fig. 5.24, which are the very same data used in Fig. 5.3. The normalization potential

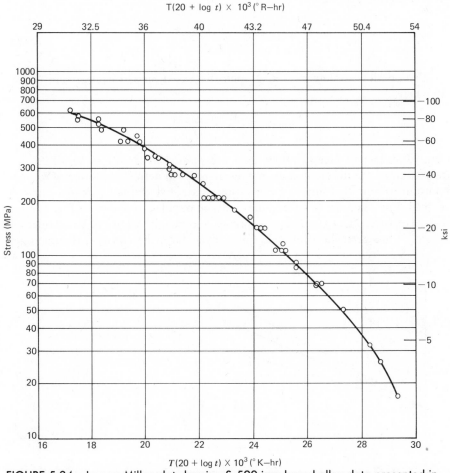

FIGURE 5.24   Larson-Miller plot showing S-590 iron based alloy data presented in Fig. 5.3.

of the Larson-Miller parameter for this material is immediately obvious. Furthermore, long-time rupture life for a given material can be estimated by extrapolating high-temperature, short rupture life response toward the more time-consuming low-temperature, long rupture life regime. It is generally found that such extrapolations to longer time conditions are reasonably accurate at higher stress levels because a smaller degree of uncertainty is associated with this portion of the Larson-Miller plot. Increased extrapolation error is found at lower stress levels where experimental scatter is greater. A comparison between predicted and experimentally determined rupture lives will be considered later in this section.

The magnitude of C for each material may be determined from a minimum of two sets of time and temperature data. Again, assuming $\Delta H / R$ to be invarient and rearranging Eq. 5-29

$$C = \frac{T_2 \log t_2 - T_1 \log t_1}{T_1 - T_2} \tag{5-30}$$

It is also possible to determine C graphically based on a rearrangement of Eq. 5.29 where

$$\log t = -C + \frac{\text{constant}}{T} \tag{5-31}$$

When experimental creep rupture data are plotted as shown in Fig. 5.25, the intersection of the different stress curves at $1/T = 0$ defines the value of C. It is important to note that not all creep rupture data give the same trends found in Fig. 5.25. For example, iso-stress lines may be parallel, as shown in Fig. 5.6, for the case of rutile (TiO_2) and other ceramics and metals. Values of C for selected materials[46] are given in Table 5.1. For convenience, the constant is sometimes

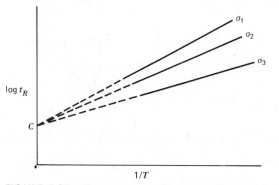

FIGURE 5.25 Convergence of iso-stress lines in plot of log t_R versus $1/T$ to determine magnitude of constant C in Larson-Miller parameter.

not determined experimentally but instead assumed equal to 20. Note that the magnitude of the material constant C does not depend on the temperature scale but only on units of time. (Since practically all data reported in the literature give both the material constant C and the rupture life in more convenient units of hours rather than in seconds—the recommended SI unit for time—test results in this section will be described in units of hours.)

TABLE 5.1   Material Constants for Selected Alloys[46]

Alloy	C Time, hr	C Time, sec
Low carbon steel	18	21.5
Carbon moly steel	19	22.5
18-8 stainless steel	18	21.5
18-8 Mo stainless steel	17	20.5
$2\frac{1}{4}$ Cr-1 Mo steel	23	26.5
S-590 alloy	20	23.5
Haynes Stellite No. 34	20	23.5
Titanium D9	20	23.5
Cr-Mo-Ti-B steel	22	25.5

In addition to being used for the extrapolation of data, the Larson-Miller parameter also serves as a figure of merit against which the elevated temperature response of different materials may be compared (e.g., in the case of alloy development studies). For example, when the curves for two materials with the same constant C are coincident, the materials obviously possess the same creep rupture behavior (Fig. 5.26a). The same conclusion does not follow, however, when the coincident curves result from materials with different values of C (Fig. 5.26b). When $C_A < C_B$, material A would be stronger of the two. (For the same

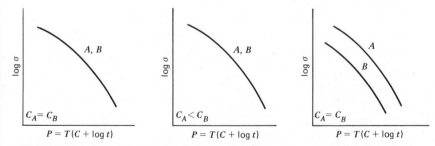

FIGURE 5.26   Parametric comparison of alloy behavior. (a) Alloy A = Alloy B; (b) and (c) Alloy A superior to Alloy B.

parameter P, and at the same test temperature, $\log t_{R_A}$ for alloy A would have to be greater than $\log t_{R_B}$ since $C_B > C_A$.) A direct comparison of material behavior is evident when C is the same but the parametric curves distinct from one another (Fig. 5.26c). Here alloy A is clearly the superior material. While such alloy comparisons for specified conditions of stress and temperature are possible using the Larson-Miller parameter (and other parameters as well), it should be understood that such parameters provide little insight into the mechanisms responsible for the creep response in a particular time-temperature regime. This is done more successfully by examining deformation maps (Section 5.4).

The Sherby-Dorn (S-D) parameter $\theta = t_R e^{-\Delta H/RT}$ (where $t = t_R$) described in Eq. 5-7 has been used to compare creep rupture data for different alloys much in the same manner as the Larson-Miller (L-M) parameter. Reasonably good results have been obtained with this parameter in correlating high temperature data of relatively pure metals[10] (Fig. 5.27). The reader should recognize that if the Sherby-Dorn parameter does apply for a given material, then when θ is constant, a plot of the logarithm of rupture life against $1/T$ should yield a series of straight lines corresponding to different stress levels. This is contrary to the response predicted by the Larson-Miller parameter, where the iso-stress lines converge when $1/T = 0$. The choice of the L-M or S-D parameters to evaluate a material's creep rupture response would obviously depend on whether the iso-stress lines converge to a common point or are parallel. In fact, the choice of a particular parameter (recall that over 30 exist) to correlate creep data for a specific alloy is a very tricky matter. Some parameters seem to provide better correlations than others for one material but not another. This may be readily seen by considering Goldhoff's tabulated results[47] for 19 different alloys (Table 5.2). Shown here are root mean square (RMS) values reflecting the accuracy of the L-M, S-D, and other parameters in predicting creep rupture life. The RMS value is defined as

$$
\text{RMS} = \left[\frac{\sum (\log \text{ actual time to rupture-log predicted time to rupture})^2}{\text{number of long-time data points}} \right]^{\frac{1}{2}}
$$

(5-32)

Note that for some metals, either the L-M or S-D parameter represented the best time-temperature parameter (TTP) of the four examined by Goldhoff and predicted actual test results most correctly. Alternately, these two parameters provided poor correlations when compared to other parameters for different materials; the use of the L-M or S-D parameters in evaluating these alloys led to significant error in the prediction of actual rupture life.

The inconsistency with which a particular TTP predicts actual creep rupture life for different alloys represents a severe shortcoming of the parametric

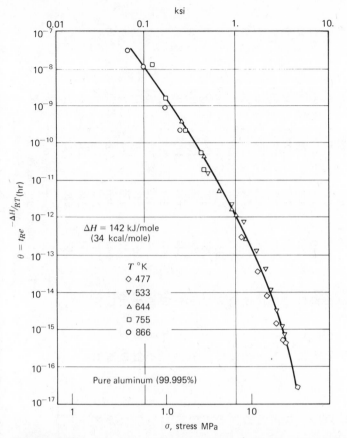

FIGURE 5.27 Correlation of stress rupture data using temperature-compensated time parameter $\theta = t_R e^{-\Delta H/RT}$ for pure aluminum. (From J. E. Dorn,[13] *Creep and Recovery*, reprinted with permission from American Society for Metals, copyright 1957.)

approach to creep design. These deficiencies may be traced in part to some of the assumptions underlying each parameter. For example, the L-M and S-D parameters are based on the assumption that the activation energy for the creep process is not a function of stress and temperature. Clearly, the test results shown in Fig. 5.8 and the extended discussion in Section 5.3 discredit this supposition. Furthermore, none of the TTP make provision for metallurgical instabilities.

Attempts are being made to standardize creep data parametric analysis procedures through the establishment of required guidelines by which an investigator arrives at the selection of a particular TTP. In this regard, the minimum commitment method (MCM)[48, 49] holds considerable promise in that it

TABLE 5.2 Comparative RMS Values Reflecting Accuracy of Different Time-Temperature Parameters[47]

Data Set	Alloy	Short-Time Data Points	Long-Time Data Points	LM[a]	MH[b]	SD[c]	MS[d]	Best TTP[e]	MCM[f]
1	Al 1100-0	53	11	.347	.377	.308	.488	.308	.260
2	Al 5454-0	68	7	.099	.166	.143	.287	.099	.081
4	Carbon steel	18	8	.456	.313	.415	.396	.313	.084
5	Cr-Mo steel	23	10	.152	.102	.056	.191	.056	.122
6	Cr-Mo-V steel	17	9	.389	.091	.162	.477	.091	.102
7A	304 stainless steel	33	19	.375	.207	.185	.309	.185	.194
7B	304 stainless steel	41	11	.454	.167	.272	.292	.167	.179
8	304 stainless steel	26	13	.334	.349	.237	.457	.237	.228
9	316 stainless steel	28	10	.244	.296	.212	.323	.212	.073
11A	347 stainless steel	18	24	.368	.203	.298	.265	.203	.123
11B	347 stainless steel	31	13	.291	.173	.267	.211	.173	.107
12	A-286	19	5	.097	.338	.089	.111	.089	.220
13	Inco 625	78	21	.343	.283	.337	.329	.283	.317
14	Inco 718	17	9	.104	.565	.110	.100	.100	.084
15	René 41	26	11	.106	.144	.139	.113	.106	.131
16	Astroloy®	21	12	.302	.343	.231	.264	.231	.107
17A	Udimet 500	65	38	.252	.342	.316	.348	.252	.268
17B	Udimet 500	93	12	.111	1.057	.247	.173	.111	.124
18A	L-605	51	49	.319	.652	.420	.261	.261	.247
18B	L-605	76	28	.374	.641	.460	.305	.305	.290
19	Al 6061-T651	74	25	.361	.382	.217	.473	.217	.311
Average of above 21 data sets				.280	.342	.244	.294	.190	.174
Average excluding B data sets				.273	.303	.228	.305	.191	.174

[a]Larson-Miller parameter.
[b]Manson-Haferd parameter.
[c]Sherby-Dorn parameter.
[d]Manson-Succop parameter.
[e]Time-temperature parameter.
[f]Minimum commitment method.

presumes initially a very general time-temperature-stress relationship. The precise form is obtained on the basis of actual test data. As such, the MCM can lead to the selection of a standard parametric relationship, such as L-M or S-D, or it may define a new parameter that can reflect the possible existence of metallurgical instabilities. Note the reduced RMS values for the MCM method as compared to the L-M, S-D, or the other two TTP evaluated by Goldhoff (Table 5.2).

5.7 MATERIALS FOR ELEVATED TEMPERATURE USE

From the previous discussions, a material suitable for high-temperature service should possess a high melting point and modulus of elasticity, and low diffusivity. As a result, alloy development has focused primarily on nickel and cobalt based superalloys,[50,51] with earlier iron based alloys being replaced because of their relatively low melting point and high diffusivity. A bar graph illustrating the comparative temperature capability of numerous nickel, cobalt, and iron based alloys is given in Fig. 5.28 for 100- and 1000-hour rupture lives at a stress of 140 MPa. From Fig. 5.29, even these cast superalloys lose their load-bearing capability rapidly at high test temperatures. Metallurgists are constantly seeking ways to improve the mechanical properties of such materials through changes in alloy chemistry and/or thermomechanical treatments. One example already referred to is the unidirectional solidification of conventional alloys to produce either a single-crystal alloy or one containing highly elongated grain boundaries, which minimize the influence of grain boundary sliding and diffusional flow processes (i.e., Nabarro-Herring and Coble type creep). By applying unidirectional solidification to alloys with different chemistries (specifically alloys of eutectic composition), it has been possible to produce eutectic composite alloys possessing properties superior to any found in conventional superalloys[53-55] (Fig. 5.30). In another recent thrust, researchers have focused attention on the development of a gas turbine engine using ceramic components. Since ceramics often possess higher melting points and moduli of elasticity and lower diffusivities than metal systems, they offer considerable potential in such applications. Unfortunately, ceramics suffer from low ductility and brittle behavior in tension (see Table 10.8). This serious problem must be resolved before the ceramic engine can become a reality.

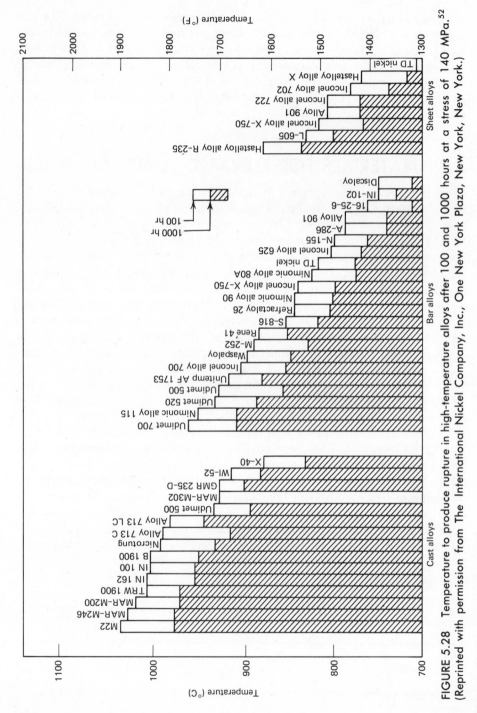

FIGURE 5.28 Temperature to produce rupture in high-temperature alloys after 100 and 1000 hours at a stress of 140 MPa.[52] (Reprinted with permission from The International Nickel Company, Inc., One New York Plaza, New York, New York.)

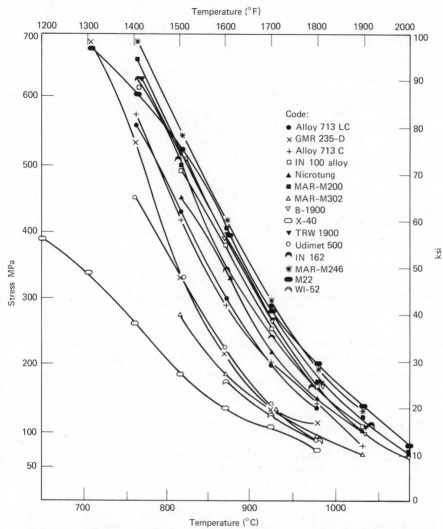

FIGURE 5.29 One hundred hour rupture strength in several cast superalloys as a function of temperature.[52] (Reprinted with permission from The International Nickel Company, Inc., One New York Plaza, New York, New York.)

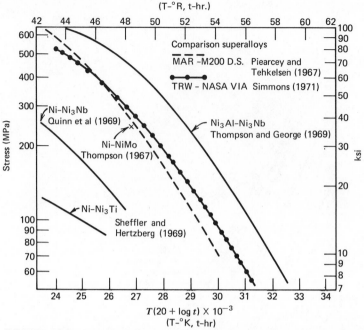

FIGURE 5.30 Larson-Miller plot for several eutectic composite and conventional superalloys. (From E. R. Thompson and F. D. Lemkey,[55] reprinted with permission from Thompson and Academic Press.)

REFERENCES

1. A. K. Mukherjee, J. E. Bird, and J. E. Dorn, *Trans.*, ASM, **62**, 1969, p. 155.
2. C. R. Barrett, W. D. Nix, and O. D. Sherby, *Trans.*, ASM, **59**, 1966, p. 3.
3. F. Garofalo, *Fundamentals of Creep and Creep-Rupture in Metals*, Macmillan, New York, 1965.
4. J. B. Conway, *Stress-Rupture Parameters: Origin, Calculation and Use*, Gordon and Breach, New York, 1969.
5. N. J. Grant and A. G. Bucklin, *Trans.*, ASM, **42**, 1950, p. 720.
6. O. D. Sherby and J. L. Lytton, *Trans.*, AIME, **206**, 1956, p. 928.
7. F. C. Monkman and N. J. Grant, *Proc.*, ASTM, **56**, 1956, p. 593.
8. A. H. Cottrell, *J. Mech. Phys. Sol.*, **1**, 1952, p. 53.
9. E. N. DaC. Andrade, *Proc. Roy. Soc.*, **A84**, 1910, p. 1.
10. R. L. Orr, O. D. Sherby, and J. E. Dorn, *Trans.*, ASM, **46**, 1954, p. 113.
11. O. D. Sherby, T. A. Trozera, and J. E. Dorn, *Trans.*, ASTM, **56**, 1956, p. 789.
12. J. E. Dorn, *Creep and Fracture of Metals at High Temperatures*, NPL Symposium, H.M.S.O., 1956, p. 89.
13. J. E. Dorn, *Creep and Recovery*, ASM, Metals Park, Ohio, 1957, p. 255.
14. O. D. Sherby, J. L. Lytton, and J. E. Dorn, *Acta Met.*, **5**, 1957, p. 219.

15. W. M. Hirthe and J. O. Brittain, *J. Amer. Ceram. Soc.*, **46**(9), 1963, p. 411.
16. S. L. Robinson and O. D. Sherby, *Acta Met.*, **17**, 1969, p. 109.
17. J. Weertman, *Trans.*, ASM, **61**, 1968, p. 681.
18. O. D. Sherby and P. M. Burke, *Prog. Mater. Sci.*, **13**, 1968, p. 325.
19. O. D. Sherby and M. T. Simnad, *Trans.*, ASM, **54**, 1961, p. 227.
20. C. R. Barrett and O. D. Sherby, *Trans.*, AIME, **230**, 1964, p. 1322.
21. J. E. Bird, A. K. Mukherjee, and J. E. Dorn, *Quantitative Relation Between Properties and Microstructure*, Israel Universities Press, Haifa, Israel, 1969, p. 255.
22. H. J. Frost and M. F. Ashby, *NTIS* Report AD-769821, August 1973.
23. F. Garofalo, *Trans.*, AIME, **227**, 1963, p. 351.
24. F. R. N. Nabarro, *Report of a Conference on the Strength of Solids*, Physical Society, London, 1948, p. 75.
25. C. Herring, *J. Appl. Phys.*, **21**, 1950, p. 437.
26. R. L. Coble, *J. Appl. Phys.*, **34**, 1963, p. 1679.
27. J. Weertman, *J. Appl. Phys.*, **28**, 1957, p. 362.
28. R. Raj and M. F. Ashby, *Met. Trans.*, **2**, 1971, p. 1113.
29. T. G. Langdon, *Deformation of Ceramic Materials*, R. C. Bradt and R. E. Tressler, eds., Plenum Press, New York, 1975, p. 101.
30. L. M. Lifshitz, *Sov. Phys. JETP*, **17**, 1963, p. 909.
31. D. K. Matlock and W. D. Nix, *Met Trans.*, **5**, 1974, p. 961.
32. D. K. Matlock and W. D. Nix, *Met Trans.*, **5**, 1974, p. 1401.
33. B. J. Piearcey and F. L. Versnyder, *Met. Prog.*, Nov. 1966, p. 66.
34. R. H. Johnson, *Met. Mater.*, **4**(9), 1970, p. 389.
35. W. A. Backofen, I. R. Turner, and D. H. Avery, *Trans.*, ASM, **57**, 1964, p. 981.
36. D. H. Avery and W. A. Backofen, *Trans.*, ASM, **58**, 1965, p. 551.
37. M. F. Ashby and R. A. Verall, *Acta Met.*, **21**, 1973, p. 149.
38. T. H. Alden, *Acta Met.*, **15**, 1967, p. 469.
39. T. H. Alden, *Trans.*, ASM, **61**, 1968, p. 559.
40. Staff Report, *Met. Prog.*, **103**(3), 1973, p. 49.
41. J. F. Hubert and R. C. Kay, *Met. Eng. Quart.*, **13**, 1973, p. 1.
42. M. F. Ashby, *Acta Met.*, **20**, 1972, p. 887.
43. M. F. Ashby, *The Microstructure and Design of Alloys*, Proceedings, Third International Conference on Strength of Metals and Alloys, Vol. 2, Cambridge, England, 1973, p. 8.
44. F. W. Crossman and M. F. Ashby, *Acta Met.*, **23**, 1975, p. 425.
45. M. R. Notis, *Deformation of Ceramic Materials*, R. C. Bradt and R. E. Tressler, eds., Plenum Press, New York, 1975, p. 1.
46. F. R. Larson and J. Miller, *Trans.*, ASME, **74**, 1952, p. 765.
47. R. M. Goldhoff, *J. Test. Eval.*, **2**(5), 1974, p. 387.
48. S. S. Manson, *Time-Temperature Parameters for Creep-Rupture Analysis*, Publication No. D8-100, ASM, Metals Park, Ohio, 1968, p. 1.
49. S. S. Manson and C. R. Ensign, *NASA Tech. Memo TM X-52999*, NASA, Washington, D.C., 1971.
50. C. T. Sims and W. C. Hagel, eds., *Superalloys*, John Wiley, New York, 1972.
51. F. J. Clauss, *Engineer's Guide to High-Temperature Materials*, Addison-Wesley, Reading, Mass., 1969.
52. *High Temperature High Strength Nickel Base Alloys*, International Nickel Co., New York, 1964.

53. *Proceedings of the Conference on In Situ Composites*, Vol. II, NMAB 308-II, Jan. 1973.
54. R. W. Hertzberg, *Composite Materials*, Vol. 1, A. G. Metcalfe, ed., Academic Press, New York, 1974, p. 329.
55. E. R. Thompson and F. D. Lemkey, *ibid.*, Vol. 4, K. G. Kreider, ed., p. 101.

PROBLEMS

5.1 (a) Two researchers have simultaneously and independently studied the activation energy for creep of Zeusalloy 300 (Tm.p. = 1000°K). Both used the Eyring rate equation to plot their data.

$$\dot{\epsilon}_s = A e^{-\Delta H / RT}$$

where $\dot{\epsilon}_s$ = minimum creep rate, T = temperature, and ΔH = activation energy for the creep process. When their data are plotted on semilog paper ($\log \dot{\epsilon}$ vs $1/T$) their results were different. Investigator I concluded from his work that one activation energy was controlling the creep behavior of the material. Investigator II, on the other hand, concluded that creep in Zeusalloy 300 was very complex since the activation energy was not constant throughout his test range. Which investigator is correct or are they both correct? Explain your answer.

(b) How might the activation energy have been determined?

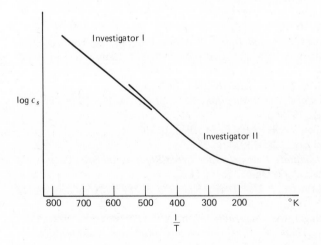

5.2 Why is the extrapolation of short-time creep rupture data to long times at the same temperatures extremely dangerous? Give a few examples to justify your answer.

5.3 For the following creep rupture data, construct a Larson-Miller plot (assuming $C = 20$). Determine the expected life for a sample tested at 650°C with a stress of 240 MPa, and at 870°C with a stress of 35 MPa. Compare these values with actual test results of 32,000 and 9000 hours, respectively.

Temp. (°C)	Stress (MPa)	Rupture Time (hr)	Temp. (°C)	Stress (MPa)	Rupture Time (hr)
650	480	22	815	140	29
650	480	40	815	140	45
650	480	65	815	140	65
650	450	75	815	120	90
650	380	210	815	120	115
650	345	2700	815	105	260
650	310	3500	815	105	360
705	310	275	815	105	1000
705	310	190	815	105	700
705	240	960	815	85	2500
705	205	2050	870	83	37
760	205	180	870	83	55
760	205	450	870	69	140
760	170	730	870	42	3200
760	140	2150	980	21	440
			1095	10	155

5.4 For the data given in the previous problem, what is the maximum operational temperature such that failure should not occur in 5000 hours at stress levels of 140 and 200 MPa, respectively.

5.5 The test results given on the next page for René 41[47] are from a different alloy from that described in the previous two problems. From a Larson-Miller plot (assume $C = 20$) estimate the rupture life associated with 650°C exposure at 655 MPa and 870°C exposure at 115 MPa. Compare these values with actual test results of 3900 and 1770 hours, respectively.

Temp. (°C)	Stress (MPa)	Rupture Time (hr)	Temp. (°C)	Stress (MPa)	Rupture Time (hr)
650	1000	12	815	460	15
650	950	56	815	395	47
650	895	165	815	345	91
650	840	479	815	260	379
650	820	319	815	220	773
650	790	504	870	260	33
705	815	23	870	215	73
705	695	110	870	180	196
705	635	254	870	145	300
705	590	478	900	170	69
760	600	28	900	140	163
760	515	62	900	105	738
760	440	291			
760	415	390			

CHAPTER
SIX
DEFORMATION
RESPONSE
OF ENGINEERING
PLASTICS

Polymers exist in nature in such forms as wood, rubber, jute, hemp, cotton, silk, wool, hair, horn, and flesh. In addition, there are countless man-made polymeric products, such as synthetic fibers, engineering plastics, and artificial rubber. In certain respects, the deformation of polymeric solids bears strong resemblance to that of metals and ceramics: polymers become increasingly deformable with increasing temperature, as witnessed by the onset of additional flow mechanisms (albeit, different ones for polymers). Also, the extent of polymer deformation is found to vary with time, temperature, stress, and microstructure consistent with parallel observations for fully crystalline solids. Furthermore, a time-temperature equivalence for polymer deformation is indicated, which is strongly reminiscent of the time-temperature parametric relationships discussed in Section 5.6. Before beginning our discussion of polymer deformation, it is appropriate to describe basic features of the polymer structure that dominate flow and fracture properties. In preparing this section, several excellent books about polymers were consulted to which the reader is referred.[1-6]

6.1 POLYMER STRUCTURE—GENERAL REMARKS

A polymer is formed by the union of two or more structural units of a simple compound. In polymeric materials used in engineering applications, the number of such unions—known as the degree of polymerization (DP)—often exceeds many thousands. Two major polymerization processes are available to the polymer chemist: addition and condensation polymerization. To examine the former, consider the polymerization of ethylene, C_2H_4, which contains a double carbon bond. If the necessary energy is introduced, the double bond may be broken, resulting in a molecule with two free radicals.

$$
\begin{array}{cc}
\text{H} \quad \text{H} & \text{H} \quad \text{H} \\
| \quad\; | & | \quad\; | \\
\text{C} = \text{C} + \text{Energy} \rightarrow -\text{C} - \text{C} - + \text{Energy} \\
| \quad\; | & | \quad\; | \\
\text{H} \quad \text{H} & \text{H} \quad \text{H}
\end{array}
\qquad (6\text{-}1)
$$

This activated molecule is now free to combine with other activated molecules to form very long chains. Since this process is exothermic, special care is needed to assure optimum temperature and pressure conditions for continuation of the polymerization process. Therefore, the initial energy input, usually in the form of heat, is needed only to initiate the reaction. To this end, the process is often aided by the addition to the monomer of an active chemical containing free radicals. These initiators break down the double bond and may then become attached to one end of the activated molecule; the other end grows by linking up with other activated molecules. The growth phase occurs very quickly, and the degree of polymerization exceeds several hundred in a fraction of a second. The key to *addition* polymerization is the availability of double or even triple carbon bonds in the monomer that may be broken if sufficient energy is provided. Examples of polymers prepared by addition polymerization include polypropylene, polystyrene, poly(vinyl chloride), and poly(methyl methacrylate).

Condensation polymerization proceeds by a different process and involves the chemical reaction between smaller molecules containing reactive groups with the associated elimination of a small molecular byproduct, usually water. For example, by combining an alcohol (R–OH) with an acid (R'–COOH)

$$
\text{R—OH} + \text{R'—COOH} \rightarrow \text{R'—COO—R} + H_2O \qquad (6\text{-}2)
$$

The hydroxyl groups have reacted to allow the two molecules to join, leaving a single molecule of water as a reaction byproduct. It should be recognized that the above reaction can proceed no further, since all active hydroxyl groups have been consumed. As in addition polymerization, the key to continued condensation polymerization is the availability of *multiple* active groups to react with other molecules. Here again the two molecules must be *bifunctional* or *trifunctional* to allow the activated molecules to react with *two* or *three* molecules and not just one as noted for the monofunctional alcohol-acid example cited above.

For the case of nylon 66, this reaction takes the form

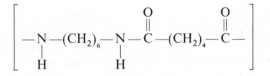

$$H—N—(CH_2)_6—N—H+OH—\overset{O}{\overset{\|}{C}}—(CH_2)_4—\overset{O}{\overset{\|}{C}}—OH→$$
$$\quad\;\; | \qquad\qquad\quad\; |$$
$$\quad\;\; H \qquad\qquad\quad H$$

$$H—N—(CH_2)_6—N—\overset{O}{\overset{\|}{C}}—(CH_2)_4—\overset{O}{\overset{\|}{C}}—OH+H_2O \quad (6\text{-}3)$$
$$\quad\;\; | \qquad\qquad\quad\; |$$
$$\quad\;\; H \qquad\qquad\quad H$$

Note that the new molecule retains H and OH groups that react shortly to continue the polymerization reaction. Consequently, long chains of nylon 66 can be formed with repeating units of

$$\left[—N—(CH_2)_6—N—\overset{O}{\overset{\|}{C}}—(CH_2)_4—\overset{O}{\overset{\|}{C}}— \right]$$
$$\quad | \qquad\qquad\quad\; |$$
$$\quad H \qquad\qquad\quad H$$

Besides nylon, polycarbonate and epoxy are other examples of condensation polymerization.

The length of a given polymer chain is determined by the statistical probability of a specific activated mer attaching itself to a particular chain. For example, a chain growing from one side (the other side being stabilized by the attachment of an initiator radical) can (1) continue to grow by the addition of successive activated mers, (2) cease to grow by the attachment of a free radical group to the free end of the chain, or (3) join with another growing chain at its free end, thus terminating the growth stage for both chains (Fig. 6.1). (Other interactions are possible but will not be considered in this book.) It is clear that some chains will be very short while others might be very long. Consequently, the reader should appreciate one of the most distinctive characteristics of a polymeric solid: the fact that there is no unique chain length for a given polymer and no specific molecular weight (MW). Instead, there is a distribution of these values. Contrast

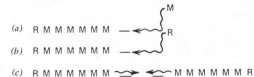

FIGURE 6.1   Different events in polymerization process: (a) continuation of chain lengthening; (b) chain termination; (c) simultaneous termination of two chains.

this with metal and ceramic solids that exhibit a well-defined lattice parameter and unit cell density. An example of the molecular weight distribution (MWD) for all the chains in a polymer is shown in Fig. 6.2; in this case, a larger number of small chains exist relative to the very long chains. The MWD will vary with the nature of the monomer and the conditions of polymerization so as to be skewed to higher or lower MW and/or made narrower or broader. For example, when the processing temperature is high and/or large amounts of initiator added to the melt, MW will be low, and vice versa.

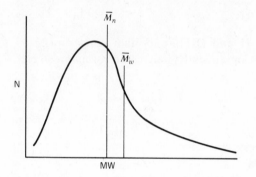

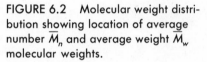

FIGURE 6.2 Molecular weight distribution showing location of average number $\overline{M}_n$ and average weight $\overline{M}_w$ molecular weights.

Rather than referring to a molecular weight distribution curve to describe the character of a polymer, it is often more convenient to think in terms of an average molecular weight $\overline{M}$. Such a value can be described in a number of ways, but is usually described either in terms of the number or weight fraction of molecules of a given weight. The *number average* molecular weight $\overline{M}_n$ is defined by

$$\overline{M}_n = \frac{\displaystyle\sum_{i=1}^{\infty} N_i M_i}{\displaystyle\sum_{i=1}^{\infty} N_i} \qquad (6\text{-}4)$$

and represents the total weight of material divided by the total number of molecules. $\overline{M}_n$ emphasizes the importance of the smaller MW chains. The *weight average* molecular weight $\overline{M}_w$ is given by

$$\overline{M}_w = \frac{\displaystyle\sum_{i=1}^{\infty} N_i M_i^2}{\displaystyle\sum_{i=1}^{\infty} N_i M_i} \qquad (6\text{-}5)$$

which reflects the weight of material of each size rather than their number. It is seen that $\overline{M}_w$ emphasizes relatively high MW fractions. As will be shown in later sections, MW exerts a very strong influence on a number of polymer physical and mechanical properties. The molecular weight distribution can be described by the ratio $\overline{M}_w / \overline{M}_n$. A narrow MWD prepared under carefully controlled conditions may have $\overline{M}_w / \overline{M}_n < 1.5$, while a broad MWD would reveal $\overline{M}_w / \overline{M}_n$ in excess of 25.

We now look more closely at a segment of the polyethylene (PE) chain just described. In the fully extended conformation, the chain assumes a zig-zag pattern, with the carbon-carbon bonds describing an angle of about 109 degrees (Fig. 6.3). With the zig-zag carbon main chain atoms lying in the plane of this page, the two hydrogen atoms are disposed above and below the paper. The chain is truly three dimensional, though often represented schematically in two-dimensional space only. Adjacent pairs of hydrogen atoms are positioned relative to one another so as to minimize their steric hindrance. That is, as rotations occur about a C–C bond (permissible as long as the bond angle remains 109 degrees), both favorable and unfavorable juxtapositions of the hydrogen atom pairs are experienced. This is perhaps more readily seen by examining the rotations about the C–C bond in ethane, C_2H_6 (Fig. 6.4), recalling that the hydrogen atoms do not lie in the plane of the page. We see that when the hydrogen atoms are located opposite one another, the potential energy of the system is maximized. Conversely, when they are staggered where $\phi = 0$, $2\pi/3$, and $4\pi/3$, the configuration has lowest potential energy. From this, it is seen that the facility by which C–C bond rotation occurs will depend on the magnitude of the energy barrier in going from one low-energy configuration to another. For the two pairs of adjacent hydrogen atoms in the PE chain, the lowest potential energy trough occurs when the hydrogen atom pair associated with one carbon atom is 180 degrees away from its neighboring hydrogen pairs (Fig. 6.3).

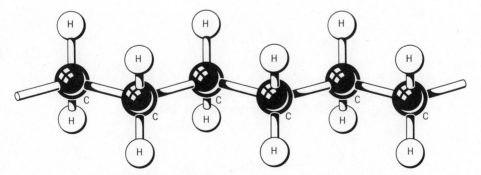

FIGURE 6.3   Extended chain of polyethylene showing coplanar zig-zag arrangement of C—C bonds with hydrogen pairs located opposite one another.

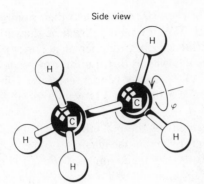

Side view

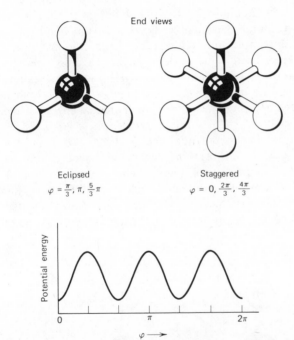

End views

Eclipsed
$\varphi = \frac{\pi}{3}, \pi, \frac{5}{3}\pi$

Staggered
$\varphi = 0, \frac{2\pi}{3}, \frac{4\pi}{3}$

Potential energy

0   π 2π

$\varphi \longrightarrow$

FIGURE 6.4   Potential energy variation associated with C—C bond rotation in ethane.[4] (Turner Alfrey and Edward F. Gurnee, *Organic Polymers* © 1967. Reprinted by permission of Prentice-Hall Inc., Englewood Cliffs, N.J.)

As might be expected, the extent of rotational freedom about the C–C bond depends on the nature of side groups often substituted for hydrogen in the PE chain. When one hydrogen atom is replaced, we have a vinyl polymer. As shown in Table 6.1, a number of different atoms or groups can be added to form a variety of vinyl polymers which all act to restrict C–C rotation to a greater or lesser degree. Generally, the bigger and bulkier the side group and the greater its polarity, the greater the resistance to rotation, since the peak-to-valley energy differences (Fig. 6.4) would be greater. Restrictions to such movement may also be effected by double carbon bonds in the main chain which rotate with much greater difficulty. Furthermore, the main chain in some polymers may contain flat cyclic groups (such as a benzene ring) which prefer to lie parallel to one another. Consequently, C–C bond rotation would be made more difficult by their presence.

TABLE 6.1 Selected Vinyl Polymers

Polymer	Repeat Unit
$\begin{bmatrix} \overset{\displaystyle H}{\underset{\displaystyle H}{-C}} - \overset{\displaystyle H}{\underset{\displaystyle H}{C}} - \end{bmatrix}$	Polyethylene
$\begin{bmatrix} \overset{\displaystyle H}{\underset{\displaystyle H}{-C}} - \overset{\displaystyle Cl}{\underset{\displaystyle H}{C}} - \end{bmatrix}$	Poly(vinyl chloride)
$\begin{bmatrix} \overset{\displaystyle H}{\underset{\displaystyle H}{-C}} - \overset{\displaystyle F}{\underset{\displaystyle H}{C}} - \end{bmatrix}$	Poly(vinyl fluoride)
$\begin{bmatrix} \overset{\displaystyle H}{\underset{\displaystyle H}{-C}} - \overset{\displaystyle CH_3}{\underset{\displaystyle H}{C}} - \end{bmatrix}$	Polypropylene
$\begin{bmatrix} \overset{\displaystyle H}{\underset{\displaystyle H}{-C}} - \overset{\displaystyle \bigcirc}{\underset{\displaystyle H}{C}} - \end{bmatrix}$	Polystyrene

Thus far, we have discussed the effect of side group size, shape, and polarity on main chain mobility. The *location* of these groups along the chain is also of critical importance, since it affects the relative packing efficiency of the polymer.

It is seen from Fig. 6.5 that the side groups can be arranged either randomly along the chain, only on one side, or on alternate sides of the chain. These three configurations are termed atactic, isotactic, and syndiotactic, respectively. Atactic polymers with large side groups (e.g., polystyrene) have low packing efficiency, with the chains arranged in a random array. Consequently, polystyrene, poly(methyl methacrylate), and, to a large extent, poly(vinyl chloride) are amorphous. In a regular and symmetric polymer (e.g., polyethylene), the chains can be packed closely together, resulting in a high degree of crystallinity. In fact, the density of a given polymer serves as a useful measure of crystallinity, the higher the density the greater the degree of crystallinity.

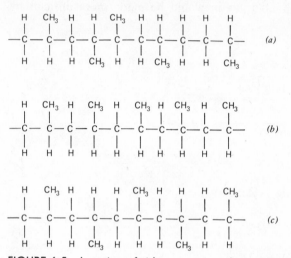

FIGURE 6.5 Location of side groups in polypropylene: (a) atactic; (b) isotactic; (c) syndiotactic.

For the data shown in Table 6.2, densities were varied by the amount of main chain branching produced during polymerization (Fig. 6.6). Extensive branching reduces the opportunity for closer packing, and little branching promotes the polymerization of higher density polyethylene. Polypropylene represents an example of a stereoregular polymer that has a high packing efficiency and resultant crystallinity. Although it is not stereoregular, the propensity for crystallinity in nylon 66 is enhanced by the highly polar nature of the nylon chain.

The $\underset{\underset{H}{|}}{N} - \overset{\overset{O}{\|}}{C}$ groups in adjacent chains have great affinity for one another with the associated hydrogen bond providing additional cause for closer packing and chain alignment.

TABLE 6.2 Relationship Between Density-Crystallinity and Ultimate Tensile Strength in Polyethylene[7]

Density (gm/cm³)	Crystallinity (%)	Ultimate Tensile Strength	
		MPa	ksi
0.92	65	13.8	2
0.935	75	17.2	2.5
0.95	85	27.6	4
0.96	87	31.0	4.5
0.965	95	37.9	5.5

$$— C — C — C — C — C — C — C — C —$$ (a)

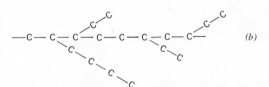

(b)

FIGURE 6.6 Degree of chain branching in polymeric solid.

To summarize, the degree to which polymers will crystallize depends strongly on the polarity, symmetry, and stereoregularity of the chain and its tendency for branching. The extent of crystallinity of several polymers is given in Table 6.3, along with other material characteristics.

6.1.1 Morphology of Amorphous and Crystalline Polymers and Their Unoccupied Free Volume

At all temperatures above absolute zero, the existing thermal energy causes the polymer chains to vibrate and wriggle about. First, small-scale vibrations are permitted. Then with increasing temperature, molecule segments begin to move more freely. Finally, at sufficiently high temperatures associated with the molten state, entire chains are free to move about. It is seen from Fig. 6.7 that these large amplitude molecular vibrations cause the polymer to become less dense. If crystallization is likely for the type of polymer described in Section 6.1, the material undergoes upon cooling a first-order transformation at B associated with the melting point T_m. Heat of fusion is liberated and the specific volume

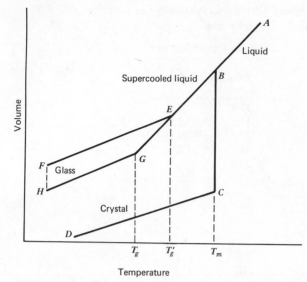

FIGURE 6.7 Change in volume as function of temperature. Crystalline melting point at T_m, glass transition temperature at T_g, and excess free volume FH.

drops abruptly to C. Further cooling involves additional change in the specific volume (D) as molecular oscillations become increasingly restricted. When crystallization does not occur in the polymer, the liquid cools beyond T_m (location B) without event. However, a point is reached where molecular motions are highly restricted and the individual chains no longer able to arrange themselves in equilibrium configurations within the supercooled liquid. Below this point (G) (called the glass transition temperature T_g), the material is relatively frozen into a glassy state. The change from a supercooled liquid to glass represents a second-order transformation that does not involve a discrete change in specific volume or internal heat. From Fig. 6.7, it is seen that the polymer in the amorphous state occupies more volume than in the crystalline form. This is to be expected, since higher density forms of a particular polymer are associated with greater crystallinity as a result of greater chain packing efficiency (Table 6.2). The relative difference in chain packing density can be described in terms of the fractional unoccupied (or free) volume given by Litt and Tobolsky[8] as

$$\bar{f} = \left[\frac{v_a - v_c}{v_a} \right] = 1.0 - \left(\frac{d_a}{d_c} \right) \tag{6-6}$$

where

$$\bar{f} = \text{fractional unoccupied free volume}$$

v_a, d_a = specific volume and density of amorphous phase

v_c, d_c = specific volume and density of crystalline phase

For many polymers, $0.01 < \bar{f} < 0.1$.

Since the glass transition occurs where molecular and segmental molecular motions are restricted, it is sensitive to cooling rate. Consequently, a polymer may not exist at its glassy equilibrium state. Instead, nonequilibrium cooling rates could preclude the attainment of the lowest possible free volume in the amorphous polymer. In Fig. 6.7, this would correspond to line *EF* with the glass transition temperature increasing to *Tg′*. Petrie[9] describes the difference between the equilibrium and actual glassy free volume as the *excess free volume* and postulates that this quantity is important in understanding the relationship between polymer properties and their thermodynamic state. Note that the free volume will differ from one polymer to another; within the same polymer, the excess free volume is sensitive to thermal history. The importance of free volume and excess free volume is considered further in Section 6.6.

The crystalline structure of polymers can be described by two factors: chain conformation and chain packing. The conformation of a chain relates to its geometrical shape. In polyethylene, the chains assume a zig-zag pattern as noted above and pack flat against one another. This is not observed in polypropylene, which has a single large methyl group, and in poly(tetrafluorethylene), which contains four large fluorine atoms. Instead, these zig-zag molecules twist about their main chain axis to form a helix. In this manner, steric hindrance is reduced. It is interesting to note that poly(tetrafluoroethylene)'s extraordinary resistance to chemical attack, mentioned in Table 6.3, is believed partly attributable to the sheathing action of the fluorine atoms that cover the helical molecule.

It is currently believed that chain packing in the crystalline polymer is effected by repeated chain folding, such that highly ordered crystalline lamellae are formed. Two models involving extensive chain folding to account for the formation of crystalline lamellae are shown in Fig. 6.8, along with an electron micrograph of a polyethylene single crystal lamellae.[10,11] The thickness of these crystals are generally about 100 to 200 Å, while its planar dimensions can be measured in micrometers. Since chains are many times longer than the observed thickness of these lamellae, chain folding is required. Consequently, chains are seen to extend across the lamellae but reverse direction on reaching the crystallite boundary. In this manner, a chain is folded back on itself many times. The loose loop model depicting less perfect chain folding (some folds occurring beyond the nominal boundary of the lamellae; see Fig. 6.8*b*) is viewed by Clark[12] as being more realistic in describing the character of real crystalline polymers. In addition to loose loops and chain ends (cilia) that lie on the surface of the lamellae, there exist tie molecules that extend from one crystal to

TABLE 6.3 Characteristics of Selected Polymers[2]

Material	Repeat Unit	Major Characteristics	Applications
Low density polyethylene	$\begin{bmatrix} \text{H} & \text{H} \\ | & | \\ -\text{C}-\text{C}- \\ | & | \\ \text{H} & \text{H} \end{bmatrix}$	Considerable branching; 55–70% crystallinity; excellent insulator; relatively cheap	Film; moldings; cold water plumbing; squeeze bottles
Polypropylene	$\begin{bmatrix} \text{H} & \text{CH}_3 \\ | & | \\ -\text{C}-\text{C}- \\ | & | \\ \text{H} & \text{H} \end{bmatrix}$	Extent of crystallinity depends on stereoregularity; can be highly oriented to form integral hinge with extraordinary fatigue behavior	Hinges; toys; fibers; pipe; sheet; wire covering
Acetal copolymer	$\begin{bmatrix} \text{H} & \text{H} \\ | & | \\ -\text{C}-\text{O}-\text{C}-\text{O}- \\ | & | \\ \text{H} & \text{H} \end{bmatrix}$	Highly crystalline; thermally stable; excellent fatigue resistance	Speedometer gears; instrument housing; plumbing valves; glands; shower heads
Nylon 66	$\begin{bmatrix} \text{H} & & \text{O} & \text{O} \\ | & & \| & \| \\ -\text{N}-(\text{CH}_2)_6-\text{N}-\text{C}-(\text{CH}_2)_4-\text{C}- \\ & & | \\ & & \text{H} \end{bmatrix}$	Excellent wear resistance; high strength and good toughness; used as plastic and fiber; highly crystalline; strong affinity for water	Gears and bearings; rollers; wheels; pulleys; power tool housings; light machinery components; fabric

Material	Repeat Unit	Major Characteristics	Applications
Poly(tetrafluoro-ethylene) (Teflon)	$\left[\begin{array}{c} F \; F \\ -C-C- \\ F \; F \end{array}\right]$	Extremely high MW; high crystallinity; extraordinary resistance to chemical attack; nonsticking	Coatings for cooking utensils; bearings and gaskets; pipe linings; insulating tape; nonstick, load-bearing pads
Poly(vinyl chloride)	$\left[\begin{array}{c} H \; Cl \\ -C-C- \\ H \; H \end{array}\right]$	Primarily amorphous; variable properties through polymeric additions; fire self-extinguishing; fairly brittle when unplasticized; relatively cheap	Floor covering; film; handbags; water pipes; wiring insulation; decorative trim; toys; upholstery
Poly(methyl methacrylate) (Plexiglas)	$\left[\begin{array}{c} \quad CH_3 \\ -C-C- \\ H \; O=C-O-CH_3 \\ H \end{array}\right]$	Amorphous; brittle; general replacement for glass	Signs; canopies; windows; windshields; sanitary ware

another[13] (Fig. 6.9). The latter provide mechanical strengthening to the crystalline aggregate as is discussed later. In the unoriented condition, crystalline polymers possess a spherulitic structure consisting of stacks of lamellae positioned along radial directions (Fig. 6.10). Since the extended chains within the lamellae are normal to the lamellae surface, the extended chains are positioned tangentially about the center of the spherulite.

For many years, the structure of amorphous polymers was presumed to consist of a collection of randomly coiled molecules surrounding a certain

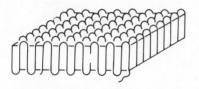

(a)

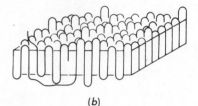

(b)

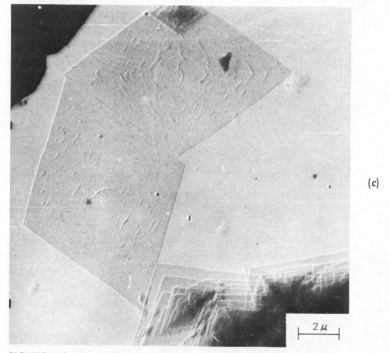

(c)

FIGURE 6.8 Crystalline lamellae in polymers. Models for regular (a) and irregular (b) chain chain folds, which produce thin crystallites;[10] (reprinted with permission from *Chem. and Eng. News, 43(33)*, copyright by the American Chemical Society). (c) Photomicrograph showing single crystals in polyethylene.[11] (Reprinted with permission from John Wiley & Sons, Inc.)

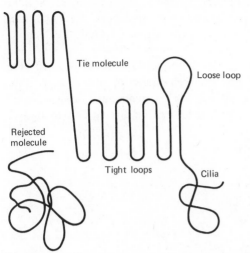

FIGURE 6.9   Schematic representation of chain-folded model containing tie molecules, loose loops, cilia (chain ends), and rejected molecules.[13] (By permission, from *Polymeric Materials,* copyright American Society for Metals, 1975.)

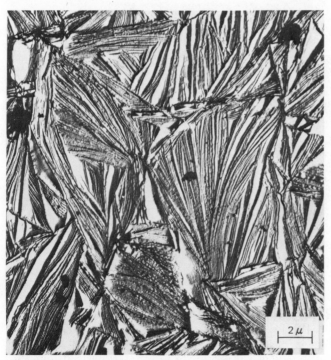

FIGURE 6.10   Sheaflike stacks of crystal lamellae in polychlorotrifluoroethylene which represent intersecting spherulites.[11] (Reprinted with permission from John Wiley & Sons, Inc.)

unoccupied volume. (A coiled molecule can be created by a random combination of C–C bond rotations along the backbone of the molecule.) Recent studies have suggested that this simple view is not correct. Instead Geil and Yeh[14–16] have proposed that seemingly amorphous polymers actually contain small domains (about 30 to 100 Å in size) in which the molecules are aligned (Fig. 6.11). More work is needed to identify the character of short-range order domains in "amorphous" polymers before the existence and role of such structural micro-details in determining properties can be proven convincingly.

FIGURE 6.11   Surface of amorphous bulk polyethylene terephthalate showing the presence of local nodular regions.[16] (Reprinted from *J. Macromol. Sci., B1*, 1967, p235, by courtesy of Marcel Dekker, Inc.)

6.2   POLYMER ADDITIONS

To this point, discussion has been confined to the structure of a pure material or homopolymer. A second monomer can be added to the starting feedstock such that the polymerization process produces a *copolymer*, designed to exhibit a

```
A A B A B B A A A B A B B A          (a)

A A A A B B B A A A B B B A A A A    (b)

A A A A A A A A A A A A A            (c)
          B
          B
          B                    FIGURE 6.12   Conformations of a
          B                    copolymer: (a) random; (b) block,
          B                    and (c) graft copolymer.
          B
```

certain set of properties. As shown in Fig. 6.12 polymer B can be added to the polymer A chain at random, in discrete blocks, or grafted to the side of a chain of polymer A. Often, commercial polymeric products contain a variety of additives that change the overall structure and associated properties. It is, therefore, appropriate to identify the major types of additives and their primary functions.

Pigments and Dyestuffs: These materials are added to impart color to the polymer.

Stabilizers: Stabilizers suppress molecular breakdown in the presence of heat, light, ozone, and oxygen. One form stabilizes the chain ends so the chains will not "unzip," thereby reversing the polymerization process. Other stabilizers act as antioxidants and antiozonants that are attached preferentially by O_2 and O_3 relative to the polymer chain.

Fillers: Various ingredients are sometimes added to the polymer to enhance certain properties. For example, the addition of carbon black to automobile tires improves their strength and abrasion resistance. Fillers also serve to lower the volume cost of the polymer-filler aggregate, since the cost of the filler is almost always much lower than that of the polymer.

Plasticizers: Plasticizers are high boiling point, low MW monomeric liquids that possess low volatility. They are added to a polymer to improve its processability and/or ductility. These changes arise for a number of reasons. Plasticizers add a low MW fraction to the melt, which broadens the MWD and shifts $\overline{M}$ to lower values. This enhances polymer processability. The liquid effectively shields chains from one another, thus decreasing their intermolecular attraction. Furthermore, by separating large chains, the liquids provide the chains with greater mobility for molecule segmental motion. The decrease in $\overline{M}$ and the lowering of intermolecular forces contribute toward improving polymer ductility and toughness. It should be recognized that these beneficial changes occur while stiffness and maximum service temperature decrease. Consequently, the extent of polymer plasticization is determined by an optimization of processability, ductility, strength, and stiffness, and service temperature requirements. It is interesting to note that nylon 66 is inadvertently plasticized by the moisture it picks up from the atmosphere (Table 6.3).

Blowing Agents: These substances are designed to decompose into gas bubbles within the polymer melt, producing stable holes. (In this way, expanded or foamed polymers are made.) The timing of this decomposition is critical. If the viscosity of the melt is too high, the bubbles will not form properly. If the viscosity is too low, the gas bubbles burst with the expanding polymer mass collapsing like an aborted soufflé.

Crosslinking Agents: The basic difference between thermoplastic and thermosetting polymers lies in the nature of the intermolecular bond. In thermoplastic materials, these bonds are relatively weak and of the Van der Waal type. Consequently, heating above T_g or T_m provides the thermal energy necessary for chains to move independently. This is not possible in thermosetting polymers, since the chains are rigidly joined with primary covalent bonds. Sulfur is a classic example of a cross-linking agent as used in the vulcanization of rubber.

From this very brief description of polymer additives, it is clear that a distinction should be made between a pure polymer and a polymer plus assorted additives; the latter is often referred to as a *plastic*, though rubbers also may be compounded. Although the terms *plastic* and *polymer* are often used synonymously in the literature, they truly represent basically different entities.

6.3 VISCOELASTIC RESPONSE OF POLYMERS AND THE ROLE OF STRUCTURE

The deformation response of many materials depends to varying degrees on both time-dependent and time-independent processes. For example, it was shown in Chapter Five that when the test temperature is sufficiently high, a test bar would creep with time under a given load. Likewise, were the same bar to have been stretched to a certain length and then held firmly, the necessary stress to maintain the stretch would gradually relax. Such response is said to be *viscoelastic*. Since T_g and T_m of most polymeric materials are not much above ambient (and in fact may be lower as in the case of natural rubbers), these materials exhibit viscoelastic creep and relaxation phenomena at room temperature. When the elastic strains and viscous flow rate are small (approximately 1 to 2% and 0.1 sec^{-1}, respectively), the viscoelastic strain may be approximated by

$$\epsilon = \sigma \cdot f(t) \tag{6-7}$$

That is, the stress-strain ratio is a function of time only. This response is called *linear* viscoelastic behavior and involves the simple addition of linear elastic and linear viscous (Newtonian) flow components. When the stress-strain ratio of a material varies with time *and stress*

$$\epsilon = g(\sigma, t) \tag{6-8}$$

the viscoelastic response is nonlinear.

On the basis of the simple creep test it is possible to define a creep modulus

$$E_c(t) = \frac{\sigma_0}{\epsilon(t)} \qquad (6\text{-}9)$$

where

$E_c(t)$ = creep modulus as a function of time

σ_0 = constant applied stress

$\epsilon(t)$ = time-dependent strain

Likewise, in a stress relaxation test where the strain ϵ_0 is fixed and the associated stress time dependent, a relaxation modulus $E_r(t)$ may be defined

$$E_r(t) = \frac{\sigma(t)}{\epsilon_0} \qquad (6\text{-}10)$$

These quantities can be plotted against log time to reveal their strong time dependence, as shown schematically in Fig. 6.13a for $E_r(t)$. It is clear that material behavior changes radically from one region to another. For very short times, the relaxation modulus approaches a maximum limiting value where the material exhibits glassy behavior associated with negligible molecule segmental motions. At longer times, the material experiences a transition to leathery behavior associated with the onset of short-range molecule segmental motions. At still longer times, complete molecule movements are experienced in the rubbery region associated with a further drop in the relaxation modulus. Beyond this point, liquid flow occurs. It is interesting to note that the same type of curve may be generated by plotting the modulus (from a simple tensile test) against test temperature (Fig. 6.13b). In this instance, the initial sharp decrease in E from its high value in the glassy state occurs at T_g. The shape of this curve can be modified by structural changes and polymer additions. For example, the entire curve is shifted downward and to the left as a result of plasticization (Fig. 6.14a). As $\overline{M}$ increases, the rubbery flow region is displaced to longer times (Fig. 6.14b), because molecular and segmental molecular movements are suppressed when chain entanglement is increased. Molecular weight has relatively little effect on the onset of the leathery region, since T_g is relatively independent of $\overline{M}$ except at low $\overline{M}$ values[17] (Fig. 6.15). The effect of $\overline{M}$ on T_g is believed to be related to the chain ends.[1] Since the ends are freer to move about, they generate a greater than average amount of free volume. Adjacent chains are then freer to move about and contribute to greater mobility of the polymer. Since the chain ends are more sensitive to $\overline{M}_n$ than $\overline{M}_w$, T_g is best correlated with the former measure of molecular weight. The leathery region is greatly retarded by cross-linking, while the flow region is completely eliminated, the latter being characteristic of thermosetting polymers (Fig. 6.16).

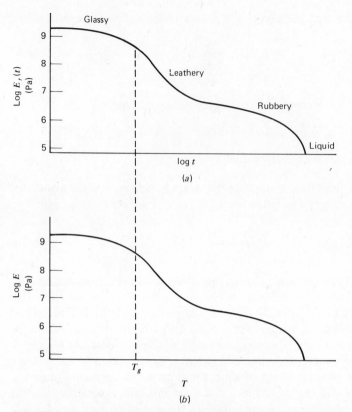

FIGURE 6.13 Time-temperature dependence of elastic modulus in thermoplastic polymeric solids: (a) change in relaxation modulus $E_r(t)$ as function of time; (b) change in tensile modulus as function of temperature.

The temperature-time (i.e., strain rate^{-1}) equivalence seen in Fig. 6.13 closely parallels similar observations made in Chapter Five. It is seen that the same modulus value can be obtained either at low temperatures and long times or at high test temperatures but short times. In fact, this equivalence is used to generate E_r versus $\log t$ curves as shown in Fig. 6.13a. The reader should appreciate that since such plots extend over 10 to 15 decades of time, they cannot be determined conveniently from direct laboratory measurements. Instead, relaxation data are obtained at different temperatures over a convenient time scale. Then, after choosing one temperature as the reference temperature, the remaining curves are shifted horizontally to longer or shorter times to generate a single master curve (Fig. 6.17). This approach was first introduced by Tobolsky and Andrews[19] and further developed by Williams et al.[20,21] Assuming that the viscoelastic response of the material is to be controlled by a single function of temperature (i.e., a single rate controlling mechanism), Williams et

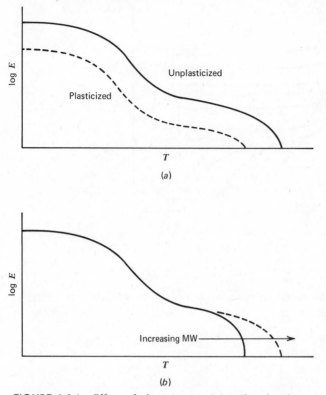

FIGURE 6.14   Effect of plasticization (*a*) and molecular weight (*b*) on elastic modulus as function of temperature.

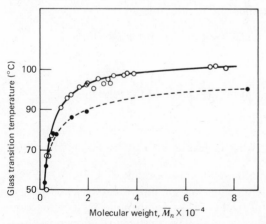

FIGURE 6.15   Glass transition temperature in PMMA (○) and polystyrene (●) as a function of $\overline{M}_n$.[17] (*The Structure of Polymers* by M. Miller © 1966 by Litton Educational Publishing by permission of Van Nostrand Reinhold.)

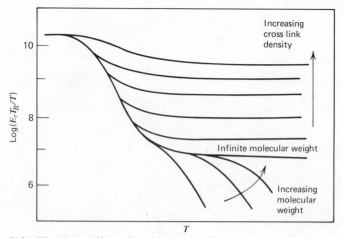

FIGURE 6.16   Effect of molecular weight and degree of cross-linking on relaxation modulus.[18] (Reprinted with permission from McGraw-Hill Book Company.)

al.[20,21] developed a semi-empirical relationship for an amorphous material, giving the time shift factor a_T as

$$\log a_T = \log \frac{t_T}{t_{T_0}} = \frac{C_1(T - T_0)}{C_2 + T - T_0} \tag{6-11}$$

where

$\quad\quad a_T =$ shift factor that is dependent on the difference

$\quad\quad\quad$ between the reference and data temperatures $T - T_0$

$\quad t_T, t_{T_0} =$ time required to reach a specific E_r at temperatures T

$\quad\quad\quad$ and T_0, respectively

$\quad C_1, C_2 =$ constants dependent on the choice of the reference

$\quad\quad\quad$ temperature, T_0

$\quad\quad\quad T =$ test temperatures where relaxation data were obtained

This relationship is found to hold in the temperature range $T_g < T < T_g + 100°C$, but is sometimes used beyond these limits on an individual basis as long as time-temperature superposition still occurs. This would indicate that the same rate controlling processes were still operative. Two reference temperatures are often used to normalize experimental data—T_g and $T_g + 50°C$—for which the constants C_1 and C_2 are given in Table 6.4.

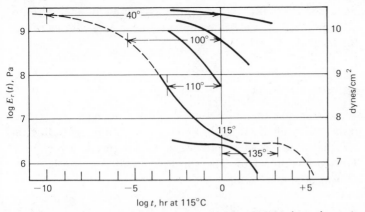

FIGURE 6.17 Modulus-time master plot for PMMA based on time-temperature superposition of data to a reference temperature of 115°C.[18] (Reprinted with permission from McGraw-Hill Book Company.)

TABLE 6.4 Constants for WLF Relationship

Reference Temperature	C_1	C_2
T_g	−17.44	51.6
$T_g + 50°C$	−8.86	101.6

The shift function may be used to normalize creep data,[4] enabling this information to be examined on a single master curve as well. Furthermore, by normalizing the creep strain results relative to the applied stress σ_0, the normalization of both axes converts individual creep-time plots into a master curve of creep compliance versus adjusted time (Fig. 6.18). These curves can be used to demonstrate the effect of MW and degree of cross-linking on polymer mechanical response much in the manner as the modulus relaxation results described in Figs. 6.14 and 6.16. Note that viscous flow is eliminated and the magnitude of

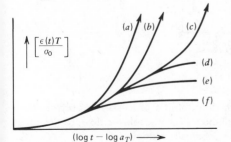

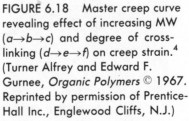

FIGURE 6.18 Master creep curve revealing effect of increasing MW ($a \rightarrow b \rightarrow c$) and degree of cross-linking ($d \rightarrow e \rightarrow f$) on creep strain.[4] (Turner Alfrey and Edward F. Gurnee, *Organic Polymers* © 1967. Reprinted by permission of Prentice-Hall Inc., Englewood Cliffs, N.J.)

the creep compliance reduced with increasing cross-linking in thermosetting polymers. For the thermoplastic materials, compliance decreases with increasing viscosity, usually the result of increased MW.

6.3.1 Mechanical Analogs

The linear viscoelastic response of polymeric solids has for many years been described by a number of mechanical models (Fig. 6.19). Many, including this author, have found that these models provide a useful physical picture of time-dependent deformation processes. The spring element (Fig. 6.19a) is intended to describe linear elastic behavior

$$\epsilon = \frac{\sigma}{E} \quad \text{and} \quad \gamma = \frac{\tau}{G} \tag{1-7}$$

such that resulting strains are not a function of time. (The stress-strain-time diagram for the spring is shown in Fig. 6.20a.) Note the instantaneous strain upon application of stress σ_0, no further extension with time, and full strain recovery when the stress is removed. The dashpot (a piston moving in a cylinder of viscous fluid) represents viscous flow (Fig. 6.19b)

$$\dot{\epsilon} = \frac{\sigma}{\eta} \quad \text{and} \quad \dot{\gamma} = \frac{\tau}{\eta} \tag{6-12}$$

where

$$\dot{\epsilon}, \dot{\gamma} = \text{tensile and shear strain rates}$$

$$\sigma, \tau = \text{applied tensile and shear stresses}$$

$$\eta = \text{fluid viscosity in units of stress-time}$$

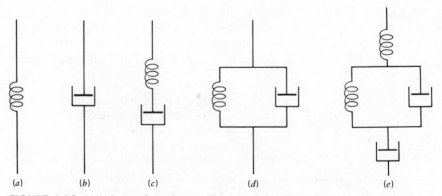

(a) (b) (c) (d) (e)

FIGURE 6.19 Mechanical analogs reflecting deformation processes in polymeric solids: (a) elastic; (b) pure viscous; (c) Maxwell model for viscoelastic flow; (d) Voigt model for viscoelastic flow; (e) four-element viscoelastic model.

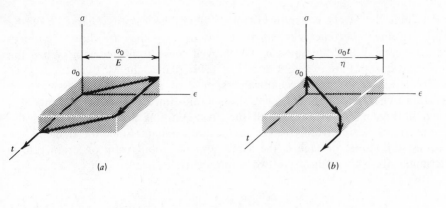

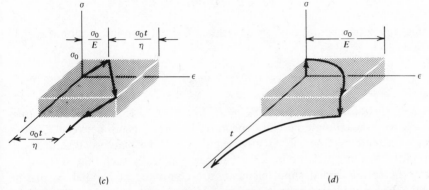

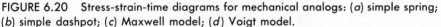

FIGURE 6.20   Stress-strain-time diagrams for mechanical analogs: (a) simple spring;
(b) simple dashpot; (c) Maxwell model; (d) Voigt model.

The viscosity η varies with temperature according to an Arrenhius-type relationship

$$\eta = Ae^{\Delta H/RT} \tag{6-13}$$

where

ΔH = viscous flow activation energy at a particular temperature

T = absolute temperature

On the basis of time-temperature equivalence, η is seen, therefore, to depend strongly on time as well. For example, at $t=0$ the viscosity will be extremely high, while at $t\to\infty$, η is small. The deformation response of a purely viscous element is shown in Fig. 6.20b. Upon loading ($t=0$), the dashpot is infinitely rigid. Consequently, there is no instantaneous strain associated with σ_0 (the same holds when the stress is removed). With time, the viscous character of the dashpot element becomes evident as strains develop that are directly proportional to time. When the stress σ_0 is removed these strains remain. When the spring and dashpot are in series, as in Fig. 6.19c (called the Maxwell model), we are able to describe the mechanical response of a material possessing both elastic and viscous components. The stress-strain-time diagram for this model is shown in Fig. 6.20c. Note that all the elastic strains are recovered, but the viscous strains arising from creep of the dashpot remain. Since the elements are in series, the stress on each is the same, and the total strain or strain rate is determined from the sum of the two components. Hence

$$\frac{d\epsilon}{dt} = \frac{\sigma}{\eta} + \frac{1}{E}\frac{d\sigma}{dt} \tag{6-14}$$

For stress relaxation conditions, $\epsilon = \epsilon_0$ and $d\epsilon/dt = 0$. Upon integration, Eq. 6-14 becomes

$$\sigma(t) = \sigma_0 e^{-Et/\eta} = \sigma_0 e^{-t/\mathcal{T}} \tag{6-15}$$

where $\mathcal{T} \equiv$ relaxation time defined by η/E. From Eq. 6-15, the extent of stress relaxation for a given material will depend on the relationship between $\mathcal{T}$ and t. When $t \gg \mathcal{T}$, there is time for viscous reactions to take place so that $\sigma(t)$ will drop rapidly. When $t \ll \mathcal{T}$, the material behaves elastically such that $\sigma(t) \approx \sigma_0$.

When the spring and dashpot elements are combined in parallel, as in Fig. 6.19d (the Voigt model), this unit predicts a different time-dependent deformation response. First, the strains in the two elements are equal, and the total stress on the pair is given by the sum of the two components

$$\epsilon_T = \epsilon_S = \epsilon_D$$
$$\sigma_T = \sigma_S + \sigma_D \tag{6-16}$$

Therefore

$$\sigma_T(t) = E\epsilon + \eta\frac{d\epsilon}{dt} \tag{6-17}$$

For a creep test, $\sigma_T(t) = \sigma_0$ and after integration

$$\epsilon(t) = \frac{\sigma_0}{E}\left(1 - e^{\frac{-t}{\mathcal{T}}}\right) \tag{6-18}$$

The strain experienced by the Voigt element is shown schematically in Fig. 6.20d. The absence of any instantaneous strain is predicted from Eq. 6-18 and is related in a physical sense to the infinite stiffness of the dashpot at $t = 0$. The creep strain is seen to rise quickly thereafter, but reach a limiting value σ_0/E associated with full extension of the spring under that stress. Upon unloading, the spring remains extended, but now exerts a negative stress on the dashpot. In this manner, the viscous strains are reversed and in the limit when both spring and dashpot are unstressed, all the strains have been reversed. Consequently, the Maxwell and Voigt models describe different types of viscoelastic response. A somewhat more realistic description of polymer behavior is obtained with a four-element model consisting of Maxwell and Voigt models in series (Fig. 6.19e). By combining Eqs. 1-7, 6-12, and 6-18, it can be readily shown that the total strain experienced by this model may be given by

$$\epsilon(t)= \frac{\sigma}{E_1} + \frac{\sigma}{E_2}(1-e^{-t/F}) + \frac{\sigma}{\eta_3}t \tag{6-19}$$

which takes account of elastic, viscoelastic, and viscous strain components, respectively (Fig. 6.21). Even this model is overly simplistic with many additional elements often required to adequately represent mechanical behavior of a polymer. For example, such a model might include a series of Voigt elements, each describing the relaxation response of a different structural unit in the molecule.

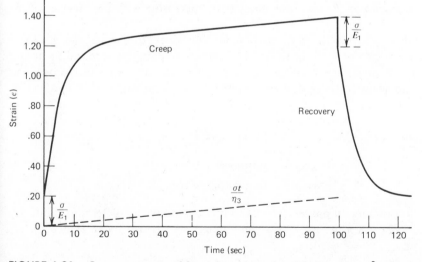

FIGURE 6.21   Creep response of four element model with $E_1 = 5 \times 10^2$ MPa, $E_2 = 10^2$ MPa, $\eta_2 = 5 \times 10^2$ MPa-sec, $\eta_3 = 50$ GPa-sec and $\sigma = 100$ MPa.[6] (*Mechanical Properties of Polymers* by L. Nielsen © 1962 by Litton Educational Publishing, Inc., reprinted by permission of Van Nostrand Reinhold.)

6.3.2 Dynamic Mechanical Testing and Energy Damping Spectra

Another method by which time-dependent moduli and energy dissipative mechanisms are examined is through the use of dynamic test methods. These studies have proven to be extremely useful in identifying the major molecular relaxation at T_g as well as secondary relaxations below T_g. It is believed that such relaxations are associated with motions of specific structural units within the polymer molecule. Two basically different types of dynamic test equipment have been utilized by researchers. One type involves the free vibration of a sample, such as that which takes place in the torsion pendulum apparatus shown in Fig. 6.22. A specimen is rotated through a predetermined angle and then released. This causes the sample to oscillate with decreasing amplitude resulting from various energy dissipative mechanisms. The extent of mechanical damping is defined by the decrement in amplitude of successive oscillations as given by

$$\Delta = \ln\frac{A_1}{A_2} = \ln\frac{A_2}{A_3} = \cdots = \ln\frac{A_n}{A_{n+1}} \tag{6-20}$$

where

$\Delta = \log\,(\text{base}\,e)$ decrement which measures the amount of damping

$A_1, A_2 =$ amplitude of successive oscillations of the freely vibrating sample.

From these same observations, stiffness of the sample is determined from the period of oscillation P, the shear modulus G increasing with the inverse square of P.

The other type of dynamic instruments introduce to the sample a forced vibration at different set frequencies. The amount of damping is found by noting the extent to which the cyclic strain lags behind the applied stress wave. The relationship between the instantaneous stress and strain values is shown in Fig. 6.23. Note that the strain vector ϵ_0 lags the stress vector σ_0 by the phase angle δ. It is instructive to resolve the stress vector into components both in phase and 90 degrees out of phase with ϵ_0. These are given by

$$\begin{aligned}\sigma' &= \sigma_0\cos\delta \quad \text{(in-phase component)}\\\sigma'' &= \sigma_0\sin\delta \quad \text{(out-of-phase component)}\end{aligned} \tag{6-21}$$

The corresponding in-phase and out-of-phase moduli are determined directly from Eq. 6-21 when the two stress components are divided by ϵ_0. Hence

$$E' = \frac{\sigma'}{\epsilon_0} = \frac{\sigma_0}{\epsilon_0}\cos\delta = E^*\cos\delta$$

$$E'' = \frac{\sigma''}{\epsilon_0} = \frac{\sigma_0}{\epsilon_0}\sin\delta = E^*\sin\delta \tag{6-22}$$

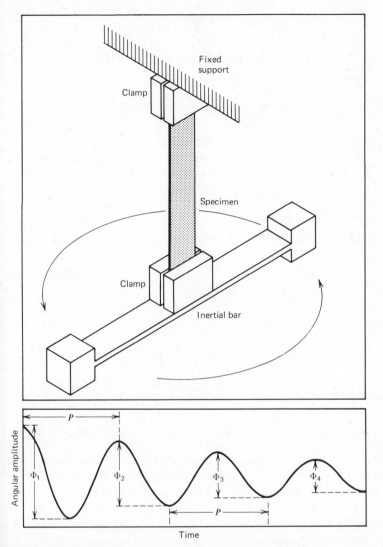

FIGURE 6.22 Simple torsion pendulum and amplitude-time curve for free decay of torsional oscillation.[6] (*Mechanical Properties of Polymers* by L. Nielsen © 1962 by Litton Educational Publishing, Inc., reprinted by permission of Van Nostrand Reinhold.)

where $E^* =$ absolute modulus $= [E'^2 + E''^2]^{\frac{1}{2}}$. E' reflects the elastic response of the material, since the stress and strain components are in phase. This part of the strain energy, introduced to the system by the application of stress σ_0, is stored but then completely released when σ_0 is removed. Consequently, E' is often referred to as the *storage* modulus. E'', on the other hand, describes the strain energy that is completely dissipated (mostly in the form of heat) and for

DEFORMATION RESPONSE OF ENGINEERING PLASTICS / **207**

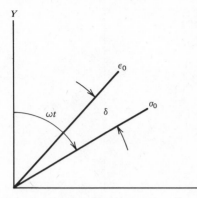

FIGURE 6.23 Forced vibration resulting in phase lag δ between applied stress σ_0 and corresponding strain ϵ_0.

this reason is called the *loss* modulus. The relative amount of damping or energy loss in the material is given by the loss tangent, $\tan\delta$

$$\frac{E''}{E'} = \frac{E^* \sin\delta}{E^* \cos\delta} = \tan\delta \tag{6-23}$$

By comparison,[6]

$$\frac{G''}{G'} \approx \frac{\Delta}{\pi} \tag{6-24}$$

with the result that

$$\Delta \approx \pi\tan\delta \tag{6-25}$$

When dynamic tests are conducted, the values of the storage and loss moduli and damping capacity are found to vary dramatically with temperature (Fig. 6.24). Note the correlation between the rapid drop in G', the rise in G'', and the corresponding damping maximum. The relaxation time associated with these changes (occurring in Fig. 6.13b in the vicinity of T_g) is considered to have an Arrhenius-type temperature dependence associated with a specific activation energy. In turn, the activation energy is then used to identify the molecular motion responsible for the change in dynamic behavior. Dynamic tests can be conducted either over a range of test temperatures at a constant frequency or at different frequencies for a constant temperature. Since the fixed frequency tests are usually more convenient to perform, most studies employ this procedure. Experiments of this type are now conducted routinely in many laboratories to characterize polymers with regard to effects of thermal history, degree of crystallinity, molecular orientation, polymer additions, molecular weight, plasticization, and other important variables. Consequently, the extant literature for such studies is enormous. Fortunately, a number of books and review articles

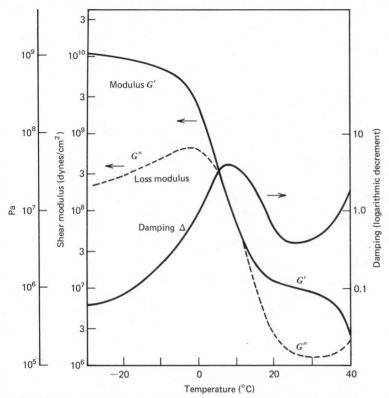

FIGURE 6.24 Dynamic mechanical response of uncross-linked styrene and butadiene copolymer revealing temperature dependence of G', G'', and Δ.[6] (*Mechanical Properties of Polymers* by L. Nielsen © 1962 by Litton Educational Publishing, Inc., reprinted by permission of Van Nostrand Reinhold.)

have been prepared[13,22–26] on the subject to which the interested reader is referred. Within the scope of this book, we can only highlight some of the major findings.

When dynamic tests are performed over a sufficiently large temperature range, multiple secondary relaxation peaks are found in addition to the T_g peak shown in Fig. 6.24. Boyer[13] has summarized some of these data in the schematic form shown in Fig. 6.25. He noted that relaxation response in amorphous and semi-crystalline polymers could be separated conveniently into four regions, as summarized in Table 6.5. Furthermore, crude temperature relationships between various damping peaks were identified (e.g., $T_m \approx 1.5\, T_g$ and the $T < T_g$ transition (β) occurring at about $0.75\, T_g$).

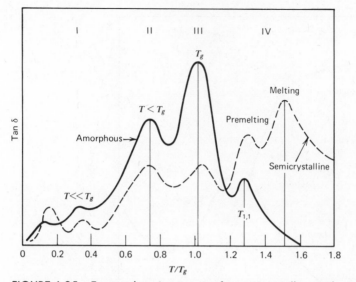

FIGURE 6.25   Energy damping spectra for semicrystalline and amorphous polymers at various temperatures normalized to T_g. Several damping peaks are found for each material.[13] (By permission, from *Polymeric Materials*, copyright American Society for Metals, 1975.)

TABLE 6.5   Transition Regions in Polymers[13]

Region	Temperature of Occurrence	Cause
I	$T \ll T_g$ (the γ peak)	Believed to be caused by movements of small groups involving only a few atoms.
II	$T < T_g$ (the β peak)	Believed related to movement of 2–3 consecutive repeat units.
III	T_g (the α peak)	Believed to be related to coordinated movements of 10–20 repeat units.
IV	$T > T_g$	Large-scale molecular motions.

6.4 DEFORMATION MECHANISMS IN CRYSTALLINE AND AMORPHOUS POLYMERS

We now consider the mechanisms by which crystalline and amorphous polymers deform. Unoriented crystalline materials are found to deform by a complex process involving initial breakdown and subsequent reorganization of crystalline regions.[27–29] After an initial stage of plastic deformation in the spherulites, the latter begin to break down. Lamellae packets oriented normal to the applied stress may separate along the amorphous boundary region between crystals, while others begin to rotate toward the stress axis (analogous to slip-plane rotation discussed in Section 3.3). The crystals themselves are now broken into smaller blocks, but the chains maintain their folded conformation. As this phase of the deformation process continues, these small bundles become aligned in tandem along the drawing direction, forming long microfibrils (Figs. 6.26 and 6.27). Note that the extended chains within each bundle are positioned parallel to the draw axis along with a large number of fully extended tie molecules. Since many tandem blocks are torn from the same lamellae, they remain connected through a number of tie molecules created by unfolding chains from the original lamellae. The combination of many more fully extended tie molecules and the orientation of the bundles within each fibril contributes toward a rapid increase in strength and stiffness. By contrast, few primary bonds join blocks in adjacent microfibrils, except those representing tie molecules from the original lamellae

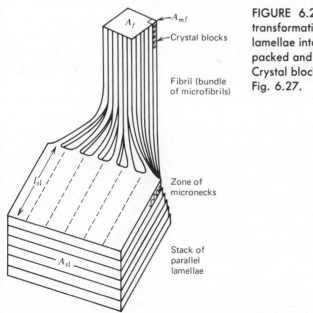

FIGURE 6.26 Model depicting transformation of a stack of parallel lamellae into a bundle of densely packed and aligned microfibrils.[28] Crystal blocks oriented as shown in Fig. 6.27.

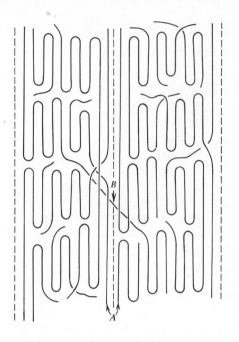

FIGURE 6.27 Alignment of crystal blocks in microfibrils. Intrafibrillar extended tie molecules shown at A with interfibrillar extended tie molecule at B.[29] (By permission, from *Polymeric Materials*, copyright American Society for Metals, 1975.)

(Fig. 6.27). It is this initial spherulite structure breakdown, followed by microfibril formation, that gives rise to the substantial hardening associated with Type V stress-strain response discussed in Section 1.2.5. Continued deformation of the microfibrillar structure is extremely difficult because of the high strength of the individual microfibrils and the increasing extension of the interfibrillar tie molecules. These tie molecules become more extended as a result of microfibril shear relative to one another. For additional comments regarding highly oriented microfibrils see Section 6.5.

Centuries ago, it was noted that pottery glazes tended to develop intricate and often beautiful arrays of fine cracks. The phenomenon was referred to as *crazing*. Analysis of crazes in glasses revealed that they usually developed on surfaces that were at right angles to the principal tensile axis. When the study and applications of polymers began to be pursued intensely, it was recognized that similarly oriented crazes in glassy polymers were not, in fact, simple cracks as in ceramics; rather, they constituted expanded material containing oriented fibrils interspersed with small (100 to 200 Å) interconnected voids[23,30-39] (Fig. 6.28). For a recent critical review of craze formation and fracture, the reader is referred to the work of Rabinowitz and Beardmore.[40] The combination of fibrils (extending across the craze thickness) and interconnected microvoids contributes toward an overall weakening of the material, though *the craze is capable of supporting some reduced stress relative to that of the uncrazed matrix*. The latter point is proven conclusively by the load-bearing capability of samples contain-

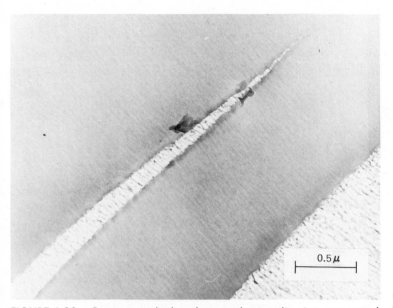

FIGURE 6.28   Crazes in polyphenylene oxide revealing interconnected microvoids and aligned fibrils.[35] (Reprinted with permission from the Polymer Chemistry Division of ACS.)

ing crazes that extend completely across the sample.[23,40] Researchers have found that crazes tend to grow along planes normal to the principal tensile stress direction with little change in craze thickness being noted (Fig. 6.29). The typical craze thickness in glassy polymers is on the order of 5 μm or less, which corresponds in some cases to plastic strains in excess of 50%.[34] Since the lateral surface contraction associated with the craze (Fig. 6.29) is negligible by comparison, it can only be concluded that the density of the craze should decrease. Indeed, Kambour[34] computed a craze density 40 to 60% that of the uncrazed matrix. The presence of the interconnected void network within the craze is consistent with this finding. Since the refractive index of the craze is lower than that of the polymer, the craze is observed readily with the unaided eye, assuming of course that it is large enough to resolve.

To explain crazing in polymers, Gent[41] proposed a theory in terms of stress-induced devitrification (change from a glassy to a weaker rubbery state) at a flaw, followed by cavitation resulting from the hydrostatic component of the stress. The following equation for the critical stress required to develop (and permit growth of) a thin band of softened material (which then can undergo cavitation with deformation characteristic of crazing) has been proposed:[41]

$$\bar{\sigma}_c = [\,\beta(T_g - T) + P\,]/k \qquad (6\text{-}26)$$

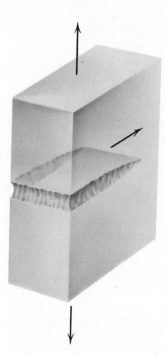

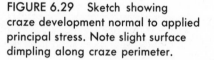

FIGURE 6.29 Sketch showing craze development normal to applied principal stress. Note slight surface dimpling along craze perimeter.

where β is a coefficient relating the effect of an applied hydrostatic pressure P on the T_g, T is the ambient temperature, and k the stress concentration factor. Thus, crazing should be favored by high temperatures and high stress concentrations, and restricted by applied hydrostatic pressure and the occurrence of creep or other types of flow that could decrease k. Regarding this point, Sternstein and co-workers[42-44] have found that under multiaxial stress conditions, glassy polymers could yield by two distinct mechanisms: normal yielding (crazing) and shear yielding (Fig. 6.30). For the biaxial stress case, shear yielding was found to depend on the mean normal stress with

$$\tau_{oct} = \tau_0 - \mu\sigma_m \tag{6-27}$$

where

τ_{oct} = octahedral yield stress

τ_0 = rate- and temperature-dependent octahedral yield stress

found for pure shear ($\sigma_m = 0$)

$\sigma_m = (\sigma_1 + \sigma_2)/2$

μ = material constant

Note that when $\sigma_m > 0$ (dilatation), τ_{oct} is reduced, with the result that shear yielding in pure shear is more difficult than in an ordinary uniaxial tensile test. Conversely, when $\sigma_m < 0$ (compressional), yielding is even more difficult than under pure shear conditions.

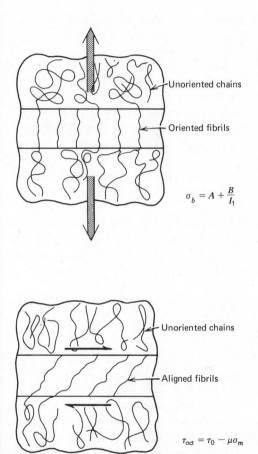

FIGURE 6.30   Deformation mechanisms in amorphous polymers: (a) normal yielding (crazing) and (b) shear yielding.

As mentioned previously, crazing develops on planes oriented normal to the principal tensile stress direction and is sensitive to the hydrostatic stress component as well. Sternstein combined these two observations and showed that

$$\sigma_b = |\sigma_1 - \sigma_2| = A(T) + \frac{B(T)}{I_1} \tag{6-28}$$

where

σ_b = stress bias given by the difference in principal stresses σ_1 and σ_2

$A(T), B(T)$ = temperature-dependent material constants

$I_1 = \sigma_1 + \sigma_2$, which must be positive to provide the necessary

dilatation for craze formation

From the above, the stress bias necessary for crazing decreases with increasing dilatation I_1. It is seen from Fig. 6.31 that the operative deformation mechanism (i.e., which yield loci is reached first) will depend on the magnitude of σ_1 and σ_2. For example, Fig. 6.31a shows that crazing will develop first. However, when a hydrostatic pressure is introduced, the stress for crazing rises relative to that for shear yielding, with the result that the latter mechanism becomes preferred over most of quadrant I as well as in the other three quadrants (Fig. 6.31b). Note also that the deformation mechanism for uniaxial tension switches from crazing to shear yielding when hydrostatic pressure is raised.

Because of its intrinsic weakness, the craze is an ideal path for crack propagation. While the presence of the craze is considered undesirable from this standpoint, the localized process of fibril and void formation does absorb considerable strain energy. For example, Berry[38] found a large difference between fracture energy and surface energy, which he attributed mainly to craze formation energy absorption. For a summary of these studies and the fracture work of others see the reviews by Rabinowitz and Beardmore[40] and Hull.[39]

6.5 STRENGTHENING OF POLYMERS

A number of approaches may be taken to strengthen a polymer. Through changes in *chemistry* some polymers can be cross-linked to lock their molecules together in rigid fashion, thereby precluding viscous flow (Section 6.2). For example, a bowling ball contains a much higher cross-link density than a handball. Strengthening through changes in chemistry can also be brought about by the introduction of large side groups and intrachain groups which restrict C–C bond rotation (Section 6.1).

FIGURE 6.31 Yield loci for normal and shear yielding in general biaxial stress space: (a) point C and D represent uniaxial tensile stresses required for crazing and shear yielding, respectively[43] (reprinted with permission from The Polymer Chemistry Division of ACS.); (b) yield loci with superimposed hydrostatic pressure. Note the inversion in points C and D from part a.[44] (Reprinted from *J. Macromol. Sci., B8*, 1973, p. 539, by courtesy of Marcel Dekker, Inc.)

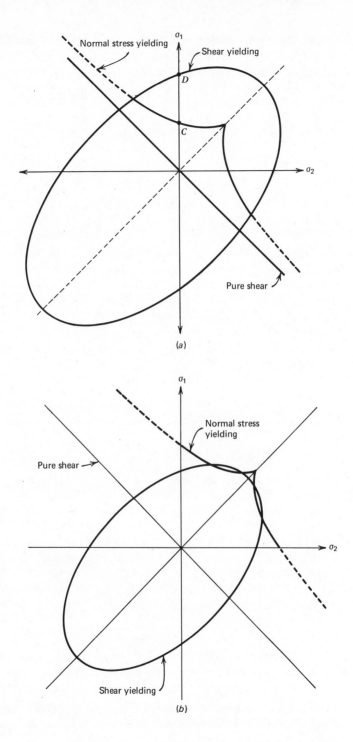

The superstructure or *architecture* of a polymer can be modified to effect dramatic changes in mechanical strength greater than those made possible by chemistry adjustments. First, mechanical properties are found to increase with molecular weight, and relationships[45] often assume the forms

$$\text{mechanical property} = A - \frac{B}{\overline{M}} \qquad (6\text{-}29a)$$

or

$$\text{mechanical property} = C + \frac{D}{\overline{M}} \qquad (6\text{-}29b)$$

An example of such data[46] is shown in Fig. 6.32. It may be argued that as chain length increases beyond a critical length, the combined resistance to flow from chain entanglement and intermolecular attractions exceeds the strength of primary bonds, which can then be broken. Consequently, once the molecular weight exceeds a critical lower limit $\overline{M}_c$, entanglement and primary bond breakage occur. At this point the mechanical property becomes less sensitive to MW. On the basis of available data it is not clear whether mechanical properties correlate best with $\overline{M}_w$ or $\overline{M}_n$.[47]

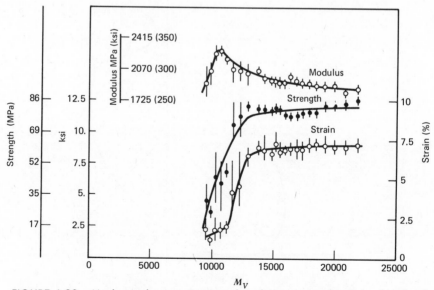

FIGURE 6.32 Mechanical properties in polycarbonate as a function of molecular weight.[46] (Reprinted with permission from John Wiley & Sons, Inc.)

Mechanical properties are improved most dramatically by molecular and molecular segment alignment parallel to the stress direction. This stands to reason, since the loads would then be borne by primary covalent bonds along the molecule rather than by weak Van der Waal forces between molecules. The alignment of molecules in an amorphous polymer is described with the aid of Fig. 6.33. Thermal energy causes the molecules in the polymer at $T > T_g$ to vibrate with relative ease in random fashion, but below T_g the randomness is "frozen in." If the material were drawn quickly at a temperature not too far above T_g (say T_1), some chain alignment could be achieved and effectively "frozen in," provided the stretched polymer were to be quenched from that temperature. The resulting material would be stronger in the direction of drawing and correspondingly weakened in the lateral direction. An example of this anisotropy is shown in Fig. 6.34 for drawn polystyrene. Note that the strength anisotropy parallels a deformation mechanism transition from crazing to shear yielding as the tensile axis approaches the draw direction. If the drawing were conducted slowly at T_1 or even at T_2, the elongation could be accommodated by viscous flow without producing chain alignment. Consequently, no strengthening would result.

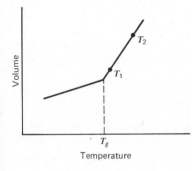

FIGURE 6.33 Rapid drawing at T_1 followed by quenching can produce molecular alignment and polymer strengthening. Viscous flow during drawing at T_2 precludes such alignment.

It should be recognized that the oriented structure is unstable and will contract upon subsequent heating above T_g. By contrast, the polymer stretched at T_2 would be dimensionally stable, since it never departed from its preferred fully random state.

In semicrystalline polymers, crystallite alignment may be produced by cold drawing spherulitic material (Section 6.4) and by forcing or drawing liquid through a narrow orifice. During the past few years, attempts have been made to extend the practice of polymer chain orientation to its logical limit—the full extension of the molecule chain—with the potential of producing a very strong and stiff fiber.[48-50] Indeed, this has been partially accomplished. Highly oriented and extended commercial fibers, such as DuPont Kevlar, possess a tensile

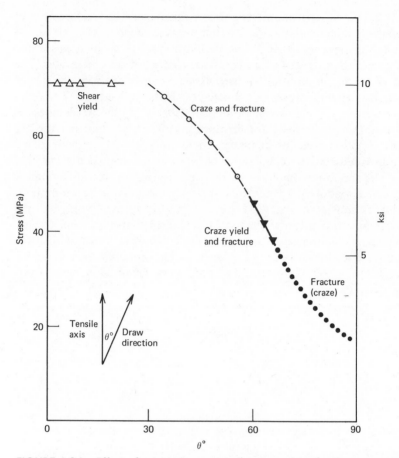

FIGURE 6.34 Effect of orientation on tensile properties of polystyrene tested at 20°C, hot drawn to a draw ratio of 2.6. Note deformation mechanism transition with test direction.[39] (By permission, from *Polymeric Materials*, Copyright American Society for Metals, 1975.)

modulus two-thirds that of steel but with a much lower density. This is truly extraordinary, since commercial plastics generally exhibit elastic moduli fully two orders of magnitude smaller than steel. By converting the folded chain conformation to a fully extended one, the applied stresses are sustained by the very strong main chain covalent bonds, which are less compliant than the much weaker intermolecular Van der Waal forces. Highly oriented fibers have been produced both by cold forming and direct spinning from the melt.

From this brief discussion, it is clear that the elastic moduli of polymeric solids are very structure sensitive, unlike those found in metals and ceramics (see

Chapter One). Specifically, it has been found[49] that the modulus of a polymer increases with: (1) thermodynamic stability of the main chain bonds; (2) percent crystallinity; (3) packing density of the chains; (4) chain orientation in the tensile direction; (5) chain end accommodation in the polymer crystal; and (6) a minimization of chain folding. There is great potential commercial use for these fibers, such as in tire cord and engineering composites including sporting equipment.

Although the formation of high-strength crystalline fibers or filaments directly from the melt is an intriguing engineering process, it is by no means a unique event in nature; rather, the materials scientist must yield to the long-recorded activities of arachnids and silkworms. For example, researchers have determined that spider silk is generated by the drawing from various glands of an amorphous protein liquid that then converts quickly to a highly oriented, very long crystalline filament with a diameter on the order of several hundred angstroms. The highly oriented and crystalline morphology of these filaments is believed responsible for their reported strengths in excess of 700 MPa.[51] In one intriguing investigation, Lucas[51] showed spider silk to have twice the tenacity, defined in g/denier*, of a 2100 MPa steel wire and four times the extension at break. These results are tabulated in Table 6.6 along with data for the man-made Kevlar fiber and others.

TABLE 6.6   Tenacity of Selected Fibers

Material	Tenacity (g/denier)	Total Elongation (%)	Ref.
Steel	~4	~8	50, 51
Glass	~10	~3	50, 51
Spider silk (drag line)	~8	~30	51
DuPont Kevlar	~25	~5	50
Monsanto X-500-G	~16	~4	50

Such high strengths, along with the apparent abundant supply of spider silk, have prompted enterprising individuals to seek commercial markets for the product of our arachnid friends. In one such feasibility study, in 1709, several pairs of stockings and gloves were woven from spider silk and presented to the French Academy of Science for their consideration.[52] It would appear that the silkworm has emerged the clear victor over the spider in the battle for the silk market.

* A denier is defined in the textile industry as the weight in grams of 9000 meters of yarn, hence, a linear density.

6.6 POLYMER TOUGHNESS

It is generally agreed that crystalline polymers tend to exhibit greater toughness than amorphous polymers owing to their folded chain conformation. There is, however, some controversy regarding the mechanism(s) responsible for whatever toughness amorphous materials possess. According to one proposal[8, 25] toughness should depend on the amount of free volume available to allow for molecule segmental motion. With enhanced motions, toughness should be higher. Litt and Tobolsky[8] found that tough amorphous polymers generally contain a fractional free volume $\bar{f}$ (Eq. 6-6) greater than 0.09. In further support, Petrie[9] found that such polymers, when aged at $T < T_g$, suffered losses in impact energy absorption which were related to corresponding decreases in excess free volume after annealing. It should be mentioned that an alternate explanation for these data has been given. Yeh[15] and Geil[14] argue that while excess free volume did decrease during aging, the change in mechanical properties could have been caused instead by the fine-scale changes in polymer structure that have been reported. In an overall sense, they contend that while free volume does exert a first-order effect on the relative toughness of a polymer, secondary factors, such as the free volume *distribution*, may be of equal importance. This point is the source of active debate at this time.

In a related finding, it has been shown that the $T < T_g(\beta)$ peak is also identified with impact toughness, the larger and broader the β peak the tougher a material generally tends to be.[22, 25, 26] For example, the β peak in polycarbonate is much greater than that shown for PMMA, consistent with the much greater toughness associated with the former material.

Finally, it is interesting to note that although a particular *polymer* possessing a low free volume and negligible β peak may be brittle, certain *plastics* based on this polymer may offer good toughness. This improvement in impact resistance is commonly achieved through the use of plasticizers. As discussed above, these high boiling point, low MW monomeric liquids serve to separate molecule chains from one another, thus decreasing their intermolecular attraction and providing chain segments with greater mobility. The ductility and toughness of plasticized polymers is found to increase, while their strength and T_g decrease. Care must be taken to add a sufficient amount of plasticizer to a particular polymer so that the material does not actually suffer a loss in toughness. This surprising reversal in material response is referred to as the *antiplasticizer* effect.[53] It has been argued that below a critical plasticizer content, the liquid serves mainly to fill some of the polymer's existing free volume with a concomitant loss in molecule chain segmental mobility.

A brittle polymer may also be toughened through the addition of a finely dispersed rubbery phase in a diameter size range of 0.1 to 10 μm. At first, it was thought that the improved toughness was due simply to the tough properties of the rubbery phase itself. More recently, researchers[23, 54–56] have come to recognize the impact energy modification as a synergistic interaction between the rubbery phase and the glassy matrix. Upon loading such a material, a myriad of fine crazes develop at the rubber particle-matrix interface (Fig. 6.35). Craze formation itself is undesirable since it is the precursor of crack formation, but the formation of many small crazes represents a large sink for strain energy release. Consequently, rubber-modified plastics such as high-impact polystyrene and acrylonitrile-butadiene-styrene (ABS) are quite tough and are useful in various engineering applications. For a further discussion of the fatigue behavior of these materials see Section 13.8.2.

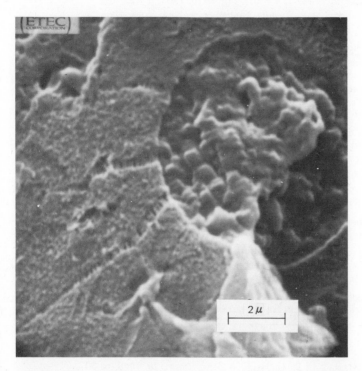

FIGURE 6.35   Matrix crazes eminating from left side of rubber particle as seen on fracture surface in high-impact polystyrene.[57] (Reprinted with permission from John Wiley & Sons, Inc.)

REFERENCES

1. R. D. Deanin, *Polymer Structure, Properties and Applications*, Cahners, Boston, Mass. 1972.
2. *Engineering Properties of Thermoplastics*, R. M. Ogorkiewicz, ed., Wiley-Interscience, London, 1970.

3. M. Kaufman, *Giant Molecules*, Doubleday, Garden City, N. Y., 1968.
4. T. Alfrey and E. F. Gurnee, *Organic Polymers*, Prentice-Hall, Englewood Cliffs, N. J., 1967.
5. S. L. Rosen, *Fundamental Principles of Polymeric Materials for Practicing Engineers*, Barnes and Noble, New York, 1971.
6. L. E. Nielsen, *Mechanical Properties of Polymers*, Reinhold, New York, 1962.
7. H. V. Boening, *Polyolefins: Structure and Properties*, Elsevier Press, Lausanne, 1966, p. 57.
8. M. H. Litt and A. V. Tobolsky, *J. Macromol. Sci. Phys.*, **B1**(3), 1967, p. 433.
9. S. E. B. Petrie, *Polymeric Materials*, ASM, Metals Park, Ohio, 1975, p. 55.
10. P. H. Geil, *Chem. Eng. News*, **43**(33), August 16, 1965, p. 72.
11. P. H. Geil, *Polymer Single Crystals*, Interscience, New York, 1963.
12. E. S. Clark, *Polymeric Materials*, ASM, Metals Park, Ohio, 1975, p. 1.
13. R. F. Boyer, *ibid*, p. 277.
14. P. H. Geil, *ibid.*, p. 119.
15. G. S. Y. Yeh, *Crit. Rev. Macromol. Sci.*, **1**, 1972, p. 173.
16. G. S. Y. Yeh and P. H. Geil, *J. Macromol. Sci.*, **B1**, 1967, p. 235.
17. M. L. Miller, *The Structure of Polymers*, Reinhold, New York, 1966.
18. F. Rodriguez, *Principles of Polymer Systems*, McGraw-Hill, New York, 1970.
19. A. V. Tobolsky and R. D. Andrews, *J. Chem. Phys.*, **13**, 1945, p. 3.
20. M. L. Williams, R. F. Landel, and J. D. Ferry, *J. Amer. Chem. Soc.*, **77**, 1955, p. 3701.
21. M. L. Williams, R. F. Landel, and J. D. Ferry, *J. Appl. Phys.*, **26**, 1955, p. 359.
22. J. Heijboer, *J. Polym. Sci.*, **16**, 1968, p. 3755.
23. R. P. Kambour and R. E. Robertson, *Polymer Science: A Materials Science Handbook*, Vol. 1, A. D. Jenkins, ed., North Holland, 1972, p. 687.
24. N. G. McCrum, B. E. Read, G. Williams, *Anelastic and Dielectric Effects in Polymeric Solids*, John Wiley, London, 1967.
25. R. F. Boyer, *Rubber Chem. Tech.*, **36**, 1963, p. 1303.
26. R. F. Boyer, *Poly. Eng. Sci.*, **8**(3), 1968, p. 161.
27. A. Peterlin, *Advances in Polymer Science and Engineering*, K. D. Pae, D. R. Morrow, and Y. Chen, eds., Plenum Press, New York, 1972, p. 1.
28. A. Peterlin, *Macromol. Chem.*, **8**, 1973, p. 277.
29. A. Peterlin, *Polymeric Materials*, ASM, Metals Park, Ohio, 1975, p. 175.
30. B. Maxwell and L. F. Rahm, *Ind. End. Chem.*, **41**, 1949, p. 1988.
31. J. A. Sauer, J. Marin and C. C. Hsiao, *J. Appl. Phys.*, **20**, 1949, p. 507.
32. C. C. Hsiao and J. A. Sauer, *J. Appl. Phys.*, **21**, 1950, p. 1071.
33. J. A. Sauer and C. C. Hsiao, *Trans.*, ASME, **75**, 1953, p. 895.
34. R. P. Kambour, *Polymer*, **5**, 1964, p. 143.
35. R. P. Kambour and A. S. Holick, *Polym. Prepr.*, **10**, 1969, p. 1182.
36. R. P. Kambour, *J. Polym. Sci.*, Part A-2, **4**, 1966, p. 17.
37. J. P. Berry, *J. Polym. Sci.*, **50**, 1961, p. 107.
38. J. P. Berry, *Fracture Processes in Polymeric Solids*, B. Rosen, ed., Interscience, 1964, p. 157.
39. D. Hull, *Polymeric Materials*, ASM, Metals Park, Ohio, 1975, p. 487.
40. S. Rabinowitz and P. Beardmore, *CRC Crit. Rev. Macromol. Sci.*, **1**, 1972, p. 1.
41. A. Gent, *J. Mater. Sci.*, **5**, 1970, p. 925.

42. S. S. Sternstein, *Polymeric Materials*, ASM, Metals Park, Ohio, 1975, p. 369.
43. S. S. Sternstein and L. Ongchin, *Polym. Prepr.*, **10**, 1969, p. 1117.
44. S. S. Sternstein and F. A. Myers, *J. Macromol. Sci.*, **B8**, 1973, p. 539.
45. P. J. Flory, *J. Amer. Chem. Soc.*, **67**, 1945, p. 2048.
46. J. H. Golden, B. L. Hammant, and E. A. Hazell, *J. Polym. Sci.*, **2A**, 1964, p. 4787.
47. J. R. Martin, J. F. Johnson, and A. R. Cooper, *J. Macromol. Sci.—Rev. Macromol. Chem.*, **C8**(1), 1972, p. 57.
48. F. C. Frank, *Proc. Roy. Soc.*, London, **A319**, 1970, p. 127.
49. R. S. Porter, J. H. Southern, and N. Weeks, *Polym. Eng. Sci.*, **15**, 1975, p. 213.
50. J. Preston, *Polym. Eng. Sci.*, **15**, 1975, p. 199.
51. F. Lucas, *Discovery*, **25**, 1964, p. 20.
52. W. J. Gertsch, *American Spiders*, Van Nostrand, New York, 1949.
53. W. J. Jackson, Jr. and J. R. Caldwell, *J. Appl. Polym. Sci.*, **11**, 1967, p. 211.
54. C. B. Bucknall and R. R. Smith, *Polymer*, **6**, 1965, p. 437.
55. C. B. Bucknall, *Brit. Plast.*, Nov. 1967, p. 118.
56. C. B. Bucknall, *J. Mater. Sci.*, **4**, 1969, p. 214.
57. J. A. Manson and R. W. Hertzberg, *J. Polym. Sci.*, Polym. Phys. Ed., **11**, 1973, p. 2483.

PROBLEMS

6.1 It is found that the viscosity of a liquid above T_g can be expressed by the relationship

$$\log\left(\frac{\eta_T}{\eta_{T_g}}\right) = \frac{-17.44(T - T_g)}{51.6 + T - T_g}$$

which is similar to Eq. 6-11. If the viscosity of a material at T_g is found to be 10^4 GPa-sec compute the viscosity of the liquid 5, 10, 50, and 100°C above T_g.

6.2 If it takes 300 seconds for the relaxation modulus to decay to a particular value at T_g, to what temperature must the material have to be raised to effect the same decay in 10 seconds?

6.3 Calculate the relaxation time for glass and comment on its propensity for stress relaxation at room temperature. $E \approx 70$ GPa and $\eta \approx 1 \times 10^{12}$ GPa-sec (10^{22} poise).

6.4 For the four-element model shown in Fig. 6.21, compute the strain-time plot over the same time span when: (a) all constants are the same except $\eta_3 = 10$ GPa-sec, and (b) all constants are the same except $E_1 = 1$ GPa-sec and $\eta_2 = 10$ GPa-sec. Compare the three plots.

SECTION TWO
FRACTURE MECHANICS OF ENGINEERING MATERIALS

CHAPTER SEVEN
FRACTURE –
AN OVERVIEW

7.1 INTRODUCTION

On January 15, 1919, something frightening happened on Commercial Street in Boston. A huge tank, 27 meters in diameter and about 15 meters high, fractured catastrophically, and over 7.5×10^6 liters (2×10^6 gallons) of molasses cascaded into the streets.

> "Without an instant's warning the top was blown into the air and the sides were burst apart. A city building nearby, where the employees were at lunch, collapsed burying a number of victims and a firehouse was crushed in by a section of the tank, killing and injuring a number of the firemen."[1]
>
> "On collapsing, a side of the tank was carried against one of the columns supporting the elevated structure [of the Boston Elevated Railway Co.]. This column was completely sheared off...and forced back under the structure...the track was pushed out of alignment and the superstructure dropped several feet.... Twelve persons lost their lives either by drowning in molasses, smothering, or by wreckage. Forty more were injured. Many horses belonging to the paving department were drowned, and others had to be shot."[2]

The molasses tank failure dramatically highlights the necessity of understanding events that contribute to premature fracture of any engineering component. For those who suspect that satisfactory solutions to such incidents surely must have been developed during the past half century, a 1973 molasses tank failure,

FIGURE 7.1   (a) Remains of a molasses tank which fractured suddenly in Bellview, New Jersey, March 22, 1973. (New York Daily News Photo.) (b) Some aspects of the cleanup operation. Note carefully the molasses dripping from the truck wheels. (New York Daily News Photo.)

shown in Fig. 7.1 along with the details of the messy clean-up operation, attests to the need for further understanding of the problem. Other man-made structures are susceptible to the same fate. For example, several bridges have fractured and collapsed in Belgium, Canada, Australia, Netherlands Antilles, and the United States during the past 40 years, resulting in the loss of many lives. In addition, numerous cargo ship failures have occurred, dating from World War II to the present (Fig. 7.2).

It is quite apparent, then, that the subject of fracture in engineering components and structures is certainly a dynamic one, with new examples being provided continuously for evaluation. You might say that things are going wrong all the time. (In all seriousness, the reader must recognize that component failures are the exception and not the rule).

7.2 THEORETICAL COHESIVE STRENGTH

Recall from Chapter Two that the theoretical shear stress necessary to deform a perfect crystal was many orders of magnitude greater than values commonly found in engineering materials. It is appropriate now to consider how high the cohesive strength σ_c might be in an ideally perfect crystal. Again, using a simple sinusoidal force-displacement law with a half period of $\lambda/2$, we see from Fig. 7.3 that the shape of the curve may be approximated by

$$\sigma = \sigma_c \sin \frac{\pi x}{\lambda/2} \qquad (7-1)$$

where σ reflects the tensile force necessary to pull atoms apart. For small atom displacements, Eq. 7-1 reduces to

$$\sigma = \sigma_c \frac{2\pi x}{\lambda} \qquad (7-2)$$

and the slope of the curve in this region becomes

$$\frac{d\sigma}{dx} = \frac{2\sigma_c \pi}{\lambda} \qquad (7-3)$$

Since Hooke's law (Eq. 1-7) applies to this region as well, the slope of the curve also may be given by

$$E = \frac{\text{stress}}{\text{strain}} = \frac{\sigma}{x/a_0} \qquad (7-4)$$

where

$$a_0 = \text{equilibrium atomic separation}$$

$$E = \text{modulus of elasticity}$$

FIGURE 7.2 (*a*) Fractured T-2 tanker, the S.S. *Schenectady*, which failed in 1941.[4] (Reprinted with permission of Earl R. Parker, *Brittle Behavior of Engineering Structures*, National Academy of Sciences, National Research Council, John Wiley & Sons, Inc., New York, 1957.) (*b*) Fractured oil barge, *Martha R. Ingram*, which failed on January 10, 1972. (With permission from the New York Times.)

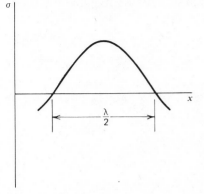

FIGURE 7.3 Simplified force versus atom displacement relationship.

Upon differentiation

$$\frac{d\sigma}{dx} = E / a_0 \qquad (7\text{-}5)$$

By combining Eqs. 7-3 and 7-5 and solving for σ_c

$$\sigma_c = \frac{E\lambda}{2\pi a_0} \qquad (7\text{-}6)$$

If we let $a_0 \approx \lambda/2$, a reasonably accurate assumption, then

$$\sigma_c \approx E / \pi \qquad (7\text{-}7)$$

It is apparent from Table 7.1 that some materials have the potential for withstanding extremely high stresses before fracture.

TABLE 7.1   Maximum Strengths in Solids[3]

Material	σ_f GPa	σ_f (psi × 10⁶)	E GPa	E (psi × 10⁶)	E/σ_f
Silica fibers	24.1	(3.5)	97.1	(14.1)	4
Iron whisker	13.1	(1.91)	295.2	(42.9)	23
Silicon whisker	6.47	(0.94)	165.7	(24.1)	26
Alumina whisker	15.2	(2.21)	496.2	(72.2)	33
Ausformed steel	3.14	(0.46)	200.1	(29.1)	64
Piano wire	2.75	(0.40)	200.1	(29.1)	73

If the energetics of the fracture process are considered, the fracture work done per unit area during fracture is given by

$$\text{fracture work} = \int_0^{\lambda/2} \sigma_c \sin \frac{\pi x}{\lambda/2}\, dx = \sigma_c \frac{\lambda}{\pi} \tag{7-8}$$

If all this work is set equal to the energy required to form two new fracture surfaces 2γ we may substitute for λ and show from Eq. 7-6 that

$$\sigma_c = \sqrt{\frac{E\gamma}{a_0}} \tag{7-9}$$

EXAMPLE 1 What is the cohesive strength of fused silica? Using a value of 1750 mJ/m² for the estimated surface energy in fused silica, 1.6 Å for the equilibrium Si-O atomic separation, and 69 to 76 GPa for the elastic modulus, the cohesive strength is found from Eq. 7-9 to be approximately 28 GPa, in agreement with experimental results from carefully prepared specimens (see Table 7.1). Note that a comparable value could have been computed from Eq. 7-7 had the modulus been the only known quantity.

Regardless of the equation used to obtain σ_c, the problem discussed in Chapter Two reappears—it is necessary to explain not the great strength of solids, but their weakness.

7.3 DEFECT POPULATION IN SOLIDS

Materials possess low fracture strengths relative to their theoretical capacity because most materials deform plastically at much lower stress levels and eventually fail by an accumulation of this irreversible damage. In addition, components and structures are not perfect. They contain myriad material defects (such as pores, slag particles, inclusions, and brittle particles), manufacturing flaws (scratches, gouges, weld torch arc strikes, weld undercutting and machining marks), and design defects (such as excessive stress concentrations resulting from inadequate fillet radii and discontinuous changes in section size). For example, the Duplessis Bridge failure in Quebec, Canada, in 1951 was traced to a preexistent crack in the steel superstructure.[4] In fact, when a crack that had been sighted before the collapse was examined, *paint* was found on the fracture surface near the crack origin. Certainly, this crack had to be present some time before failure.

In light of such findings, is it not reasonable and even conservative to assume that an engineering component will fail as a consequence of preexistent defects and that this hypothesis provides the basis for fracture control design planning? To a first approximation, then, the problem reduces to one of statistics. How many defects are present in the component or structure, how big are they, and where are they located with respect to the highly stressed portions of the part? Certainly, component size should have some bearing on the propensity for premature failure if for no other reason than the fact that larger pieces of material should contain more defects than smaller ones. Indeed, Leonardo da Vinci used the simple wire testing apparatus shown in Fig. 7.4 400 years ago to demonstrate that short iron wires were stronger than long sections. In one of his manuscripts we find the following passage:[5]

"The object of this test is to find the load an iron wire can carry. Attach an iron wire 2 braccia [about 1.3 m] long to something that will firmly support it, then attach a basket or any similar container to the wire and feed into

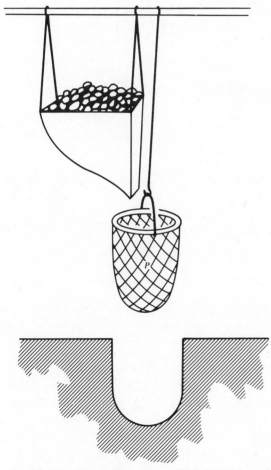

FIGURE 7.4 Sketch from the notebook of Leonardo da Vinci illustrating tensile test apparatus for iron wires.

the basket some fine sand through a small hole placed at the end of a hopper. A spring is fixed so that it will close the hole as soon as the wire breaks. The basket is not upset while falling, since it falls through a very short distance. The weight of sand and the location of the fracture of the wire are to be recorded. The test is repeated several times to check the results. Then a wire of one-half the previous length is tested and the additional weight it carries is recorded; then a wire of one-fourth length is tested and so forth, noting each time the ultimate strength and the location of the fracture."

The fact that Leonardo repeated his experiments several times to verify his results reflects his concern with the statistical nature of the problem. Modern research has also demonstrated the so-called size effect, wherein larger sections often possess fracture properties inferior to smaller sections. As we will see in Chapter Eight, the probability argument is not the complete explanation of the size effect in fracture.

7.4 THE STRESS CONCENTRATION FACTOR

By analyzing a plate containing an elliptical hole, Inglis[6] was able to show that the applied stress σ_a was magnified at the ends of the major axis of the ellipse (Fig. 7.5) so that

$$\frac{\sigma_{max}}{\sigma_a} = 1 + \frac{2a}{b} \tag{7-10}$$

where

σ_{max} = maximum stress at the end of the major axis

σ_a = applied stress applied normal to the major axis

a = half major axis

b = half minor axis

Since the radius of curvature ρ at the end of the ellipse is given by

$$\rho = \frac{b^2}{a} \tag{7-11}$$

Eqs. 7-10 and 7-11 may be combined so that

$$\sigma_{max} = \sigma_a \left[1 + 2\sqrt{a/\rho} \, \right] \tag{7-12}$$

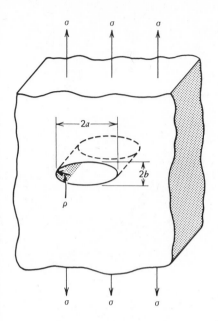

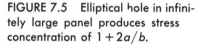

FIGURE 7.5 Elliptical hole in infinitely large panel produces stress concentration of $1 + 2a/b$.

In most cases $a \gg \rho$, therefore

$$\sigma_{max} \approx 2\sigma_a \sqrt{a/\rho} \qquad (7\text{-}13)$$

The term $2\sqrt{a/\rho}$ is defined as the stress concentration factor K_t and describes the effect of crack geometry on the local crack tip stress level. Many textbooks and standard handbooks describe stress concentrations in components with a wide range of crack configurations; a few are listed in the bibliography.[7,8] Although the exact formulations vary from one case to another, they all reflect the fact that K_t increases with increasing crack length and decreasing crack radius. Therefore, all cracks, if present, should be kept as small as possible. One way to accomplish this is by periodic inspection and replacement of components that possess cracks of dangerous length. Alternately, once a crack has developed, the relative severity of the stress concentration can be reduced by drilling a hole through the crack tip. In this way, ρ is increased from a curvature associated with a natural, sharp crack tip to that of the hole radius. In fact, it was suggested by Wilfred Jordan that the crack in the Liberty Bell "might have been stopped by boring a very small hole through the metal a short distance beyond the former termination of the first crack [However], it was never done, with the result that the original fracture, known to millions, has more than doubled in length and has extended up and around the shoulder of the Bell."[9]

The latter procedure is used occasionally in engineering practice today but should be employed with caution. Obviously, a crack is still present after drilling

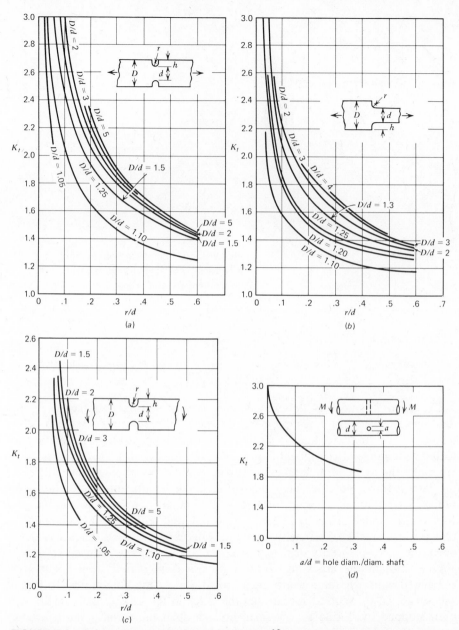

FIGURE 7.6 Selected stress concentration factors:[10] (*a*) Axial loading of notched bar; (*b*) axial loading of bar with fillet; (*c*) bending of notched bar; (*d*) bending of shaft with transverse hole; (*e*) axial loading of bar with transverse hole; (*f*) straight portion of shaft keyway in torsion. (From *Metals Engineering; Design* by American Society of Mechanical Engineers. Copyright © 1953 by the American Society of Mechanical Engineers. Used with permission of McGraw-Hill Book Company.)

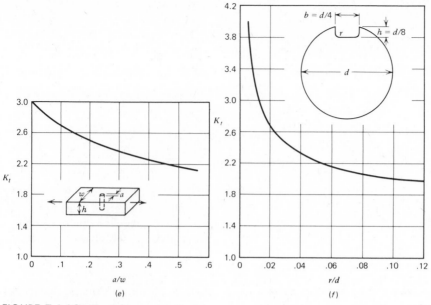

FIGURE 7.6 (Continued)

and may continue to grow beyond the hole after possible reinitiation. Also, field failures have occurred earlier than expected, simply because the blunting hole was introduced *behind* the crack tip, and the sharp crack tip radius was not eliminated.

Stress concentrations also may be defined for component configuration changes such as those associated with section size changes. As shown in Fig. 7.6, K_t increases for a number of design configurations whenever there is a large change in cross-sectional area and/or where the associated fillet radius is small. For every book written on the analysis of stress concentrations no doubt many others could be filled with case histories of component failures attributable to either ignorance or neglect of these same factors.

The completely elastic stress-strain material response, denoted in Chapter One as Type I behavior, is affected greatly by the presence of stress concentrations within the sample. As shown schematically in Fig. 7.7, the maximum stress and strain level that any component may support decreases with increasing K_t. This explains, in part, why many ceramic materials are much weaker in tension than in compression. The higher the stress concentration, the easier it is to break the sample. This fact is put to good use regularly by glaziers, who first score and then bend glass plate to induce fracture with minimal effort and along a desired path.

Fortunately, stress concentrations in most materials will not result in the escalation of the local crack tip stress to dangerously high levels. Instead, this

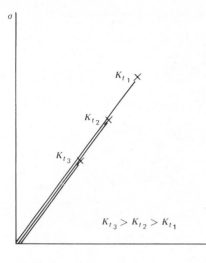

FIGURE 7.7 Effect of stress concentration K_t on allowable stress and strain in completely elastic material.

In the figure: σ (vertical axis), ϵ (horizontal axis), K_{t_1}, K_{t_2}, K_{t_3}, and $K_{t_3} > K_{t_2} > K_{t_1}$.

potentially damaging stress elevation is avoided by plastic deformation processes in the highly stressed crack tip region. As a result, the local stress does not greatly exceed the material's yield strength level as the crack tip blunts, thereby reducing the severity of the stress concentration. The ability of a component to plastically deform in the vicinity of a crack tip is the saving grace of countless engineering structures.

7.5 NOTCH STRENGTHENING

When an appreciable amount of plastic deformation is possible, an interesting turn of events may occur with regard to the fracture behavior of notched components. We saw in Chapter One that plastic constraint is developed in the necked region of a tensile bar as a result of a triaxial stress state; the unnecked regions of the sample experience a lower true stress than the necked section and, therefore, restrict the lateral contraction of the material in the neck. Similar stress conditions exist in the vicinity of a notch in a round bar. When the net section stress reaches the yield strength level, the material in the reduced section attempts to stretch plastically in the direction parallel to the loading axis. Since conservation of volume is central to the plastic deformation process, the notch root material seeks to contract also, but is constrained by the bulk of the sample still experiencing an elastic stress. The development of tensile stresses in the other two principal directions—the constraining stresses—makes it necessary to raise the axial stress to initiate plastic deformation. The deeper the notch, the greater is the plastic constraint and the higher the axial stress must be to deform the sample. Consequently, the yield strength of a notched sample may be *greater* than the yield strength found in a smooth bar tensile test. The data shown in Table 7.2 demonstrate the "notch strengthening" effect in 1018 steel bars, notched to reduce the cross-sectional area by up to 70%.

TABLE 7.2 Notch Strengthening in 1018 Steel

Reduction of Area in Notched Sample	Yield Strength, $\dfrac{\text{Notched Bar}}{\text{Smooth Bar}}$
0	1.00
20	1.22
30	1.36
40	1.45
50	1.64
60	1.85
70	2.00

A laboratory demonstration is suggested to illustrate a seemingly contradictory test response in two different materials. First, austenitize and quench a high-strength steel, such as AISI 4340, to produce an untempered martensite structure, and then perform a series of notched tensile tests. You will note that the net section stress *decreases* with increasing notch depth because of the increasing magnitude of the stress concentration factor. Now conduct notch tests with a ductile material such as a low carbon steel or aluminum alloy. In this case, note that the net section stress will *increase* with increasing notch depth as a result of the increased plastic constraint. In this manner, you may prove to yourself that materials with limited deformation capacity will *notch weaken*, and highly ductile materials will *notch strengthen*.

EXAMPLE 2 Two 0.5 cm diameter rods of 1020 steel ($\sigma_{ts}=395$ MPa) are to be joined with a silver braze alloy 0.025 cm thick ($\sigma_{ys}=145$ MPa) to produce one long rod. What will be the ultimate strength of this composite? The response of this bar may be equated to that of a notched rod of homogeneous material. In this instance, preferential yielding in the weaker braze material would be counterbalanced by a constraining triaxial stress field similar to that found in a notched bar of homogeneous material. As such, the strength of the joint will depend to a great extent on the geometry of the joint. Specifically, it would be expected that braze joint constraint would increase with increasing rod diameter and decreasing joint thickness. The experimental results by Moffatt and Wulff[11] (Fig. 7.8) reflect the importance of these two geometrical variables on the composite strength $\bar{\sigma}$. Accordingly, $\bar{\sigma}$ is found to be approximately 345 MPa.

Two factors need to be emphasized when discussing the observed notch strengthening effect. First, even though the notched component may have a

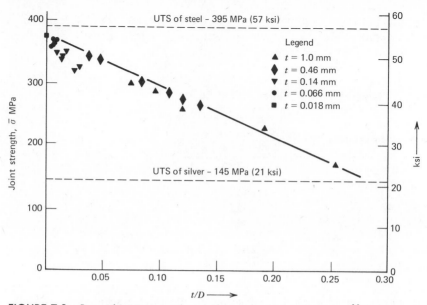

FIGURE 7.8   Brazed joint strength as a function of joint geometry.[11] (Reprinted with permission from Metallurgical Society of AIME.)

higher net section stress, it requires a *lower* load for fracture than does the smooth sample when based on the *gross* cross-sectional area. I trust that this should temper the enthusiasm of any overzealous student who might otherwise race about, hacksaw in hand, with the intent of "notch strengthening" all the bridges in town. Second, there is a limit to the amount of notch strengthening that a material may exhibit. From theory of plasticity considerations, it is shown that the net section stress in a deformable material may be elevated to $2\frac{1}{2}$ to 3 times the smooth bar yield strength value. It would appear, then, that the brazed joint system described in the previous example represented an optimum matching of material properties. Using a higher strength steel with the same braze alloy would not have made the joint system stronger.

7.6   EXTERNAL VARIABLES AFFECTING FRACTURE

As mentioned in Section 7.4, the damaging effect of an existing stress concentration depends strongly on the material's ability to yield locally and thereby blunt the crack tip. Consequently, anything that affects the deformation capacity of the material will affect its fracture characteristics as well. Obviously, any metallurgical strengthening mechanism designed to increase yield strength will simultaneously suppress plastic deformation capacity and the ability of the material to blunt the crack tip. For any given material, there are, in addition, a trilogy of external test conditions that contribute to premature fracture: notches, reduced temperatures, and high strain rates (Fig. 7.9). The more the flow curve is raised for a given material, the more likely brittle fracture becomes. As we saw in Section 7.5, the presence of a notch acts to plastically constrain the material in the reduced section and serves to elevate the stress necessary for yielding. Likewise, lowering the test temperature and/or increasing strain rate will elevate the yield strength (the magnitude of this change depends on the material). For example, yield strength temperature and strain rate sensitivity varies with the Peierls stress component of the overall yield strength of the material (Section 2.3). In BCC alloys, such as ferritic steels, the Peierls stress increases rapidly with decreasing temperature, thereby causing these materials to exhibit a sharp rise in yield strength at low temperatures. This can and often does precipitate premature fracture in structures fabricated from these materials. In other materials, such as FCC alloys, the Peierls stress component is small. Consequently, yield strength temperature sensitivity is small, and these materials may be used under cryogenic conditions, provided they possess other required mechanical properties.

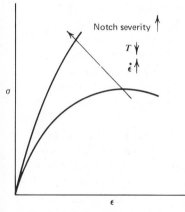

FIGURE 7.9   Effect of temperature, strain rate, and plastic constraint on flow curve.

7.7 NOMENCLATURE OF THE FRACTURE PROCESS

Many words and phrases have been used to characterize failure processes in engineering materials. Since these terms are born of different disciplines, each having its own relatively unique jargon, confusion exists along with some incorrect usage. In an attempt to simplify and clarify this situation, I have found it convenient to describe the fracture of a test sample or engineering component in terms of three general characteristics: energy of fracture, macroscopic fracture path and texture, and microscopic fracture mechanisms.

7.7.1 Energy to Break

As discussed in Chapter One, the toughness of a given material is a measure of the energy absorbed before and during the fracture process. The area under the tensile stress-strain curve would provide a measure of toughness where

$$energy = \int_0^{\epsilon_f} \sigma \, d\epsilon \qquad (1\text{-}25)$$

If the energy is high, (such as for Curve C in Fig. 1.13) the material is said to be tough or possess high fracture toughness.

Conversely, if the energy were low, the material (e.g., Curves A and B in Fig. 1.13) would be described as brittle. In notched samples, a determination of toughness is more complicated, and will be considered in more detail in Chapter Eight. For the time being, the relative toughness or brittleness of a material may be estimated by noting the extent of plasticity surrounding the crack tip. Since the stress concentration at a crack tip often will elevate the applied stress above the level necessary for irreversible plastic deformation, a zone of plastically deformed material will be found at the crack tip, embedded within an elastically deformed media. Since much more energy is dissipated during plastic flow than during elastic deformation, the toughness of a notched sample should increase with the potential volume of the crack tip plastic zone. As shown in Fig. 7.10, when the plastic zone is small just before failure, the overall toughness level of the sample is low and the material is classified as *brittle*. On the other hand, were plasticity to extend far from the crack tip to encompass the specimen's unbroken ligament, the energy to break would be high and the material would be *tough*.

An additional measure of toughness in notched samples is that reflected by the notched to smooth bar tensile strength ratio. A low ratio reflects considerable notch sensitivity and low toughness, and a high ratio—say, greater than two —depicts high toughness behavior and minimal notch sensitivity. As will be discussed in Chapter Eleven, the notch test can reveal metallurgical embrittlement, with brittle behavior becoming apparent through changes in the notched to smooth bar tensile strength ratio as a function of heat treatment.

7.7.2 Macroscopic Fracture Mode and Texture

The macroscopic crack path may also provide useful information about the toughness exhibited by a component prior to failure. This is particularly true in the case of sheet or plate type components where the level of toughness can be related to the relative amounts of flat and slant fracture (Fig. 7.11). For a given material, the fracture toughness values would be higher in the thin sample exhibiting a full slant fracture condition than the thick sample exhibiting a

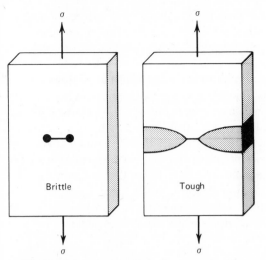

FIGURE 7.10 Extent of plastic zone development (shaded region) at fracture for brittle and tough fractures.

FIGURE 7.11 Fractured surfaces of aluminum test specimens revealing flat and slant-type failure. Toughness level increases with increasing relative amount of slant fracture.

completely flat fracture mode. Correspondingly, samples with mixed-mode (i.e., part slant and part flat fracture) would reflect an intermediate toughness level. The significance of the macroscopic fracture mode transition is discussed in greater detail in Chapter Eight.

When we approach the task of describing the character of the fracture surface texture, we find a plethora of terms to excite a sympathetic image in the mind of the reader. What do you visualize when you hear the words "rock-candy like," "woody," "glassy"? If you pictured fractured surfaces consisting of small shiny angular facets, a rough splintered surface, and a smooth featureless surface, respectively, the terms were apt descriptions of the fracture texture. As unscientific as these words may sound, they perform their descriptive function successfully. However, although such terms may create a useful mental picture of the macroscopic texture of the fracture surface, they do not provide a precise description of crack propagation through the material's microstructure.

7.7.3 Microscopic Fracture Mechanisms

As recently as 10 years ago, the light microscope was the tool most often used in the microscopic examination of fracture processes. Because of the very shallow depth of focus, examination of the fracture surface was not possible except at very low magnifications. Consequently, the fracture surface analysis procedure entailed the examination of a metallographic section containing a profile of the fracture surface. Using this technique, it was possible to obtain important information about the fracture path. For example, by comparing the path of the fracture with the metallographic grain structure, it was possible to determine whether the failure was of transcrystalline or intercrystalline nature (Fig. 7.12). This point is often more easily clarified when secondary cracks are present in the sectioned component, thereby revealing profiles of mating fracture surfaces. Since the condition of the profile edge is critical for proper failure analysis, precautions are often taken to preserve the sharpness of the fracture profile. To this end, fracture surfaces usually are plated with nickel to protect the specimen edge from rounding caused by the metallographic polishing procedure. The most widely used procedures for metallographic specimen preparation are described in a standard metallographic text.[12]

The understanding of fracture mechanisms in materials increased dramatically when the electron microscope was developed. Because its depth of field and resolution were superior to those of the light microscope, many topographical fracture surface features were observed for the first time. Many of these markings have since been applied to current theories of fracture. Until recently, much of the fractographic work had been conducted on transmission electron microscopes (TEM). Since the penetrating power of electrons is quite limited, fracture surface observations in a TEM require the preparation of a replica of the fracture surface that allows transmission of the high-energy electron beam.

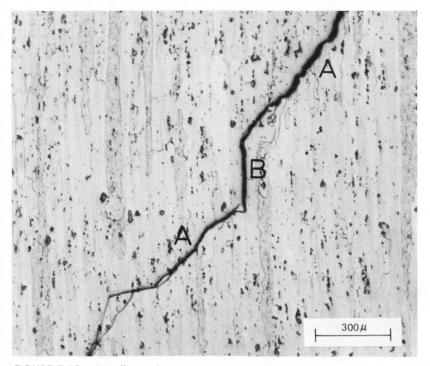

FIGURE 7.12 Metallographic section revealing transcrystalline crack propagation at (A) and intercrystalline crack growth at (B).

During the past few years, encouraging progress also has been made in the utilization of scanning electron microscopy in failure analysis. A major advantage of the scanning electron microscope (SEM) for some examinations is that the fractured sample may be viewed directly in the instrument, thereby obviating the need for replica preparation. When it is not possible to cut the fractured component to fit into the viewing chamber, replicas must be used instead. At present, the resolution capability of scanning electron microscopes is less than that of transmission electron microscopes. It is anticipated that later models of the SEM will be more competitive with respect to this specification, and both instruments may soon be needed in a laboratory committed to failure analysis.

Before one can proceed with an interpretation of fracture surface markings, it is necessary to review replication techniques and electron image contrast effects. To this end, the reader is referred to Appendix A. In addition, the *Electron Fractography Handbook*[13] and Volume 9 of the ASM *Metals Handbook*[14] are two excellent sources that contain both discussions of techniques and thousands of electron fractographs. For the purposes of the present discussion, it is important only to recognize that the most commonly employed replication technique for the TEM leads to a reversal in the "apparent" fracture surface morphology. That is, electron images may suggest that the fracture surface consists of mounds or hillocks when in reality, it is composed of troughs or depressions. *Everything that looks up is really down and vice versa.* On the other hand, SEM images do not possess this height deception. For completeness and comparative purposes, all microscopic fracture mechanisms discussed in this book are described with both TEM and SEM electron images.

7.7.3.1 MICROVOID COALESCENCE

An important fracture mechanism, common to most materials regardless of fundamental differences in crystal structure and alloy composition, is microvoid coalescence. In fact, amorphous polymers also experience failure by this mechanism. It is believed that stress-induced fracture of brittle particles, particle-matrix interface failure, and, perhaps, complex dislocation interactions lead to the formation of microcracks or pores within the stressed component. These mechanically induced micropores should not be confused with preexistent microporosity sometimes present as a result of casting or powder sintering procedures. At increasing stress levels, the voids grow and finally coalesce into a broad crack front. When this growing flaw reaches critical dimensions, total failure of the component results. Even after the point of instability, the unstable crack often grows by a repetitive process of void formation and subsequent coalescence with the main crack front.

Three distinct processes for void formation and coalescence can be envisioned, depending on the state of stress.[15] Under simple uniaxial loading conditions, the microvoids will tend to form in association with fractured particles and/or interfaces and grow out in a plane generally normal to the stress axis. (This occurs in the fibrous zone of the cup-cone failure shown in Section 1.2.2.4.) The resulting micron-sized "equiaxed dimples" are generally spherical, as shown in Fig. 7.13a,b. Since the growth and coalescence of these voids involves a plastic deformation process, it is to be expected that total fracture energy should be related in some fashion to the size of these dimples. In fact, it has been shown in laboratory experiments that fracture energy does increase with increasing depth and width of the observed dimples.[16, 17]

When failure is influenced by shear stresses, the voids that nucleate in the manner cited above grow and subsequently coalesce along planes of maximum shear stress. Consequently, these voids tend to be elongated and result in the formation of parabolic depressions on the fracture surface, as shown in Fig. 7.14a,b (such voids are found in the shear walls of the cup-cone failure). If one were to compare the orientation of these "elongated dimples" from matching fracture faces, one would find that the voids are elongated in the direction of the shear stresses and point in opposite directions on the two matching surfaces.

Finally, when the stress state is one of combined tension and bending, the resulting tearing process produces "elongated dimples," which can appear on

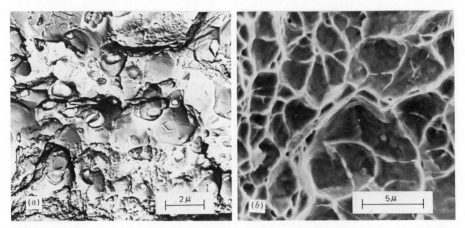

FIGURE 7.13 Microvoid coalescence under tensile loading, which leads to "equiaxed dimple" morphology; (a) TEM fractograph shows "dimples" as mounds, (b) SEM fractograph shows "dimples" as true depressions.

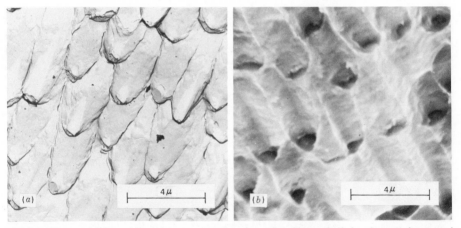

FIGURE 7.14 Microvoid coalescence under shear loading, which leads to "elongated dimple" morphology; (a) TEM fractograph shows "dimples" as raised parabolas, (b) SEM fractograph shows "dimples" as true elongated troughs.

gross planes normal to the direction of loading. The basic difference between these "elongated dimples" and those produced by shear is that the tear dimples point in the same direction on both halves of the fracture surface. It is important to note that these dimples point back toward the crack origin. Consequently, when viewing a replica that contains impressions of tear dimples, the dimples may be used to direct the viewer to the crack origin. A schematic diagram illustrating the effect of stress state on microvoid morphology is presented in Fig. 7.15.

It may be desirable to determine the chemical composition of the particle responsible for the initiation of the voids. By selected area diffraction in the TEM of particles extracted from replicas or by X-ray detector instrumentation in the SEM, it often is possible to identify the composition of particles responsible for microvoid initiation. With this information, it may be possible to select a

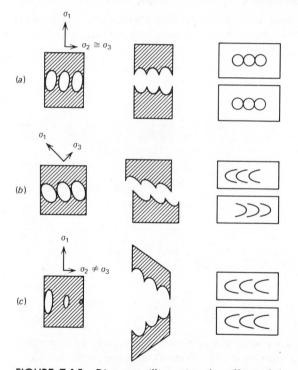

FIGURE 7.15 Diagrams illustrating the effect of three stress states on microvoid morphology; (a) tensile stresses produce equiaxed microvoids; (b) pure shear stresses generate microvoids elongated in the shearing direction (voids point in opposite directions on the two fracture surfaces); (c) tearing associated with nonuniform stress, which produces elongated dimples on both fracture surfaces that point back to crack origin.[15] (Reprinted with permission from the American Society of Mechanical Engineers.)

different heat treating procedure and/or select an alloy of higher purity so as to suppress the void formation initiation process.

7.7.3.2 CLEAVAGE

The process of cleavage involves transcrystalline fracture along specific crystallographic planes and is usually associated with low-energy fracture. This mechanism is observed in BCC, HCP, and ionic and covalently bonded crystals, but occurs in FCC metals only when they are subjected to severe environmental conditions. Cleavage facets are typically flat, although they may reflect a parallel plateau and ledge morphology (Fig. 7.16a). Often these cleavage steps appear as "river patterns" wherein fine steps are seen to merge progressively into larger ones (Fig. 7.16b). It is generally believed that the "flow" of the "river pattern" is in the direction of microscopic crack propagation (from right to left in Fig. 7.16b). The sudden appearance of the "river pattern" in Fig. 7.16b was probably brought on by the movement of a cleavage crack across a high-angle grain boundary, where the splintering of the crack plane represents an accommodation process as the advancing crack reoriented in search of cleavage planes in the new grain. It is also possible that the cleavage crack traversed a low-angle twist boundary, and the cleavage steps were produced by the intersection of the cleavage crack with screw dislocations.[18]

In some materials, such as ferritic steel alloys, the temperature and strain rate regime necessary for cleavage formation is similar to that required to activate deformation twinning (see Chapter Four). Fine-scale height elevations (so-called tongues) seen in Fig. 7.17a,b provide proof of deformation twinning during or immediately preceding failure. In BCC iron, etch pit studies have verified that

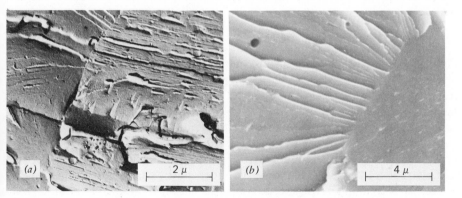

FIGURE 7.16 Cleavage fracture in a low carbon steel. Note parallel plateau and ledge morphology and river patterns reflecting crack propagation along many parallel cleavage planes; (a) TEM, (b) SEM.

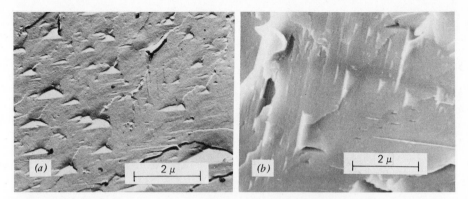

FIGURE 7.17 Cleavage facets revealing fine-scale height elevations caused by localized deflection of the cleavage crack along twin matrix interfaces; (a) TEM, (b) SEM.

these fracture surfaces consist of {100} cleavage facets and {112} tongues; the latter representing failure along twin matrix interfaces.

Little information may be obtained from cleavage facets for use in failure analyses. However, one may learn something about the phase responsible for failure by noting the shape of the facet and comparing it to the morphology of different phases in the alloy. Furthermore, in materials that undergo a fracture mechanism transition (e.g., void coalescence to cleavage failure), it is possible to relate the presence of the cleavage mechanism to a general set of external conditions. In most mild steel alloys (which undergo the above fracture mechanism transition), the observation of cleavage indicates that the component was subjected to some combination of low temperature, high strain rate, and/or a high tensile triaxial stress condition. This point is discussed further in Chapter Nine.

7.7.3.3 INTERGRANULAR FRACTURE

Perhaps the most readily recognizable microscopic fracture mechanism is that of intergranular failure wherein the crack prefers to follow grain surfaces. The resulting fracture surface morphology (Fig. 7.18a,b) immediately suggests the three-dimensional character of the grains that comprise the alloy microstructure. Intergranular failure can be induced by several factors. For example, it will occur whenever the material is heat treated improperly and develops a brittle phase distributed as a grain boundary film (see Chapter Eleven). Also, if the material has an insufficient number of independent slip systems (see Chapter Three) to accommodate plastic deformation between contiguous grains, grain boundary separation may occur. Finally, if the grain boundary regions are weakened by high-temperature deformation processes, or intrusion of aggravating liquids and gases (Chapter Eleven), intergranular failure may occur.

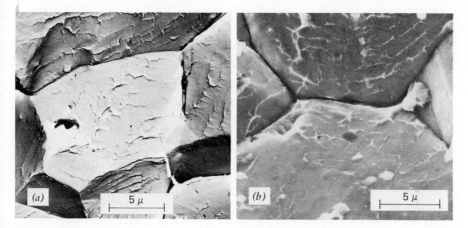

FIGURE 7.18 Intergranular fracture in maraging steel tested in a gaseous hydrogen environment; (a) TEM, (b) SEM.

REFERENCES

1. *Scientific American*, **120** (February 1, 1919), p. 99.
2. *Engineering News-Record*, **82**(20) May 15, 1919, p. 974.
3. W. J. McGregor Tegart, *Elements of Mechanical Metallurgy*, Macmillan, New York, 1966.
4. E. R. Parker, *Brittle Behavior of Engineering Structures*, John Wiley, New York, 1957.
5. W. B. Parsons, *Engineers and Engineering in the Renaissance*, Williams and Wilkens, Baltimore, 1939, p. 72.
6. C. E. Inglis, *Proceedings*, Institute of Naval Architects, **55**, 1913, p. 219
7. R. E. Peterson, *Stress Concentration Design Factors*, John Wiley, New York, 1953.
8. H. Neuber, *Kerbspannungslehre*, Springer, Berlin; English translation available from Edwards Bros., Ann Arbor, Mich., 1959.
9. W. Jordan, *Proceedings*, American Numismatical and Antiquarian Society, **27**, 1915, p. 109.
10. *ASME Handbook—Metals Engineering—Design*, McGraw-Hill, New York, 1953.
11. W. Moffatt and J. Wulff, *J. Met.*, April 1957, p. 440.
12. G. L. Kehl, *Principles of Metallographic Laboratory Practice*, McGraw-Hill, New York, 1949.
13. A. Phillips, V. Kerlins and B. V. Whiteson, *Electron Fractography Handbook*, AFML TDR-64-416, WPAFB, Ohio, 1965.
14. *Metals Handbook*, Volume 9, American Society of Metals, 1974.
15. C. D. Beachem, *Trans.*, ASME, *J. Basic Eng.*, Series D, **87**, 1965, p. 299.
16. A. J. Birkle, R. P. Wei and G. E. Pellissier, *Trans.* ASM, **59**, 1966, p. 981.
17. D. E. Passoja and D. C. Hill, *Met. Trans.*, **5**, 1974, p. 1851.
18. J. J. Gilman, *Trans. Met. Soc.* AIME, **212**, 1958, p. 310.

PROBLEMS

7.1 For a given engineering alloy, it is found that the notched tensile strength decreased with increasing notch depth (assuming a constant notch root radius) to a point beyond which σ_{net} began to increase. Explain this behavior.

7.2 Demonstrate to yourself that the intersection of a cleavage crack with a screw dislocation will produce a step on the fracture surface and that no step will result from intersection of the crack with an edge dislocation.

7.3 For most of the configurations shown in Fig. 7.6, K_t increases toward infinity as the notch root radius to minor diameter ratio approaches zero. In practice, such high K_t values are never experienced. Why?

7.4 Since cleavage in BCC crystals is preferred along (100) planes, consider what might happen to the fracture energy of a randomly oriented polycrystalline sample if the grain size were reduced several fold. (Trace the crack path across the grain boundaries.)

7.5 Might the molasses tank that fractured in Boston have remained intact had it been erected in the tropical zone?

7.6 It has been shown that carefully prepared rods of silicon can withstand extremely high stresses before fracturing. When failure does occur, the sample explodes into powder. Why?

CHAPTER
EIGHT
ELEMENTS
OF FRACTURE
MECHANICS

As outlined in the previous chapter, the fracture behavior of a given structure or material will depend on stress level, presence of a flaw, material properties, and the mechanism(s) by which the fracture proceeds to completion. The purpose of this chapter is to develop quantitative relationships between some of these factors. With knowledge of these relationships, fracture phenomena may be better understood and design engineers more equipped to anticipate and thus prevent structural deficiencies. In addition to several sample problems given in this chapter, other real case histories are discussed in Chapter Fourteen which bear upon the material discussed below.

8.1 GRIFFITH CRACK THEORY

The quantitative relationships that engineers and scientists use today in determining the fracture of cracked solids were initially stated some 50 years ago by A. A. Griffith.[1] Griffith noted that when a crack is introduced to a stressed plate of elastic material, a balance must be struck between the decrease in potential energy (related to the release of stored elastic energy and work done by movement of the external loads) and the increase in surface energy resulting from the presence of the crack. Likewise, an existing crack would grow by some increment if the necessary additional surface energy were supplied by the system. This "surface energy" arises from the fact that there is a nonequilibrium configuration of nearest neighbor atoms at any surface in a solid. For the configuration seen in Fig. 8.1, Griffith estimated the surface energy term to be the product of the total crack surface area $(2a \cdot 2 \cdot t)$, and the specific surface energy γ_S, which has units of energy/unit area. He then used the stress analysis of Inglis[2] for the case of an infinitely large plate containing an elliptical crack and computed the decrease in potential energy of the cracked plate to be $(\pi\sigma^2 a^2 t)/E$. Hence, the change in potential energy of the plate associated with the introduction of a crack may be given by

$$U - U_0 = -\frac{\pi\sigma^2 a^2 t}{E} + 4at\gamma_S \qquad (8\text{-}1)$$

where

$$U = \text{potential energy of body with crack}$$

$$U_0 = \text{potential energy of body without crack}$$

$$\sigma = \text{applied stress}$$

$$a = \text{one-half crack length}$$

$$t = \text{thickness}$$

$$E = \text{modulus of elasticity}$$

$$\gamma_S = \text{specific surface energy}$$

By rewriting Eq. 8-1 in the form

$$U = 4at\gamma_S - \frac{\pi\sigma^2 a^2 t}{E} + U_0 \qquad (8\text{-}2)$$

and determining the condition of equilibrium by differentiating the potential energy U with respect to the crack length and setting equal to zero

$$\frac{\partial U}{\partial a} = 4t\gamma_S - \frac{2\pi\sigma^2 at}{E} = 0 \qquad (8\text{-}3)$$

$(\partial U_0 / \partial a = 0$, since U_0 accounts for the potential energy of the body without a crack and does not vary with crack length). Therefore

$$2\gamma_S = \frac{\pi\sigma^2 a}{E} \qquad (8\text{-}4)$$

which represents the equilibrium condition.

The nature of the equilibrium condition described by Eqs. 8-3 and 8-4 is

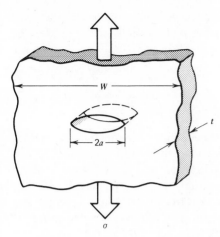

FIGURE 8.1 Through thickness crack in a large plate.

determined by the second derivative, $\partial^2 U / \partial a^2$. Since

$$\frac{\partial^2 U}{\partial a^2} = -\frac{2\pi\sigma^2 t}{E} \tag{8-5}$$

and is negative, the equilibrium condition described by Eq. 8-3 is unstable, and the crack will always grow.

Griffith rewrote Eq. 8-4 in the form

$$\sigma = \sqrt{\frac{2E\gamma_S}{\pi a}} \tag{8-6}$$

for the case of plane stress (biaxial stress conditions), and

$$\sigma = \sqrt{\frac{2E\gamma_S}{\pi a(1-\nu^2)}} \tag{8-7}$$

for the case of plane strain (triaxial stress conditions associated with the suppression of strains in one direction).

Since Poisson's ratio ν is approximately 0.25 to 0.33 for many materials, the difference in allowable stress level in a given material subjected to plane strain or plane stress conditions does not appear to be large. However, major differences do arise as will be discussed in Section 8.6.

It is important to recognize that the Griffith relationship was derived for an elastic material containing a very sharp crack. Although Eqs. 8-6 and 8-7 do not explicitly involve the crack tip radius ρ, as was the case for the stress concentration in Eq. 7-13, the radius is assumed to be very sharp. As such, the Griffith relationship, as written, should be considered necessary but not sufficient for failure. The crack tip radius also would have to be atomically sharp to raise the local stress above the cohesive strength.

8.1.1 Verification of the Griffith Relationship

The other half of Griffith's classic paper was devoted to experimentally confirming the accuracy of Eqs. 8-6 and 8-7. Thin round tubes and spherical bulbs of soda lime silica glass were deliberately scratched or cracked with a sharp instrument, annealed to eliminate any residual stresses associated with the cracking process, and fractured by internal pressure. By recording the crack size and stress at fracture for these glass samples, Griffith was able to compute values of $\sigma\sqrt{a}$ in the range of 0.25 to 0.28 MPa$\sqrt{m}$ (0.23 to 0.25 ksi$\sqrt{in.}$),* which correspond to values of $\sqrt{2\gamma_S E/\pi}$ (Eq. 8-6). From experimentally determined surface tension values γ_S for glass fibers between 745°C and 1110°C, a room-temperature value was obtained by extrapolation (risky business, but reasonable for a first approximation). By multiplying this value, 0.54 N/m, by the modulus of elasticity, 62 GPa, the value of $\sqrt{2\gamma_S E/\pi}$ determined from material properties was found to be 0.15 MPa$\sqrt{m}$. It should be recognized that the exceptional agreement between theoretical and experimental values may be somewhat fortuitous in light of some inaccuracies contained in the original development by Griffith.[1] Nevertheless, the Griffith equation for the case of ideally brittle solids is considered to be valid.

Since plastic deformation processes in amorphous glasses are very limited, the difference in surface energy and fracture energy values is not expected to be great. This is not true for metals and polymers, where the fracture energy is found to be several orders of magnitude greater than the surface energy of a given material. Orowan[3] recognized this fact and suggested that Eq. 8-6 be modified to include the energy of plastic deformation in the fracture process so that

$$\sigma = \sqrt{\frac{2E(\gamma_S + \gamma_P)}{\pi a}} = \sqrt{\frac{2E\gamma_S}{\pi a}\left(1 + \frac{\gamma_P}{\gamma_S}\right)} \tag{8-8a}$$

where γ_P = plastic deformation energy and $\gamma_P \gg \gamma_S$.

Under these conditions

$$\sigma \approx \sqrt{\frac{2E\gamma_S}{\pi a}\left(\frac{\gamma_P}{\gamma_S}\right)} \tag{8-8b}$$

*To convert from ksi$\sqrt{in.}$ to MPa$\sqrt{m}$ multiply by 1.099.

The applicability of Eqs. 8-6 or 8-8 in describing the fracture of real materials will depend on the sharpness of the crack and the relative amount of plastic deformation. The following relationship reveals these two factors to be related. By combining Eqs. 7-9 and 7-13 and letting $\sigma_{max} = \sigma_c$, we see that the applied stress σ_a for fracture will be

$$\sigma_a = \frac{1}{2}\sqrt{\frac{E\gamma_S}{a}\left(\frac{\rho}{a_0}\right)} \quad \text{or} \quad \sqrt{\frac{2E\gamma_S}{\pi a}\left(\frac{\pi\rho}{8a_0}\right)} \qquad (8\text{-}8c)$$

The similarity between Eqs. 8-8b and 8-8c is obvious and suggests a correlation between γ_P/γ_S and $\pi\rho/8a_0$; that is, plastic deformation can be related to a blunting process at the crack tip—ρ will increase with γ_P. From Eqs. 8-8b and 8-8c, it is seen that the Griffith relation (Eq. 8-6) is valid for sharp cracks with a tip radius in the range of $(8/\pi)a_0$. Equation 8-6 is believed to be applicable also where $\rho < (8/\pi)a_0$, since it would be unreasonable to expect the fracture stress to approach zero as the crack root radius became infinitely small. When $\rho > (8/\pi)a_0$, Eqs. 8-8b or 8-8c would control the failure condition where plastic deformation processes are involved.

At the same time, Irwin[4] also was considering the application of Griffith's relation to the case of materials capable of plastic deformation. Instead of developing an explicit relationship in terms of the energy sink terms, γ_S or $\gamma_S + \gamma_P$, Irwin chose to use the energy source term (i.e., the elastic energy per unit crack length increment $\partial U/\partial a$). Denoting $\partial U/\partial a$ as $\mathcal{G}$, Irwin showed that

$$\sigma = \sqrt{\frac{E\mathcal{G}}{\pi a}} \qquad (8\text{-}9)$$

which is one of the most important relationships in the literature of fracture mechanics. By comparison of Eqs. 8-8 and 8-9, it is seen that

$$\mathcal{G} = 2(\gamma_S + \gamma_P) \qquad (8\text{-}10)$$

At the point of instability, the elastic energy release rate $\mathcal{G}$ (also referred to as the crack driving force) reaches a critical value $\mathcal{G}_c$, whereupon fracture occurs. This critical elastic energy release rate may be interpreted as a material parameter and can be measured in the laboratory with sharply notched test specimens.

8.1.2 Energy Release Rate Analysis

In the previous section, the elastic energy release rate $\mathcal{G}$ was related to the release of strain energy and the work done by the boundary forces. The significance of these two terms will now be considered in greater detail. For an elastically loaded body containing a crack of length a (Fig. 8.2), the amount of

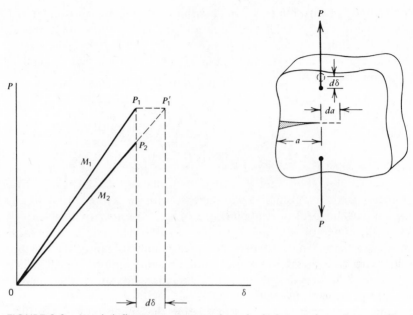

FIGURE 8.2 Load-deflection response of cracked plate such as shown in Fig. 8.1 for case where crack length increases by da. OP_2 corresponds to fixed grip condition while OP_1' corresponds to fixed load case.

stored elastic strain energy is given by

$$V = \tfrac{1}{2} P\delta \qquad \text{or} \qquad \tfrac{1}{2}\frac{P^2}{M_1} \tag{8-11}$$

where

$$V = \text{stored strain energy}$$

$$P = \text{applied load}$$

$$\delta = \text{load displacement}$$

$$M_1 = \text{body stiffness for crack length } a$$

If the crack extends by an amount da, the necessary additional surface energy is obtained from the work done by the external body forces $P\,d\delta$ and the release of strain energy dV.[5] As a result

$$\mathcal{G} = \frac{dU}{da} = P\frac{d\delta}{da} - \frac{dV}{da} \tag{8-12}$$

with the stiffness of the body decreasing to M_2. Whether the body was rigidly gripped such that incremental crack growth would result in a load drop from P_1 to P_2 or whether the load was fixed such that crack extension would result in an increase in δ by an amount $d\delta$, the stiffness of the plate M would decrease. For the fixed grip case, both P and M would decrease but the ratio P/M would remain the same, since from Fig. 8.2

$$\delta_1 = \delta_2 = \frac{P_1}{M_1} = \frac{P_2}{M_2} \tag{8-13}$$

The elastic energy release rate would be

$$\left(\frac{\partial U}{\partial a}\right)_\delta = \frac{1}{2}\left[\frac{2P}{M}\frac{\partial P}{\partial a} + P^2 \frac{\partial(1/M)}{\partial a}\right] \tag{8-14}$$

By differentiating Eq. 8-13 to obtain

$$\frac{1}{M}\frac{\partial P}{\partial a} + P\frac{\partial(1/M)}{\partial a} = 0 \tag{8-15}$$

and substituting the result into Eq. 8-14

$$\left(\frac{\partial U}{\partial a}\right)_\delta = -\frac{1}{2}P^2 \frac{\partial(1/M)}{\partial a} \tag{8-16a}$$

It may be shown[5] that under fixed load conditions

$$\left(\frac{\partial U}{\partial a}\right)_P = \frac{1}{2}P^2 \frac{\partial(1/M)}{\partial a} \tag{8-16b}$$

Note that in both conditions, the elastic energy release rate is the same (only the sign is reversed), reflecting the fact that $\mathcal{G}$ is independent of the type of load application (e.g., fixed grip, constant load, combinations of load change and displacement, and machine stiffness). At instability, then, the critical strain energy release rate is

$$\mathcal{G}_c = \frac{P_{max}^2}{2} \frac{\partial(1/M)}{\partial a} \tag{8-17}$$

where $1/M$ is the compliance of the cracked plate, which depends on the crack size. Once the compliance versus crack length relationship has been established for a given specimen configuration, $\mathcal{G}_c$ can be obtained by noting the load at fracture, provided the amount of plastic deformation at the crack tip is kept to a minimum.

8.2 STRESS ANALYSIS OF CRACKS

The fracture of flawed components also may be analyzed by a stress analysis based on concepts of elastic theory. By assuming solutions (Airy stress functions) to the equilibrium and compatibility requirements for various sample configurations and stress patterns,[6] Westergaard[7] was able to determine the nature of crack tip stress distributions. With some introductory knowledge of theory of elasticity and some familiarity with complex variable mathematical functions, Westergaard's paper can be digested properly by the reader and provides a useful insight into the derivation of the stress equations discussed below. For this reason, this paper is recommended for these readers with the necessary background who are interested in pursuing this point in greater depth.

The stress fields surrounding a crack tip can be divided into three major modes[8] of loading that involve different crack surface displacements, as shown in Fig. 8.3.

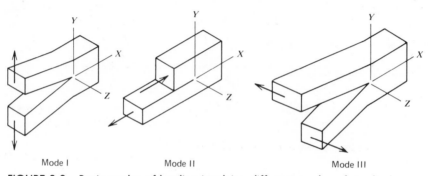

| Mode I | Mode II | Mode III |

FIGURE 8.3 Basic modes of loading involving different crack surface displacements.

Mode I   Opening or tensile mode, where the crack surfaces move directly apart.

Mode II   Sliding or in-plane shear mode, where the crack surfaces slide over one another in a direction perpendicular to the leading edge of the crack.

Mode III   Tearing or antiplane shear mode, where the crack surfaces move relative to one another and parallel to the leading edge of the crack.

Mode I loading is encountered in the overwhelming majority of actual engineering situations involving cracked components. Consequently, considerable attention has been given to both analytical and experimental methods designed to quantify Mode I stress-crack length relationships. Mode II is found less frequently and is of little engineering importance. One example of mixed Mode I–II loading involves axial loading (in the Y direction) of a crack inclined as a result of rotation about the Z axis (Fig. 8.4). Even in this instance, analytical methods[8] show the Mode I contribution to dominate the crack tip stress field when $\beta > 60$ degrees. Mode III may be regarded as a pure shear problem such as that involving a notched round bar in torsion.

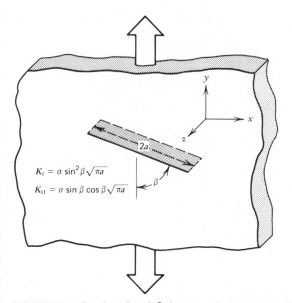

$$K_I = \sigma \sin^2 \beta \sqrt{\pi a}$$
$$K_{II} = \sigma \sin \beta \cos \beta \sqrt{\pi a}$$

FIGURE 8.4 Crack inclined β degrees about z axis. Mode I dominates when $\beta > 60°$.

For the notation shown in Fig. 8.5, the crack tip stresses are found to be

$$\sigma_y = \frac{K}{\sqrt{2\pi r}} \cos\frac{\theta}{2} \left[1 + \sin\frac{\theta}{2} \sin\frac{3\theta}{2} \right]$$

$$\sigma_x = \frac{K}{\sqrt{2\pi r}} \cos\frac{\theta}{2} \left[1 - \sin\frac{\theta}{2} \sin\frac{3\theta}{2} \right] \qquad (8\text{-}18)$$

$$\tau_{xy} = \frac{K}{\sqrt{2\pi r}} \left[\sin\frac{\theta}{2} \cos\frac{\theta}{2} \cos\frac{3\theta}{2} \right]$$

It is apparent from Eq. 8-18 that these local stresses could rise to extremely high levels as r approaches zero. As pointed out earlier in the chapter, this circumstance is precluded by the onset of plastic deformation at the crack tip. Since

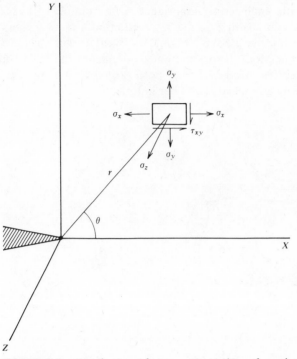

FIGURE 8.5 Distribution of stresses in vicinity of crack tip.

this plastic enclave is embedded within a large elastic region of material and is acted upon by either biaxial $(\sigma_y + \sigma_x)$ or triaxial $(\sigma_y + \sigma_x + \sigma_z)$ stresses, the extent of plastic strain within this region is suppressed. For example, if a load were applied in the Y direction, the plastic zone would develop a positive strain ϵ_y and attempt to develop corresponding negative strains in the X and Z direction, thus achieving a constant volume condition required for a plastic deformation process $(\epsilon_y + \epsilon_z + \epsilon_x = 0)$. However, σ_x acts to restrict the plastic zone contraction in the X direction, while the negative ϵ_z strain is counteracted by an induced tensile stress σ_z. Since there can be no stress normal to a free surface, the through-thickness stress σ_z must be zero at both surfaces but may attain a relatively large value at the midthickness plane. At one extreme, the case for a thin plate where σ_z cannot increase appreciably in the thickness direction, a condition of *plane stress* dominates, so

$$\sigma_z \approx 0 \tag{8-19}$$

In thick sections, however, a σ_z stress is developed, which creates a condition of triaxial tensile stresses acting at the crack tip and severely restricts straining in the z direction. This condition of *plane strain* can be shown to develop a through-thickness stress

$$\sigma_z \approx \nu(\sigma_x + \sigma_y) \qquad (8\text{-}20)$$

The distribution of σ_z stress through the plate thickness is sketched in Fig. 8.6 for conditions of plane stress and plane strain.

An important feature of Eq. 8-18 is the fact that the stress distribution around any crack in a structure is similar and depends only on the parameters r and θ. The difference between one cracked component and another lies in the magnitude of the stress field parameter K, defined as the *stress intensity factor*. In essence, K serves as a scale factor to define the magnitude of the crack tip stress field. From Westergaard's paper we see that

$$K = f(\sigma, a) \qquad (8\text{-}21)$$

where the functionality depends on the configuration of the cracked component and the manner in which the loads are applied. Many functions have been determined for various specimen configurations and are available from the fracture mechanics literature.[8-10] In recent years, stress intensity factor functions have been determined by mathematical procedures other than the Airy stress function approach used by Westergaard. Several solutions are shown in Fig. 8.7 for both commonly encountered cracked component configurations and standard laboratory test sample shapes, where the function is defined by $Y(a/W)$. Consistent with Eq. 8-21, the stress intensity factor is most often found to be a function of stress and crack length. [Situations where K is independent of crack

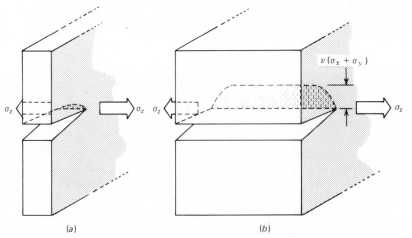

(a)   (b)

FIGURE 8.6   Through-thickness stress σ_z in (a) thin sheets under plane stress state and (b) thick plates under plane strain conditions.

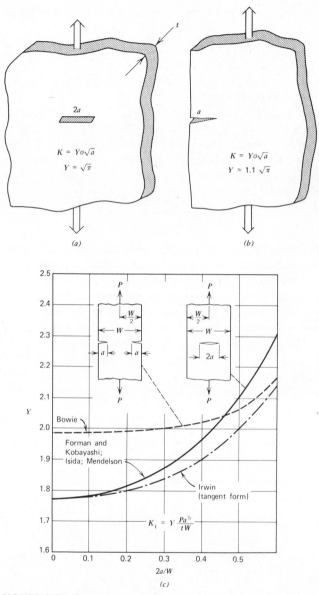

FIGURE 8.7 Stress intensity factor solutions for several specimen configurations.[17] (Reprinted by permission of the American Society for Testing and Materials from copyright material.)

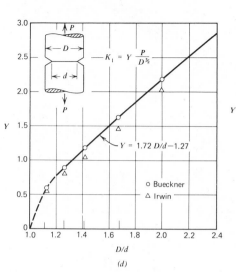

$K_1 = Y \dfrac{P}{D^{3/2}}$

$Y = 1.72 \, D/d - 1.27$

○ Bueckner
△ Irwin

D/d

(d)

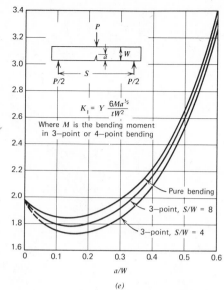

$K_1 = Y \dfrac{6Ma^{1/2}}{tW^2}$

Where M is the bending moment
in 3–point or 4–point bending

Pure bending

3–point, $S/W = 8$

3–point, $S/W = 4$

a/W

(e)

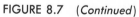

$K_1 = Y \dfrac{Pa^{1/2}}{tW}$

a/W

(f)

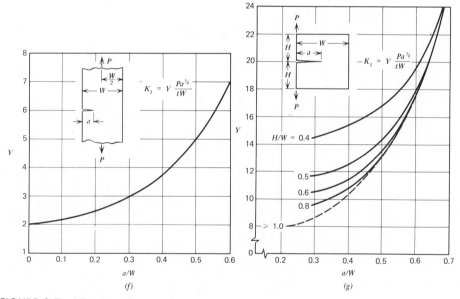

$K_1 = Y \dfrac{Pa^{1/2}}{tW}$

$H/W = 0.4$

0.5

0.6

0.8

$\geqslant 1.0$

a/W

(g)

FIGURE 8.7 (*Continued*)

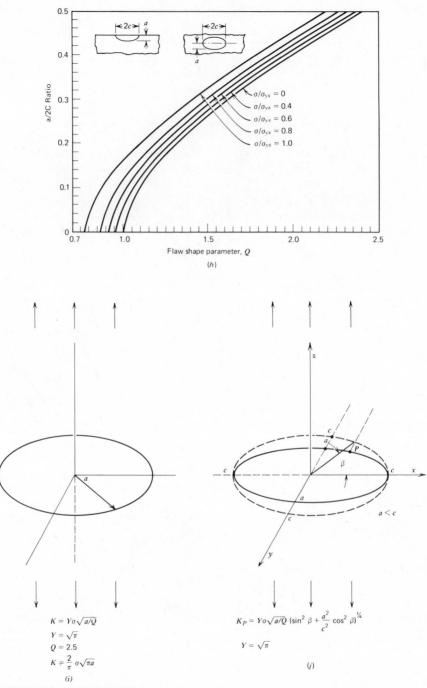

$$K = Y\sigma\sqrt{a/Q}$$
$$Y = \sqrt{\pi}$$
$$Q = 2.5$$
$$K = \frac{2}{\pi}\sigma\sqrt{\pi a}$$

(i)

$$K_P = Y\sigma\sqrt{a/Q}\left(\sin^2\beta + \frac{a^2}{c^2}\cos^2\beta\right)^{1/4}$$

$$Y = \sqrt{\pi}$$

(j)

FIGURE 8.7 (*Continued*)

length or varies inversely with a (e.g., Eq. 13-4) are reported elsewhere.[8–10] At this point, it is informative to compare the stress intensity factor K and the stress concentration factor K_t introduced in Chapter Seven. Although K_t accounts for the geometrical variables, crack length and crack tip radius, the stress intensity factor K incorporates *both geometrical terms* (the crack length appears explicitly, while the crack tip radius is assumed to be very sharp) *and the stress level*. As such, the stress intensity factor provides more information than does the stress concentration factor.

Once the stress intensity factor for a given test sample is known, it is then possible to determine the maximum stress intensity factor that would cause failure. This critical value K_c is described in the literature as the *fracture toughness* of the material.

A useful analogy may be drawn between stress and strength, and the stress intensity factor and fracture toughness. A component may experience many levels of stress, depending on the magnitude of load applied and the size of the component. However, there is a unique stress level that produces permanent plastic deformation and another stress level that causes failure. These stress levels are defined as the yield *strength* and fracture *strength*. Similarly, the stress intensity level at the crack tip will vary with crack length and the level of load applied. That unique stress intensity level that causes failure is called the critical stress intensity level or the *fracture toughness*. *Therefore, stress is to strength as the stress intensity factor is to fracture toughness.*

Any specimen size and shape may be used to determine the fracture toughness of a given material, provided the stress intensity factor calibration is known. Obviously, some samples are more convenient and cheaper to use than others. For example, when the nuclear power plant manufacturers set out to test the steels to be used in nuclear reactors, they chose a small sample like the one shown in Fig. 8.7g so that fracture studies of neutron irradiated samples could be carried out in relatively small environmental chambers. You would use a similar sample in your laboratory if you had a limited amount of material available for the test program or if your testing machine had limited loading capacity. The notched bend bar (Fig. 8.7e) with a long span S also would be an appropriate sample to use when laboratory load capacity is limited. Of course, this sample would require much more material than the compact tension sample (Fig. 8.7g).

There is an important connection between the stress intensity correction factors given in Figs. 8.7h and 8.7i. Note that as $a/2c$ approaches 0.5, Q approaches 2.5, where $\sqrt{1/Q}$ equals $2/\pi$, which is the solution for an embedded circular flaw or a semicircular surface flaw. An embedded elliptical or semielliptical surface flaw will grow such that $a/2c$ always increases to a limiting value of 0.5. (For the semielliptical surface flaw, the equilibrium $a/2c$ is closer to 0.36, because an additional K correction associated with the free surface is present.) This results from a variation in K along the surface of the ellipse (Fig. 8.7*j*). When $\beta = 90°$, K is maximized but is smallest where $\beta = 0°$. As a result, the crack will always grow fastest where $\beta = 90°$ until the crack assumes a circular configuration. At this point, the K level is the same along the entire crack perimeter, with additional crack growth maintaining a circular shape. A corollary to this case is that elliptical flaws will grow first into a circular configuration before appreciably increasing the crack size in the direction of the major axis of the ellipse.

8.3   DESIGN PHILOSOPHY

The interaction of material properties, such as the fracture toughness, with the design stress and crack size controls the conditions for fracture in a component. For example, it is seen from Fig. 8.7*a* that the fracture condition for an infinitely large cracked plate would be

$$K = K_c = \sigma \sqrt{\pi a} \tag{8-22}$$

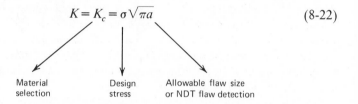

| Material selection | Design stress | Allowable flaw size or NDT flaw detection |

   This relationship may be used in one of several ways to design against a component failure. For example, if you are to build a system that must withstand the ravages of a liquid metal environment, such as in some nuclear reactors, one of your major concerns is the selection of a suitable corrosion-resistant material. Once done, you have essentially fixed K_c. In addition, if you allow for the presence of a relatively large stable crack—one that can be readily detected and repaired—the design stress is fixed and must be less than $K_c / \sqrt{\pi a}$.

   A second example shows another facet of the fracture control design problem. A certain aluminum alloy was chosen for the wing skin of a military aircraft because of its high strength and light weight; hence, K_c was fixed. The design stress on the wing was then set at a high level to increase the aircraft's payload capacity. Having fixed K_c and σ, the allowable flaw size was defined by Eq. 8-22 and beyond the control of the aircraft designers. In one case history, a fatigue crack grew out from a rivet hole in one of the aluminum wing plates and progressed to the point where the conditions of Eq. 8-22 were met. The result—fracture. What was most unfortunate about this particular failure was the fact that the allowable flaw size that could be tolerated by the material under the applied stress was smaller than the diameter of the rivet head covering the hole. Consequently, it was impossible for maintenance and inspection people to know that a crack was growing from the rivet hole until it was too late. This situation could have been avoided in several ways. Had it been recognized beforehand that the wing plate should have tolerated a crack greater than the diameter of the rivet head, the stresses could have been reduced and/or a tougher material selected for the wing plates. It is worth noting that one of the difficulties leading to the early demise of the British Comet jet transport was the selection of a lower toughness 7000 series aluminum alloy for application in critical areas of the aircraft.

The significance of Eq. 8-22 lies in the fact that you must first decide what is most important about your component design: certain material properties, the design stress level as affected by many factors such as weight considerations, or the flaw size that must be tolerated for safe operation of the part. Once such a priority list is established, certain critical decisions can be made. However, once any combination of two of three variables (fracture toughness, stress, and flaw size) is defined, the third factor is fixed.

Another design philosophy based on fracture mechanics considerations is the leak-before-break concept[11] as applied to pressure vessels. This approach to the design of a vessel would allow for the growth of a surface flaw (such as the one sketched in Fig. 8.7h) through the thickness without creating an unstable condition that would result in the rupture of the vessel. Instead, the through-thickness flaw would allow fluid or gas to escape and, thereby, alert technicians to a potentially serious problem. The leak-before-break condition would exist when a crack of length equal to at least twice the vessel wall thickness was stable under the prevailing stresses.

8.4 RELATIONSHIP BETWEEN ENERGY RATE AND STRESS FIELD APPROACHES

Thus far, two approaches to the relationship between stresses, flaw sizes, and material properties in the fracture of materials have been discussed. At this point, it is appropriate to demonstrate the similarity between the two. If Eq. 8-9 is rearranged so that

$$\sigma\sqrt{\pi a} = \sqrt{E \mathcal{G}} \qquad (8\text{-}23)$$

it is seen from Eq. 8-22 that

$$K = \sqrt{E \mathcal{G}} \quad \text{(plane stress)} \qquad (8\text{-}24)$$

and

$$K = \sqrt{\frac{E \mathcal{G}}{(1 - \nu^2)}} \quad \text{(plane strain)} \qquad (8\text{-}25)$$

for plane strain. This relationship between K and $\mathcal{G}$ is not merely fortuitous but, rather, can be shown to be valid based on an analysis credited to Irwin.[12]

Consider the energy needed to reclose part of a crack that had formed in a solid. In reverse manner, then, once this energy was removed the crack should reopen. With the notation shown in Fig. 8.8, the work done per unit area (unit thickness) to close the crack by an amount α is given by

$$\mathcal{G} = \frac{2}{\alpha} \int_0^\alpha \frac{\sigma_y V(x)}{2} \, dx \tag{8-26}$$

The constant "2" accounts for the total closure distance, since $V(x)$ is only half of the total crack opening displacement; $1/\alpha$ relates to the average energy released over the total closure distance α; $\sigma_y V(x)/2$ defines the energy under the load deflection curve. From Eq. 8-18 where $\theta = 0$,

$$\sigma_y = \frac{K}{\sqrt{2\pi x}} \tag{8-27}$$

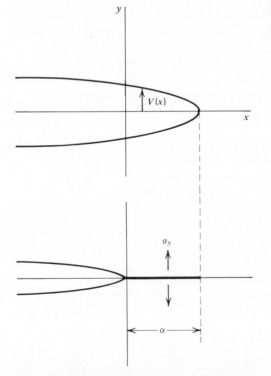

FIGURE 8.8 Diagram showing partial closure of crack over distance α.

while it has been shown[8] that

$$V = \frac{2K}{E}\sqrt{\frac{2(\alpha - x)}{\pi}} \tag{8-28}$$

Combining Eqs. 8-27 and 8-28 with 8-26, note that

$$\mathcal{G} = \frac{2K^2}{\alpha E \pi} \int_0^\alpha \sqrt{\frac{\alpha - x}{x}}\, dx \tag{8-29}$$

Note that $\mathcal{G}$ represents an average value taken over the increment α, while K will vary with α because K is a function of crack length. These difficulties can be minimized by shrinking α to a very small value so as to arrive at a more exact solution for $\mathcal{G}$. Therefore, taking the limit of Eq. 8-29, where α approaches zero, and integrating, it is seen that

$$\mathcal{G} = \frac{K^2}{E} \text{ (plane stress)} \tag{8-30}$$

and

$$\mathcal{G} = \frac{K^2}{E}(1 - \nu^2) \text{ (plane strain)} \tag{8-31}$$

which is the same result given by Eqs. 8-24 and 8-25.

8.5 CRACK TIP PLASTIC ZONE SIZE ESTIMATION

As you know by now, a region of plasticity is developed near the crack tip whenever the stresses described by Eq. 8-18 exceed the yield strength of the material. An estimate of the size of this zone may be obtained in the following manner. First, consider the stresses existing directly ahead of the crack where $\theta = 0$. As seen in Fig. 8.9, the elastic stress $\sigma_y = K/\sqrt{2\pi r}$ will exceed the yield strength at some distance r from the crack tip, thereby truncating the elastic stress at that value. By letting $\sigma_y = \sigma_{ys}$ at the elastic-plastic boundary

$$\sigma_{ys} = \frac{K}{\sqrt{2\pi r}} \tag{8-32}$$

and the plastic zone size is computed to be $K^2/2\pi\sigma_{ys}^2$. Since the presence of the plastic region makes the material behave as though the crack was slightly longer than actually measured, the "apparent" crack length is assumed to be the actual crack length plus some fraction of the plastic zone diameter. As a first approximation, Irwin[13] set this increment equal to the plastic zone radius, so that

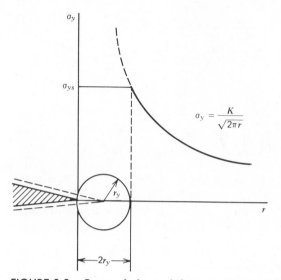

FIGURE 8.9   Onset of plastic deformation at the crack tip. "Effective" crack length taken to be initial crack length plus the plastic zone radius.

the apparent crack length is increased by that amount. In effect, the plastic zone diameter is a little larger than $K^2/2\pi\sigma_{ys}^{\,2}$ as a result of load redistributions around the zone and is estimated to be twice that value. Therefore

$$r_y \approx \frac{1}{2\pi}\frac{K^2}{\sigma_{ys}^{\,2}} \text{ (plane stress)} \tag{8-33}$$

For conditions of plane strain where the triaxial stress field suppresses the plastic zone size, the plane strain plastic zone radius is smaller and has been estimated[14] to be

$$r_y \approx \frac{1}{6\pi}\frac{K^2}{\sigma_{ys}^{\,2}} \text{ (plane strain)} \tag{8-34}$$

By comparing Eqs. 8-33 and 8-34 it is seen that the size of the plastic zone varies along the crack front, being largest at the two free surfaces and smallest at the midplane.

The reader should recognize that the size of the plastic zone also varies with θ. If the plastic zone size is determined for the more general case by the distortion energy theory, where σ_x, σ_y, and σ_z are described in terms of r and θ, it can be shown that

$$r_y = \left(\frac{K^2}{2\pi\sigma_{ys}^{\,2}}\right)\cos^2\frac{\theta}{2}\left(1+3\sin^2\frac{\theta}{2}\right) \text{ (plane stress)} \tag{8-35}$$

where the zone assumes a shape as drawn in Fig. 8.10a. Hahn and Rosenfield[15] have confirmed this plastic zone shape by way of etch pit studies in an iron–silicon alloy (see Fig. 8.10b). Note that when $\theta = 0$, Eq. 8-35 reduces to Eq. 8-33. Eqs. 8-33 and 8-34 may now be used to determine the effective stress intensity level K_{eff}, based on the effective or apparent crack length, so that

$$K_{eff} \approx Y\left(\frac{a+r_y}{W}\right)\sigma\sqrt{a+r_y} \qquad (8\text{-}36)$$

Since the plastic zone size is itself dependent on the stress intensity factor, the value of K_{eff} must be determined by an iterative process that may be truncated at any given level to achieve the desired degree of exactness for the value of K_{eff}. For example, the iteration may be terminated when $(a+r_y)_2 - (a+r_y)_1 \leqslant X$, where X is arbitrarily chosen by the investigator. A special case is an infinite plate with a small central notch, where the stress intensity factor is defined by Eq. 8-22. Iteration is not necessary in this case, and K_{eff} may be determined directly. Substituting Eq. 8-33 into Eq. 8-22 yields

$$K_{eff} = \sigma\sqrt{\pi\left[a + \frac{1}{2\pi}\left[\frac{K_{eff}^{2}}{\sigma_{ys}^{2}}\right]\right]} \qquad (8\text{-}37)$$

Upon rearranging Eq. 8-37, it is seen that

$$K_{eff} = \frac{\sigma\sqrt{\pi a}}{\left[1 - \frac{1}{2}\left(\frac{\sigma}{\sigma_{ys}}\right)^{2}\right]^{\frac{1}{2}}} \qquad (8\text{-}38)$$

so that K_{eff} will always be greater than $K_{applied}$, although the difference may be very small under low stress conditions.

EXAMPLE 1 A plate of steel with a central through-thickness flaw of length 16 mm is subjected to a stress of 350 MPa normal to the crack plane. If the yield strength of the material is 1400 MPa, what is the plastic zone size and the effective stress intensity level at the crack tip?

Assuming the plate to be infinitely large, r_y may be determined from Eqs. 8-22 and 8-33 so that

$$r_y \approx \frac{1}{2\pi}\left[\frac{350^2\pi(0.008)}{1400^2}\right] \approx 0.25 \text{ mm}$$

Since r_y/a is very small, it would not be expected that K_{eff} would greatly exceed $K_{applied}$. In fact, from Eq. 8-38

$$K_{eff} = \frac{350\sqrt{\pi(0.008)}}{\left[1 - \frac{1}{2}(350/1400)^2\right]^{\frac{1}{2}}} = 56.4 \text{ MPa}\sqrt{m}$$

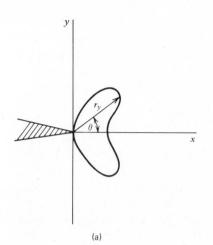

(a)

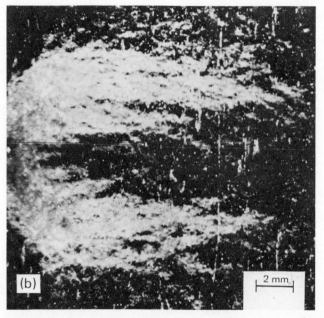

(b)

2 mm

FIGURE 8.10   Crack tip plastic zone. (a) Variation in plastic zone boundary as function of θ; (b) etch-pitted Fe-Si revealing similar zone shape. (After Hahn and Rosenfield,[15] Fig. 8.10b reprinted with permission of Hahn, *Acta Metall.*, **13** (1965), Pergamon Publishing Company.)

which is only about 2% greater than $K_{applied}$. When the plastic zone is relatively small in relation to the overall crack length, the plastic zone correction to the stress intensity factor is usually ignored in practice. This occurs often under fatigue crack propagation conditions, where the applied stresses are well below the yield strength of the material.

If, on the other hand, a second plate of steel with the same crack size and applied stress level were heat treated to provide a yield strength of 385 MPa, the plasticity correction would be substantially larger. The plastic zone size would be

$$r_y = \frac{1}{2\pi}\left[\frac{350\sqrt{\pi(0.008)}}{385}\right]^2 = 3.3 \text{ mm}$$

or one-fifth the size of the total crack length. Correspondingly, the effective stress intensity factor would be considerably greater than the applied level, wherein

$$K_{eff} = \frac{350\sqrt{\pi(0.008)}}{\left[1-\frac{1}{2}(350/385)^2\right]^{\frac{1}{2}}} = 72.4 \text{ MPa}\sqrt{m}$$

which represents a 30% correction. When the computed plastic zone becomes an appreciable fraction of the actual crack length, as found above, and generates a large correction for the stress intensity level, the entire procedure of applying the plasticity correction becomes increasingly suspect. When such a large plasticity correction is made to the elastic solution, the assumptions of a dominating elastic stress field become tenuous.

8.5.1   Dugdale Plastic Strip Model

Another model of the crack tip plastic zone has been proposed by Dugdale[16] for the case of plane stress. As shown in Fig. 8.11, Dugdale considered the plastic regions to take the form of narrow strips extending a distance R from each crack tip. For purposes of the mathematical analysis, the internal crack of length $2c$ is allowed to extend elastically to a length $2a$; however, an internal stress is applied in the region $|c| < |x| < |a|$ to reclose the crack. It may be shown that this internal stress must be equal to the yield strength of the material such that $|c| < |x| < |a|$ represents local regions of plasticity. By combining the internal stress field surrounding the plastic enclaves with the external stress field associated with a stress σ acting on the crack, Dugdale demonstrated that

$$c/a = \cos\left(\frac{\pi}{2}\frac{\sigma}{\sigma_{ys}}\right) \tag{8-39}$$

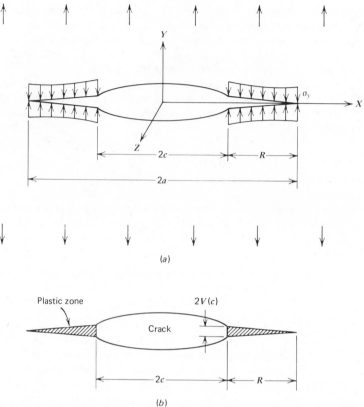

FIGURE 8.11 Dugdale plastic zone strip model for nonstrain hardening solids. Plastic zones R extend as thin strips from each end of the crack. (After Hahn and Rosenfield,[15] reprinted with permission of Hahn, *Acta Metall.*, **13** (1965), Pergamon Publishing Company.)

or, since $a = c + R$

$$R/c = \sec\left(\frac{\pi}{2}\frac{\sigma}{\sigma_{ys}}\right) - 1 \qquad (8\text{-}40)$$

When the applied stress $\sigma \ll \sigma_{ys}$, Eq. 8-40 reduces to

$$R/c \approx \frac{\pi^2}{8}\left(\frac{\sigma}{\sigma_{ys}}\right)^2 \qquad (8\text{-}41)$$

By rearranging Eq. 8-41 in the form of $D(K/\sigma_{ys})^2$, it is encouraging to note the reasonably good agreement between Eqs. 8-41 and 8-33 (i.e., $D = \pi/8 \approx 1/\pi$). (Note that $2\,r_y$ is used in Eq. 8-33 for comparison with the Dugdale zone.)

8.6 FRACTURE MODE TRANSITION: PLANE STRESS VERSUS PLANE STRAIN

As discussed in Section 8.5, the plastic zone size depends on the state of stress acting at the crack tip. When the sample is thick in a direction parallel to the crack front, a large σ_z stress can be generated that will restrict plastic deformation in that direction. As shown by Eqs. 8-33 and 8-34, the plane strain plastic zone size is correspondingly smaller than the plane stress counterpart. Since the fracture toughness of a material will depend on the volume of material capable of plastically deforming prior to fracture, and since this volume depends on specimen thickness, it follows that the fracture toughness K_c will vary with thickness as shown in Fig. 8.12. When the sample is thin (for example, at t_1) and the degree of plastic constraint acting at the crack tip minimal, plane stress conditions prevail and the material exhibits maximum toughness. (Note that if the sample were made thinner the toughness would gradually decrease because less material would be available for plastic deformation energy absorption.) Alternately, when the thickness is increased to bring about plastic constraint and plane strain conditions at the crack tip, the toughness drops sharply to a level that may be one-third (or less) that of the plane stress value. One very important aspect of this lower level of toughness (the *plane strain fracture toughness K_{IC}*) is that it does not decrease further with increasing thickness, thereby making this value a conservative lower limit of material toughness in any given engineering application. Once K_{IC} is determined in the laboratory for a given material with a sample at least as thick as t_2 (Fig. 8.12), an engineering component much thicker than t_2 should exhibit the same toughness. To summarize, the plane stress fracture toughness K_c is related to both metallurgical and specimen geometry, while the plane strain fracture toughness K_{IC} depends only on metallurgical factors. Consequently, the best way to compare materials of different thickness

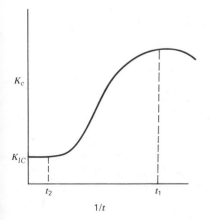

FIGURE 8.12 Variation in fracture toughness with plate thickness.

on the basis of their respective intrinsic fracture toughness levels should involve a comparison of K_{IC} values, since thickness effects may be avoided.

Since stress state effects on fracture toughness are affected by the size of the plastic enclave in relation to the sheet thickness, it is informative to consider the change in stress state in terms of the ratio, r_y/t, where r_y is computed arbitrarily with the plane stress plastic zone size relationship as given by Eq. 8-33. Experience has shown that when $r_y/t \geqslant 1$, plane stress conditions prevail and toughness is high. At the other extreme, plane strain conditions will exist when $r_y/t < \frac{1}{10}$. In either case, the necessary thickness to develop a plane stress or plane strain condition will depend on the yield strength of the material, since this will control r_y at any given stress intensity level. Therefore, if the yield strength of a material were increased by a factor of two by some thermo-mechanical treatment (TMT), the thickness necessary to achieve a plane strain condition for a given stress intensity level could be reduced by a factor of four, assuming, of course, that K_{IC} was not altered by the TMT. Clearly, very thin sections can still experience plane strain conditions in high yield strength material, whereas very large sections of low yield strength material may never bring about a full plane strain condition.

Another feature of the fracture toughness-stress state dependency is the commonly observed fracture mode transition mentioned in the previous chapter. As shown in Fig. 8.13, the relative degree of flat and slant fracture depends on the crack tip stress state. When plane stress conditions prevail and $r_y \geqslant t$, the fracture plane often assumes a ± 45-degree orientation with respect to the load axis and sheet thickness (Fig. 8.14a). This may be rationalized in terms of failure

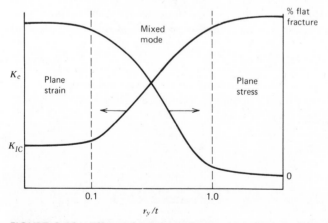

FIGURE 8.13 Effect of relative plastic zone size to plate thickness on fracture toughness and macroscopic fracture surface appearance. Plane stress state associated with maximum toughness and slant fracture. Plane strain state associated with minimum toughness and flat fracture.

occurring on those planes containing the maximum resolved shear stress. (Since $\sigma_z = 0$ in plane stress, a Mohr circle construction will show that the planes of maximum shear will lie along ± 45-degree lines in the Y–Z plane.) In plane strain, where $\sigma_z \approx \nu(\sigma_y + \sigma_x)$ and $r_y \ll t$, the plane of maximum shear is found in the X–Y plane (Fig. 8.14b), (σ_y, σ_x, and σ_z may be computed from Eqs. 8-18 and 8-20, where it may be shown, for example, that when $\theta = 60°$, $\sigma_y > \sigma_z > \sigma_x$). Apparently, the fracture plane under plane strain conditions lies midway between the two maximum shear planes. This compromise probably also reflects the tendency for the crack to remain in a plane containing the maximum net section stress.

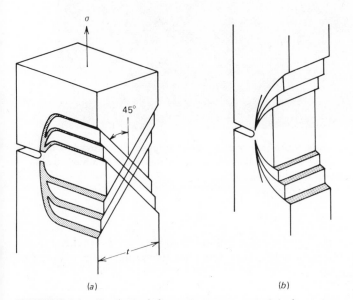

(a)   (b)

FIGURE 8.14   Crack tip deformation patterns in (a) plane stress, and (b) plane strain. (After Hahn and Rosenfield,[15] reprinted with permission of Hahn, *Acta Metall.*, **13** (1965), Pergamon Publishing Company.)

The existence of a fracture mode transition in many engineering materials such as aluminum, titanium, and steel alloys, and a number of polymers makes it possible to estimate the relative amount of energy absorbed by a component during a fracture process. (Beryllium is one material, however, that does not seem to fit this pattern.) When the fracture surface is completely flat (Fig. 8.15c), plane strain test conditions probably prevail, and the observed fracture toughness is low. If the fracture is completely of the slant or shear type (Fig.

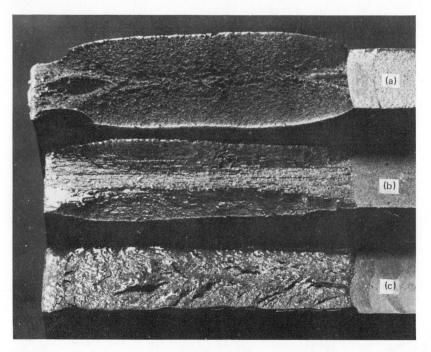

FIGURE 8.15 Fracture mode transition in alloy steel induced by change in test temperature. (*a*) Slant fracture at high temperature, (*b*) mixed mode at intermediate temperature, (*c*) flat fracture at low temperature.

8.15*a*), plane stress conditions probably dominate to produce a tougher failure. Obviously, a mixed fracture appearance (Fig. 8.15*b*) would reflect an intermediate toughness condition.

8.7 PLANE STRAIN FRACTURE TOUGHNESS TESTING

Since the plane strain fracture toughness K_{IC} is such an important material property in fracture prevention, it is appropriate to consider the procedures by which this property is measured in the laboratory. Accepted test methods have been set forth by the American Society for Testing and Materials under Standard E399-72. Although the reader should examine this standard for precise details, the most important features of K_{IC} testing are summarized in this section.

A recommended test sample [as of this writing, a three-point bend bar* (Fig. 8.7e) and compact tension sample† (Fig. 8.7g) are considered acceptable for this purpose] is initially fatigue-loaded to extend the machined notch a prescribed amount. A clip gage is then placed at the mouth of the crack to monitor its displacement when the specimen load is applied. Typical load-displacement records for a K_{IC} test are shown in Fig. 8.16. From such curves, two important questions should be answered. First, what is the apparent plane strain fracture toughness value for the material? Second, is this value *valid* in the sense that a thicker or bigger sample might not produce a lower K_{IC} number for the same material? If a lower toughness level is achieved with a thicker sample, then the value obtained initially is not a valid number. Brown and Srawley[17] examined the fracture toughness of several high-strength alloys and found empirically that a valid plane strain fracture toughness test is performed when the specimen thickness and crack length are both greater than a certain minimum value. Specifically

$$t \text{ and } a \geqslant 2.5 \left(\frac{K_{IC}}{\sigma_{ys}} \right)^2 \tag{8-42}$$

The ratio $(K_{IC}/\sigma_{ys})^2$ suggests that the required sheet thickness and crack length are related to some measure of the plastic zone size, since Eqs. 8-33 and 8-34 are of the same form. Using the plastic zone sizes determined from Eqs. 8-33 and 8-34 and substituting into Eq. 8-42, it is seen that the criteria for plane strain conditions reflect a condition where

$$a \text{ and } t \geqslant 50 r_y \text{ plane strain} \tag{8-43}$$

or

$$a \text{ and } t \geqslant 17 r_y \text{ plane stress} \tag{8-44}$$

Under these conditions, K_{eff} as defined in Eq. 8-36 would not be significantly different from $K_{applied}$, such that the plastic zone correction would be unnecessary.

To arrive at a valid K_{IC} number, it is necessary to first calculate a tentative number K_Q, based on a graphical construction on the load-displacement test record. If K_Q satisfies the conditions of Eq. 8-42, then $K_Q = K_{IC}$. From ASTM Specification E-399-72, the graphical construction involves the following procedures:[18] On the load-deflection test record, draw a secant line OP_5 through the origin with a slope that is 5% less than the tangent OA to the initial part of the curve. For the three-point bend bar and the compact tension sample, a 5% reduction in slope is approximately equal to a 2% increase in the effective crack

$$*K = \frac{PS}{tW^{3/2}} [2.9 \left(\frac{a}{W} \right)^{\frac{1}{2}} - 4.6 \left(\frac{a}{W} \right)^{\frac{3}{2}} + 21.8 \left(\frac{a}{W} \right)^{\frac{5}{2}} - 37.6 \left(\frac{a}{W} \right)^{\frac{7}{2}} + 38.7 \left(\frac{a}{W} \right)^{\frac{9}{2}}]$$

$$†K = \frac{P}{tW^{1/2}} [29.6 \left(\frac{a}{W} \right)^{\frac{1}{2}} - 185.5 \left(\frac{a}{W} \right)^{\frac{3}{2}} + 655.7 \left(\frac{a}{W} \right)^{\frac{5}{2}} - 1017 \left(\frac{a}{W} \right)^{\frac{7}{2}} + 638.9 \left(\frac{a}{W} \right)^{\frac{9}{2}}]$$

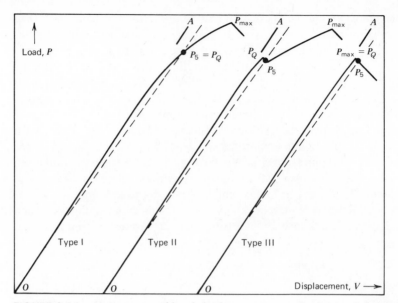

FIGURE 8.16 Major types of load-displacement records obtained during K_{IC} testing.[18] (Reprinted by permission of the American Society for Testing and Materials from copyright material.)

length of the sample, a level reflecting minimal crack extension and plasticity correction.[17] P_5 is defined as the load at the intersection of the secant line OP_5 with the original test record. (Note the similarity between this graphical method and the one described in Chapter One for the determination of the 0.2% offset yield strength.) The load P_Q, which will be used to calculate K_Q, is then determined as follows: "if the load at every point on the record which precedes P_5 is lower than P_5, then P_Q is P_5 (Fig. 8.16, Type I); if, however, there is a maximum load preceding P_5 which exceeds it, then this maximum load is P_Q [Fig. 8.16, Types II and III]."[18] If the ratio of P_{max}/P_Q is less than 1.1, it is then permissible to compute K_Q with the aid of the appropriate K calibration. If K_Q satisfies Eq. 8-42, then K_Q is equal to K_{IC}. If not, then a thicker and/or more deeply cracked sample must be prepared for additional testing so that a valid K_{IC} may be determined.

8.8 FRACTURE TOUGHNESS OF ENGINEERING ALLOYS

Typical K_{IC} values for various steel, aluminum, and titanium alloys are listed in Table 8.1 along with associated yield strength levels. The table also provides a listing of critical flaw sizes for each material, based on a hypothetical service condition involving a through-thickness center notch of length $2a$ embedded in an infinitely large sheet sufficiently thick to develop plane strain conditions at the crack tip. If it is assumed that the operating design stress is taken to be one-half the yield strength, then the critical crack length would equal $(1/\pi)$ $(K_{IC}/(\sigma_{ys}/2))^2$. One basic data trend becomes immediately obvious: the fracture

TABLE 8.1 Plane Strain Fracture Toughness of Selected Engineering Alloys

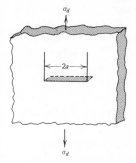

$$\sigma_d = \frac{\sigma_{ys}}{2} \; ; \; K_{IC} = \sigma_d\sqrt{\pi a}$$

$$a_c = \frac{1}{\pi}\left(\frac{K_{IC}}{\sigma_{ys}/2}\right)^2$$

Material	K_{IC}		σ_{ys}		a_c	
	MPa$\sqrt{\text{m}}$	ksi$\sqrt{\text{in.}}$	MPa	ksi	mm	in.
2014-T651	24.2	22	455	66	3.6	0.14
2024-T3	~44.	~40	345	50	~21.	~0.82
2024-T851	26.4	24	455	66	4.3	0.17
7075-T651	24.2	22	495	72	3.0	0.12
7178-T651	23.1	21	570	83	2.1	0.08
7178-T7651	33.	30	490	71	5.8	0.23
Ti-6Al-4V	115.4	105	910	132	20.5	0.81
Ti-6Al-4V	55.	50	1035	150	3.6	0.14
4340	98.9	90	860	125	16.8	0.66
4340	60.4	55	1515	220	2.	0.08
4335+V	72.5	66	1340	194	3.7	0.15
17-7PH	76.9	70	1435	208	3.6	0.14
15-7Mo	49.5	45	1415	205	1.5	0.06
H-11	38.5	35	1790	260	<0.6	<0.02
H-11	27.5	25	2070	300	0.23	0.009
350 Maraging	55.	50	1550	225	1.6	0.06
350 Maraging	38.5	35	2240	325	<0.4	<0.02
52100	~14.3	~13	2070	300	~0.06	<0.002

toughness and allowable flaw size of a given material decreases, often precipitously, when the yield strength is elevated. Consequently, there is a price to pay when one wishes to raise the strength of a material. More will be said about this in Chapter Ten.

EXAMPLE 2 Assume that a component in the shape of a large sheet is to be fabricated from 0.45C-Ni-Cr-Mo steel. It is required that the critical flaw size be greater than 3 mm, the resolution limit of available flaw detection procedures. A design stress level of one-half the tensile strength is indicated. To save weight, an increase in the tensile strength from 1520 MPa to 2070 MPa is suggested. Is such a strength increment allowable? (Assume plane strain conditions in all computations.)

The answer to this question bears heavily upon the changes in fracture toughness of the material resulting from the increase in tensile strength. At the 1520 MPa strength level, it is found that the K_{IC} value is 66 MPa$\sqrt{m}$, while at 2070 MPa K_{IC} drops sharply to 33 MPa$\sqrt{m}$ (Fig. 8.17). For a large sheet the stress intensity factor is determined from Eq. 8-22, where the design stress is $\sigma_{ts}/2$. For the alloy heat treated to the 1520 MPa

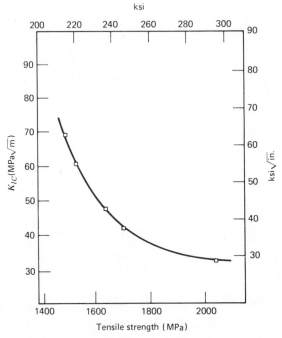

FIGURE 8.17 Fracture toughness in 0.45C-Ni-Cr-Mo steel containing 0.045% S. (After Birkle, et al.,[19] copyright American Society for Metals 1966.)

strength level

$$66 \text{ MPa}\sqrt{\text{m}} = 760 \text{ MPa}\sqrt{\pi a}$$

$$2a = 4.8 \text{ mm}$$

which exceeds the minimum flaw size requirements. At the 2070 MPa strength level, however,

$$33 \text{ MPa}\sqrt{\text{m}} = 1035 \text{ MPa}\sqrt{\pi a}$$

$$2a = 0.65 \text{ mm}$$

which is five times smaller than the minimum flaw size requirement and approximately eight times smaller than the maximum flaw tolerated at the 1520 MPa strength level. Therefore, it is not possible to raise the strength of the alloy to 2070 MPa to save weight and still meet the minimum flaw size requirement. Furthermore, using the flaw size found in the 1520 MPa material for the 2070 MPa alloy would necessitate a decrease in design stress from 1035 to 380 MPa.

$$\sigma = \frac{33 \text{ MPa}\sqrt{\text{m}}}{\sqrt{\pi(0.0024)}} \approx 380 \text{ MPa}$$

Therefore, under similar flaw size conditions, the allowable stress level in the stronger alloy could be only half that in the weaker alloy, resulting in a two-fold *increase* in the weight of the component!

The fracture toughness data for the aluminum alloys deserves additional comment at this time. It is seen that the allowable flaw size in 7075-T651 is only one-eighth that for the 2024-T3. Accordingly, a design engineer would be alerted to the greater propensity for brittle fracture in the 7075-T651 alloy. In actual engineering structures, such as the wing skins of commercial and military aircraft, the relative difference between the two materials is even greater. Since wing skins are about 1 to 2 cm thick, it may be shown by Eq. 8-42 that 7075-T651 would experience approximately plane strain conditions, while 2024-T3 would be operating in a plane stress environment, where K_c is about two or three times as large as K_{IC}. Consequently, there may be a 50-fold difference in allowable flaw size between the two materials, taking into account both the metallurgical and stress state factors.

EXAMPLE 3 A plate of Zeusalloy 100, a steel alloy with a yield strength of 415 MPa, has been found to exhibit a K_{IC} value of 132 MPa$\sqrt{\text{m}}$. The material is available in various gages up to 250 mm but will be used in very wide 100 mm thick plates for a given application. If the plate is subjected to a stress of 100 MPa how large can a crack grow from a hole in the middle of the plate before catastrophic failure occurs?

You would be correct in assuming a central crack configuration so that Eq. 8-22 could be used but would be wrong if you substituted the K_{IC}

value of 132 MPa$\sqrt{m}$ into the equation. In fact, not enough information has been provided to answer the question properly. To use K_{IC} in Eq. 8-22 presumes that plane strain conditions exist in the component. It is seen from Eq. 8-42 that a plate 250 mm thick is required for plane strain conditions to apply. Since the plate in question is only 100 mm thick, plane *strain* conditions would not prevail. (In all likelihood, the valid K_{IC} value was obtained from a 250 mm plate). It would be possible to determine K_c for Zeusalloy 100 if data such as shown in Fig. 8.12 were available. In addition, an estimate of K_c may be obtained from the K_{IC} value in the region near plane strain by an empirical relationship shown by Irwin,[20] where

$$K_c^2 = K_{IC}^2(1 + 1.4\beta_{IC}^2)$$ (8-45)

where

$$\beta_{IC} = \frac{1}{t}\left(\frac{K_{IC}}{\sigma_{ys}}\right)^2$$

t = thickness

8.9 PLANE STRESS FRACTURE TOUGHNESS TESTING

The graphical procedures set forth in the last section for the determination of K_{IC} are sometimes confounded by the deformation and fracture response of the material. With rising load conditions, it is possible for the material to experience slow stable crack extension prior to failure, which makes it difficult to determine the maximum stress intensity level at failure because the final crack length is uncertain. During the early days of fracture toughness testing, a droplet of recorder ink was placed at the crack tip to follow the course of such slow crack extension and stain the fracture surface, thereby providing an estimate of the final stable crack length. Since the fracture properties of some high-strength materials are adversely affected by aqueous solutions (see Chapter Eleven), this laboratory practice has been abandoned. In addition to uncertainties associated with the measurement of the stable crack growth increment Δa, a plastic zone may develop in tougher materials which also must be accounted for in the stress intensity factor computation. Consequently, different methods must be employed to determine the fracture toughness value of a material when the final crack length is not clearly defined but is greater than the initial value.

Irwin[21] proposed that crack instability would occur in accordance with the requirements of the Griffith formulation. That is, failure should occur when the rate of change in the elastic energy release rate $\partial \mathcal{G} / \partial a$ equals the rate of change in material resistance to such crack growth $\partial R/\partial a$. The material's resistance to fracture R is expected to increase with increasing plastic zone development and strain hardening. Consequently, both $\mathcal{G}$ and R increase with increasing stress level. The Griffith instability criterion is depicted graphically in Fig. 8.18 as the point of tangency between the $\mathcal{G}$ and R curves when plotted against crack length. Knowing the material's R curve and using the correct stress and crack length dependence of $\mathcal{G}$ for a given specimen configuration, it would then be possible to determine $\mathcal{G}_c$ or K_c. This fracture toughness value is generally designated as the *plane stress* fracture toughness level, since r_y is no longer much smaller than the sheet thickness and the criterion set forth in Eq. 8-42 is not met.

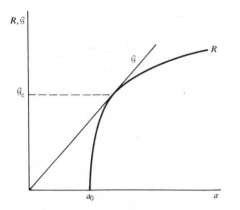

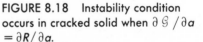

FIGURE 8.18   Instability condition occurs in cracked solid when $\partial \mathcal{G}/\partial a = \partial R/\partial a$.

One should recognize that the fracture toughness level will vary with the planar dimensions of the specimen. For example, it is seen from Fig. 8.19a that for a given material the fracture toughness value will depend on the initial crack length, since the tangency point is displaced slightly when the starting crack length is changed. However, this effect is not too large and is minimized when large samples are used, since K_c reaches a limiting value with increasing initial crack length. A considerably larger potential change in K_c values is evident when samples of different planar dimensions are used. Since $\mathcal{G}$ and K depend on specimen configuration, the shape of the $\mathcal{G}$ curve is different for each case. Consequently, the point of tangency and, hence, $\mathcal{G}_c$ will change for a given material. As shown in Fig. 8.19b, the $\mathcal{G}_c$ value increases with increasing sample width and in the limit (the case of an infinitely large panel, where $Y(a/W) = \sqrt{\pi}$ and $\mathcal{G} = \sigma^2 \pi a/E$) reaches a maximum value. Again, it should be fully recognized that the plane stress fracture toughness of a material is dependent on both metallurgical factors and specimen geometry, while the plane strain fracture toughness relates only to metallurgical variables.

Although the $\mathcal{G}$ curve can be determined from known analytical relationships,

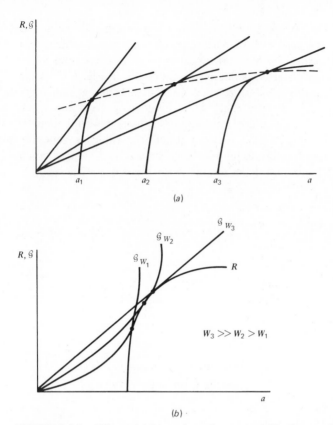

(a)

(b)

$W_3 \gg W_2 > W_1$

FIGURE 8.19   Effect of (a) initial crack size and (b) plate width on fracture toughness.

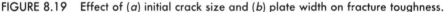

such as those provided in Fig. 8.7, the R curve is determined by graphical means. In one accepted procedure, the initial step involves the construction of a series of secant lines on the load-displacement record from a test sample (Fig. 8.20a). The compliance values δ/P from these secant lines are used in Fig. 8.20b (the compliance curve for the given specimen configuration) to determine the associated a/W values that reflect the effective crack length $a_{\text{eff}} = a_0 + \Delta a + r_y$. Each a_{eff}/W value is then used to determine a respective K_{eff} value from the calibration curve (Fig. 8.20c). Finally, the $\mathcal{G}$ or K^2/E curve is constructed from these K_{eff} and associated a_{eff} values (Fig. 8.20d). Once the R curve for a given material is determined from this graphical procedure, it becomes part of the

FIGURE 8.20   Graphical procedure to determine R curve. (a) Secant lines drawn on P-δ test record and plotted on compliance curve; (b) to obtain effective a/W values; (c) calibration factors determined from a_{eff}/W and used to calculate K_{eff}^2/E values for; (d) R curve construction.

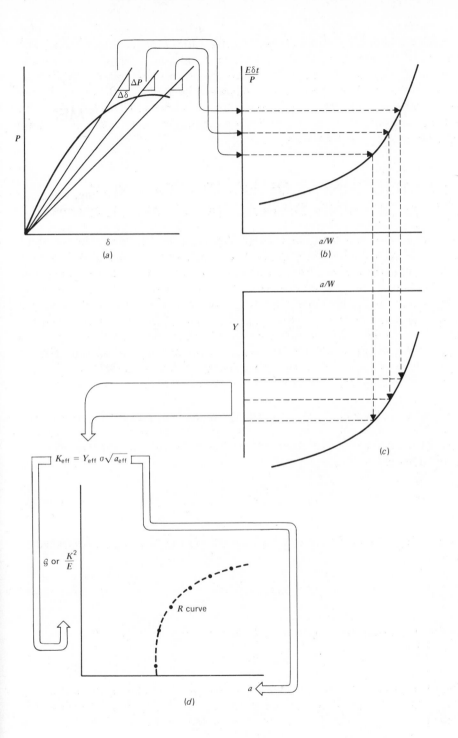

information about the material. As such, the same R curve may be used with samples of different crack length and planar dimension, as discussed above and depicted in Figs. 8.19a and 8-19b. Once the R curve is known, the fracture toughness may be determined for an engineering component with any configuration, as long as an analytical expression for $\mathcal{G}$ is known. (Recall that $\mathcal{G}_c$ also may be determined from Eq. 8-17 even when the stress and crack length dependence of $\mathcal{G}$ is not known.)

8.10 TOUGHNESS DETERMINATION FROM CRACK OPENING DISPLACEMENT MEASUREMENT

As demonstrated earlier in this chapter, there is a limit to the extent to which K_{eff} may be adjusted for crack tip plasticity. Whether one uses Eqs. 8-33 or 8-41 in the computation of r_y, it becomes increasingly inaccurate to determine K_{eff} from Eq. 8-36 when r_y/a becomes large. Since the allowable loads or stresses in a component rapidly approach a limiting value upon general yielding, while the associated strains or crack opening displacements (COD) (i.e., how much the crack opens under load) increase continually to the point of failure, it is more reasonable to monitor the latter in a general yielding situation. Accordingly, the concept of a critical crack opening displacement near the crack tip has been introduced to provide a fracture criteria and an alternate measure of fracture toughness.[22,23]

By using the Dugdale crack tip plasticity model[16] for plane stress, it has been possible to conveniently compute the magnitude of the crack opening displacement $2V(c)$ defined at the elastic-plastic boundary (Fig. 8.11b). Goodier and Field[24] demonstrated that

$$2V(c) = \frac{8\sigma_{ys}c}{\pi E} \ln \left[\sec \frac{\pi\sigma}{2\sigma_{ys}} \right] \qquad (8\text{-}46)$$

When $\sigma \ll \sigma_{ys}$, $\ln x \approx x - 1$, so that Eqs. 8-40, 8-41, and 8-46 can be combined to yield

$$2V(c) = \frac{K^2}{E\sigma_{ys}} = \frac{\mathcal{G}}{\sigma_{ys}} \qquad (8\text{-}47)$$

The fracture toughness of a material is then shown to be

$$\mathcal{G}_c \approx 2\sigma_{ys} V^*(c) \qquad (8\text{-}48)$$

where $V^*(c) =$ critical crack opening displacement. One important aspect of Eq. 8-47 is that $V(c)$ can be computed for conditions of both elasticity and plasticity, while $\mathcal{G}$ may be defined only in the former situation. The COD concept, therefore, bridges the elastic and plastic fracture conditions. It is important to recognize, however, that the strain fields and crack opening displacements associated with the crack tip will vary with specimen configuration. Consequently, a single critical crack opening displacement value for any given material cannot be defined, since it will be affected by the geometry of the specimen used in the test program.[25]

The physical significance of Eq. 8-48 warrants additional comment. Since the crack opening displacement $2V(c)$ is related to the extent of plastic straining in the plastic zone, Eq. 8-48 is analogous to the measurement of toughness from the area under the stress-strain curve in a uniaxial tensile test. As discussed in Chapter One, maximum toughness is achieved by an optimum combination of stress and strain. Since fracture toughness most often varies *inversely* with yield strength (see Table 8.1 and Chapter Ten), Eq. 8-48 would appear to predict toughness-yield strength trends incorrectly. This is not the case, since an increase in σ_{ys} of an alloy resulting from any thermomechanical treatment is offset normally by a proportionately greater *decrease* in $2V(c)$. The important point to keep in mind is that σ_{ys} and $2V(c)$ are interrelated. As a final note, some uncertainty exists concerning the correct value of yield strength to use in Eq. 8-48. When plane strain conditions dominate at the crack tip, should not the tensile yield strength value be elevated to account for crack tip triaxiality? Certainly, this would be consistent with a similar previous justification to adjust the plastic zone size estimate for plane stress and plane strain conditions (Eqs. 8-33 and 8-34). On this basis, Eq. 8-48 would be modified so that

$$\mathcal{G}_c \approx n\sigma_{ys} 2V^*(c) \tag{8-49}$$

where $1 \leqslant n \leqslant 1.5 - 2.0$. Consequently, when plane stress conditions are prevalent, $n = 1$, and n increases with increasing plane strain.

8.11 FRACTURE TOUGHNESS DETERMINATION WITH THE J INTEGRAL

Another approach is being developed to determine fracture energy for conditions involving both elastic and plastic deformation. Rice et al.[25,26,27] have defined the "J integral" which, in a physical sense, characterizes the stress-strain conditions existing near a crack tip in an elastic-plastic solid. By taking the load-displacement records from the same material for two different crack lengths and determining the change in potential energy for the incremental crack length change, we see from Fig. 8.21 that

$$J = \frac{\partial (U/B)}{\partial a} \tag{8-50}$$

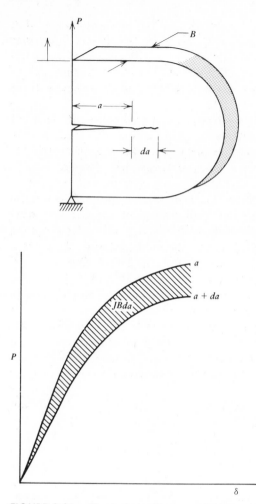

FIGURE 8.21 Physical significance of J integral. (After Bucci, et al.,[26] reprinted by permission of the American Society for Testing and Materials from copyright material.)

where

$$U/B = \text{potential energy per unit thickness}$$

$$a = \text{crack length}$$

It may be seen that for the purely elastic case, Eq. 8-50 reduces to Eq. 8-12 except for the $1/B$ term.

This approach considers J to be the general fracture energy release rate and in the limit equivalent to $\mathcal{G}$, the elastic energy release rate for the purely elastic case. In fact, Begley and Landes[28] have used small, fully plastic specimens to determine J_{IC} values. These values were found to be in good agreement with $\mathcal{G}_{IC}$ data obtained from large elastic specimens that satisfied plane strain fracture toughness test requirements. Certainly this approach justifies much additional research, since the laboratory costs for small specimen J_{IC} testing (which allows for extensive plastic deformation) are much lower than comparable $\mathcal{G}_{IC}$ experimentation (which provides useful data only in the elastic regime). In this connection, it will be necessary to evaluate the potential limitation of this approach because of imposed restrictions on subcritical crack growth prior to fracture and prefracture unloading events. The interested reader is referred to an additional article concerning this important emerging method for the measurement of fracture toughness.[29]

REFERENCES

1. A. A. Griffith, *Transactions, Royal Society of London*, Vol. 221, 1920. (This article has been republished with additional commentary in *Trans.*, ASM, **61**, 1968, p. 871.)
2. C. E. Inglis, *Proceedings*, Institute of Naval Architects, Vol. 55, 1913, p. 219.
3. E. Orowan, *Fatigue and Fracture of Metals*, MIT Press, Cambridge, 1950, p. 139.
4. G. R. Irwin, *Fracturing of Metals*, ASM, Cleveland, Ohio, 1949, p. 147.
5. G. R. Irwin and J. A. Kies, *Weld. J. Res. Suppl.*, **33**, 1954, p. 193s.
6. J. P. Den Hartog, *Advanced Strength of Materials*, McGraw-Hill, New York, 1952, p. 174.
7. H. M. Westergaard, *Trans.*, ASME, *J. Appl. Mech.*, **61**, 1939, p. 49.
8. P. C. Paris and G. C. M. Sih, ASTM *STP 381*, 1965, p. 30.
9. H. Tada, P. C. Paris, and G. R. Irwin, "The Stress Analysis of Cracks Handbook," Del Research, Hellertown, Pa. 1973.
10. G. C. M. Sih, "Handbook of Stress Intensity Factors," Lehigh University, 1973.
11. G. R. Irwin, *Appl. Mater. Res.*, **3**, 1964, p. 65.
12. G. R. Irwin, *Trans.*, ASME, *J. Appl. Mech.*, **24**, 1957, p. 361.
13. G. R. Irwin, *Handbuch der Physik*, Vol. VI, Springer, Berlin, 1958, p. 551.
14. F. A. McClintock and G. R. Irwin, ASTM *STP 381*, 1965, p. 84.
15. G. T. Hahn and A. R. Rosenfield, *Acta Met.*, **13**, 1965, p. 293.
16. D. S. Dugdale, *J. Mech. Phys. Sol.*, **8**, 1960, p. 100.
17. W. F. Brown, Jr. and J. E. Srawley, ASTM *STP 410*, 1966.
18. ASTM Standard E399-72, *Annual Book of ASTM Standards*, 1972.
19. A. J. Birkle, R. P. Wei, and G. E. Pellissier, *Trans.*, ASM, **59**, 1966, p. 981.
20. G. R. Irwin, NRL Report 6598, Nov. 21, 1967.
21. G. R. Irwin, *ASTM Bulletin*, Jan. 1960, p. 29.
22. A. A. Wells, *Brit. Weld. J.*, **13**, 1965, p. 2.
23. A. A. Wells, *Brit. Weld. J.*, **15**, 1968, p. 221.
24. J. N. Goodier and F. A. Field, *Fracture of Solids*, Interscience, New York, 1963, p. 103.
25. D. C. Drucker and J. R. Rice, *Eng. Fract. Mech.*, **1**, 1970, p. 577.
26. R. J. Bucci, P. C. Paris, J. D. Landes, and J. R. Rice, ASTM *STP 514*, 1972, p. 40.

27. J. R. Rice, P. C. Paris, and J. G. Merkle, ASTM *STP 536*, 1973, p. 231.
28. J. A. Begley and J. D. Landes, ASTM *STP 514*, 1972, p. 1.
29. J. D. Landes and J. A. Begley, ASTM, *STP 560*, 1974, p. 170.

PROBLEMS

8.1 A compact tension test specimen ($H/W = 0.6$), is designed and tested according to the ASTM E399-72 procedure. Accordingly, a Type I load versus displacement (P vs. δ) test record was obtained and a measure of the maximum load P_{max} and a critical load measurement point P_Q were determined. The specimen dimensions were determined as $W = 10$ cm, $t = 5$ cm, $a = 5$ cm, the critical load point measurement point $P_Q = 100$ kN and $P_{max} = 105$ kN. Assuming all other E399 requirements regarding the establishment and sharpness of the fatigue starter crack were met, determine the critical value of stress intensity. Does it meet conditions for a valid K_{IC} test if the material yield stress is 700 MPa? If it is 350 MPa?

8.2 An infinitely large sheet is subjected to a gross stress of 350 MPa. There is a central crack $5/\pi$ cm long and the material has a yield strength of 500 MPa.
(a) Calculate the stress intensity factor at the tip of the crack.
(b) Calculate the plastic zone size at the crack tip.
(c) Comment upon the validity of this plastic zone correction factor for the above case.

8.3 A sharp penny-shaped crack with a diameter of 2.5 cm is completely embedded in a solid. Catastrophic fracture occurs when a stress of 700 MPa is applied.
(a) What is the fracture toughness for the material? (Assume that this value is for plane strain conditions.)
(b) If a sheet (0.75 cm thick) of this material is prepared for fracture toughness testing ($t = 0.75$, $a = 3.75$ cm), would the fracture toughness value be a valid test number (the yield strength of the material is given to be 1100 MPa)?
(c) What would be a sufficient thickness for valid K_{IC} determination?

8.4 Determine the angle θ (Fig. 8.10a) associated with the maximum dimension of the plastic zone.

8.5 Calculate the leak-before-break criteria in terms of flaw size and section thickness.

8.6 Prove to yourself that the ASTM criteria for a valid plane strain fracture toughness test (Eq. 8-42) reflects the conditions associated with Eqs. 8-43 and 8-44.

8.7 For the Ti-6 Al-4 V alloy test results given in Table 8.1, determine the sizes of the largest elliptical surface flaws ($a/2c \approx 0.2$) that would be stable when the design stress is 75% of σ_{ys}.

CHAPTER
NINE
TRANSITION TEMPERATURE APPROACH TO FRACTURE CONTROL

9.1 TRANSITION TEMPERATURE PHENOMENON AND THE CHARPY IMPACT SPECIMEN

Before fracture mechanics concepts were developed, engineers sought laboratory-sized samples and suitable test conditions with which to simulate field failures without resorting to the forbidding expense of destructively testing full-scale engineering components. To anticipate the worst possible set of circumstances that might surround a potential failure, these laboratory tests employed experimental conditions that could suppress the capacity of the material to plastically deform by elevating the yield strength: low test temperatures, high strain rates, and a multiaxial stress state caused by the presence of a notch or defect in the sample. Of considerable importance in pressure vessel, and bridge and ship structure applications was the fact that in body-centered cubic metals, such as ferritic alloys, the yield strength is far more sensitive to temperature and strain rate changes than it is in face-centered cubic metals, such as aluminum, nickel, copper, and austenitic steel alloys. As pointed out in Chapter Two, this increased sensitivity in BCC alloys can be related to the temperature-sensitive Peierls-Nabarro stress contribution to yield strength, which is much larger in BCC metals than in FCC metals.

To a first approximation, the relative notch sensitivity of a given material may be estimated from the yield to tensile strength ratio. When the ratio is low, the plastic constraint associated with a biaxial or triaxial stress state at the crack tip will elevate the entire stress-strain curve and allow for a net section stress greater than the smooth bar tensile strength value. Recall from Chapter Seven that a $2\frac{1}{2}$- to 3-fold increase in net section strength is possible in ductile materials that "notch strengthen" (Fig. 9.1). On the other hand, in materials that have less ability for plastic deformation, the stress concentration at a notch root is not offset by the necessary degree of crack tip plasticity needed to blunt the crack tip. Consequently, the notch with its multiaxial stress state raises the local stress to a high level and suppresses what little plastic deformation capacity the material possesses, and brittle failure occurs (Fig. 9.2).

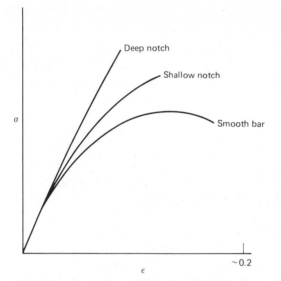

FIGURE 9.1   Plastic constraint resulting from triaxial stresses at notch root produces elevation of flow curve in ductile material.

The Charpy specimen (Fig. 9.3) and associated test procedure provides a relatively severe test of material toughness. The notched sample is loaded at very high strain rates because the material must absorb the impact of a falling pendulum and is tested over a range of temperatures. Considerable data can be obtained from the impact machine reading and from examination of the broken sample.

First, the amount of energy absorbed by the notched Charpy bar can be measured by the maximum height to which the pendulum rises after breaking the sample (Fig. 9.4). If a typical 325-J (240-ft-lb)* machine were used, the

*To convert from foot-pounds to joules, multiply by 1.356.

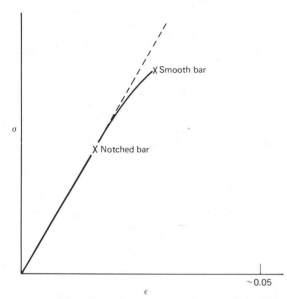

FIGURE 9.2 With little intrinsic plastic flow capacity, introduction of sharp crack induces premature brittle failure.

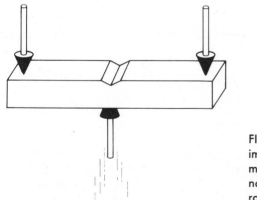

FIGURE 9.3 Standard three-point, impact loaded Charpy specimen (10 mm × 10 mm × 50 mm) with 45° notch 2 mm deep and with 0.25 mm root radius.

extreme final positions of the pendulum would be either at the same height of the pendulum before it was released (indicating no energy loss in breaking the sample), or at the bottom of its travel (indicating that the specimen absorbed the full 325 joules of energy). The dial shown in Fig. 9.4 provides a direct readout of the energy absorbed by the sample. Typical impact energy versus test temperature for several materials is plotted in Fig. 9.5. It is clearly evident from this plot that some materials show a marked change in energy absorption when a wide

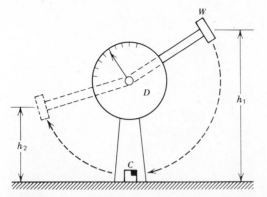

FIGURE 9.4 Diagram showing impact hammer W dropping from height h_1, impacting sample at C and rising to maximum final height h_2. Energy absorbed by sample, related to height differential $h_2 - h_1$, is recorded on dial D.

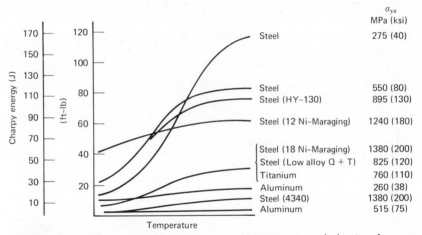

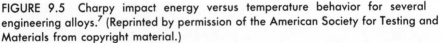

	σ_{ys} MPa (ksi)
Steel	275 (40)
Steel	550 (80)
Steel (HY–130)	895 (130)
Steel (12 Ni-Maraging)	1240 (180)
Steel (18 Ni-Maraging)	1380 (200)
Steel (Low alloy Q + T)	825 (120)
Titanium	760 (110)
Aluminum	260 (38)
Steel (4340)	1380 (200)
Aluminum	515 (75)

FIGURE 9.5 Charpy impact energy versus temperature behavior for several engineering alloys.[7] (Reprinted by permission of the American Society for Testing and Materials from copyright material.)

range of temperatures is examined. In fact, this sudden shift or transition in energy absorption with temperature has suggested to engineers the possibility of designing structural components with an operating temperature above which the component would not be expected to fail.

The effect of temperature on the energy to fracture has been related in low-strength ferritic steels to a change in the microscopic fracture mechanism: cleavage at low temperatures and void coalescence at high temperatures. The onset of cleavage and brittle behavior in low-strength ferritic steels is so closely related that "cleavage" and "brittle" often are used synonymously in the fracture literature. This is unfortunate since, in Chapter Seven, brittle is defined as a low level of fracture *energy* or limited crack tip plasticity, while cleavage describes a failure micro-*mechanism*. Confusion arises since brittle behavior can occur without cleavage, as in the fracture of high-strength aluminum alloys; alternately, you can have 4% elongation (reflecting moderate energy absorption) in a tungsten-25 a/o rhenium alloy specimen and still have a cleavage fracture.[1] Since a direct correlation does not always exist between a given fracture mechanism and the magnitude of fracture energy, it is best to treat the two terms separately.

Unless the fracture energy changes discontinuously at a given temperature, some criterion must be established to *define* the "transition temperature." Should it be defined at the 13.5, 20, or 27 J (10, 15, or 20 ft-lb)* level as it is sometimes done or at some fraction of the maximum or shelf energy? The answer depends on how well the defined transition temperature agrees with the service experience of the structural component under study. For example, Charpy test results

TABLE 9.1a   Transition Temperature Data for Selected Steels[3]

Material	$\dfrac{\sigma_{ys}, \text{MPa}}{\sigma_{ts}, \text{MPa}}$	Transition Temperature, °C		
		20 J	0.38 mm	50% fibrous
Hot rolled C-Mn steel	$\dfrac{210}{442}$	27	17	46
Hot rolled, low alloy steel	$\dfrac{385}{570}$	-24	-22	12
Quenched and tempered steel	$\dfrac{618}{688}$	-71	-67	-54

TABLE 9.1b   Transition Temperature Data for Selected Steels[3]

Material	$\dfrac{\sigma_{ys}, \text{ksi}}{\sigma_{ts}, \text{ksi}}$	Transition Temperature, °F		
		15 ft-lb	15 mil	50% fibrous
Hot rolled C-Mn Steel	$\dfrac{30.5}{64.1}$	80	62	115
Hot rolled, low alloy steel	$\dfrac{55.9}{82.6}$	-12	-7	53
Quenched and tempered steel	$\dfrac{89.7}{99.8}$	-95	-88	-66

*Dual units are retained for reference since the specific foot-pound energy levels cited above represent long-standing design criteria.

for steel plate obtained from failures of Liberty ships revealed that plate failures never occurred at temperatures greater than the 20 J (15 ft-lb) transition temperature. Unfortunately, the transition temperature criterion based on such a specific energy level is not constant but varies with material. Specifically, Gross[2] has found for several steels with strengths in the range of 415 to 965 MPa that the appropriate energy level for the transition temperature criterion should increase with increasing strength.

The same problem arises when the transition temperature is estimated from other measurements. For example, if the amount of lateral expansion on the compression side of the bar is measured, (Fig. 9.6), it is found that it, too, undergoes a transition from small values at low temperature to large values at high temperature. (This increase in observed plastic deformation is consistent with the absorbed energy-temperature trend.) Whether the correct transition temperature conforms to an absolute or relative contraction depends on the material. Finally, transitional behavior is found when the amount of fibrous or cleavage fracture on the fracture surface is plotted against temperature. A typical series of fracture surfaces produced at different temperatures is shown in Fig. 9.7a. Here again, the appropriate percentage of cleavage or fibrous fracture (based on comparison with a standard chart such as in Fig. 9.7b or measured

(a)

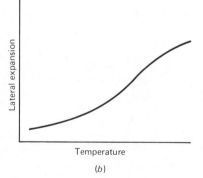

Temperature

(b)

FIGURE 9.6 (a) Measurement of lateral expansion at compression side of Charpy bar; (b) schematic of temperature dependence of lateral expansion revealing transition behavior.

directly as in Fig. 9.7c) to use to define the transition temperature will depend on the material as well as other factors. To make matters worse, transition temperatures based on either energy absorption, ductility, or fracture appearance criteria do not agree even for the same material. As shown in Table 9.1, the transition temperature defined by a 20 J energy criterion or by a 0.38 mm (15 mil) lateral expansion are in reasonably good agreement but are consistently lower than the 50% fibrous fracture transition temperature. Which transition temperature to use "is a puzzlement!"

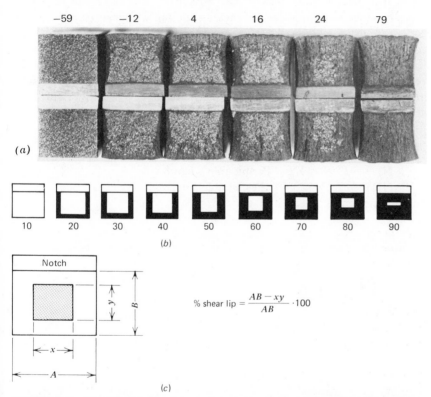

FIGURE 9.7 Transition in fracture surface appearance as function of test temperature. (a) Actual fracture series for A36 steel tested in the transverse direction; (b) standard comparison chart showing percent shear lip; (c) computation for percent shear lip.

$$\% \text{ shear lip} = \frac{AB - xy}{AB} \cdot 100$$

9.2 ADDITIONAL FRACTURE TEST METHODS

In addition to transition temperature determinations from Charpy data, related critical temperatures may be obtained from other laboratory samples, such as the drop-weight and Robertson crack arrest test procedures. The drop-weight sample[4] (Fig. 9.8) consists of a flat plate, one surface of which contains a notched bead of brittle weld metal. After reaching a desired test temperature, the plate is placed in a holder, weld-bead face down, and impacted with a falling weight. Since a crack can begin to run at the base of the brittle weld-bead notch with very little energy requirement, the critical factor is whether the base plate can withstand this advancing crack and not break. According to ASTM Standard E208, the nil ductility temperature (NDT) is defined as that temperature below which the plate "breaks" (Fig. 9.8a) but above which it does not (Fig. 9.8b). (The specimen is considered to be "broken" if a crack grows to one or both edges on the tension surface. Cracking on the compression side is not required to establish the "break" condition.) Therefore, NDT reflects a go, no go condition associated with a negligible level of ductility.

NDT test results have been used in the design of structures made with low-strength ferritic steels. For example, allowable minimum service temperatures (T_{min}) for structures containing sharp cracks have been defined[5] (but may be a function of plate thickness):

1. $T_{min} \geqslant NDT$: Permissible when applied stress σ is less than 35–55 MPa
2. $T_{min} \geqslant NDT + 17°C$ (30°F): Permissible when $\sigma \leqslant \sigma_{ys}/2$
3. $T_{min} \geqslant NDT + 33°C$ (60°F): Permissible when $\sigma \leqslant \sigma_{ys}$
4. $T_{min} \geqslant NDT + 67°C$ (120°F): Permissible since failure will not occur below the ultimate tensile strength of the material

These criteria represent specific conditions along the curve marked a_5 in Fig. 9.9, corresponding to a large flaw size. For progressively smaller flaw sizes, the allowable stress levels are seen to increase for a given minimum operating temperature. Information of this type has been collected and averaged by Pellini and Puzak[6] and presented in the form of fracture analysis diagrams (FAD) such as the one shown in Fig. 9.9.

Although some success has been achieved by using this diagram for low-strength steel applications, certain inherent deficiencies should be pointed out.[7] First, it is not altogether clear which stress, local, or gross section should be used. Also, the FAD treats all steels of the same strength level alike and ignores the possibility of toughness differences from one steel grade to another. The allowable flaw size to be tolerated by a structure is strongly dependent on the toughness of the material and a reasonable estimate of the stress level; if the latter quantities are not defined, the FAD cannot provide very reliable quantitative information. Nevertheless, these diagrams do represent an attempt to bridge the differences between the transition temperature and fracture mechanics philosophies of fracture control design.

The Robertson sample[8] is designed to measure a crack arrest condition. As seen in Fig. 9.10, this sample contains a saw cut at one side of the plate and is subjected to a thermal gradient across the plate width such that the starter notch is at the lowest temperature, while the right side of the plate is considerably

(*a*) Break   (*b*) No break

FIGURE 9.8 Drop weight test sample with notched weld bead. Used to measure nil ductility temperature (NDT). (*a*) Break, (*b*) no break. (Courtesy Dr. A. W. Pense, Lehigh University.)

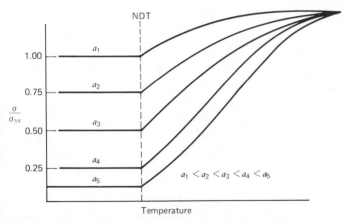

FIGURE 9.9 Fracture analysis diagram revealing allowable stresses as a function of flaw size and operating temperature. Curve a_5 corresponds to crack arrest temperature as measured from the Robertson test.

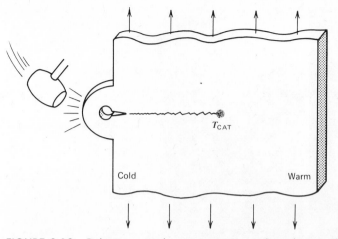

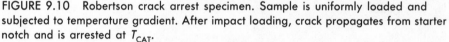

FIGURE 9.10 Robertson crack arrest specimen. Sample is uniformly loaded and subjected to temperature gradient. After impact loading, crack propagates from starter notch and is arrested at T_{CAT}.

warmer. After a uniform load is applied normal to the starter crack plane, the plate is impacted on the cold side, causing an unstable crack to grow from the cold starter notch root. The crack will run across the plate until it encounters a plate temperature at which the material offers too much resistance to further crack growth; this is defined as the crack arrest temperature (CAT). From such tests, it has been shown that CAT depends on the material, the magnitude of the applied stress, and the specimen thickness. As a final note, the CAT, when determined over a range of stress levels, provides the information for the construction of the lower curve (marked a_5) in the FAD shown in Fig. 9.9.

9.3 LIMITATIONS OF THE TRANSITION TEMPERATURE PHILOSOPHY

It is important to recognize some limitations in the application of the transition temperature philosophy to component design. First, the absolute magnitude of the experimentally determined transition temperature, as defined by any of the previously described methods (energy absorbed, ductility, and fracture appearance), depends on the thickness of the specimen used in the test program. This is due to the potential for a plane strain-plane stress, stress state transition when sample thickness is varied. In evaluating this effect, McNicol[3] found that the transition temperature in several steels, based on energy, ductility, and fracture appearance criteria, increased with increasing Charpy bar thickness t. Figure 9.11 shows temperature-related changes in energy absorbed per 2.5-mm sample thickness and percent shear fracture as a function of sample thickness for A283, a hot rolled carbon manganese steel. It is clear from this figure that the transition temperature increased with increasing thickness. Moreover, the transition temperature was different for the two criteria. With increasing sample thickness, it would be expected that the transition temperature would rise to some limiting value as full plane strain conditions were met. This condition is inferred from Fig. 9.12, which shows the transition temperature reaching a maximum level with increasing thickness for three different steel alloys.

It is clear, then, that the defined transition temperature will depend not only on the measurement criteria but also on the thickness of the test bar. Therefore, laboratory results may bear no direct relation to the transition temperature characteristics of the engineering component if the component's thickness is different from that of the test bar. To overcome this difficulty, the dynamic tear test (DT)[5] and drop weight tear test (DWTT)[9] were developed wherein the sample thickness was increased to the full thickness of the plate. As seen in Fig. 9.13, both tests involve three-point bending of a notched bar. The basic difference between the two is the notch detail: the DWTT contains a shallow notch

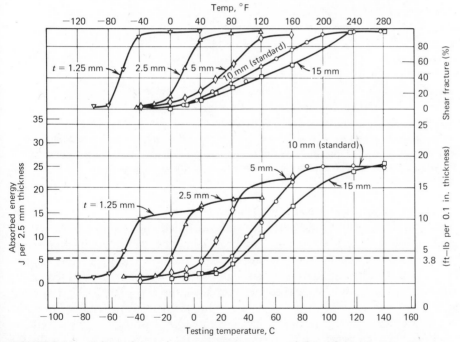

FIGURE 9.11   Adjusted energy-temperature curves and shear fracture-temperature curves for 38 mm thick plate of A283 steel tested with Charpy V-notch specimens of various thickness. Absorbed energy defined at 5.2 J/2.5 mm (3.8 ft-lb/0.1 in.) of specimen thickness.[3] (Reprinted from *Welding Journal* by permission of the American Welding Society.)

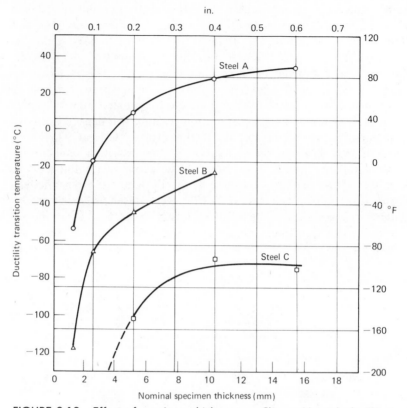

FIGURE 9.12   Effect of specimen thickness on Charpy V-notch ductility transition temperature of steels A, B, and C. The ductility transition temperature was selected with the same relative energy/unit thickness ratio given in Fig. 9.11.[3] (Reprinted from *Welding Journal* by permission of the American Welding Society.)

(5 mm deep) which is pressed into the edge of the sample with a sharp tool, while the DT notch is deeper and embedded within a titanium embrittled electron beam weld. These samples are broken in either pendulum or dropweight machines that are calibrated to measure the fracture energy of the sample. Hence, energy absorption versus test temperature plots can be obtained in the same manner as with Charpy specimens. As such, the DWTT and DT specimens may be considered to be oversized Charpy samples. The big difference lies in the fact that these samples are much thicker and wider than the Charpy specimen, resulting in much greater plastic constraint at the notch root. As a result, the transition temperature is shifted dramatically to higher temperatures (Fig. 9.14).[10] It is important to note from Fig. 9.14 that the DT, Robertson crack arrest test, and the drop weight NDT test results all indicate brittle material response at about −20°C for this material while the Charpy test indicates very

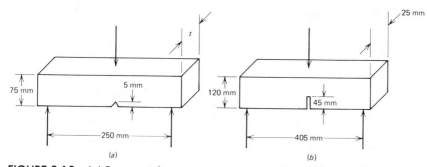

FIGURE 9.13 (*a*) Drop weight tear test specimen (DWTT) with shallow notch pressed into bar; (*b*) dynamic tear test specimen with machined slot introduced into titanium embrittled electron beam weld.

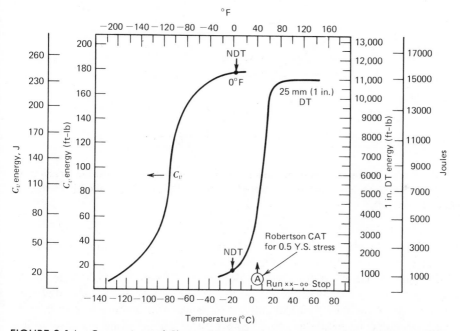

FIGURE 9.14 Comparison of Charpy V-notch, dynamic tear, drop-weight NDT, and Robertson crack arrest test results for A541 (Class 6 steel at 580 MPa yield strength). Note that the C_v test indicates a very high level of toughness at the NDT temperature that, in reality, corresponds to brittle behavior as indicated by the DT and Robertson test results.[10] (Reprinted by permission of the American Society for Testing and Materials from copyright material.)

tough behavior. Such sharp contrasts in test results are most disturbing when engineering design decisions must be made.

In addition to transition temperature-thickness effects, there are uncertainties relating to crack length effects as well. This may be seen by considering implications of a graphical representation of Eq. 8-22 (Fig. 9.15). We see the general relationship between flaw size and allowable stress level for a material with a given toughness level. The solid line represents the material toughness K_c,

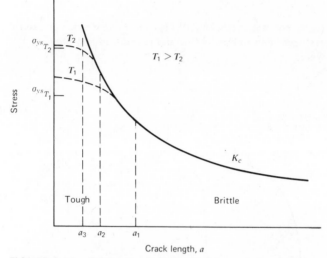

FIGURE 9.15 Schematic diagram showing relationship between allowable stress level and flaw size. Solid line represents material fracture toughness K_c; dashed lines show effect of plasticity.

assuming ideally elastic conditions, and the dashed portion of the curve reflects the reality of crack tip plasticity. On the basis of the necessary energy to break a component, brittle conditions would be associated with the right side of the curve, while tough behavior would be found under conditions associated with the left side of the plot. Consequently, it is seen that a notched bar with crack length a_1 would be brittle at room temperature, but the same material with a crack length a_3 would exhibit tough behavior. If a sample with intermediate crack length, say a_2, were tested, the material also would be tough at room temperature. Since the brittle region of this curve is truncated by the onset of plastic deformation when the applied stress reaches the material yield strength, the brittle domain can be expanded simply by lowering the test temperature. Consequently, if the test temperature were to be reduced from T_1 to T_2, the sample response with crack length a_2 would change from tough to brittle, and the sample with crack length a_3 would still exhibit high toughness. An additional temperature reduction would be necessary for this sample to exhibit brittle behavior.

From the above discussion, it becomes apparent that a wide range of "transition temperatures" can be obtained simply by changing the specimen thickness and/or crack length of the test bar. For this reason, transition temperature values obtained in the laboratory bear little relation to the performance of the full-scale component, thereby necessitating a range of correction factors as discussed earlier.

As mentioned above, the onset of brittle fracture is not always accompanied by the occurrence of the cleavage microscopic fracture mechanism. Rather, it should be possible to choose a specimen size for a given material, and tailor both thickness and planar dimensions such that a temperature-induced transition in energy to fracture, amount of lateral contraction, and macroscopic fracture appearance would occur *without the need for a microscopic mechanism transition*. Figure 9.16, from the work of Begley,[11] is offered as proof of this statement. Substandard thickness Charpy bars of 7075-T651 aluminum alloy were tested and shown to exhibit a temperature-induced transition in impact energy and fracture appearance. As noted in Fig. 9.5, no such transition was observed when standard Charpy specimens of an aluminum alloy were broken.

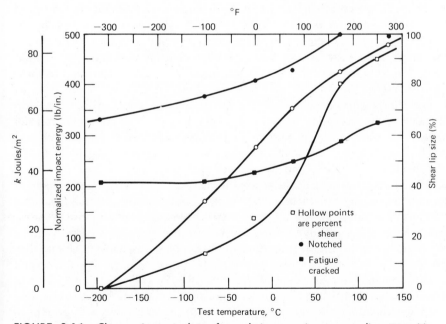

FIGURE 9.16 Charpy impact data for subsize specimen revealing transition temperature response in 7075-T651 aluminum alloy.[11] (Courtesy James A. Begley, Westinghouse Electric Co.)

9.4 IMPACT ENERGY-FRACTURE TOUGHNESS CORRELATIONS

Although handicapped by the inability to bridge the size gap between small laboratory sample and large engineering component, the Charpy test sample method does possess certain advantages, such as ease of preparation, simplicity of test method, speed, low cost in test machinery, and low cost per test. Recognizing these factors, many researchers have attempted to modify the test procedure to extract more fracture information and seek possible correlations between Charpy data and fracture toughness values obtained from fracture mechanics test samples. In one such approach. Orner and Hartbower[12] pre-cracked the Charpy sample so that the impact energy for failure represented energy for crack propagation but not energy to initiate the crack.

$$E_T = E_i + E_p \qquad (9\text{-}1)$$

where

$$E_T = \text{total fracture energy}$$

$$E_i = \text{fracture initiation energy}$$

$$E_p = \text{fracture propagation energy}$$

They found that a correlation could be made between the fracture toughness of the material $\mathcal{G}_c$ and the quantity W/A, where W is the energy absorbed by the precracked Charpy test piece and A the cross-sectional area broken in the test Although promising results have been observed for some materials (for example, see Fig. 9.17), the applicability of this test method should be restricted to those materials that exhibit little or no strain rate sensitivity, since dynamic Charpy data are being compared with static fracture toughness values. Also, the neglect of kinetic energy absorption by the broken samples as part of the energy transfer process from the loading pendulum to the specimen makes it impossible to develop good data in brittle materials where the kinetic energy component is no longer negligible.[13] Orner and Hartbower did point out, however, that the precracked Charpy sample could be used to measure the strain rate sensitivity of a given material by conducting tests under both impact and slow bending conditions. Barsom and Rolfe[14] have verified this hypothesis with a direct comparison of static and dynamic test results from precracked Charpy V-notch (CVN) and plane strain fracture toughness samples, respectively. First, they established the strain rate induced shift in transition temperature for several steel alloys in the strength range of 275 to 1725 MPa (Figs. 9.18 and 9.19 and Table 9.2). They noted that the greatest transition temperature shift was found in the low-strength steels and no apparent strain rate sensitivity was present in alloys with yield strengths in excess of 825 MPa. When these same materials were tested to determine their plane strain fracture toughness value, a corresponding shift was noted as a function of strain rate.

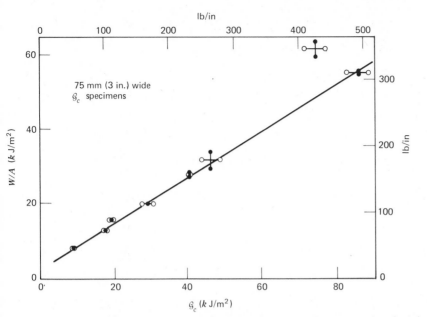

FIGURE 9.17 Relationship between fatigue-cracked V-notch Charpy slow bend and $\mathcal{G}_c$ in a variety of 3.2 mm (0.125-in.) thick aluminum alloys.[12] (Reprinted from *Welding Journal* by permission of the American Welding Society.)

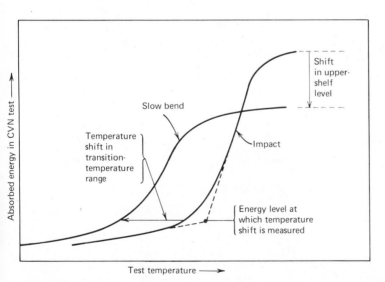

FIGURE 9.18 Diagram of impact energy versus test temperature revealing shift in transition temperature due to change in strain rate. (Note the higher shelf energy resulting from dynamic loading conditions, which may be related to a strain rate induced elevation in yield strength.)[14] (Reprinted by permission of the American Society for Testing and Materials from copyright material.)

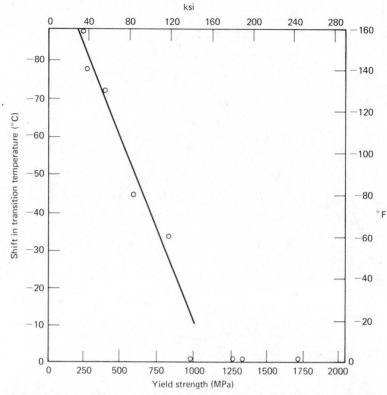

FIGURE 9.19 Effect of yield strength on shift in transition temperature between impact and slow bend CVN tests.[14] (Reprinted by permission of the American Society for Testing and Materials from copyright material.)

TABLE 9.2 Transition Temperature Shift Related to Change in Loading Rate[14]

| Steel | σ_{ys} | | Shift in Transition Temperature | |
	MPa	(ksi)	°C	(°F)
A36	255	(37)	−89	(−160)
ABS-C	269	(39)	−78	(−140)
A302B	386	(56)	−72	(−130)
HY-80	579	(84)	−44	(−80)
A517-F	814	(118)	−33	(−60)
HY-130	945	(137)	0	(0)
10 Ni-Cr-Mo-V	1317	(191)	0	(0)
18 Ni (180)	1241	(180)	0	(0)
18 Ni (250)	1696	(246)	0	(0)

Figures 9.20a and 9.21 show static (K_{IC}) and dynamic (K_{ID}) plane strain fracture toughness values plotted as a function of test temperature. One additional point should be made with regard to these data. Although K_{IC} increased gradually with temperature for the high-strength steels, a dramatic transition to higher values was observed for the low- and intermediate-strength alloys. It should be emphasized that this transition was not associated with the plane strain to plane stress transition, since all the data reported represented valid plane strain conditions. A similar transition in plane strain ductility (measured with a thin, wide sample) occurred in the same temperature region, but no such transition developed in axisymmetric ductility (measured with a conventional round tensile bar; see Fig. 9.20b). This tentative correlation between the K_{IC} and plane strain ductility transitions was strengthened with the observation that both transitions were associated with a fracture mechanism transition from cleavage at low temperatures to microvoid coalescence at high test temperatures.[15, 16]

From these data, Barsom and Pellegrino[16] suggested that the fracture strain near the crack tip could be given by the ductility associated with the plane strain ductility test sample. Assuming the plane strain ductility to be related to the crack opening displacement

$$\epsilon_{ps} = \alpha \delta^m \tag{9-2}$$

where

$$\epsilon_{ps} = \text{plane strain ductility}$$

$$\delta = \text{crack opening displacement}$$

$$m = \text{material parameter}$$

$$\alpha = \text{proportionality factor}$$

From Chapter Eight

$$\delta = 2V(c) = \frac{K^2}{E\sigma_{ys}} \tag{8-47}$$

which may be combined with Eq. 9-2 to yield

$$K^2 = \frac{E\sigma_{ys}\epsilon_{ps}^{1/m}}{\alpha^{1/m}} \tag{9-3}$$

or

$$K = A\sqrt{\sigma_{ys}}\left(\epsilon_{ps}^{1/2m}\right) \tag{9-4}$$

where

$$A = \text{material parameter}$$

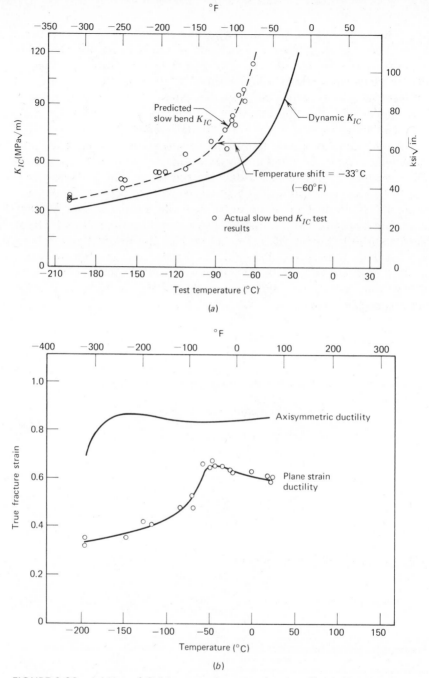

FIGURE 9.20 (a) Use of CVN test results to predict the effect of loading rate on K_{IC} for A517-F steel.[14] (Reprinted by permission of the American Society for Testing and Materials from copyright material.) (b) Variation of plane-strain ductility and axisymmetric ductility with temperature.[15] (Reprinted with permission from S. Rolfe, *Eng. Fract. Mech.*, 2(4), 1971, Pergamon Press.)

At fracture

$$K_{IC} = A \sqrt{\sigma_{ys}} \left(\epsilon_{f_{ps}}^{1/2m} \right) \tag{9-5}$$

By comparing data in a number of ferrous alloys ranging in strength from 550 to 1725 MPa, m was found to be about 0.25 such that

$$K_{IC} \approx A \sqrt{\sigma_{ys}} \left(\epsilon_{f_{ps}}^{2} \right) \tag{9-6}$$

Note that after combining Eqs. 9-2 and 9-6, the latter relationship has the same form as Eq. 8-47.

It is seen that the toughness levels of both strain-rate sensitive and insensitive materials increased with increasing temperature (Figs. 9.20a and 9.21). Of significance is the fact that the predicted static K_{IC} values (broken line), obtained by applying the appropriate temperature shift (Fig. 9.19) to the dynamic test results (solid line) were confirmed by experimentation. Since dynamic plane strain fracture toughness testing procedures are more complex and beyond the capability of many laboratories, estimation of K_{ID} from more

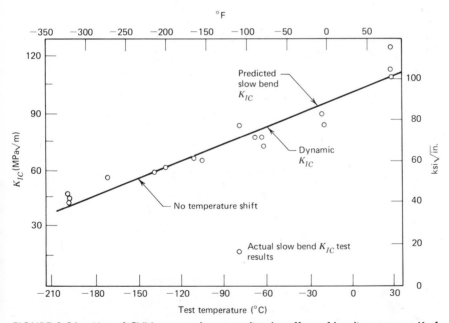

FIGURE 9.21   Use of CVN test results to predict the effect of loading rate on K_{IC} for 18 Ni-(250) maraging steel.[14] (Reprinted by permission of the American Society for Testing and Materials from copyright material.)

easily determined K_{IC} values represents a potentially greater application of the strain rate induced temperature shift in the determination of fracture properties.

Finally, empirical correlations have been developed between impact energy absorbed in DT[10] and Charpy[14] specimens and K_{IC} values such as shown in Figs. 9.22 and 9.23. Whereas the correlation shown in Fig. 9.23 is applicable in the upper shelf region, Barsom and Rolfe[15] proposed an additional K_{IC}–CVN correlation for use in the transition temperature region:

$$\frac{K_{IC}^{2}}{E} = 2(CVN)^{3/2} \text{ (English units)} \qquad (9\text{-}7)$$

More recently Sailors and Corten[17] reexamined this relationship in light of additional findings and recommended that the K_{IC} value for low- to medium-

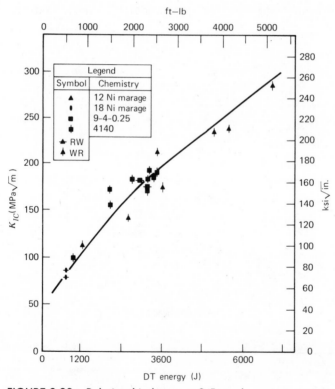

FIGURE 9.22   Relationship between 2.5 cm dynamic tear energy and K_{IC} values of various high-strength steels.[10] (Reprinted by permission of the American Society for Testing and Materials from copyright material.)

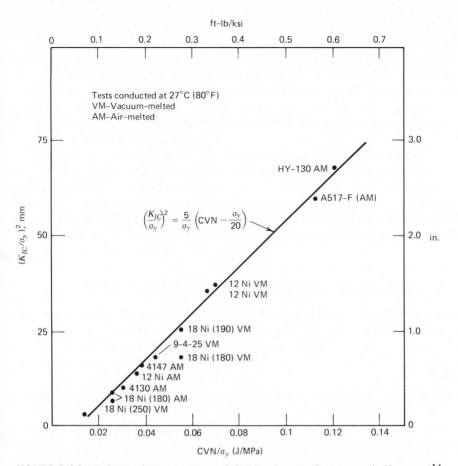

FIGURE 9.23 Relation between K_{IC} and CVN values in the upper shelf region.[14] (Reprinted by permission of the American Society for Testing and Materials from copyright material.)

strength steels in the transition temperature region be given by

$$\frac{K_{IC}^2}{E} = 8(CVN) \text{ (English units)} \qquad (9\text{-}8)$$

The reader is cautioned to fully appreciate that Eq. 9-7 and 9-8 are empirical correlations defined for specific steels and may not apply in situations involving other alloys.* Furthermore, test scatter associated with CVN results is usually much greater than that found for K_{IC} measurements, thereby reducing the

---

*Barsom showed recently the K_{IC}–CVN relationship for bridge steels was $K_{IC}^2/E \approx 5(CVN)$ (English units). See J. M. Barsom, *Eng. Fract. Mech.*, **7**, 1975, p. 605.

confidence one should place in K_{IC} estimations derived from these empirical K_{IC}–CVN correlations. In addition, some basic problems must not be overlooked. For example, the K_{IC}–CVN correlation implies that you can directly compare data from blunt and sharp notched samples and data from statically and dynamically loaded samples, respectively. The latter difficulty may not be too important for the materials shown in Fig. 9.23, since they all have yield strengths greater than 825 MPa (except A517-F) where strain rate effects are minimized (see Fig. 9.19 and Table 9.2). The same probably holds true for the DT–K_{IC} data in Fig. 9.22, since only high-strength materials are shown.

In summary, the empirical correlations between notched impact and fracture mechanics test data are useful in that they provide an estimate of K_{IC} as determined from simpler test procedures. However, they do possess a certain "apples versus bananas" incompatibility and, as such, should be considered strictly as *empirical* guidelines.

9.5 INSTRUMENTED CHARPY IMPACT TEST

In recent years, considerable attention has been given to instrumenting the impact hammer in the Charpy machine pendulum so as to provide more information about the load-time history of the sample during the test.[18, 19] A typical load-time trace from such a test is shown in Fig. 9.24. A curve of this type can provide information concerning the general yield load, maximum and fracture loads, and time to the onset of brittle fracture. To determine the fracture energy of the sample requires integration of a load-*displacement* record. However, it is possible to calculate the fracture energy from a load-*time* curve if

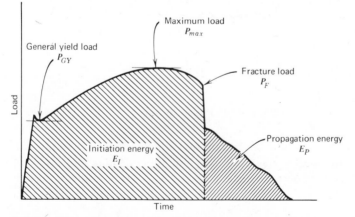

FIGURE 9.24 Drawing of load-time curve from an instrumented Charpy test.[20] (Reprinted by permission of John Wiley & Sons, Inc.)

the pendulum velocity is known. Assuming this velocity to be constant throughout the test, the fracture energy is computed to be

$$E_1 = V_0 \int_0^t P \, dt \qquad (9\text{-}9)$$

where

E_1 = total fracture energy based on constant pendulum velocity

V_0 = initial pendulum velocity

P = instantaneous load

t = time

In reality, the assumption of a constant pendulum velocity V is not valid. Instead, V decreases in proportion to the instantaneous load on the sample. From the work of Augland,[21] we find that

$$E_t = E_1 (1 - \alpha) \qquad (9\text{-}10)$$

where

E_t = total fracture energy

$$E_1 = V_0 \int_0^t P \, dt$$

$$\alpha = \frac{E_1}{4 E_0}$$

E_0 = initial pendulum energy

When computed total fracture energy values are compared[22] with conventionally determined results based on final pendulum position (direct read out from the impact machine), an almost one-to-one correlation is obtained (Fig. 9.25). On the basis of such good agreement, it has been possible for researchers to use Eq. 9-10 to separately compute the initiation and propagation fracture energies at any given test temperature. The ability to provide such information along with data relating to yielding, maximum, and fracture loads, has enabled materials engineers to more clearly identify the various stages in the fracture process. In addition, the instrumented Charpy test provides a relatively inexpensive screening test to compare material properties.

By precracking the Charpy sample and introducing side grooves (to enhance conditions of plain strain at the notch root) it was the hope of some engineers that instrumented Charpy testing could be used to determine G_{ID} and K_{ID}

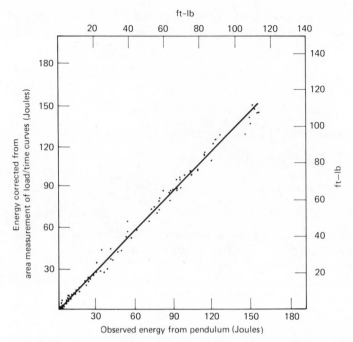

FIGURE 9.25 Comparison between fracture energy measured from final pendulum height and from load-time record.[22] (Reprinted by permission of the American Society for Testing and Materials from copyright material.)

values. In fact, G_{ID} could be determined in three separate ways:

1. By relating Charpy energy absorption, defined by pendulum position, to a G_{ID} level.
2. By relating Charpy energy absorption, defined from Eq. 9-10, to a G_{ID} level.
3. Computing G_{ID} from the appropriate K calibration equation for three-point bending using the maximum load from a load-time trace.

Unfortunately, this is not a realistic goal for many materials, since the Charpy specimen dimensions are generally too small to satisfy the ASTM specimen size requirement for crack length and sample thickness (Eq. 8-42).

REFERENCES

1. P. L. Raffo, NASA *TND-4567*, May 1968, Lewis Research Center, Cleveland, Ohio.
2. J. Gross, ASTM *STP 466*, 1970, p. 21.
3. R. C. McNicol, *Weld. Res. Suppl.*, Sept. 1965, p. 385s.

4. *1971 Annual Book of ASTM Standards*, Part 31, E208-69, p. 590.
5. W. S. Pellini, NRL Report 1957, U.S. Naval Research Laboratory, Sept. 23, 1969.
6. W. S. Pellini and P. Puzak, NRL Report 5920, U.S. Naval Research Laboratory, 1963.
7. W. T. Matthews, ASTM *STP 466*, 1970, p. 3.
8. T. S. Robertson, *Engineering*, **172**, 1951, p. 445.
9. *1971 Annual Book of ASTM Standards*, Part 31, E436-71T, p. 1005.
10. E. A. Lange and F. J. Loss, ASTM *STP 466*, 1970, p. 241.
11. J. A. Begley, "Fracture Transition Phenomena," Ph.D. Diss., Lehigh University, 1970.
12. G. M. Orner and C. E. Hartbower, *Weld. Res. Suppl.*, 1961, p. 405s.
13. W. F. Brown, Jr. and J. F. Srawley, ASTM *STP 410*, 1966.
14. J. M. Barsom and S. T. Rolfe, ASTM *STP 466*, 1970, p. 281.
15. J. M. Barsom and S. T. Rolfe, *Eng. Fract. Mech.*, **2**(4), 1971, p. 341.
16. J. M. Barsom and J. V. Pellegrino, *Eng. Fract. Mech.*, **5**(2), 1973, p. 209.
17. R. H. Sailors and H. T. Corten, ASTM *STP 514*, Part II, 1972, p. 164.
18. C. E. Turner, ASTM *STP 466*, 1970, p. 93.
19. R. A. Wullaert, ibid., p. 418.
20. T. R. Wilshaw and A. S. Tetelman, *Techniques of Metals Research*, Volume 5, Part 2, R. F. Bunshah, ed., Wiley-Interscience, New York, 1971, p. 103.
21. B. Augland, *Brit. Weld. J.*, **9**(7), 1962, p. 434.
22. G. D. Fearnehough and C. J. Hoy, *JISI*, **202**, 1964, p. 912.

PROBLEMS

9.1 Summarize the relative advantages and disadvantages of the transition temperature approach in analyzing the fracture of solids.

9.2 Multiaxial stress conditions may be beneficial or detrimental. Give examples of both situations and discuss the role of material properties.

9.3 Some investigators have established correlations between Charpy *CVN* and K_{IC} values. Discuss the potential pitfalls of such correlations.

9.4 What would happen to the relative position of the Charpy impact energy curves for the 275 MPa steel and the 1380 MPa 4340 steels shown in Fig. 9.5 if the specimens were tested slowly in bending?

9.5 Speculate as to why some materials exhibit a sharply defined tough-brittle transition temperature while others do not. Consider both macroscopic and microscopic factors in your evaluation.

9.6 Describe the effect of thickness and stress level on the CAT in the Robertson crack arrest sample (Fig. 9.10).

CHAPTER
TEN
MICROSTRUCTURAL
ASPECTS
OF FRACTURE
TOUGHNESS

Lest the reader become too enamored with the continuum mechanics approach to fracture control, it should be noted that the profession of metallurgy predates to a considerable extent the mechanics discipline. To wit: "And Zillah she also bore Tubal-cain, the forger of every cutting instrument of brass and iron."*

10.1   SOME USEFUL GENERALITIES

Before considering specific microstructural modifications that improve fracture toughness properties in engineering materials, it is appropriate to cite certain aspects of the material's structure that have a fundamental influence on fracture resistance. It has been pointed out that the deformation and fracture characteristics of a given material will depend on the nature of the electron bond, the crystal structure, and degree of order in the material.[1] The extent of brittle behavior based on these three factors is summarized in Table 10.1 for different types of materials.

It is seen that the more rigidly fixed the valence electrons, the more brittle the material is likely to be. Since covalent bonding involves sharing of valence electrons between an atom and its nearest neighbors only, materials such as diamond, silicon, carbides, nitrides, and silicates tend to be very brittle. Ionic bonding is less restrictive to the location of valence electrons; the electrons are

*Genesis, Chapter 4:22.

TABLE 10.1 Relationship Between Basic Structure of Solids and Their Effect on Brittle Behavior[1]

Basic Characteristic	Increasing Tendency for Brittle Fracture →		
Electron bond	Metallic	Ionic	Covalent
Crystal structure	Close-packed crystals	Low symmetry crystals	
Degree of order	Random solid solution	Short-range order	Long-range order

simply transferred from an electropositive anion to an electronegative cation. Furthermore, greater deformation capability is usually found in monovalent rather than multivalent ionic compounds. As mentioned in Chapter Three, plastic flow in ionic materials is also limited by the number of allowable slip systems that do not produce juxtaposition of like ions across the slip plane after a unit displacement. Metallic bonding provides the least restriction to valence electron movement; valence electrons are shared equally by all atoms in the solid. These materials generally have the greatest deformation capability.

As seen in Table 10.1, brittle behavior is more prevalent in materials of low crystal symmetry where slip is more difficult. On the other hand, considerable plastic deformation is possible in close-packed metals of high crystal symmetry. Finally, the ability of a given material to plastically deform generally will decrease as the degree of order of atomic arrangement increases. Consequently, the addition of a solute to a metal lattice will cause greater suppression of plastic flow whenever the resulting solid solution changes from that of a random distribution to that of short-range order and finally to long-range order.

Additional trends appear when one considers the propensity for brittle fracture based on fundamental engineering properties, such as yield and tensile strength and tensile ductility. Recall from Chapter One that toughness was defined by the area under the stress-strain curve; consequently, toughness would be highest when an optimum combination of strength and ductility is developed. Again, in Chapter Eight we saw that fracture toughness of a notched specimen depended on an optimum combination of yield strength and crack opening displacement [$V(c)$ (Eq. 8-48)]. Since $V(c)$ decreases sharply with increasing strength, a basic trend of decreased toughness with increased strength has been identified (see Table 8.1). It is apparent, then, that one is faced with a dilemma; metallurgists can limit a material's ability to deform by various strengthening procedures that enhance load-bearing capacity, but almost always to the detriment of the fracture toughness. As such, it is not difficult to raise the fracture toughness level of a material; one only needs to alter the thermomechanical treatment to lower strength and toughness increases as a consequence (Fig. 10.1). This approach is often impractical, however, since material requirements would be increased as a result of the material's lower load-bearing capacity. Of

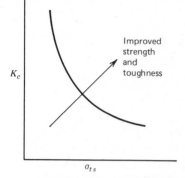

FIGURE 10.1 Diagram showing inverse relationship between fracture toughness and strength. Optimization of alloy properties would involve shifting curve in direction of the arrow.

course, one might be satisfied with lower alloy strength to achieve higher toughness if prevention of low-energy fracture were of paramount importance (for example, in the case of nuclear energy generating facilities).

The most desirable approach would involve shifting the curve in Fig. 10.1 up and toward the right so that the material might exhibit both higher strength and toughness for a given metallurgical condition. These improvements may be effected in several ways such as by:

1. Improved alloy chemistry and melting practice to remove or make innocuous undesirable tramp elements that degrade toughness.
2. Development of optimum microstructures and phase distributions to maximize toughness.
3. Microstructural refinement.

It would appear that these variables are extremely important to K_{IC}, since toughness levels can vary widely for the same material and in the same strength range (Fig. 10.2).

10.2 TOUGHNESS AND MICROSTRUCTURAL ANISOTROPY

Before discussing toughness improvement based on the approaches cited above, it is convenient to consider another approach to toughness improvement. This approach is concerned with means by which cracks may be deflected from their normal plane and direction of growth—that is, by taking advantage of material anisotropy.

As we saw from Eqs. 8-18 to 8-20, a triaxial tensile stress state is developed at the crack tip when plane strain conditions are present. Since fracture toughness was shown to increase with decreasing tensile triaxiality (for example, with thin sections where $\sigma_z \approx 0$), some potential for improved toughness is indicated if

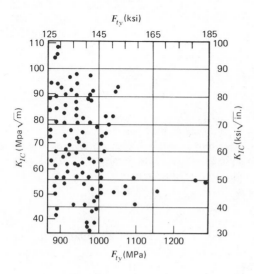

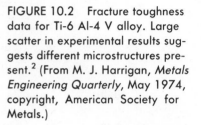

FIGURE 10.2 Fracture toughness data for Ti-6 Al-4 V alloy. Large scatter in experimental results suggests different microstructures present.[2] (From M. J. Harrigan, *Metals Engineering Quarterly*, May 1974, copyright, American Society for Metals.)

ways could be found to reduce the crack tip induced σ_x and/or σ_z stresses. One way to reduce the σ_x stress would involve the generation of an internally free surface perpendicular to σ_x and the direction of crack propagation. This can be accomplished by providing moderately weak interfaces perpendicular to the anticipated direction of crack growth, which could be pulled apart by the σ_x stress in advance of the crack tip (Fig. 10.3).[3] Since there can be no stress normal to a free surface, σ_x would be reduced to zero at this interface. In addition to reducing the crack tip triaxiality by generation of the internally free surface, the crack becomes blunted when it reaches the interface (Fig. 10.3b). Both conditions make it difficult for the crack to reinitiate in the adjacent layer with the result that toughness is improved markedly.

Embury et al.[4] conducted laboratory experiments to demonstrate the dramatic improvement in toughness arising from delamination, which can effectively arrest crack propagation. These investigators soldered together a number of thin, mild steel plates to produce a standard-sized Charpy impact specimen with an "arrester" orientation (Fig. 10.4a). As seen in Fig. 10.5a, the transition temperature for the "arrester" sample was found to be more than 130°C lower than that exhibited by homogeneous samples of the same steel. Additional confirmation of such favorable material response was reported by Leichter[6] who observed 163 J and 326 J Charpy impact energy absorption in "arrester" laminates of high-strength titanium and maraging steel alloys, respectively. Such energies are much higher than values expected from homogeneous samples of the same materials (Fig. 9.5).

The benefits of the crack arrester geometry have been utilized for many years in a number of component designs. For example, one steel fabricator has developed a procedure for on-site construction of large pressure vessels using a

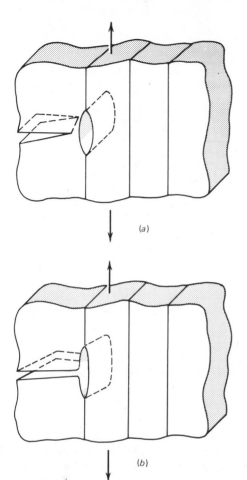

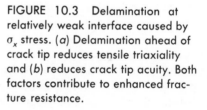

FIGURE 10.3 Delamination at relatively weak interface caused by σ_x stress. (a) Delamination ahead of crack tip reduces tensile triaxiality and (b) reduces crack tip acuity. Both factors contribute to enhanced fracture resistance.

number of tightly wrapped and welded concentric shells of relatively thin steel plate. This approach, though expensive to construct, boasts several advantages. First, only a *thin* layer of corrosion-resistant (more expensive) material would be needed (if at all) to contain an aggressive fluid within the vessel, as opposed to a full thickness vessel of the more expensive alloy. Second, the tightly wrapped layers are designed to create a favorable residual compressive stress on the inner layers, thus counteracting the hoop stresses of the pressure vessel. Third, the free surfaces between the layers act as crack arresters to a crack that might otherwise penetrate the vessel thickness. Finally, the metallurgical structure of thin plates (especially for low hardenability steels) is generally superior to that of thicker sections.

In another example of crack arrester design, large gun tubes often contain one

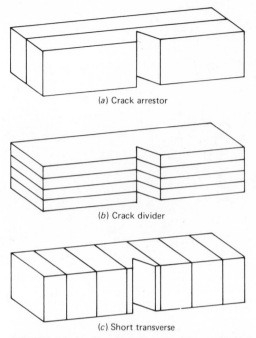

(a) Crack arrestor

(b) Crack divider

(c) Short transverse

FIGURE 10.4   Specimens containing relatively weak interfaces; (a) arrester; (b) divider; and (c) short transverse configurations.[5] (Reprinted with permission from American Society of Mechanical Engineers.)

or more sleeves shrunk fit into the outer jacket of the tube. Here, again, the procedure was originally developed to produce a favorable residual compressive stress in the inner sleeve(s), but serves also to introduce an internal surface for possible delamination and crack arrest. In one actual case history, a fatigue crack was found to have initiated at the inner bore of the sleeve, propagated to the sleeve-jacket interface, and proceeded around the interface, but *not* across the interface into the jacket itself. A similar crack "arrester" response is found in conventional materials given thermomechanical treatments that produce layered microstructures. McEvily and Bush[7] showed that a Charpy specimen made from ausformed steel (warm rolled above the martensite transformation temperature) completely stopped a 325 J (240 ft-lb) impact hammer when the carbide embrittled former austenite grain boundaries were oriented normal to the direction of crack propagation (Fig. 10.6).

Triaxiality can also be reduced by relaxing σ_z stresses brought about by de-lamination of interfaces positioned normal to the thickness direction. When delamination occurs, the effective thickness of the sample is reduced, since σ_z decreases to zero at each delamination. Consequently, the specimen acts like a series of thin-plane stress samples instead of one thick-plane strain sample. For

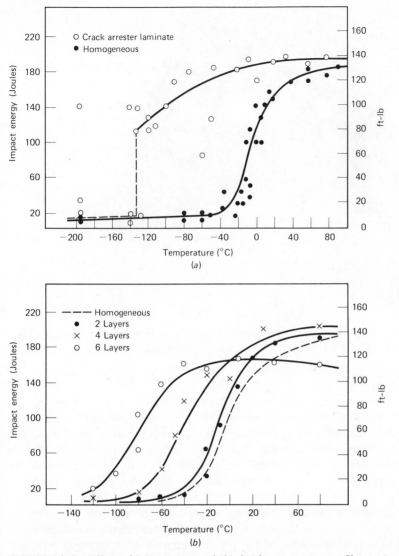

FIGURE 10.5   Effect of (*a*) arrester and (*b*) divider geometry on Charpy impact energy temperature transition.[4] (Reprinted with permission from American Institute of Mining, Metallurgical and Petroleum Engineers.)

FIGURE 10.6 Extensive delaminations in ausformed steel with "arrester" orientation. Specimen absorbed 325 J energy.[7] (Copyright American Society for Metals, 1962.)

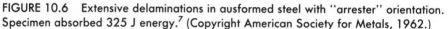

this reason, the resulting shift in transition temperature will depend on the number of weak planes introduced in the specimen—the more planes introduced, the thinner the delaminated segments will be and the greater the tendency for plane stress response. Embury et al.[4] conducted such tests with laminated samples in the "divider" orientation (Fig. 10.4b) and found the transition temperature to decrease with increasing number of weak interfaces (Fig. 10.5b). Note that at sufficiently low temperatures all the samples exhibited minimal toughness, indicating that even the thinnest layers were experiencing essentially plane strain conditions at these low temperatures. Leichter[6] confirmed the beneficial character of the divider orientation in raising toughness. For example, the fracture toughness of a titanium alloy was improved six- to seven-fold for a laminated sample over that of the full thickness sample made from the same material. It is interesting to note that the homogeneous samples exhibited an extensive amount of flat fracture, while each delaminated layer showed 100% full shear. As we saw in Chapter Eight, this difference in fracture mode appearance also reflects the improvement in fracture toughness.

The strength of the interface represents an important parameter in the delamination-induced toughening process. On one hand, the interface should not be so weak that the sample slides apart like a deck of playing cards. On the other hand, if the interface is too strong, delamination will not occur. In considering this point, Kaufman[8] demonstrated that the toughness of multi-layered, adhesive-bonded panels of 7075-T6 aluminum alloy was significantly greater than that shown by homogeneous samples with the same total thickness. Conversely, when similar multilayered panels of the same material were metallurgically bonded (resulting in strong interfaces), no improvement in toughness was observed over that of the homogeneous sample.

We must also consider the third possible orientation of weakened interfaces relative to the stress direction. As one might expect, fracture toughness properties are lowest in the short transverse orientation. It is as though all the positive increments in toughness associated with crack "arrester" and "divider" orientations are derived at the expense of short transverse properties. Because of the anisotropy of wrought products, fracture toughness values may be expected to vary with the type of specimen used to measure K_c or K_{IC}. This is not related to the specimen shape per se in terms of the K calibration but rather to the material anisotropy. For example, a rolled plate containing a surface flaw (arrester orientation) might be expected to exhibit somewhat higher toughness than the same material prepared in the form of an edge-notched plate, where the crack would propagate parallel to the rolling direction. To illustrate this behavior, the fracture toughness anisotropy in a number of wrought aluminum alloys is shown in Table 10.2 with the fracture toughness data given as a function of fracture plane orientation (first letter in code) and crack propagation direction (second letter in code) (Fig. 10.7). Additional data revealing fracture toughness anisotropy in aluminum, steel, and titanium alloys are given in Table 10.8 at the end of this chapter.

TABLE 10.2a Plane Strain Fracture Toughness Anisotropy in Wrought, High-Strength Aluminum Alloys[9]

Alloy and Temper Designation	Product	K_{IC}(MPa$\sqrt{m}$)		
		L-T	T-L	S-T
2014-T651	127 mm plate	22.9	22.7	20.4
7075-T651	45 mm plate	29.7	24.5	16.3
7079-T651	45 mm plate	29.7	26.3	17.8
7075-T6511	90 × 190 extruded bar	34.0	22.9	20.9
7178-T6511	90 × 190 extruded bar	25.0	17.2	15.4

A potentially dangerous condition—lamellar tearing—can develop because of the poor short transverse properties often found in rolled plate. Consider the consequences of a large T-joint weld such as the one shown in Fig. 10.8. After the weld is deposited, large shrinkage stresses are developed that act in the

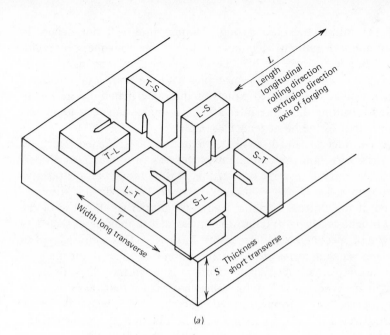

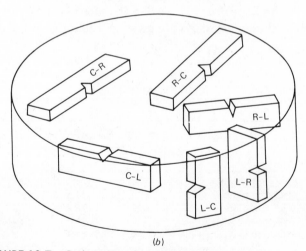

FIGURE 10.7 Code system for specimen orientation and crack propagation direction in (a) plate and (b) round bar stock.

TABLE 10.2b   Plane Strain Fracture Toughness Anisotropy in Wrought, High-Strength Aluminum Alloys[9]

Alloy and Temper Designation	Product	$K_{IC}(\text{ksi}\sqrt{\text{in}})$		
		L-T	T-L	S-T
2014-T651	5 in. plate	20.8	20.6	18.5
7075-T651	1 3/4 in. plate	27.0	22.3	14.8
7079-T651	1 3/4 in. plate	27.0	23.9	16.2
7075-T6511	3 1/2×7 1/2 in. extruded bar	30.9	20.8	19.0
7178-T6511	3 1/2×7 1/2 in. extruded bar	22.7	15.6	14.0

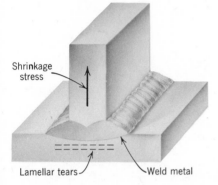

FIGURE 10.8   Lamellar tears generated along rolling planes as a result of weld shrinkage stresses.

thickness direction of the bottom plate. These stresses can be large enough to cause numerous microfissures at inclusion-matrix interfaces, which were aligned during the rolling operation. Clusters of these short transverse cracks can seriously degrade the weld joint efficiency and should be minimized if at all possible.

10.3   IMPROVED ALLOY CLEANLINESS

Although certain elements are added to alloys to develop the best microstructures and properties, other (tramp) elements serve no such useful purpose and are, in fact, often very deleterious. For example, we see from Fig. 10.9 that small amounts of oxygen have a severely embrittling effect on the fracture toughness of diffusion-bonded Ti-6 Al-4 V alloy.[10] Also, hydrogen in solid solution is known to produce hydrogen embrittlement in a number of high-strength alloys and their weldments. The latter problem is examined in the next chapter. For the moment we focus on those elements that contribute to the formation of

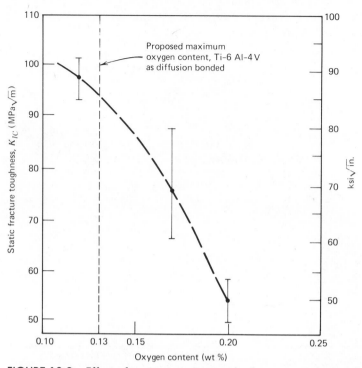

FIGURE 10.9 Effect of oxygen content on the fracture toughness of diffusion bonded Ti-6 Al-4 V.[10] (Copyright *Aviation Week & Space Technology.*)

undesirable second phases which serve as crack nucleation sites. Edelson and Baldwin[11] demonstrated convincingly that second-phase particles act to reduce alloy ductility (Fig. 10.10). The severe effect of sulfide inclusions on toughness in steel is shown in Fig. 10.11, where the Charpy V-notch shelf energy drops appreciably as sulfur content increases. Since the yield strength of this material is greater than 965 MPa, it would be interesting to compute the fracture toughness level K_{IC} for these alloys with different sulfur content, using the Barsom-Rolfe relationship described in Chapter Nine (Fig. 9.23). This will be left to the student as an interesting exercise. By using K_{IC} as the measure of toughness, Birkle et al.[13] demonstrated the deleterious effect of sulfur content at all tempering temperatures in a Ni-Cr-Mo steel (Fig. 10.12).

The task, then, is to remove sulfur, phosphorous, and gaseous elements (such as hydrogen, nitrogen, and oxygen) from the melt before the alloy is processed further. This has been done with a number of more sophisticated melting techniques developed in recent years. For example, melting in a vacuum rather than in air has contributed to a dramatic reduction in inclusion count and in the amount of trapped gases in the solidified ingot. To obtain a still better quality

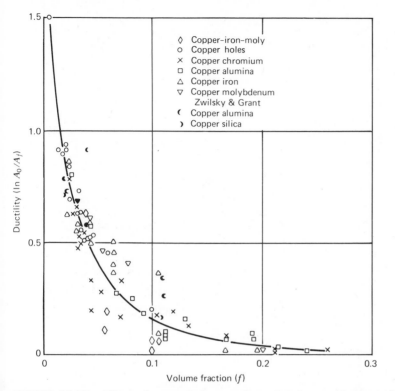

FIGURE 10.10 Effect of second-phase volume fraction on fracture ductility.[11] (Copyright American Society for Metals, 1962.)

steel, steels are vacuum arc remelted (VAR). In this process, the electrode (the steel to be refined) is remelted from the heat generated by the arc and the molten metal collected in a water-cooled crucible. Electroslag remelting (ESR) represents a variation of the consumable electrode remelting process; when the steel electrode is remelted, the molten metal droplets must first filter through a slag blanket floating above the molten metal pool. By carefully controlling the chemistry of the slag layer, various elements contained within the molten drops may be selectively removed.

As one might expect, removal of tramp elements increases the cost of the product. Although these costs are justifiable in terms of improved alloy behavior, the price of the final product may not be competitive in the marketplace. The task is to devise inexpensive means by which the tramp elements are rendered more harmless. One truly excellent example that we may discuss relates to correction of inferior transverse fracture properties in hot rolled, low alloy steels. As we saw in the previous section, the alignment of inclusions during rolling develops a considerable anisotropy in fracture toughness. In these alloys,

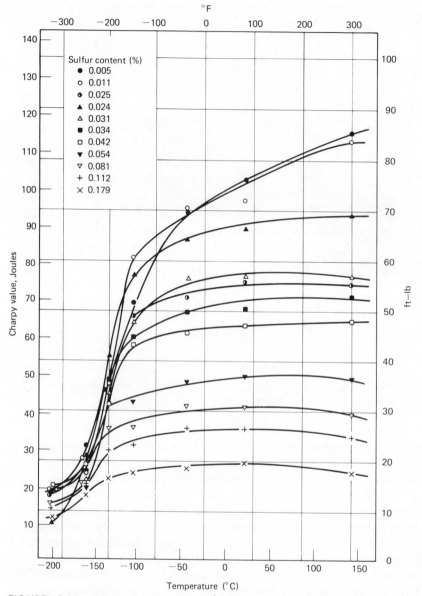

FIGURE 10.11   Effect of sulfur content on Charpy impact energy in steel plate (30 R_c).[12] (Reprinted with permission from American Institute of Mining, Metallurgical, and Petroleum Engineers.)

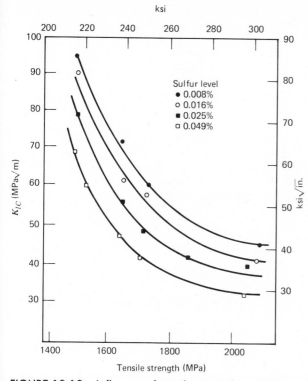

FIGURE 10.12 Influence of tensile strength and sulfur content on plane strain fracture toughness of 0.45 C-Ni-Cr-Mo steels.[13] (Copyright American Society for Metals, 1966.)

the objectionable particles are manganese sulfide inclusions that become soft at the hot rolling temperature and, consequently, smear out on the rolling plane and in the rolling direction (Fig. 10.13a). The result: very poor transverse fracture properties. Since these alloys are used in automotive designs where components are bent in various directions for both functional and aesthetic reasons, poor transverse bending properties severely restrict the use of these materials. The objective, then, was to suppress the tendency for softening of the sulfide at the hot rolling temperature and thereby preclude its smearing out on the rolling plane. This was accomplished by very small additions of rare earth metals to the melt. (Certainly rare earth elements are not cheap, but the small amounts necessary result in limited additional cost per ton of steel produced.) Luyckx et al.[14] found that the manganese in the sulfide was replaced by rare earth elements (mostly cerium), which produced more stable and higher melting point sulfides. Since these did not deform during hot rolling, they maintained their globular shape (Fig. 10.13b), thus giving rise to greater isotropy in fracture

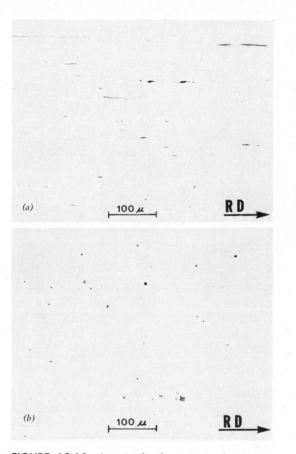

FIGURE 10.13 Longitudinal sections shown (*a*) elongated manganese sulfide inclusions in quenched and tempered steel without inclusion shape control; (*b*) globular rare earth inclusions found in hot rolled, low alloy steel with inclusion shape control.[15]

properties. This is demonstrated in Fig. 10.14 by noting the rise in transverse Charpy shelf energy with increasing cerium/sulfur ratio. Recent plane stress fracture toughness results comparing K_c anisotropy in a quenched and tempered steel that has no inclusion shape control with corresponding values from a rare earth modified, hot rolled, low alloy steel that has inclusion shape control are given in Table 10.3.[15] The K_c anisotropy ratio is obviously much higher for the steel without inclusion shape control.

No one would argue with the desirability of removing sulfides from the microstructure or at least making them more harmless. However, removal of carbides presents a more serious problem, since carbon both in solid solution and in carbides serves as a potent hardening agent in ferrous alloys. In addition,

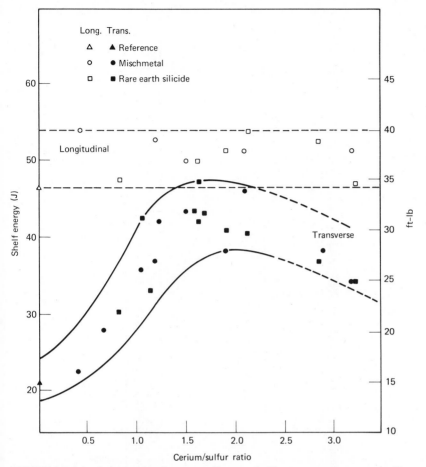

FIGURE 10.14　Relationship between Charpy shelf energy and cerium/sulfur ratio in longitudinal and transverse oriented half-sized impact specimens of VAN-80 steel.[14] (Copyright American Society for Metals, 1970.)

carbon provides the most effective means by which steel hardenability may be raised. And, yet, carbides provide the nucleation sites for many cracks. In a painstaking study designed to identify the origin of microcracks in high-purity iron, McMahon[16] demonstrated that almost every microcrack found could be traced to the fracture of a carbide particle (Table 10.4). This was found true even for the alloy that contained less carbon than the solubility limit. Of particular importance in this study was a critical evaluation of the relative importance of various mechanisms of microcrack nucleation proposed by a number of investigators.

TABLE 10.3 Fracture Toughness Anisotropy in Steels With and Without Inclusion Shape Control[15]

	Shape Control	σ_{ys} MPa	(ksi)	σ_{ts} MPa	(ksi)	% Uniform Elong.	K_c (long.) MPa$\sqrt{\text{m}}$	(ksi$\sqrt{\text{in.}}$)	K_c (trans.) MPa$\sqrt{\text{m}}$	(ksi$\sqrt{\text{in.}}$)
VAN-80	Yes	550	(80)	725	(105)	15	255–275	(230–250)	225	(206)
Q+T	No	910	(132)	1000	(145)	6	305	(275)	145	(132)

Alloy	C	Mn	S	P	Si	Cr	Ni	Cu	Zr	Al	N_2	V	Ce
VAN-80	0.15	1.40	0.005	0.004	0.52	—	—	—	—	0.052	0.018	0.096	0.02–0.03
Q+T	0.26	1.16	0.018	0.012	0.06	0.068	0.013	0.023	0.01	0.007	0.008	<0.005	—

TABLE 10.4   Initiation Sites of Surface Microcracks in Ferrite[16]

Material	0.035% C	0.035% C	0.005% C
Test temperature	$-140°C$	$-180°C$	$-170°C$
Total microcracks per 10^4 grains	66	43	17
Microcracks originating at cracked carbides	63	42	12
Microcracks probably originating at cracked carbides	3	1	4
Microcracks possibly originating at twin matrix interface	0	0	1

As we see from Fig. 10.15, McMahon found that cracks were much more likely to occur at embrittled grain boundaries and brittle second-phase particles than as a result of twin and/or slip band interactions. Consequently, although dislocation models proposed to account for microcrack formation, such as those shown in Fig. 10.16, appear to be valid for certain ionic materials,[19] they are precluded by other crack nucleation events in metals. Indeed, McMahon stated that "there appears to be no direct evidence of initiation of cleavage by slip band blockage in metals."[16]

Because of the negative side effects of carbon solid solution and carbide strengthening in ferrous alloys, attempts have been made to develop alloys that derive their strength instead by precipitation hardening processes involving

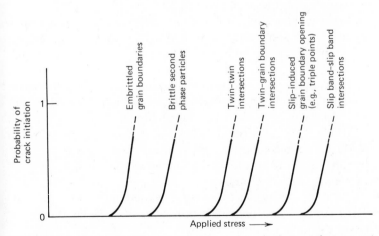

FIGURE 10.15   Probability as a function of applied stress that a particular microcrack formation mechanism will be operative.[16] (Reprinted with permission of Plenum Publishing Corporation.)

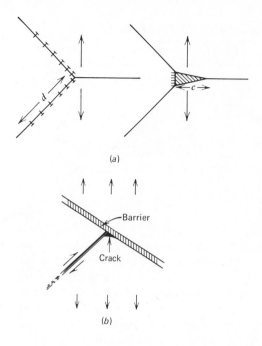

(a)

(b)

FIGURE 10.16 Dislocation models for crack nucleation: (a) Cottrell model[17] (reprinted with permission from American Institute of Mining, Metallurgical and Petroleum Engineers); and (b) Zener model[18] (reprinted from *Fracturing of Metals*, copyright, American Society for Metals, 1948.)

various intermetallic compounds. Maraging steels represent such a class of very low carbon, high alloy steels; they are soft upon quenching but harden appreciably after a subsequent aging treatment. It is seen from Fig. 10.17 that for all strength levels the toughness of maraging steels is superior to that of AISI 4340 steel, a conventional quenched and tempered steel. The chemistry of AISI 4340 steel and a typical maraging steel is given in Table 10.5.

It is felt that the lower carbon levels in maraging steels are partly responsible for their improved toughness and resistance to hydrogen, neutron, and temper embrittlement (see Chapter Eleven). Additional factors are discussed in Section 10.4.

Striking improvements in the fracture toughness of aluminum alloys also have been achieved by eliminating undesirable second-phase particles. Since precipitation hardening is achieved by dislocation interaction with closely spaced submicron-sized particles, the very large, dark inclusions (e.g., Al_7Cu_2Fe, $(Fe,Mn)Al_6$ and Mg_2Si), or secondary microconstituents (depending on your point of view), seen in Fig. 10.18 provide no strengthening increment. Instead, they provide sites for early crack nucleation.

Piper et al.[21] conducted an exhaustive study to determine how various elements affected the strength and toughness properties in 7178-T6 aluminum alloy.* Their results are summarized in Table 10.6.

*7178 aluminum alloy: 1.6–2.4 Cu, 0.70 max. Fe, 0.50 max Si, 0.30 Mn, 2.4–3.1 Mg, 6.3–7.3 Zn, 0.18–0.40 Cr, 0.20 max Ti.

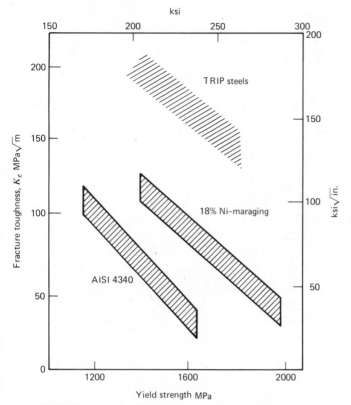

FIGURE 10.17   Fracture toughness-tensile strength behavior in AISI 4340, 18% Ni maraging, and TRIP steels.[20] (Reprinted with permission from V. F. Zackay and Elsevier Sequoia S. A.)

TABLE 10.5   Nominal Chemistry of Typical High-Strength Steels

Material	C	Ni	Cr	Mo	Si	Mn	Co	Ti	Al
					Composition				
AISI 4340	0.40	1.65–2.00	0.70–0.90	0.20–0.30	0.2–0.35	0.60–0.80	—	—	—
Maraging Steel	0.03 max.	18	—	5	(0.20 max.)	(0.20 max.)	8	0.4	0.1

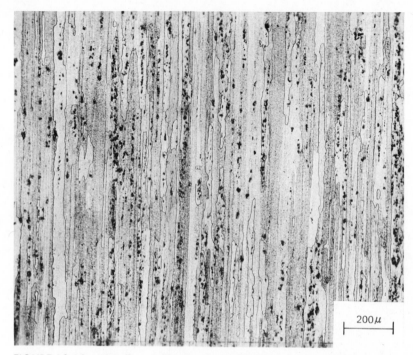

FIGURE 10.18 Metallographic section in 2024-T3 aluminum alloy revealing typically large number of Al_7Cu_2Fe second-phase particles.

It is seen that some elements (copper and magnesium) provide a solid solution strengthening component to alloy strength, while zinc and magnesium contribute a precipitation hardening increment. By comparing data from other investigators, Piper et al.[21] determined that strengthening of this alloy caused an expected reduction in fracture toughness. After examining strength and fracture toughness data in 18 different alloys, all reasonably close to the composition of 7178, they were able to isolate the strength–toughness relationship for the major alloying additions. Zinc was found to degrade K_c less for a given strength increment than that associated with the average response of the alloy. This would indicate that zinc is a desirable alloy strengthening addition. Although yield strength increments associated with copper and magnesium produced about average degradation in K_c, iron was found to degrade fracture toughness by the greatest amount ($3\frac{1}{2}$ times that associated with zinc additions). As expected, reduction in iron content brought about a significant improvement in alloy toughness and an associated reduction in the number of insoluble large particles. More recent results have confirmed the deleterious effect of iron *and* silicon content on fracture toughness (Figs. 10.19 and 10.20).[22–24] This has lead to the development of new alloys possessing the same general chemistry as

Element	Function
Zinc	Found in Guinier-Preston zones and subsequently found in $MgZn_2$ precipitates. Element acts as precipitation hardening agent.
Magnesium	Some Mg_2Si formation but mostly found in $MgZn_2$ precipitates and in solid solution.
Copper	Exists in solid solution, in $CuAl_2$ and Cu-Al-Mg type precipitates, and in Al_7Cu_2Fe intermetallic compounds.
Iron	Initially reacts to form Al-Fe-Si intermetallic compounds. Copper later replaces Si to form Al_7Cu_2Fe (the large black particles seen in Fig. 10.18).
Silicon	Initially reacts to form Al-Fe-Si compound prior to being replaced by Cu. Also forms Mg_2Si.
Manganese	Exact role not clear.
Chromium	Combines with Al and/or Mg to form fine precipitates, which serve to grain refine.

previous ones with the exception that iron and silicon contents are kept to an absolute minimum. Examples of these new materials include 2124 (the counterpart of 2024) and 7475 (the counterpart of 7075), which have the same strength as the older alloys but enhanced toughness (see Fig. 10.20 and Table 10.8).

The role of inclusions in initiating microvoids in a wide variety of aluminum alloys has been examined recently by Broek,[25] who showed that microvoid dimple size was directly related to inclusion spacing (Fig. 10.21). Large particles that fractured at low stress levels allowed for considerable void growth prior to final failure, but smaller particles nucleated and grew spontaneously to failure. Using an analysis similar to that proposed by McClintock[26] and supported by the work of Edelson and Baldwin[11] (Fig. 10.10), Broek suggested that the fracture strain was related to some function of the volume fraction of particles or voids. Hence $\epsilon_f \propto f(1/V)$, where ϵ_f = fracture strain and V = volume fraction of the second phase. Consequently, the toughness of these alloys would be expected to rise with decreasing particle content. Therefore, one would predict that had Broek examined low iron and silicon alloys in his investigation, he would have found these materials to reveal larger microvoids and fracture strains. Indeed, Kaufman[23] confirmed larger microvoids in the tougher, cleaner aluminum alloys he examined. (The effect of inclusion content on fracture toughness is discussed further in Section 10.6.)

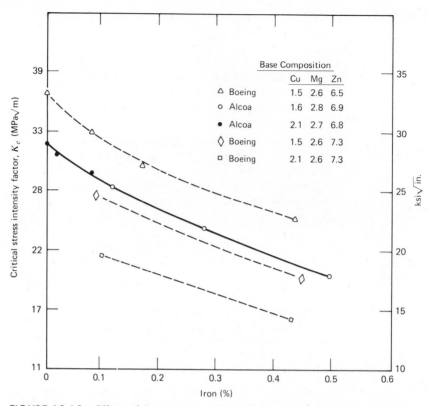

FIGURE 10.19 Effect of Fe content on fracture toughness of compositions within 7178 limits.[23] (Reprinted with permission from J. G. Kaufman, Alcoa Laboratories, Aluminum Company of America.)

10.4 OPTIMIZING MICROSTRUCTURES FOR MAXIMUM TOUGHNESS

10.4.1 Ferrous Alloys

Numerous studies have been conducted to determine which alloying elements and microstructures provide a given steel alloy with the best combination of strength and toughness. Since these structure-property correlations were established with many different properties (strength, ductility, impact energy, ductile-brittle transition temperature, and fracture toughness), it is difficult to make immediate data comparisons. There are, however, certain general statements that can be made with regard to the role of major alloying elements in optimizing mechanical properties. These are summarized in Table 10.7.

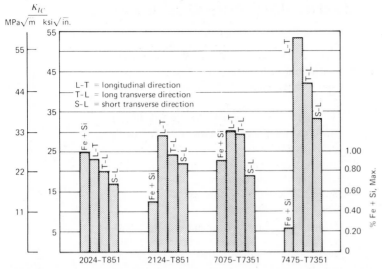

FIGURE 10.20   High-purity metal (low iron and silicon) and special processing techniques used to optimize toughness of 2xxx and 7xxx aluminum alloys.[24] (From R. Seng and E. Spuhler, *Metal Progress*, March 1975, copyright, American Society for Metals.)

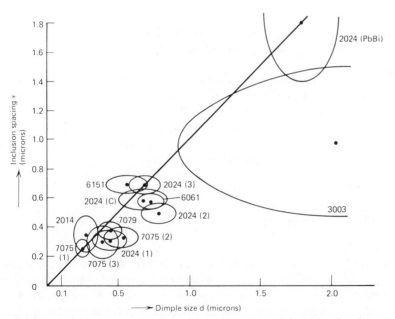

FIGURE 10.21   Observed relation between microvoid size and inclusion spacing. Numbers represent aluminum alloy designations; ellipses indicate scatter.[25] (Reprinted with permission from D. Broek, *Engineering Fracture Mechanics*, 1973, Pergamon Press.)

TABLE 10.7   Role of Major Alloying Elements in Steel Alloys

Element	Function
C	Extremely potent hardenability agent and solid solution strengthener; carbides also provide strengthening but serve to nucleate cracks.
Ni	Extremely potent toughening agent; lowers transition temperature; hardenability agent; austenite stabilizer.
Cr	Provides corrosion resistance in stainless steels; hardenability agent in quenched and tempered steels; solid solution strengthener; strong carbide former.
Mo	Hardenability agent in quenched and tempered steels; suppresses temper embrittlement; solid solution strengthener; strong carbide former.
Si	Deoxidizer; increases σ_{ys} and transition temperature when found in solid solution.
Mn	Deoxidizer; forms MnS, which precludes hot cracking caused by grain boundary melting of FeS films; lowers transition temperature; hardenability agent.
Co	Used in maraging steels to enhance martensite formation and precipitation hardening kinetics.
Ti	Used in maraging steels for precipitation hardening; carbide and nitride former.
V	Strong carbide and nitride former.
Al	Strong deoxidizer; forms AlN, which pins grain boundaries and keeps ferrite grain size small. AlN formation also serves to remove N from solid solution, thereby lowering lattice resistance to dislocation motion and lowering transition temperature.

The beneficial role of nickel in improving toughness and lowering the transition temperature (Fig. 10.22) has been recognized for many years and used in the development of new alloys with improved properties. For example, high nickel steels are presently being evaluated as candidates for cryogenic applications, such as in the construction of liquified natural gas storage tanks. The explanation for the improved response of nickel steels remains unclear despite

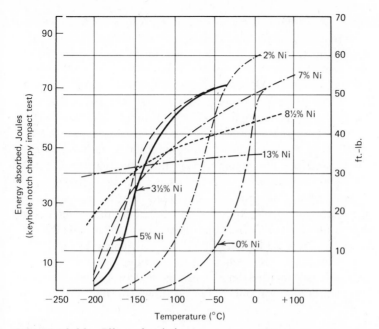

FIGURE 10.22 Effect of nickel content on transition temperature in steel.[27] (Reprinted with permission from the International Nickel Company, Inc., One New York Plaza, New York.)

the efforts of many researchers. In a recent review of the subject, Leslie[28] examined several theories and found all to be incomplete and/or incapable of rationalizing a number of exceptions to the proposed rule(s). One interesting result of Leslie's review, however, was the identification of platinum as an even better toughening agent than nickel when found in iron solid solutions. Obviously, the exorbitant cost of platinum will preclude its use as an alloying addition in steels.

Low[29] examined the effect of typical alloy steel microconstituents on toughness and concluded that the finer ones, namely lower bainite and martensite, provided greater fracture resistance than the coarser high-temperature transformation products such as ferrite, pearlite, and upper bainite. (The question of structural refinement is discussed further in Section 10.5.) More recently, Cox and Low[30] sought explanation(s) for the toughness difference between two types of important commercial steel alloys—quenched and tempered (for example, AISI 4340) and maraging steels (Fig. 10.17). One factor already mentioned (Section 10.4) was the beneficial effect of lower carbon content in the maraging steel. Proceeding further, they found that voids were nucleated in both alloys— nucleated by fracture of Ti (C,N) inclusions in maraging steels and at the interfaces between MnS inclusions and the matrix in AISI 4340 steel. However, a critical difference between the two alloys was noted in the crack growth and coalescence stage. In maraging steels, these voids grew until impingement caused coalescence and final failure. By contrast, the growth of the initial large voids in AISI 4340 was terminated prematurely by the development of void sheets—consisting of small voids—that linked the large voids (Fig. 10.23). This difference was verified by fractographic observations, which revealed uniform and relatively large microvoids in the maraging steel, but a duplex void size distribution in the 4340 steel. The small voids found in AISI 4340 were attributed to fracture of coarse carbide particles found along martensite lath boundaries. Since the strengthening intermetallic precipitates in the maraging steel are much finer and more resistant to fracture than the corresponding carbides in the quenched and tempered steel, Cox and Low concluded that the superior toughness shown by the maraging steels could be attributed to their much lower tendency to form void sheets. Correspondingly, they suggested that the quenched and tempered AISI 4340 steel could be toughened by a thermomechanical treatment resulting in refinement of the carbides.

So far, our discussion has focused on the mechanical properties associated with alloy steel transformation products. Let us now consider the fracture

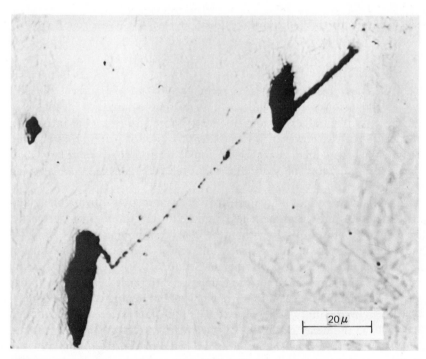

FIGURE 10.23 Large voids in AISI 4340 linked by narrow void sheets consisting of small microvoids.[30] (Copyright American Society for Metals, 1974.)

behavior of the parent austenite phase. As we shall see, this is both a complex and intriguing task. For one thing, the stability of the austenite phase can be increased through judicious alloying so as to completely stabilize this high-temperature phase at very low cryogenic temperatures or partially stabilize it at room temperature. Low-temperature stability of austenite (γ) is highly beneficial in light of the general observation that austenitic steels are tougher than ferritic (α) or martensitic (α') steels because of the intrinsically tougher austenite FCC crystal structure. In a study of AFC 77, a high-strength steel alloy containing both martensite and austenite microconstituents, Webster[31] showed that the fracture toughness level increased with increasing amount of retained austenite in the microstructure (Fig. 10.24). It is believed that the retained austenite phase in this alloy serves as a crack arrester or crack blunter, since it is softer and tougher than the martensite phase. By sharp contrast, retained γ in high carbon steels can damage overall material response when it undergoes an *ill-timed*, stress-induced transformation to untempered martensite, a much more brittle microconstituent. Several research groups have been experimenting with certain alloys that will undergo (in a carefully controlled manner) this mechanically induced phase transformation. The result has been the development of high-

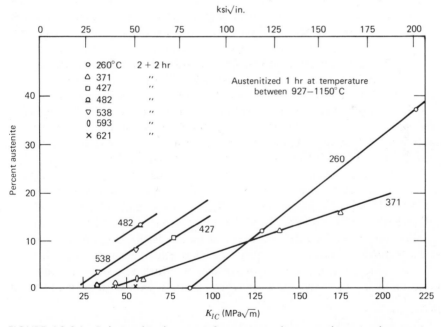

FIGURE 10.24 Relationship between fracture toughness and retained austenite content in AFC 77 high-strength alloy. Retained austenite content varies with austenitizing and tempering temperatures.[31] (Copyright American Society for Metals, 1968.)

strength steels possessing remarkable ductility and toughness brought about by transformation-induced plasticity in the material. These TRIP[32] (an acronym for *tr*ansformation-*i*nduced *p*lasticity) steels compare most favorably with both quenched and tempered and maraging steels (Fig. 10.17).

How can this be? How can you transform a tough phase γ into a brittle phase α' and produce a tougher alloy? Antolovich[33] argued that a considerable amount of energy is absorbed by the system when the $\gamma \rightarrow \alpha'$ transformation takes place. Assuming for the moment that the fracture energy of γ and α' is the same, the total fracture energy of the system would be the elastic and plastic energies of fracture for each phase plus the energy required for the transformation itself. Since the fracture energy of α' is lower than that of the γ phase, the toughness of the unstable γ alloy would be greater than that of a stable γ alloy, so long as the energy of transformation more than made up the loss in fracture energy associated with fracture of α' rather than the γ phase. Obviously, the toughness of the TRIP steel would be enhanced whenever the toughness difference between the two phases was minimized. An additional rationalization for the TRIP effect has been given, based on the 3% volumetric expansion associated with the $\gamma \rightarrow \alpha'$ transformation. It has been argued[34] that this expansion would provide for some stress relaxation within the region of tensile triaxiality at the crack tip. Bressanelli and Moskowitz[35] pointed out that the *timing* of the transformation was extremely critical to alloy toughness. Transformation to the more brittle α' phase was beneficial only if it occurred during incipient necking. That is, if martensite formed at strains where plastic instability by necking was about to occur, the γ matrix could be strain hardened and, thereby, resist neck formation. If the $\gamma \rightarrow \alpha'$ transformation occurred at lower stress levels prior to necking because the alloy was very unstable, brittle α' would be introduced too soon with the result that the alloy would have lower ductility and toughness. Note in this connection that prestraining these alloys at room temperature would be very detrimental. At the other extreme, if alloy stability were too high, the transformation would not occur and the material would not be provided with the enhanced strain hardening capacity necessary to suppress plastic necking. Some success has been achieved in relating the fracture toughness of a particular TRIP alloy to the relative stability of the austenite phase.[36,37]

Although the TRIP process offers considerable promise as a means to improve alloy toughness, engineering usage of materials utilizing this mechanism must await further studies to identify optimum alloy chemistry and thermomechanical treatments. Equally important will be efforts to reduce the unit cost of these materials to make them more competitive with existing commercial alloys.

10.4.2 Nonferrous Alloys

Microstructural effects are also important when attempting to optimize the toughness of titanium based alloys. For example, it has been found that toughness depends on the size, shape, and distribution of different phases that are present (Fig. 10.25). We see, for example, that metastable β (BCC phase) alloys possess the highest toughness with α (HCP phase) + β alloys generally being inferior. Furthermore, in these mixed phase alloys, acicular α rather than equiaxed α within a β matrix is found to provide superior toughness. It is quite probable that the large amount of scatter in K_{IC} values shown in Fig. 10.2 for Ti-6 Al-4 V (an $\alpha + \beta$ alloy) was largely a result of variations in the microstructures just described.

Let us now reexamine the observation made in Chapter Eight (Table 8.1) that the 2024-T3 aluminum alloy possessed higher toughness than the 7075-T6 sister alloy. Although this is true when comparison is based on the *different* strength levels designated for these alloys, the 7075-T6 alloy actually is the tougher material when compared at the same strength level (Fig. 10.26).

Nock and Hunsicker[40] demonstrated that the superiority of 7000 series alloys was attributable to a relatively small amount of insoluble intermetallic phase and to a reduced precipitate size, which would be less likely to fracture (25 to 50 Å in 7075-T6 versus 500 to 1000 Å in 2024-T86). Recent metallurgical studies

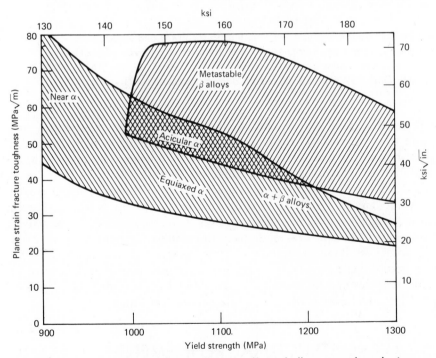

FIGURE 10.25 Schematic diagram showing effect of alloy strength and microconstituents on toughness in titanium alloys.[38] (Reprinted with permission from A. Rosenfeld and A. J. McEvily, Jr., NATO AGARD Report 610, Dec. 1973, p. 23.)

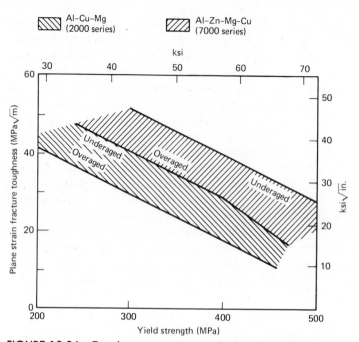

FIGURE 10.26 Toughness versus strength data for 2000 and 7000 series aluminum alloys revealing superior toughness in the latter at any given strength level.[38] (Reprinted with permission from A. Rosenfeld and A. J. McEvily, Jr., NATO AGARD Report 610, Dec. 1973, p. 23, based on data from Develay.[39])

have been concerned with optimizing the fracture toughness, strength, and resistance to environmental attack of various aluminum alloys. For example, it has been shown[22,23,38] that while strength decreases, toughness is improved when the material is underaged with a somewhat smaller improvement being associated with the overaged condition (Fig. 10.27). However, the overaged alloy, with its greater resistance to stress corrosion cracking, is preferred, even though the toughness level is somewhat lower than that obtained in the under-aged condition. Attempts are being made to combine mechanical deformation with variations in aging procedures to optimize material response.[23,41–43] Pre-liminary results have shown that toughness increases with decreasing size of Al_2CuMg particles and minimization of $Al_{12}Mg_2Cr$ dispersoids.[43] More studies of important engineering alloys are expected in the near future.

10.5 MICROSTRUCTURAL REFINEMENT

Microstructural refinement represents a unique opportunity by which the material may be *both* strengthened and toughened (Fig. 10.28). This represents a

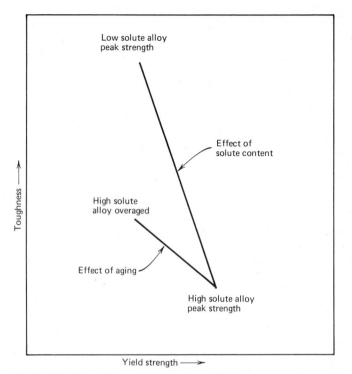

Low solute alloy
peak strength

Effect of
solute content

High solute
alloy overaged

Toughness

Effect of aging

High solute alloy
peak strength

Yield strength

FIGURE 10.27 Effect of decreasing solute content and overaging on the toughness-yield strength relationship of 7xxx alloy products.[23] (Reprinted with permission from J. G. Kaufman, Alcoa Research Laboratories.)

particularly attractive strengthening mechanism in view of the generally observed inverse relationship between strength and toughness (Figs. 10.1, 10.12, 10.17, 10.25, 10.26, and 10.27). The toughness and strength superiority of fine-grained materials has been recognized for many years, as evidenced by the well-accepted view that quenched and tempered steel alloy microstructures are superior to those associated with the normalizing process. (Quenched and tempered steels contain the finer transformation products, such as lower bainite and martensite, while normalizing produces coarser aggregates of proeutectoid ferrite and pearlite.) One beneficial effect of grain refinement is revealed by a reduction in the ductile-brittle transition temperature, as shown in Fig. 10.29. In addition to illustrating the beneficial effect of grain refinement on transition temperature, this figure reveals a shift to lower transition temperatures in the "controlled" as opposed to the "standard" specimens. Kapadia et al.[45] demonstrated that the superior behavior of the controlled group of samples was attributable to enhanced delamination and associated stress relaxation in divider type samples which resulted from a thermomechanical treatment designed to

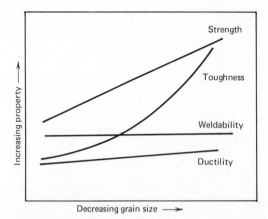

FIGURE 10.28 Diagram revealing simultaneous improvement in alloy strength and toughness with decreasing grain size. Ductility and weldability are not impaired.[44] (Reprinted with permission from American Society of Agricultural Engineers.)

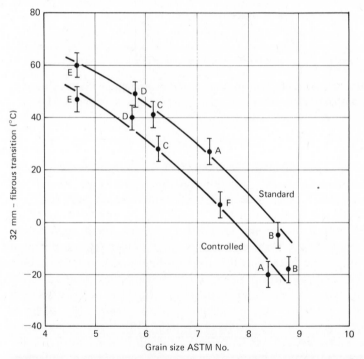

FIGURE 10.29 Effect of grain refinement in reducing transition temperature in hot rolled steel. "Controlled" specimens exhibited more delaminations in crack "divider" configuration to account for superior behavior.[45] (Copyright American Society for Metals, 1962.)

accentuate mechanical fibering. More recent data reveal improvements also in K_{IC} levels with reduced grain size. Mravic and Smith[46] reported that a 0.3 C-0.9 Mn-3.2 Ni-1.8 Cr-0.8 Mo steel with a prior austenite ultra-fine grain size of ASTM No. 15 (produced by multiple-cycle rapid austenitizing), exhibited K_{IC} values in the range of 100 to 110 MPa$\sqrt{m}$ at a strength level of 1930 to 2000 MPa. By comparison, for the same strength level, 4340 steel with a conventional grain size of about ASTM No. 7 exhibits a K_{IC} value of about 55 MPa$\sqrt{m}$.

To explain the beneficial role of structural refinement on toughness it may be argued that a microcrack will be stopped by an effective barrier (the grain boundary) more often the finer the grain size. As a result, the crack is forced to reinitiate repeatedly, and considerable energy is expended as it alters direction in search of the most likely propagation plane in the contiguous grain. Recall from Chapter Seven that this twisting of the crack front at the boundary gives rise to "river patterns" on cleavage fracture surfaces. One may argue, too, that fine-grained structures produce smaller potential flaws, thereby increasing the stress necessary for fracture (Eq. 8-6).

A number of investigators have attempted to describe the role of grain size in cleavage fracture for materials that undergo a temperature-sensitive fracture mechanism transition (see Chapter Nine). Cottrell[17,47] and Petch[48] used dislocation theory to independently develop similar relationships that could account for the effect temperature and various metallurgical factors have on the likelihood for cleavage failure. They found that the fracture stress could be given by

$$\sigma_f \approx \frac{4G\gamma_m}{k_y} d^{-1/2} \tag{10-1}$$

where

σ_f = fracture stress

G = shear modulus

γ_m = plastic work done around a crack as it moves through the crystal

k_y = dislocation locking term from Hall-Petch relationship (Eq. 3-16)

d = grain size

The increase in σ_f with decreasing grain size parallels a similar increase in yield strength with grain refinement. The familiar Petch-Hall relationship for yield strength is given by

$$\sigma_{ys} = \sigma_i + k_y d^{-1/2} \tag{3-16}$$

where

σ_{ys} = yield strength

σ_i = lattice resistance to dislocation movement resulting

from various strengthening mechanisms and intrinsic

lattice friction (Peierls stress)

k_y = dislocation locking term

d = grain size

As seen in Fig. 10.30, Low[49] demonstrated σ_f to be more sensitive to grain size than the associated yield strength σ_{ys}. There are some important implications to be drawn from these data. First, the intersection of the yield strength and fracture strength curves represents a transition in material response. For large grains (greater than the critical size), failure must await the onset of plastic flow; hence, fracture occurs when $\sigma = \sigma_{ys} = \sigma_f$. For grains smaller than the critical size, yielding occurs first and is followed by eventual failure after a certain amount of plastic flow—the amount increasing with decreasing grain size. The latter situation reflects greater toughness with an increasing ratio σ_f / σ_{ys}. Since σ_f and

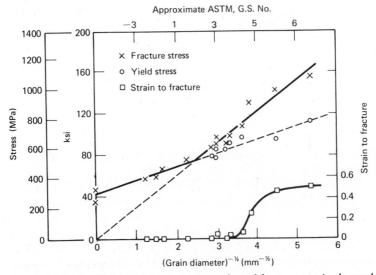

FIGURE 10.30 Yield and fracture strength and fracture strain dependence on grain size in low carbon steel at $-196°C$.[49] (Reprinted from *Relation of Properties to Microstructure*, copyright, American Society for Metals, 1954.)

σ_{ys} are temperature-sensitive properties, the critical grain size for the fracture transition would be expected to vary with test temperature. Consequently, the transition temperature is shown to decrease strongly with decreasing grain size (Fig. 10.31). As such, grain refinement serves to increase yield strength (Eq. 3-16) and fracture strength (Eq. 10-1), while lowering the ductile-brittle transition temperature (Fig. 10.31).

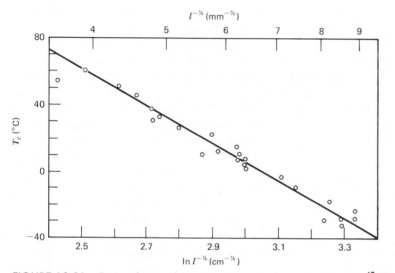

FIGURE 10.31   Dependence of transition temperature on grain size.[48] (Reprinted with permission from MIT Press.)

The significance of the terms in Eq. 10-1 has been treated at greater length by Tetelman and McEvily.[50] They argued that γ_m should increase with increasing number of unpinned dislocation sources, temperature, and decreasing crack velocity. Obviously, the more dislocations that can be generated near the crack tip the more blunting can take place and the tougher the material will be. However, when these sources are pinned by solute interstitials, such as nitrogen and carbon in the case of steel alloys, or highly immobile, as in ionic or covalent materials, because of a high Peierls stress, γ_m and σ_f are reduced. The beneficial effect of increased test temperature may be traced to a reduction in the Peierls stress and an increase in dislocation velocity. As we saw in Chapter Three, dislocation velocity was found to depend on the applied shear stress

$$v = \left(\frac{\tau}{D}\right)^m \tag{3-9}$$

where

$$v = \text{dislocation velocity}$$

$$\tau = \text{applied shear stress}$$

$$D, m = \text{material properties}$$

With increasing temperature, D decreases so that for the same applied stress, dislocation velocity will increase, thereby enabling dislocations to move more rapidly to blunt the crack tip. In related fashion, a slower crack velocity will provide more time for dislocations to glide to the crack tip to produce blunting. In short, anything that enhances the number of mobile dislocations, their mobility and speed, and the time allowed for such movement will increase γ_m and σ_f and contribute to improved toughness. By comparing Eqs. 10-1 and 3-16, it should be noted that strengthening mechanisms (such as solid solution strengthening, precipitation hardening, dispersion hardening, and strain hardening) that restrict the number of free dislocations and their mobility contribute toward increasing σ_i, while at the same time reducing the magnitude of γ_m. Therefore, attempts to increase yield strength by increasing σ_i are counterproductive, since γ_m and σ_f decrease. Likewise, k_y can be adjusted to improve σ_f or σ_{ys} but only at the expense of the other. Enhanced dislocation locking will increase σ_{ys} but will decrease σ_f directly (Eq. 10-1) and indirectly (since the number of mobile dislocations and γ_m decrease). In this regard, the more brittle nature of nitrogen bearing steel is attributed to its stronger dislocation locking character (Fig. 10.32).

We may summarize this discussion by stating that the only way to improve σ_{ys}, σ_f, and toughness simultaneously is not by changing γ_m, σ_i, or k_y but rather by decreasing grain size. By using the transition temperature as a measure of toughness (higher toughness corresponding to a lower transition temperature) the diagrams in Fig. 10.33 illustrate that only by grain refinement can you have both high yield strength and good toughness. Equations 10-1 and 3-16 are instructive in identifying those parameters that affect the temperature of the fracture mechanism transition (from void coalescence to cleavage). However, the reader should recognize that these relationships are not applicable for materials that do not cleave but which, however, may undergo a stress-state controlled fracture energy transition (Chapter Nine).

10.6 K_{IC}-MECHANICAL PROPERTY CORRELATIONS

Since K_{IC} testing is fairly complex and expensive by most standards, it would be extremely valuable to be able to determine K_{IC} on the basis of more readily obtained mechanical properties, such as those associated with a tensile test. One

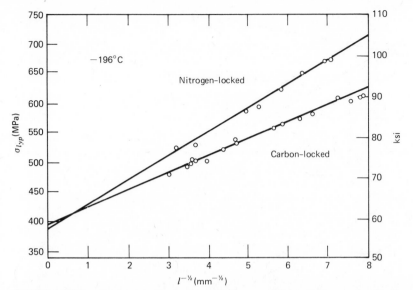

FIGURE 10.32 Grain size dependence of lower yield point in steel reflecting greater dislocation locking k_y with nitrogen interstitial.[48] (Reprinted with permission from MIT Press.)

such model—that relating plane strain fracture ductility to K_{IC}—was discussed previously in Chapter Nine. In another study, Hahn and Rosenfield[51] related the critical strain to the crack opening displacement at fracture and showed that K_{IC} could be estimated from tensile test results according to Eq. 10-2.

$$K_{IC} \approx n\sqrt{2E\sigma_{ys}\epsilon_f/3} \qquad (10\text{-}2)$$

where

$n =$ strain hardening coefficient

$E =$ modulus of elasticity

$\sigma_{ys} =$ tensile yield strength

$\epsilon_f =$ true fracture strain in uniaxial tension

The $\epsilon_f/3$ term in Eq. 10-2 was introduced to reflect plane strain conditions, with plane strain fracture ductility estimated as one-third the value associated with uniaxial stress conditions. Computed values of K_{IC} based on this equation were within ±30% of actual values for several materials. Recent efforts[52,53] to quantify the relationship between K_{IC} and alloy cleanliness (Section 10.3) have

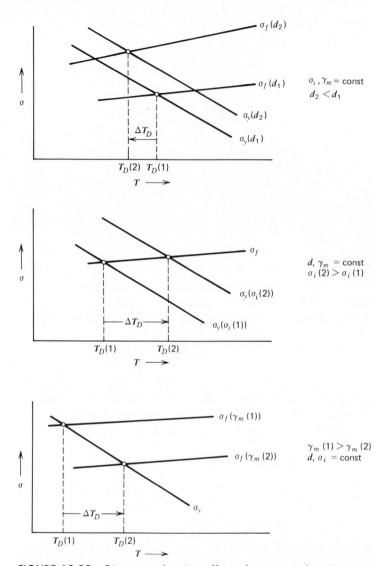

FIGURE 10.33 Diagrams showing effect of γ_m, σ_i, and grain size on σ_{ys}, σ_f, and transition temperature. Only grain refinement produces simultaneous increase in σ_{ys}, σ_f, and reduction in transition temperature.[50] (Reprinted with permission from John Wiley & Sons, Inc.)

lead to another equation of the form

$$K_{IC} \approx \left[2\sigma_{ys} E \left(\frac{\pi}{6} \right)^{1/3} D \right]^{1/2} f_c^{-1/6} \qquad (10\text{-}3)$$

where

D = diameter of cracked particles

f_c = volume fraction of second phase particles

Equation 10-3 is based on a model involving fracture of large second phase particles (with a size D), and initial growth of the associated voids and their subsequent linkup by rupture of the intervening ligaments. (Recall Fig. 10.23.) The ligament length between broken particles is found to depend on f_c. The relationship between Eqs. 10-2 and 10-3 is clear when one recalls the test results shown in Fig. 10.10 and the discussion associated with Fig. 10.21.

Another model developed by Krafft[54] assumed fracture to occur within a small ligament ahead of the crack tip when the strain in this region equaled the smooth bar tensile instability strain (Fig. 10.34). Using the analysis from Chapter One, where $\sigma = K\epsilon^n$, the instability strain was then taken to be the strain hardening coefficient and K_{IC} given by

$$K_{IC} \approx En\sqrt{2\pi d_T} \qquad (10\text{-}4)$$

where

E = modulus of elasticity

n = strain hardening coefficient

d_T = tensile ligament distance over which $\epsilon \geqslant n$ (this

distance is referred to by Krafft as the "process

zone size")

Within d_T, Krafft postulated the existence of a somewhat smaller elemental fracture cell whose size d_F might correspond to the diameter of an equiaxed microvoid, though a specific relationship between d_T and d_F was not given. In confirmation of this model, it has been found that K_{IC} values tend to increase with increasing microvoid size (that is, increasing inclusion spacing, Fig. 10.21) for a controlled set of conditions.[13] Furthermore, Yoder[55] observed a series of parallel ridges and troughs of constant spacing on the fracture surfaces of high-strength steels (Fig. 10.35) and equated the ridge to ridge distance with Krafft's process zone size, d_T. Values of K_{IC} computed from Eq. 10-4 based on

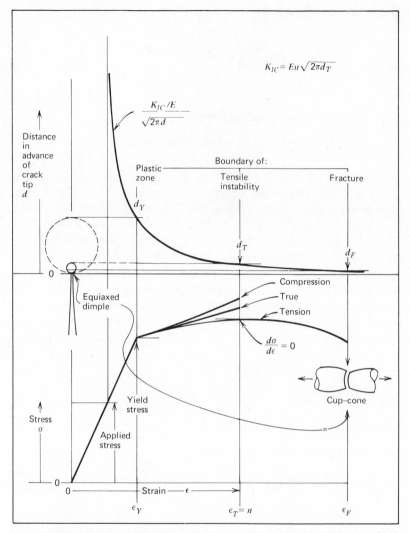

FIGURE 10.34 Tensile ligament instability model developed by Krafft.[54] Fracture will occur when small region d_T experiences $\epsilon_T \geqslant n$. (Reprinted with permission from IPC Industrial Press, Ltd., London.)

this assumption were in reasonably good agreement with experimental values of K_{IC}.[55]

It should be noted that the earlier Krafft model given by Eq. 10-4 was based on smooth bar tensile properties. As pointed out by Barsom and Pellegrino[57] (Eq. 9-6) and Hahn and Rosenfeld[51] (Eq. 10-2), the fracture toughness of a material should be more strongly dependent on the plane strain ductility than on

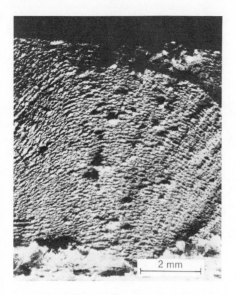

FIGURE 10.35 Constant spacing ridges on steel fracture surface related to Krafft's process zone size.[56] (Copyright American Society for Metals, 1973.)

the axisymmetric ductility, since the former reflects more closely the stress-state conditions at the crack tip. (Recall from Chapter Nine that the sharp rise in K_{IC} values with increasing temperatures for low and intermediate-strength steels could be anticipated by a similar transition in plane strain ductility but not by the axisymmetric ductility data.) Additional support for this hypothesis has been given by Weiss,[58] who demonstrated for several materials that fracture strains decreased markedly (by a factor or five or more) when test samples were subjected to a balanced biaxial stress state ($\sigma_1 = \sigma_2, \sigma_3 = 0$) as opposed to simple uniaxial tension ($\sigma_2 = \sigma_3 = 0$).

Other attempts have been made to develop K_{IC} data based on certain fractographic measurements in conjunction with the fracture toughness model that includes crack opening displacement considerations. From Fig. 10.36, the sharp crack produced by cyclic loading [each line represents one load cycle (Region A)] is blunted by a stretching open process (Region B) prior to crack instability where microvoid coalescence occurs (Region C). Some investigators[13,59] have postulated that the width of the "stretch zone" reflects the amount of crack opening displacement, but Broek[60] claims the depth of this zone to be the more relevant dimension. Broek argues that the "stretch zone" width is related, instead, to the process zone size from Eq. 10-4. In either case, these researchers have contended that K_{IC} values may be estimated from Eq. 8-48, with the crack opening displacement being estimated by some measure of the "stretch zone." Although there is some merit to this postulate, this author does not feel that much confidence can be placed in K_{IC} values determined on the basis of "stretch zone" widths, since these measurements are difficult to make and vary widely along the crack front.

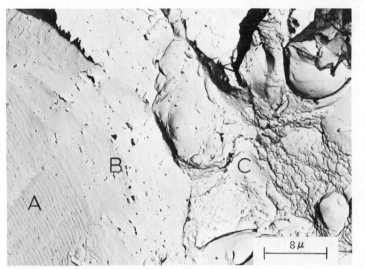

FIGURE 10.36 Precracked sample subsequently loaded to fracture. Region A represents fatigue precracking zone (see Section 13.3); Region B is the "stretch zone" representing crack blunting prior to crack instability; Region C is overload fracture region revealing microvoid coalescence.

The models discussed above possess some merit, but it is obvious that further work is necessary before one is able to determine K_{IC} from more readily obtained mechanical properties.

10.7 ADDITIONAL DATA

Additional K_{IC} data are provided in Table 10.8 to enable the reader to become more familiar with the fracture properties of a number of commercial alloys, ceramics, and polymers. The relatively brittle response of ceramic solids, which was anticipated in Table 10.1, is verified convincingly by the tabulated results.

TABLE 10.8a   Strength and Fracture Toughness Data for Selected Materials[61]

Alloy	Material Supply	Specimen Orientation	Test Temperature (°C)	σ_{ys} (MPa)	K_{IC} (MPa$\sqrt{\text{m}}$)
Aluminum Alloys					
2014-T651	Plate	L-T	21–32	435–470	23–27
"	"	T-L	"	435–455	22–25
"	"	S-L	24	380	20
2014-T6	Forging	L-T	"	440	31
"	"	T-L	"	435	18–21
2020-T651	Plate	L-T	21–32	525–540	22–27
"	"	T-L	"	530–540	19
2024-T351	"	L-T	27–29	370–385	31–44
"	"	T-L	"	305–340	30–37
2024-T851	"	L-T	21–32	455	23–28
"	"	T-L	"	440–455	21–24
2124-T851	"	L-T	"	440–460	27–36
"	"	T-L	"	450–460	24–30
2219-T851	"	L-T	"	345–360	36–41
"	"	T-L	"	340–345	28–38
7049-T73	Forging	L-T	"	460–510	31–38
"	"	T-L	"	460–470	21–27
7050-T73651	Plate	L-T	"	460–510	33–41
"	"	T-L	"	450–510	29–38
"	"	S-L	"	430–440	25–28
7075-T651	Plate	L-T	21–32	515–560	27–31
"	"	T-L	"	510–530	25–28
"	"	S-L	"	460–485	16–21
7075-T7351	"	L-T	"	400–455	31–35
"	"	T-L	"	395–405	26–41
7475-T651	"	T-L	"	505–515	33–37
7475-T7351[a]	"	T-L	"	395–420	39–44
7079-T651	"	L-T	"	525–540	29–33
"	"	T-L	"	505–510	24–28
7178-T651	"	L-T	"	560	26–30
"	"	T-L	"	540–560	22–26
"	"	S-L	"	470	17

TABLE 10.8a   (Continued)

Alloy	Material Supply	Specimen Orientation	Test Temperature (°C)	σ_{ys} (MPa)	K_{IC} (MPa$\sqrt{m}$)
Ferrous Alloys					
4330V(275°C temper)	Forging	L-T	21	1400	86–94
4330V(425° C temper)	"	L-T	"	1315	103–110
4340(205°C temper)	"	L-T	"	1580–1660	44–66
4340(260°C temper)	Plate	L-T	"	1495–1640	50–63
4340(425°C temper)	Forging	L-T	"	1360–1455	79–91
D6AC(540°C temper)	Plate	L-T	"	1495	102
D6AC(540°C temper)	"	L-T	−54	1570	62
9-4-20(550°C temper)	"	L-T	21	1280–1310	132–154
18 Ni(200)(480°C 6 hr)	Plate	L-T	21	1450	110
18 Ni(250)(480°C 6 hr)	"	L-T	"	1785	88–97
18 Ni(300)(480°C)	"	L-T	"	1905	50–64
18 Ni(300)(480°C 6 hr)	Forging	L-T	"	1930	83–105
AFC77 (425°C temper)	"	L-T	24	1530	79
Titanium Alloys					
Ti-6 Al-4 V	(Mill anneal plate)	L-T	23	875	123
"	"	T-L	"	820	106
"	(Recryst. anneal plate)	L-T	22	815–835	85–107
"	"	T-L	22	825	77–116
Ceramics					
Mortar[62]	—	—	—	—	0.13–1.3
Concrete[62]	—	—	—	—	0.23–1.43
Al_2O_3[63-65]	—	—	—	—	3–5.3
SiC[63]	—	—	—	—	3.4
Si_3N_4[64]	—	—	—	—	4.2–5.2
Soda lime silicate glass[64]	—	—	—	—	0.7–0.8
Electrical porcelain ceramics[66]	—	—	—	—	1.03–1.25
WC(2.5–3µm)—3 w/o Co[67]	—	—	—	—	10.6
WC(2.5–3µm)–9 w/o Co[67]	—	—	—	—	12.8
WC(2.5–3.3µm)–15 w/o Co[67, 68]	—	—	—	—	16.5–18
Indiana limestone[69]	—	—	—	—	0.99
Polymers					
PMMA[70]	—	—	—	—	0.8–1.75[b]
PS[71]	—	—	—	—	0.8–1.1[b]
Polycarbonate[72]	—	—	—	—	2.75–3.3[b]

[a]Special processing.   [b]K_{IC} is f (crack speed).

TABLE 10.8b Strength and Fracture Toughness Data for Selected Materials[61]

Alloy	Material Supply	Specimen Orientation	Test Temperature (°F)	σ_{ys} (ksi)	K_{IC} (ksi$\sqrt{\text{in}}$)
Aluminum Alloys					
2014-T651	Plate	L-T	70–89	63–68	21–24
"	"	T-L	"	63–66	20–22
"	"	S-L	75	55	18
2014-T6	Forging	L-T	"	64	28
"	"	T-L	"	63	16–19
2020-T651	Plate	L-T	70–89	76–78	20–25
"	"	T-L	"	77–78	17–18
2024-T351	"	L-T	80–85	54–56	28–40
"	"	T-L	"	44–49	27–34
2024-T851	"	L-T	70–89	66	21–26
"	"	T-L	"	64–66	19–21
2124-T851	"	L-T	"	64–67	25–33
"	"	T-L	"	65–67	22–27
2219-T851	"	L-T	"	50–52	33–37
"	"	T-L	"	49–50	26–34
7049-T73	Forging	L-T	"	67–74	28–34
"	"	T-L	"	67–68	19–25
7050-T73651	Plate	L-T	"	67–74	30–37
"	"	T-L	"	65–74	26–35
"	"	S-L	"	62–64	22–26
7075-T651	Plate	L-T	70–89	75–81	25–28
"	"	T-L	"	74–77	23–26
"	"	S-L	"	67–70	15–19
7075-T7351	"	L-T	"	58–66	28–32
"	"	T-L	"	57–59	24–37
7475-T651	"	T-L	"	73–75	30–33
7475-T7351[a]	"	T-L	"	57–61	35–40
7079-T651	"	L-T	"	76–78	26–30
"	"	T-L	"	73–74	22–25
7178-T651	"	L-T	"	81	23–27
"	"	T-L	"	78–81	20–23
"	"	S-L	"	68	15

TABLE 10.8b (*Continued*)

Alloy	Material Supply	Specimen Orientation	Test Temperature (°F)	σ_{ys} (ksi)	K_{IC} (ksi$\sqrt{\text{in}}$)
Ferrous Alloys					
4330V(525°F temper)	Forging	L-T	70	203	78–85
4330V(800°F temper)	"	L-T	"	191	94–100
4340(400°F temper)	"	L-T	"	229–241	40–60
4340(500°F temper)	Plate	L-T	"	217–238	45–57
4340(800°F temper)	Forging	L-T	"	197–211	72–83
D6AC(1000°F temper)	Plate	L-T	"	217	93
D6AC(1000°F temper)	Plate	L-T	−65	228	56
9-4-20(1025°F temper)	Plate	L-T	70	186–190	120–140
18 Ni(200)(900°F 6 hr)	Plate	L-T	70	210	100
18 Ni(250)(900°F 6 hr)	Plate	L-T	"	259	80–88
18 Ni(300)(900°F)	"	L-T	"	276	45–58
18 Ni(300)(900°F 6 hr)	Forging	L-T	"	280	75–95
AFC77 (800°F temper)	"	L-T	75	222	72
Titanium Alloys					
Ti-6 Al-4 V	(Mill anneal plate)	L-T	74	127	112
"	"	T-L	"	119	96
"	(Recryst. anneal plate)	L-T	72	118–121	77–97
"	"	T-L	72	120	70–105
Ceramics					
Mortar[62]	—	—	—	—	0.12–1.15
Concrete[62]	—	—	—	—	0.21–1.30
Al_2O_3[63-65]	—	—	—	—	2.7–4.8
SiC[63]	—	—	—	—	3.1
Si_3N_4[64]	—	—	—	—	3.8–4.7
Soda lime silicate glass[64]	—	—	—	—	0.64–0.73
Electrical porcelain ceramics[66]	—	—	—	—	0.94–1.14
WC(2.5–3 μm)—3 w/o Co[67]	—	—	—	—	9.6
WC(2.5–3 μm)–9 w/o Co[67]	—	—	—	—	11.6
WC(2.5–3.3 μm)—15 w/o Co[67,68]	—	—	—	—	15–16.4
Indiana limestone[69]	—	—	—	—	0.9
Polymers					
PMMA[70]	—	—	—	—	0.73–1.6[b]
PS[71]	—	—	—	—	0.73–1.0[b]
Polycarbonate[72]	—	—	—	—	2.5–3.0[b]

[a]Special processing. [b]K_{IC} is f (crack speed).

REFERENCES

1. R. A. Jaffee and G. T. Hahn, *Symposium on Design with Materials That Exhibit Brittle Behavior*, Vol. 1, MAB-175-M, National Materials Advisory Board, Washington, D.C., Dec. 1960, p. 126.
2. M. J. Harrigan, *Met. Eng. Quart.*, May 1974, p. 16.
3. J. Cook and J. E. Gordon, *Proc. Roy. Soc.*, **A282**, 1964, p. 508.
4. J. D. Embury, N. J. Petch, A. E. Wraith, and E. S. Wright, *Trans.*, AIME, **239**, 1967, p. 114.
5. F. A. Heiser and R. W. Hertzberg, *Trans.* ASME, *J. Basic Eng.*, **93**, 1971, p. 71.
6. H. I. Leichter, *J. Spacecr. and Rockets*, 3(7), 1966, p. 1113.
7. A. J. McEvily, Jr. and R. H. Bush, *Trans.*, ASM, **55**, 1962, p. 654.
8. J. G. Kaufman, *Trans.*, ASME, *J. Basic Eng.*, **89**(3), 1967, p. 503.
9. J. G. Kaufman, P. E. Schilling, and F. G. Nelson, *Met. Eng. Quart.*, **9**(3), 1969, p. 39.
10. W. S. Hieronymus, *Aviat. Week Space Tech.*, July 26, 1971, p. 42.
11. B. I. Edelson and W. M. Baldwin, Jr., *Trans.* ASM, **55**, 1962, p. 230.
12. J. M. Hodge, R. H. Frazier, and F.W. Boulger, *Trans.*, AIME, **215**, 1959, p. 745.
13. A. J. Birkle, R. P. Wei, and G. E. Pellesier, *Trans.* ASM, **59**, 1966, p. 981.
14. L. Luyckx, J. R. Bell, A. McClean, and M. Korchynsky, *Met. Trans.*, **1**, 1970, p. 3341.
15. R. W. Hertzberg and R. Goodenow, *Microalloying 1975*, Oct. 1975, Washington, D.C.
16. C. J. McMahon, Jr., *Fundamental Phenomena in the Material Sciences*, Plenum, New York, Vol. 4, 1967, p. 247.
17. A. H. Cottrell, *Trans. Met. Soc.*, AIME, **212**, 1958, p. 192.
18. C. Zener, *Fracturing of Metals*, ASM, Cleveland, 1948, p. 3.
19. T. L. Johnston, R. J. Stokes, and C. H. Li, *Phil. Mag.*, **7**, 1962, p. 23.
20. V. F. Zackay, E. R. Parker, J. W. Morris, Jr., and G. Thomas, *Mater. Sci. Eng.*, **16**, 1974, p. 201.
21. D. E. Piper, W. E. Quist and W. E. Anderson, *Application of Fracture Toughness Parameters to Structural Metals*, Vol. 31, Metallurgical Society Conference, 1966, p. 227.
22. R. E. Zinkham, H. Liebowitz, and D. Jones, *Mechanical Behavior of Materials*, Vol. 1, Proceedings of the International Conference on Mechanical Behavior of Materials, The Society of Materials Science, Kyoto, Japan, 1972, p. 370.
23. J. G. Kaufman, Agard Meeting of the Structures and Materials Panel, Apr. 15, 1975.
24. R. R. Senz and E. H. Spuhler, *Met. Prog.*, **107**(3), 1975, p. 64.
25. D. Broek, *Eng. Fract. Mech.*, **5**, 1973, p. 55.
26. F. A. McClintock, *Int. J. Fract. Mech.*, **2**, 1966, p. 614.
27. H. E. McGannon, *The Making, Shaping and Treating of Steel*, 9th Ed., United States Steel Corporation, Pittsburgh, 1971.
28. W. C. Leslie, *Met. Trans.*, **3**, 1972, p. 5.
29. J. R. Low, Jr., *Fracture*, Technology Press MIT and John Wiley, New York, 1959, p. 68.
30. T. B. Cox and J. R. Low, Jr., *Met. Trans.*, **5**, 1974, p. 1457.
31. D. Webster, *Trans.*, ASM, **61**, 1968, p. 816.
32. V. F. Zackay, E. R. Parker, D. Fahr, and R. Busch, *Trans.*, ASM, **60**, 1967, p. 252.

33. S. Antolovich, *Trans. Met. Soc.*, AIME, **242**, 1968, p. 237.
34. E. R. Parker and V. F. Zackay, *Eng. Fract. Mech.*, **5**, 1973, p. 147.
35. J. P. Bressanelli and A. Moskowitz, *Trans.*, ASM, **59**, 1966, p. 223.
36. W. W. Gerberich, G. Thomas, E. R. Parker, and V. F. Zackey, *Proceedings of the Second International Conference on the Strength of Metals and Alloys*, Asilomar, Calif., 1970, p. 894.
37. D. Bhandarkar, V. F. Zackay, and E. R. Parker, *Met. Trans.*, **3**, 1972, p. 2619.
38. A. R. Rosenfield and A. J. McEvily, Jr., NATO AGARD Report No. 610, Dec. 1973, p. 23.
39. R. Develay, *Met. Mater.*, **6**, 1972, p. 404.
40. J. A. Nock, Jr. and H. Y. Hunsicker, *J. Met.*, Mar. 1963, p. 216.
41. D. S. Thompson, S. A. Levy, and D. K. Benson, *Third International Conference on Strength of Metals and Alloys*, Paper 24, 1973, p. 119.
42. N. E. Paton and A. W. Sommer, *op. cit.*, Paper 21, 1973, p. 101.
43. J. T. Staley, AIME Spring Meeting, Alcoa Report, May 23, 1974.
44. M. Korchynsky, American Society of Agricultural Engineers, Paper No. 70-682, Dec. 1970.
45. B. M. Kapadia, A. T. English and W. A. Backofen, *Trans.*, ASM, **55**, 1962, p. 389.
46. B. Mravic and J. H. Smith, "Development of Improved High-Strength Steels for Aircraft Structural Components," AFML-TR-71-213, Oct. 1971.
47. A. H. Cottrell, *Fracture*, Technology Press MIT and John Wiley, New York, 1959, p. 20.
48. N. J. Petch, *op. cit.*, p. 54.
49. J. R. Low, Jr., *Relation of Properties to Microstructure*, ASM, Metals Park, Ohio, 1954, p.163.
50. A. S. Tetelman and A. J. McEvily, Jr., *Fracture of Structural Materials*, John Wiley, New York, 1967.
51. G. T. Hahn and A. R. Rosenfield, ASTM *STP 432*, 1968, p. 5.
52. G. T. Hahn and A. R. Rosenfield, *Met. Trans.*, **6A**, 1975, p. 653.
53. R. H. Van Stone and J. A. Psioda, *Met. Trans.*, **6A**, 1975, p. 668.
54. J. M. Krafft, *Appl. Mater. Res.*, **1**, 1964, p. 88.
55. G. R. Yoder, *Met. Trans.*, **3**, 1972, p. 1851.
56. C. D. Beachem and G. R. Yoder, *Met. Trans.*, **4**, 1973, p. 1145.
57. J. M. Barsom and J. V. Pellegrino, *Eng. Fract. Mech.*, **5**, 1973, p. 209.
58. V. Weiss, *Proceedings of the International Conference on Mechanical Behavior of Materials*, **1**, 1972, p. 458.
59. R. C. Bates and W. G. Clark, Jr., *Trans.*, ASM, **62**, 1969, p. 380.
60. D. Broek, *Eng. Fract. Mech.*, **6**, 1974, p. 173.
61. J. E. Cambell, W. E. Berry and C. E. Feddersen, *Damage Tolerant Design Handbook*, *MCIC-HB-01*, Sept. 1973.
62. D. J. Nans, G. B. Batson and J. L. Lott, *Fracture Mechanics of Ceramics*, Vol. 2, R. C. Bradt, D. P. H. Hasselman, and F. F. Lange, eds., Plenum Press, New York, 1974, p. 469.
63. R. F. Pabst, *ibid.*, p. 555.
64. S. M. Wiederhorn, *ibid.*, p. 613.
65. S. W. Freiman, K. R. McKinney, and H. L. Smith, *ibid.*, p. 659.
66. W. G. Clark, Jr. and W. A. Logsdon, *ibid.*, p. 843.

67. R. C. Lueth, *ibid.*, p. 791.
68. N. Ingelström and H. Nordberg, *Eng. Fract. Mech.*, **6**, 1974, p. 597.
69. R. A. Schmidt, *Closed Loop*, **5**, November 1975, p. 3.
70. G. P. Marshall and J. G. Williams, *J. Mater. Sci.*, **8**, 1973, p. 138.
71. G. P. Marshall, L. E. Culver and J. G. Williams, *Int. J. Fract.*, **9**(3), 1973, p. 295.
72. J. C. Radon, *J. Appl. Polym. Sci.*, **17**, 1973, p. 3515.

PROBLEMS

10.1 Why would you expect a steel refined by the Bessemer process (air blown through the melt) to exhibit inferior fracture properties to a steel refined in a BOF (oxygen blown through the melt)?

10.2 Discuss three ways in which the toughness of a material may be increased.

10.3 For a stress level of 240 MPa compute the maximum radius of a semicircular surface flaw in 7075-T651 aluminum alloy plate when loaded in the L-T, T-L, and S-L orientations. Assume plane strain conditions.

10.4 For the room-temperature CVN test results shown in Fig. 10.11, estimate the K_{IC} level based on the K_{IC}-CVN upper shelf correlation given in Fig. 9.23. Assume $\sigma_{ys} = 1000$ MPa. Plot these estimates of K_{IC} versus sulfur content and then superimpose the data from Fig. 10.12. Discuss your results.

10.5 Discuss the overall virtues of fine grain microstructures with regard to room-temperature and elevated-temperature behavior.

10.6 An 8-cm diameter extruded rod of 7178-T651 aluminum alloy is to be machined into a closed-end, hollow cylinder with a 7-cm bore. If a fluid is introduced into the bore and compressed by a piston, calculate the largest semicircular surface flaw (oriented along the axis of the bore) that could withstand a fluid pressure of 50 MPa.

CHAPTER
ELEVEN
ENVIRONMENT ASSISTED CRACKING AND METALLURGICAL EMBRITTLEMENT

"The image of stress corrosion I see
Is that of a huge unwanted tree,
Against whose trunk we chop and chop,
But which outgrows the chips that drop;

And from each gash made in its bark
A new branch grows to make more dark
The shade of ignorance around its base,
Where scientists toil with puzzled face"

*On Stress Corrosion, S. P. Rideout**

Much attention was given in preceding chapters to the importance of the plane strain fracture toughness parameter K_{IC} in material design considerations. It was argued that this value represents the lowest possible material toughness corresponding to the maximum allowable stress intensity factor that could be applied short of fracture. And, yet, failures are known to occur when the *initial* stress intensity factor level is considerably below K_{IC}. How can this be? These failures arise because cracks are able to grow to critical dimensions with the initial stress intensity level increasing to the point where $K = K_{IC}$ (Eq. 8-22). Such crack extension can occur by a number of processes. Subcritical flaw growth

*Reprinted with permission from *Fundamental Aspects of Stress Corrosion Cracking*, 1969, National Association of Corrosion Engineers.

mechanisms involving a cooperative interaction between a static stress and the environment include stress corrosion cracking (SCC), hydrogen embrittlement (HE), and liquid metal embrittlement (LME). The subject of fatigue and corrosion fatigue is examined in Chapter Thirteen, while SCC, HE, and LME are considered in this chapter. The literature dealing with these topics is as staggering as is the history and significance of the problem. It is not within the scope of this book to cover this material in depth, especially considering the complexity of the environmentally induced embrittlement phenomena itself. Indeed, as Staehle[1] has pointed out, "A general mechanism for stress corrosion cracking···seems to be an unreasonable and unattainable goal. Specific processes appear to operate under specific sets of metallurgical and environmental conditions." A montage of some major SCC mechanisms is shown in Fig. 11.1, representing the cumulative results of many researchers. In addition to Ref. 1, the interested reader should find the several dozen other papers in this volume of particular interest regarding the specifics of SCC. In addition, the reader is referred to the review by Uhlig.[2] Rather than deal with the problem on such a precise level, the author has chosen to expose the reader only to certain common phenomenological characteristics of SCC, HE, and LME, which suggest that these mechanisms are related in an overall sense to a single general phenomenon. For simplicity, this phenomenon shall be referred to as environment assisted cracking (EAC).

In addition to embrittlement brought about by *external* aggressive gaseous and liquid environments, a material may suffer significant loss of intrinsic toughness as a result of various interactions between point, line, surface, and volume defects within the lattice. Examples of these *internal* embrittling processes include 300°C embrittlement, temper embrittlement, and irradiation damage embrittlement. These material conditions represent a special group of problems not considered previously in Chapter Ten.

The objective of this chapter is to provide an overview of externally and internally induced embrittlement that will enable the reader to better appreciate some of the major problems that befall many engineering materials.

11.1 ENVIRONMENT ASSISTED CRACKING (EAC)

As already mentioned, the various manifestations of EAC have been long recognized. Consequently, different approaches to "solving" the problem have been developed, along with "standard" specimen types. For example, stress corrosion cracking studies of engineering materials had often made use of smooth test bars which were stressed in various aggressive environments. Here the *nucleation* kinetics of cracking, as well as its character (transgranular versus intergranular), were examined closely. Most often these studies focused on the nature of anodic dissolution in the vicinity of the crack tip. Recently, more

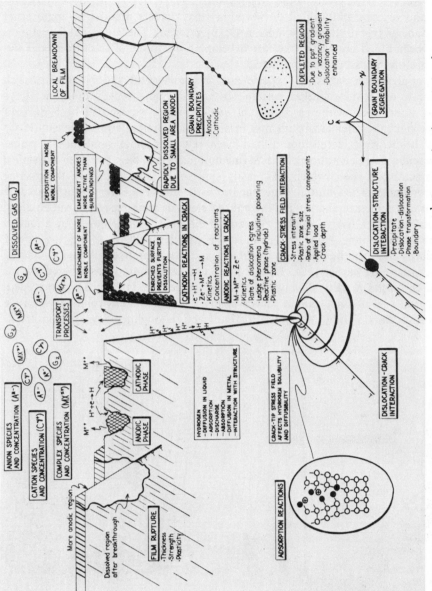

FIGURE 11.1   Montage of important stress corrosion cracking processes.[1] (Reprinted with permission from R. W. Staehle, *Fundamental Aspects of Stress Corrosion Cracking*, 1969, National Association of Corrosion Engineers.)

attention has been given to the *propagation* stage, reflecting the more conservative and realistic philosophical viewpoint[3] that defects preexist in engineering components (recall the discussions in Chapter Seven). These propagation studies have been aided greatly by the fast developing discipline of fracture mechanics and are the focus of attention here.

In a dramatic series of experiments, researchers at the Naval Research Laboratories[4-6] showed that certain *precracked*, high-strength titanium alloys failed under load within a matter of *minutes* when exposed to both distilled water and salt water environments. In all tests, the initial stress intensity levels were below K_{IC}. Heretofore it had been felt that these same alloys would represent a new generation of submarine hull materials, based on their resistance to general corrosion, which was vastly superior to steel alloys in these same environments. These initial experiments made use of the very simple loading apparatus shown in Fig. 11.2a. Precracked samples were placed in the environmental chamber and stressed in bending at different initial K levels by a loaded scrub bucket hung from the end of the cantilever beam. Note the strong similarity between the NRL test apparatus and the diagram attributed to Galileo some 400 years earlier (Fig. 11.2b). For each test condition associated with a different initial K value (always less than K_{IC}) the time to failure was recorded. A typical plot of such data is shown in Fig. 11.3 for the environment sensitive Ti-8 Al-1 Mo-1 V alloy. With an apparent fracture toughness level of about 100

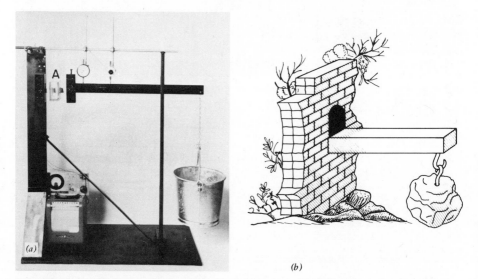

FIGURE 11.2   (*a*) Environment assisted cracking test stand. Specimen is placed in environment chamber at A and loaded by weights placed in scrub bucket.[5] (Reprinted with permission from B. F. Brown and C. D. Beachem, *Corrosion Science*, **5**, 1965, Pergamon Press.) (*b*) Cantilever beam arrangement (adapted from Galileo).

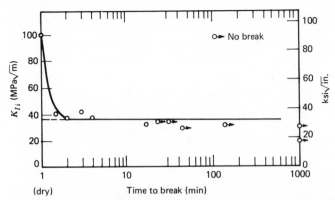

FIGURE 11.3 Initial stress intensity level plotted versus time to break for Ti-8 Al-1 Mo-1 V alloy in 3.5% NaCl solution. Note threshold behavior.[4] (Reprinted by permission of the American Society for Testing and Materials from copyright work.)

$MPa\sqrt{m}$, test failures occurred at initial K levels of only 40 $MPa\sqrt{m}$ after a few minutes of exposure to a 3.5% NaCl solution. At slightly lower K levels, the time to failure increased rapidly, suggesting the existence of a threshold K level, originally designated K_{ISCC},[4] below which stress corrosion cracking would not occur. To be consistent with the philosophical viewpoint expressed in this chapter, K_{ISCC} shall be redefined hereafter as K_{IEAC} where "I" represents Mode I opening and "EAC" represents environment assisted cracking. As a result, a new safe lower limit of the applied stress intensity value (time-dependent in this case) was identified with fracture occurring according to the following criteria:

1. $K < K_{IEAC}$ No failure expected even after long exposures under stress to aggressive environments.
2. $K_{IEAC} < K < K_{IC}$ Subcritical flaw growth with fracture occurring after a certain loading period in an aggressive environment.
3. $K > K_{IC}$ Immediate fracture upon initial loading.

One should keep in mind that for the test conditions associated with Fig. 11.3 (criteria 2, above), stable crack extension causes the initial stress intensity level to increase to the point where failure occurs when K approaches K_{IC}. Stated differently, the fracture toughness of the material is not affected by the environment; instead, small cracks grow under sustained loads to the point where the critical stress intensity factor level is reached (Fig. 11.4).

Determination of K_{IEAC} values is not an easy matter. Since the environmental threshold level depends on how long one chooses to conduct the test, K_{IEAC} values may vary from one laboratory to another, depending on the patience of the investigator. Currently, it is suggested that such tests be conducted for at least 1000 hr. However, we see from Table 11.1 that the apparent K_{IEAC} value for a 1240 MPa steel in a sea water environment continues to decrease sharply

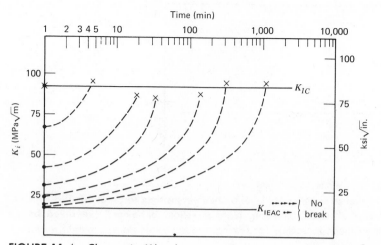

FIGURE 11.4 Change in K level associated with subcritical flaw growth. Regardless of $K_{initial}$, failure in any sample occurs at $K = K_{IC}$.[5] (Reprinted with permission from B. F. Brown and C. D. Beachem, *Corrosion Science*, **5**, 1965, Pergamon Press.)

TABLE 11.1 Influence of Cutoff Time on Apparent K_{IEAC} for High-Strength Steel (1240 MPa) in Synthetic Sea Water at Room Temperature[7]

Test Time (hr)	Apparent K_{IEAC}	
	MPa$\sqrt{m}$	(ksi$\sqrt{in}$)
100	187	(170)
1,000	127	(115)
10,000	28	(25)

with increasing test time beyond that point. It may be that K_{IEAC} test times will have to be determined for each material-environment system on an individual basis.

Some materials, such as high-strength steels and titanium alloys, exhibit a rather well-defined K_{IEAC} limit after a reasonable test time period, but in aluminum alloys this does not appear to be the case. Instead, K_{IEAC} values in high-strength aluminum alloys tend to decrease with increasing patience of the investigator. Consequently, K_{IEAC} data must be used carefully, especially in the design of engineering components that will be stressed in an aggressive atmosphere for time periods longer than those associated with generation of the K_{IEAC} data.

During the past few years, different specimen configurations have been developed to make determinations of K_{IEAC} both cheaper and easier. One such test method utilizes a bolt-loaded compact tension sample.[8] As seen in Fig. 11.5, a screw, engaged in the top half of the sample, bears against the bottom crack surface. This produces a crack opening displacement corresponding to some initial load. In this manner, the specimen is self-stressed and does not require a test machine for application of loads. As the crack extends by environment assisted cracking, the load and, hence, the K level drops under the prevailing constant displacement condition. The crack finally stops when the K level drops below K_{IEAC}. Consequently, only one specimen is needed to determine K_{IEAC}. Such a test is very easy to conduct and very portable, since the self-stressed sample can be carried to any environment rather than vice versa. All one needs to do is to engage the screw thread to produce a given crack opening displacement and place the specimen in the environment. Samples are examined periodically to determine when the crack stops growing. The K_{IEAC} value is then defined by the residual applied load remaining after the crack has ceased growing and the final crack length as seen on the fracture surface. The major advantages of the modified compact tension sample relative to the precracked cantilever beam are:

1. The need for one sample versus 8 to 10 in determining K_{IEAC}.
2. The specimen is self-stressed and highly portable.
3. The method is less costly.
4. K_{IEAC} is determined directly by the arrest characteristics of the sample because of the continual decrease in K with increasing crack length. By comparison, the K_{IEAC} value determined with the precracked cantilever beam samples represents an interpolated value between the highest K level at which EAC does not occur and the lowest K level where failures still occur.
5. The need for a sharp notch is not as great, since K is initially high, which results in early crack growth. By contrast, a poorly prepared notch in the cantilever beam specimen would involve a considerable period of time for crack initiation, especially at low K levels.

Although the modified compact tension sample represents an improvement in the method by which K_{IEAC} data are obtained, one must still contend with the fact that K_{IEAC} may not represent a true material property. As noted above, threshold values are often found to be a function of the length of the test (i.e., the patience of the technician). Recognizing this limitation (particularly in aluminum alloys), engineers and researchers have sought to characterize the kinetics of the crack growth rate process by monitoring the rate of crack advance da/dt as a function of the instantaneous stress intensity level. From the work of Wiederhorn[9,10] on the static crack growth of glass and sapphire in water, a $\log da/dt$-K relationship was determined, which took the form shown schematically in Fig. 11.6. Three distinct crack growth regimes are readily identified. In Region I, $(da/dt)_{\text{I}}$ is found to depend strongly on the prevailing

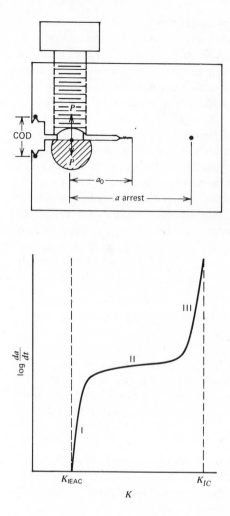

FIGURE 11.5 Modified compact tension sample with threaded bolt bearing on loading pin. Initial crack opening displacement determined by extent to which bolt is engaged.

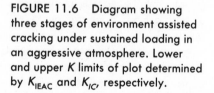

FIGURE 11.6 Diagram showing three stages of environment assisted cracking under sustained loading in an aggressive atmosphere. Lower and upper K limits of plot determined by K_{IEAC} and K_{IC}, respectively.

stress intensity level, along with temperature, pressure, and the environment. For some materials, the slope of this part of the curve is so steep as to allow for an alternate definition of K_{IEAC}; that is, the K level below which da/dt becomes vanishingly small. For aluminum alloys that do not appear to exhibit a true threshold level and which have a shallower Stage I slope, a "K_{IEAC}" value can be defined at a specific da/dt level much the same as the yield strength of a material exhibiting Type II behavior (Section 1.2.2.1) is defined by the 0.2% offset method. Environment assisted crack growth is essentially independent of the prevailing K level in Region II, but it is still affected strongly by temperature, pressure, and the environment. Finally, Region III reflects a second regime where da/dt varies strongly with K. In the limit, crack growth rates become unstable as K approaches K_{IC}.

In addition to the three *steady-state* crack growth regimes just described, a number of additional *transient* growth regions have been identified followed by a dormant or incubation period prior to steady-state growth.[7] Consequently, the total time to fracture is the summation of transient, incubation, and steady-state cracking periods. Hence

$$t_T = t_{tr} + t_{inc} + t_s \tag{11-1}$$

where

t_T = total time to failure

t_{tr} = time during transient crack growth

t_{inc} = incubation time

t_s = time of Region I, II, and III steady-state crack growth.

The transient time t_{tr} is usually small relative to t_{inc} and t_s and is often ignored in life computations. The relative importance of the other two regimes in affecting total life is shown schematically in Fig. 11.7. Note that the incubation time decreases rapidly with increasing initial K level.[7,11-13] (Higher test temperatures also decrease t_{inc}.[12]) Since the initial K level of the bolt-loaded, constant displacement type K_{IEAC} test sample is large, this configuration is preferred over the cantilever beam geometry. As noted previously, the very long times to failure at low K values suggest a threshold condition. It should be recognized, however, that the incubation period represents a large part of the time to failure. As a

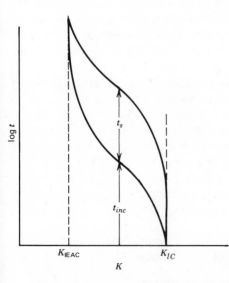

FIGURE 11.7 Diagram showing time for crack incubation and steady-state growth as function of applied K level. Note smaller incubation time at higher K values.

result, initial crack growth rate readings often are abnormally low, suggesting the existence of an erroneously high K_{IEAC} level. These data, therefore, should be used with extreme caution.

11.1.1 Major Variables Affecting Environment Assisted Cracking

The degree to which materials are subject to EAC depends on a number of factors, including alloy chemistry and thermomechanical treatment, the environment itself, temperature, and pressure (for the case of gaseous atmospheres). The effect of these important variables on the cracking process will now be considered.

Alloy Chemistry and Thermomechanical Treatment: As one might expect, many studies have been conducted to examine the relative EAC propensity of different families of alloys and specific alloys thermomechanically treated to different specifications. For example, the $\log da/dt$-K plot for several high-strength aluminum alloys exposed to alternate immersion in a 3.5% NaCl solution reveals the 7079-T651 alloy to be markedly inferior relative to the response of the other alloys (Fig. 11.8).[14] For example, Stage II cracking in the 7079 alloy occurs at a rate 1000 times greater than in the 7178-T651 alloy and corresponds to a Stage II crack growth rate of greater than 3 cm in 1 hr. No engineering component would be expected to resist final failure for long at that growth rate. During the past 10 years, many investigations have been conducted to improve the EAC resistance of these materials. These studies indicate that overaging is the most effective way to accomplish this objective.[14,15] Coincidentally, toughness is also improved while strength decreases as a result of the overaging process (see Fig. 10.27). The effect of overaging (denoted by the -T7 temper designation) on 7079 and 7178 aluminum alloys is shown in Fig. 11.9. Although Stage I in the 7079 alloy is shifted markedly to higher K levels, reflecting a sharp increase in K_{IEAC}, the growth rates associated with Stage II cracking remain relatively unchanged. Consequently, the major problem of very high Stage II cracking rates in this material remains even after overaging. By contrast, preliminary data for the 7178 alloy show a marked decrease in Stage II crack growth rate with increasing aging time, while Stage I cracking is shifted to a much lesser extent.[14,15] Note the dramatic six order of magnitude difference in Stage II crack growth rates between these two alloys in the overaged condition. Upon reflection, it would be most desirable to have the overaging treatment effect a *simultaneous* lowering of the Stage II cracking rate and a displacement of the Stage I regime to higher K levels. This may prove to be the case in other alloy systems.

In general, K_{IEAC} values tend to be greater in materials possessing higher K_{IC} levels and lower yield strength. For example, we see from Fig. 11.10 that K_{IEAC}

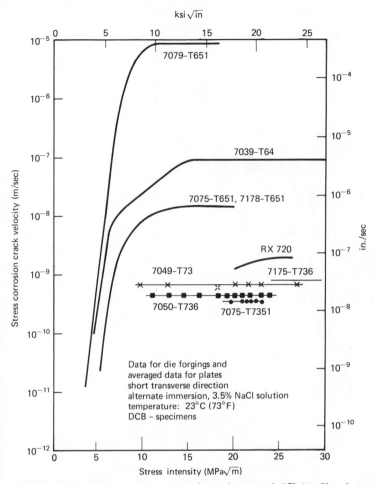

FIGURE 11.8 Environment assisted cracking in 3.5% NaCl solution for several aluminum alloys and heat treatments.[14] (Reprinted with permission of the American Society for Metals.)

values decrease rapidly with increasing yield strength in a 4340 steel.[16] However, Speidel[14, 15] has noted exceptions to this rule for the case of aluminum alloys. If one defines the relative degree of susceptibility to environment assisted cracking by the ratio K_{IEAC}/K_{IC}, the generally observed trend is for K_{IEAC}/K_{IC} to decrease with increasing alloy strength. That is, K_{IEAC} values drop faster than K_{IC} values with increasing strength.

Most often, hydrogen and stress corrosion cracks follow an intergranular path in the important high strength steel, titanium, and aluminum alloys (see Section 7.7.3.3). This is particularly true for aluminum alloys, which suffer stress

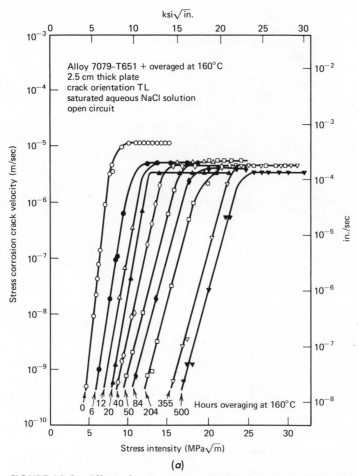

FIGURE 11.9   Effect of overaging on EAC (salt water) in 7xxx series aluminum alloys: (a) 7079 alloy shows pronounced shift of Stage I behavior to higher K levels while $(da/dt)_{II}$ remains relatively constant; (b) 7178 alloy shows sharp drop in $(da/dt)_{II}$.[15] (With permission of Markus O. Speidel, Brown Bovari Co.)

corrosion cracking almost exclusively along grain boundary paths. Consequently, environment assisted cracking in wrought alloys is usually of greater concern in the short transverse direction than in other orientations. As such, EAC orientation sensitivity parallels K_{IC} orientation dependence as described in Section 10.2.

K_{IEAC} data for selected materials is listed in Table 11.2. Note the sharp disagreement in results for several aluminum alloys. Since Speidel's values are based on K levels associated with crack velocities less than about 10^{-10} m/s and were conducted over a long time period, they provide a more representative and

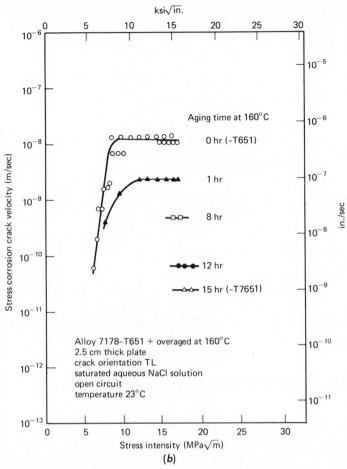

ksi√in.

Aging time at 160°C

0 hr (-T651)

1 hr

─□□─ 8 hr

─●●●─ 12 hr

─△△─ 15 hr (-T7651)

Alloy 7178-T651 + overaged at 160°C
2.5 cm thick plate
crack orientation TL
saturated aqueous NaCl solution
open circuit
temperature 23°C

Stress corrosion crack velocity (m/sec)

in./sec

Stress intensity (MPa√m)

(b)

FIGURE 11.9 (*Continued*)

conservative estimate of the material's environmental sensitivity. Additional K_{IEAC} information is provided in Ref. 17 which also contains numerous log $da/dt\text{-}K$ plots for aluminum, steel, and titanium alloys.

Environment: As one might expect, the kinetics of crack growth and the threshold K_{IEAC} level depend on the material-environment system. In fact, the reality of this situation is reflected by the formalization of various cracking processes, such as stress corrosion cracking (generally involving aqueous solutions), hydrogen gas embrittlement, and liquid metal embrittlement. As mentioned at the beginning of this chapter, the complex aspects of the material-environment interaction can be greatly simplified by treating the problem from a phenomenological viewpoint in terms of a single mechanism, environment assisted cracking. This concept is supported by Speidel's results shown in Fig.

TABLE 11.2 Selected K_{IEAC} Data[17]

Metal	Environment	Test Orientation	Yield Strength MPa	Yield Strength ksi	K_{IC} or $(K_{IX})^a$ MPa√m	K_{IC} or $(K_{IX})^a$ ksi√in.	K_{IEAC} MPa√m	K_{IEAC} ksi√in.	Test Time, Hours
Aluminum Alloys									
2014-T6	Synth. seawater	S-L	420	61	21	19	18	16	—
2014-T6	NaCl solution	S-L	—	—	—	—	≈8	≈7	≈10,000^b
2024-T351	3 1/2% NaCl	S-L	325	47	(55)	(50)	11	10	—
2024-T351	NaCl solution	S-L	—	—	—	—	≈9	≈8	≈10,000^b
2024-T852	Seawater	S-L	370	54	19	17.6	15	14	—
2024-T852	NaCl solution	S-L	—	—	—	—	≈17	≈15	≈10,000^b
2024-T851	Dist. water	L-T	410	59	21	18.6	24	22	—
7075-T6	3 1/2% NaCl	S-L	505	73	25	23	21	19	—
7075-T6	NaCl solution	S-L	—	—	—	—	≈8	≈7	≈10,000^b
7075-T7351	3 1/2% NaCl	S-L	360	52	26	24	23	21	—
7075-T7351	NaCl solution	S-L	—	—	—	—	≤22	≤20	≈10,000^b
7075-T7351	3 1/2% NaCl	T-L	365	53	32	29	26	24	—
7175-T66	3 1/2% NaCl	—	525	76	32	29	≤6.6	≤6	—
7175-T66	NaCl solution	S-L	—	—	—	—	7	6	≈10,000^b
7175-T736	NaCl solution	—	455	66	27	25	21	19	>1029
Steel Alloys									
18 Ni(300)-maraging	"	T-L	1960	284	80	72	8	7.5	>150
4340	"	T-S	1335	194	79	72	9	8.5	>333
4340	"	L-T	1690	245	56	51	17	15	>58
4340	Seawater	T-L	1550	225	(69)	(63)	6	5	>20
"	"	"	1380	200	(65)	(59)	11	10	—
"	"	"	1205	175	(83)	(75)	30	27	—
"	"	"	1035	150	(94)	(85)	65	59	—
"	"	"	860	125	(98)	(89)	77	70	—
300M	3.5% NaCl	L-S	1735	252	70	64	22	20	—
"	"	T-L	1725	250	61	56	20	18	—

TABLE 11.2 (*Continued*)

Metal	Environment	Test Orientation	Yield Strength MPa	Yield Strength ksi	K_{IC} or (K_{IX})[a] MPa$\sqrt{m}$	K_{IC} or (K_{IX})[a] ksi$\sqrt{in}$	K_{IEAC} MPa$\sqrt{m}$	K_{IEAC} ksi$\sqrt{in}$	Test Time, Hours
Titanium Alloys									
Ti-6 Al-4 V	3.5% NaCl	L-T	890	129	104	95	39 ± 10	35 ± 9	—
"	"	L-S	890	129	99	90	45 ± 8	41 ± 7	—
Ti-8 Al-1 Mo-1 V	"	T-S	825	120	97	88	25	23	—
"	"	"	745	108	123	112	31	28	—
"	Water	T-L	855	124	(105)	(95)	29	26	—
"	Methanol	"	855	124	(105)	(95)	15	14	—
"	CCl$_4$	"	855	124	(105)	(95)	22	20	—
"	Water + 21000 ppm Chloride	"	1035	150	(74)	(67)	15	14	—
"	Water + 100 ppm Chloride	"	1035	150	(65)	(59)	23	21	—
"	Water + 0.1 ppm Chloride	"	1035	150	(65)	(60)	27	24	—

[a]Numbers in parenthese are invalid K_{IC} values which do not satisfy Eq. 8-42

[b]M. O. Speidel and M. W. Hyatt, *Advances in Corrosion Science and Technology*, 2, Plenum Press, New York, 1972, p. 115.

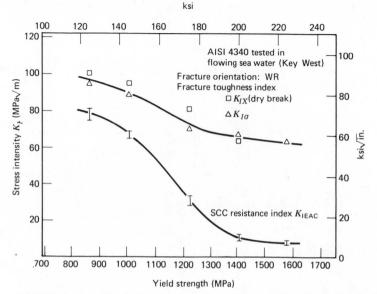

FIGURE 11.10   Effect of yield strength on K_{IC} and K_{IEAC} (in water) in 4340 steel.[16] (Reprinted with permission from M. H. Peterson, B. F. Brown, R. L. Newbegin, and R. E. Grover, *Corrosion*, **23**, 1967, National Association of Corrosion Engineers.)

11.11, which reveal parallel Stage I and II responses for the 7075 aluminum alloy in liquid mercury and aqueous potassium iodide environments.[14] Obviously, the liquid metal represents a more severe environment for this aluminum alloy (some five orders of magnitude difference in Stage II cracking rate), but the phenomenology is the same. Furthermore, we see that the alloy in the overaged condition is more resistant to the liquid metal EAC, as was the case for the salt solution results discussed above.

Environment assisted cracking in dry gases does not appear to occur in aluminum alloys.[14] However, with increasing moisture content, cracking develops with increasing speed (Fig. 11.12). Consequently, EAC in aluminum alloys may take the form of stress corrosion cracking and liquid metal embrittlement but not gaseous hydrogen embrittlement.*

By contrast, HE plays an important role in the fracture of steel and titanium alloys. In fact, hydrogen embrittlement in steel alloys can take several forms. For simplicity, these can be described as being of internal or external origin. Before vacuum degassing techniques were developed, large steel castings were subject to a phenomenon called hydrogen flaking, wherein dissolved hydrogen

---

*There is some debate, however, as to whether aqueous stress corrosion cracking in aluminum is related to hydrogen embrittlement. See M. O. Speidel, *Hydrogen in Metals*, I. M. Bernstein and A. W. Thompson, eds., ASM, 1974, p. 249.

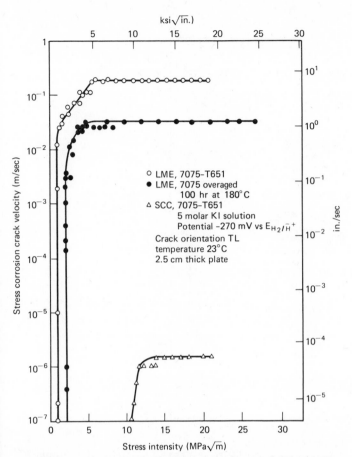

FIGURE 11.11 Environment assisted cracking with liquid mercury and aqueous iodide solution in 7075 aluminum alloy.[15] (With permission of Markus O. Speidel, Brown Bovari Co.)

in the molten metal would form entrapped gas pockets upon solidification. The large, localized pressures associated with these gas pockets generated many sharp cracks which, when located near the casting surface, caused chunks of steel to be spalled away. A service failure of a large rotor forging caused by this type of defect is discussed in Chapter Fourteen. Hydrogen can also be picked up from the electrode cover material or from residual water during welding. After diffusing into the base plate while the weld is hot, embrittlement occurs upon cooling by a process referred to as cold cracking in the weld heat affected zone.[18] Hydrogen may also enter the material as a result of electroplating (i.e., cathodic charging) which contributes to early failure. It is ironic that the electroplating process, designed to protect a material against aqueous environ-

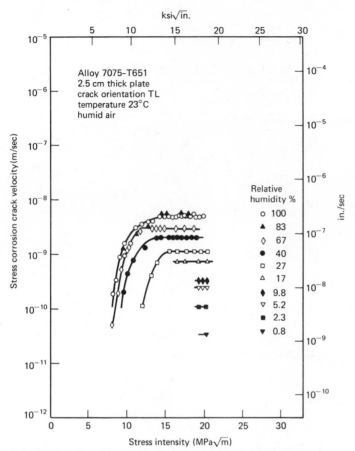

FIGURE 11.12 Effect of humidity on EAC in 7075-T651 aluminum alloy.[15] (With permission of Markus O. Speidel, Brown Bovari Co.)

ments and SCC, actually undermines fracture resistance of the component by simultaneously introducing a different cracking process.* Finally, hydrogen embrittlement may result from the formation of metal hydrides in such materials as titanium, vanadium, and zirconium.

Hydrogen embrittlement in high-strength steel and titanium alloys can also occur without the need for internal charging and may occur whenever a sample

---

*It is possible to overcome many of the problems associated with cathodic charging by subjecting the electroplated material to a baking treatment. This involves heating the metal to a moderate temperature for a sufficient period of time to drive the hydrogen out of solid solution. Furthermore, weld related cold cracking is suppressed by preheating the sample. This has the effect of lowering the postweld cooling rate, thereby allowing more time for the hydrogen to diffuse away from the weld zone.

under stress is exposed to a hydrogen gas atmosphere. It should be noted that embrittlement does not occur as a result of prior exposure to hydrogen gas in the absence of stress.[19,20] The dramatic difference in cracking behavior in H-11 steel for oxygen, argon, and hydrogen gases and water is illustrated in Fig. 11.13.[21] Note the severe effect of dry hydrogen and moisture on the cracking rate, while oxygen causes total crack arrest. It is presently believed that molecular hydrogen is dissociated by chemisorption on iron,[22] allowing liberated atomic hydrogen to diffuse internally and embrittle the metal. Likewise, it has been shown that hydrogen is a product of the corrosion reaction between iron and water; this hydrogen then follows the same path as the chemisorped hydrogen to the metal interior.[23] On the basis of this latter observation, it has been suggested that SCC and HE in steels are one and the same mechanism.[24] Apparently, oxygen has a greater affinity for iron and forms a protective oxide barrier to block the chemisorption process.[19,21] It is believed that once the oxygen is removed, hydrogen can reduce the oxide layer and thereby react again with a clean iron surface.

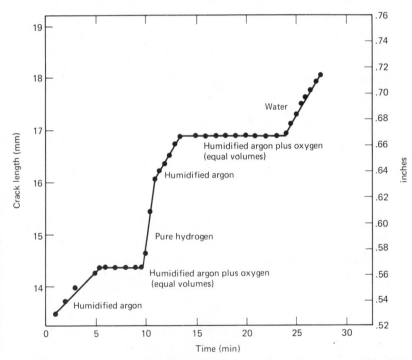

FIGURE 11.13   Fast crack growth of high-strength steel in water and hydrogen, but crack arrest in oxygen.[21] (Reprinted with permission of the American Institute of Mining, Metallurgical, and Petroleum Engineers.)

The hydrogen embrittling process, therefore, depends on three major factors: (1) the original location and form of the hydrogen (internally charged versus atmospheric water or gaseous hydrogen); (2) the transport reactions involved in moving the hydrogen from its source to the locations where it reacts with the metal to cause embrittlement; and (3) the embrittling mechanism itself. We may now ask what that embrittling mechanism is. Unfortunately, the answer is not a simple one, as evidenced by the number of theories that have been proposed. According to one model, called the "planar pressure mechanism," the high pressures developed within internal hydrogen gas pores of charged material cause cracking.[25,26] Although this mechanism appears valid for hydrogen charged steels, it can not be operative for the embrittlement of steel by low-pressure hydrogen atmospheres. In the latter situation, there would be no thermodynamic reason for a low gas pressure external atmosphere to produce a high gas pressure within the solid. Troiano and co-workers[27,28] have argued that hydrogen diffuses under the influence of a stress gradient to regions of high tensile triaxiality which then interacts with the metal lattice to lower its cohesive strength. This model resembles the one proposed to account for liquid metal embrittlement (Fig. 11.14).[29] The latter model adopts the Kelly et al.[30] failure criterion, with brittle fracture occurring when the cohesive strength to shear strength ratio is reduced. If it is assumed that liquid metal atoms reduce the cohesive strength of the cleavage plane near the contact surface but not internally along the slip plane (since the chemisorbed liquid-metal atom would not be expected to penetrate far into the lattice), then the cohesive strength σ is reduced relative to the metal's shear strength τ. Westwood et al.[29] argued that this condition would bring about LME.

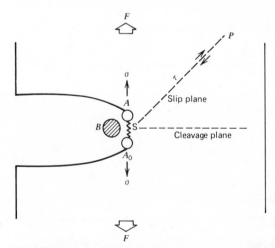

FIGURE 11.14 Diagram showing interaction of liquid metal atom B with solid metal atoms at crack tip.[29] (Reprinted with permission of the American Society for Metals.)

A third model to explain HE was proposed by Petch and Stables,[31] who suggested that hydrogen acts to reduce the surface energy of the metal at internally free surfaces. Which of the above-mentioned theories best describes the HE process will be determined only after additional study. To this end, the reader is referred to the proceedings of a recent HE conference.[32]

Temperature and Pressure: Since EAC processes involve chemical reactions, it is to be expected that temperature and pressure would be important variables. Test results, such as those shown in Fig. 11.15a, for hydrogen cracking in a titanium alloy show the strong effect of temperature on the Stage II cracking rate.[33] These data can be expressed mathematically in the form

$$\left(\frac{da}{dt}\right)_{II} \propto e^{-\Delta H/RT} \tag{11-2}$$

where

$$\Delta H = \text{activation energy for the rate controlling process}$$

The apparent activation energy may then be compared with other data to suggest the nature of the rate controlling process. Recall from Chapter Five that at $T_h > 0.5$, ΔH_{creep} was approximately equal to the activation energy for self-diffusion in many materials. In similar fashion, it has been found that the apparent activation energy for the cracking of high-strength steel in water and humidified gas are both about 38 kJ/mole,[3] which corresponds to the activation energy for hydrogen diffusion in the steel lattice.[34] On the other hand, recent studies have shown that the apparent activation energy for Stage II cracking in the presence of gaseous hydrogen is only 16 to 17 kJ/mole.[11,35] Since the embrittling mechanism appears to be the same for the two environments[35] [e.g., the fracture path is intergranular in both cases (see Section 7.7.3.3)], the change in ΔH probably reflects differences in the rate controlling hydrogen transport process. In this regard, note that the cracking rate in gaseous hydrogen is higher than that in water (Fig. 11.13).

The increase in Stage II crack growth rate with increasing pressure noted in Fig. 11.15b can be described mathematically in the form

$$\frac{da}{dt}_{II} \propto P^n \tag{11-3}$$

In all likelihood, increased pressure enhances hydrogen transport, which in turn increases the cracking rate. At present, there is considerable debate regarding the magnitude of the exponent n.

11.1.2 Life and Crack Length Calculations

Kinetic crack growth data can be integrated to provide estimates of component life and crack length as a function of time.[7,33] For reactions occurring in parallel

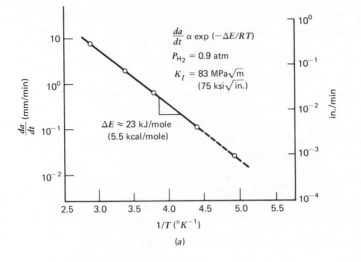

(a)

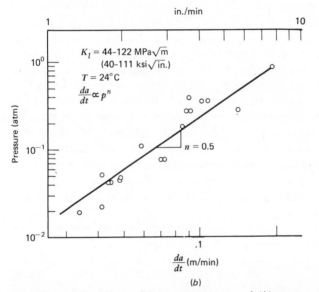

(b)

FIGURE 11.15   Effect of (a) temperature and (b) pressure on hydrogen-induced cracking in Ti-5 Al-2.5 Sn in Region II.[33] (From "A New Criterion for Failure of Materials by Environment-Induced Cracking" by Dell P. Williams, *International Journal of Fracture*, **9**, 1973, pp. 63-74, published by Noordhoff, Leyden, The Netherlands.)

(Eq. 5-9), the effective steady-state cracking rate is controlled by the slowest process acting in regions I, II, and III. If one ignores the contribution of $(\frac{da}{dt})_{III}(= \dot{a}_{III})$, then the controlling crack growth rate is given by

$$\frac{1}{\dot{a}_T} = \frac{1}{\dot{a}_I} + \frac{1}{\dot{a}_{II}} \tag{11-4}$$

or

$$\dot{a}_T = \frac{\dot{a}_I \dot{a}_{II}}{\dot{a}_I + \dot{a}_{II}} \tag{11-5}$$

Upon rearrangement of terms, the time devoted to steady-state cracking is

$$t = \int_0^t dt = \int_{a_0}^{a_i} \frac{\dot{a}_I + \dot{a}_{II}}{\dot{a}_I \dot{a}_{II}} \, da \tag{11-6}$$

To solve Eq. 11-6 expressions for $\dot{a}_{I,II}$ are needed in terms of K and the crack length a. In Stage I

$$\dot{a}_I = f(K, T, P, \text{environment}) \tag{11-7}$$

where

$$K = \text{stress intensity factor}$$

$$T = \text{temperature}$$

$$P = \text{pressure}$$

Since $\log da/dt - K$ plots are often linear, Williams[33] has suggested that

$$\dot{a}_I = C_1 e^{mK} \tag{11-8}$$

where C_1 and m are independent of K but may depend on T, P, and environment. For Region II

$$\dot{a}_{II} = f(T, P, \text{environment}) = C_2 \tag{11-9}$$

Note the lack of K dependence in $\dot{a}_{II}$ and the fact that C_2 depends on T, P, and environment. From the previous discussion in Section 11.1.1, C_2 can be evaluated by combining Eqs. 11-2 and 11-3, such that

$$\dot{a}_{II} = C_3 P^n e^{-\Delta H/RT} \tag{11-10}$$

By combining Eqs. 11-8 and 11-10 into Eq. 11-6, it is possible to calculate the length of a crack at any given time, once the various constants are determined from experimental data. The value of Eq. 11-6 lies in its potential to estimate failure times for conditions of T, P, and environments beyond those readily examined in a test program. One additional subtle point should be made regarding the life computation. It should be recognized that the life of a component or test specimen will depend on the rate of change of the stress intensity factor with crack length dK/da. Consequently, for the same initial K level, the sample with the lowest dK/da characteristic will have the longest life. That is, changing specimen geometry would alter the time to failure. Much work is needed to determine the best method of loading and the best specimen geometry so that one may establish a standard EAC test procedure.

11.2   METALLURGICAL EMBRITTLEMENT

Attention will now be focused on several undesirable circumstances that lead to serious loss of a material's fracture toughness. As will be shown, these changes can be brought about by alterations in microstructure and/or solute redistribution as produced by improper heat treatment or prolonged exposure to neutron irradiation.

11.2.1   300 to 350°C Embrittlement

Metallurgists have long recognized the potentially embrittling effects of tempering martensitic steels at about 300 to 350°C. Evidence for embrittlement has been found in this tempering temperature range by noting decreases in notched impact energy, ductility, and tensile strength[36] and a reduction in smooth bar tensile properties when unnotched samples are tested at subzero temperatures[37] (Fig. 11.16). This trough in material properties has been attributed to the precipitation of cementite films along martensite plates which lie along prior austenite grain boundaries. These plate boundaries then act as easy crack paths.[38,39] However, more recent studies have shown this type of embrittlement to be dependent also on alloy chemistry, with pure steels not subject to this problem[40] (Fig. 11.17). In this context, there exists a basic similarity between 300°C embrittlement and temper embrittlement, which will be discussed shortly.

Short of preparing high-purity (but expensive) alloys, the most obvious way to avoid 300°C embrittlement is simply to avoid tempering at that temperature. Usually this involves tempering at a higher temperature but with some sacrifice in strength. However, there are material applications that arise where the higher strengths associated with tempering at 300°C are desired. Fortunately, it has been found possible to obtain the strength levels associated with a 300°C temper while simultaneously suppressing the embrittling kinetics. This optimization of properties has been achieved through the addition of 1.5 to 2% silicon to the alloy steel.[41] It is believed that the silicon suppresses the kinetics of the martensite tempering process with the result that the embrittling reaction shifts to a higher temperature (about 400°C).

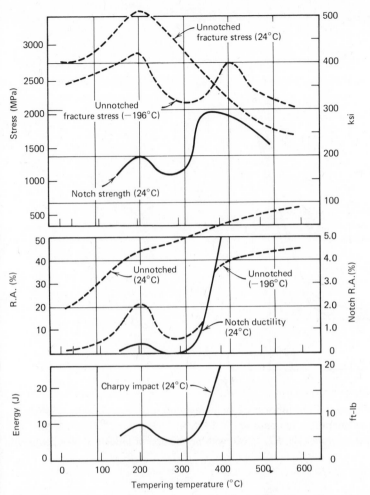

FIGURE 11.16 Notched and unnotched tensile properties at room and low temperatures for SAE 1340 steel, quenched and tempered at various temperatures. Poor properties associated with tempering in range of 300°C.[36] (Reprinted with permission of the American Society for Metals.)

11.2.2 Temper Embrittlement

Temper embrittlement (TE) develops in alloy steels when cooled slowly or isothermally heated in the temperature range of 400 to 600°C. The major consequence of TE is found to be an increase in the tough-brittle transition temperature and is associated with intercrystalline failure along prior austenite grain boundaries. Using the change in transition temperature as the measure of TE, the kinetics of the embrittlement process are found to exhibit a C-curve

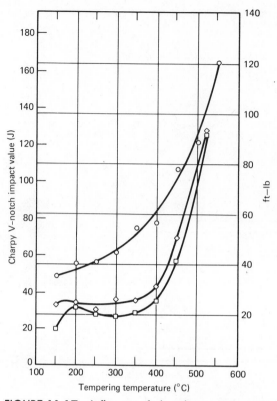

FIGURE 11.17 Influence of phosphorus and antimony on room-temperature impact energy as function of tempering temperature in $1\frac{1}{2}\%$ Ni-Cr-Mo steel.[40] (From J. M. Capus and G. Mayer, *Metallurgia*, **62**, 1960, with permission of Industrial Newspapers Ltd.)

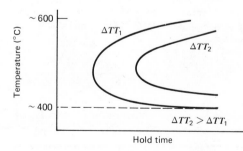

FIGURE 11.18 Isoembrittlement lines (fixed shift in tough-brittle transition temperature) as function of exposure temperature and hold time.

response, with isoembrittlement lines depicting maximum embrittlement in the shortest hold time at intermediate temperatures in the 400 to 600°C range (Fig. 11.18). It is important to note that TE can be largely reversed by reheating the steel above 600°C. Although TE has been recognized for over 75 years (e.g., see the reviews by Holloman,[42] Woodfine,[43] Low,[44] and McMahon[45]), it is by no means under control. For example, the catastrophic failure in 1969 of two forged alloy steel discs from the Hinkley Point nuclear power station steam turbine rotor offers dramatic proof that additional understanding of the TE process is needed.[46] In this instance, failure was attributed to a combination of two factors: TE resulting from slowly cooling the discs during manufacture through the critical temperature range; and EAC resulting from the entrapment of condensate in the keyways of the discs.

Balajiva and co-workers[47,48] contributed much to our current understanding of temper embrittlement. They demonstrated that TE occurred only in alloy steels of commercial purity but not in comparable alloys of high purity (Fig. 11.19). The most potent embrittling elements in order of severity were found to be antimony, phosphorus, tin, and arsenic. These results have been verified by others,[50] along with the additional finding that for a given impurity level, Ni-Cr alloy steels are embrittled more than alloys containing nickel or chromium alone. It has generally been thought that embrittlement resulted from the segregation of impurity elements at prior austenite grain boundaries as a result of exposure to the 400 to 600°C temperature range. This has since been verified

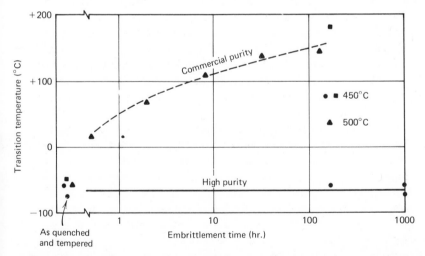

FIGURE 11.19   Effect of 450 and 500°C exposure on temper embrittlement in commercial and high-purity nickel-chromium steels. (Data from Woodfine[43] and Steven, et al.[48])[49] (Reprinted by permission of the American Society for Testing and Materials from copyright work.)

using Auger electron spectroscopy[51-53]—a technique by which the chemistry of the first few atomic layers of a material's surface is analyzed. Marcus and Palmberg[51,52] found, in a modified AISI 3340 steel alloy, antimony on the fracture surface (along prior austenite grain boundaries) in amounts exceeding 100 times that of the bulk concentration (0.03 atomic percent). Furthermore, the high antimony concentration layer was very shallow, extending only one to two atomic layers below the fracture surface.

The actual mechanism for TE remains unclear; how the embrittling elements segregate to the prior austenite grain boundaries and the element's role in reducing fracure properties are yet to be determined. Regarding the first point, there is some question as to whether the concentration of the impurity elements at prior austenite grain boundaries represents an equilibrium or nonequilibrium transient condition. In this regard, Rellick and McMahon[54] proposed recently that the impurity element segregation was caused by rejection of these elements in advance of growing carbide particles. They concluded that the relative stability of this transient concentration would depend on the diffusivity of that atom specie in the iron matrix. The explanation for the embrittlement process itself is believed to be related to an impurity element-induced reduction in the cohesive energy of the carbide-ferrite interface. Certainly, more work will be needed to clarify both these points.

11.2.3   Neutron Irradiation Embrittlement

Proper selection of alloy chemistry and heat treatment may provide a material with adequate toughness at the outset of service life for nuclear applications. There exists, however, the potential for mechanical property degradation after exposure to neutron irradiation. For example, we see from Fig. 11.20 a sharp reduction in fracture toughness after neutron irradiation, especially in the region where K_{IC} values normally rise rapidly with test temperature.[55] As a result, the initial K_{IC} level of 200 MPa$\sqrt{m}$ anticipated for this material at room temperature is reduced drastically to about 45 MPa$\sqrt{m}$, representing a 20-fold *decrease* in the allowable flaw size for a given applied stress. As one might expect, considerable attention has been devoted to determine the extent of embrittlement in a number of steels being used or considered for use in reactor pressure vessels (Table 11.3). Earlier studies[56,57] had focused on documenting changes in tensile properties (for example, strength and ductility) and Charpy impact energy absorption. To a large extent, these test procedures have continued to the present with precious few fracture mechanics studies undertaken to date. Researchers have found that neutron irradiation raises the yield strength (mainly as a result of an increase in the lattice friction component σ_i in the Petch-Hall relationship) and the tough-brittle transition temperature. At the same time, there is a corresponding decrease in tensile ductility and Charpy impact shelf energy. A diagram showing irradiation-induced changes in Charpy impact response is given in Fig. 11.21, where higher fluence* levels are seen to cause greater embrittlement. Although the cause of this embrittlement is not clearly understood, it is believed to be related to the interaction of dislocations with defect aggregates, such as solute atom-vacancy clusters that are generated by neutron bombardment. As one might expect, the greater the fluence, the greater the number of defect aggregates and the greater the elevation in yield strength and transition temperature rise.

Studies have shown that the extent of irradiation damage depends strongly on the irradiation temperature, with more damage accompanying low-temperature neutron exposures.[58] Figure 11.22 shows a larger transition temperature shift for a given neutron fluence level when exposed at lower temperatures. Steele[59] reported that embrittlement in this temperature range was attributable to nitrogen-defect cluster aggregates that impeded dislocation motion in the lattice. It is of some comfort to note that actual service temperatures in current nuclear reactors are in the range of 260 to 288°C (500 to 550°F) where damage is less extensive. Preliminary test results at 307°C (see Fig. 11.22) show further reductions in the degree of embrittlement. It may be argued that since the defect clusters probably are being annealed out more rapidly at higher temperatures, it would be reasonable to expect minimal irradiation damage at even higher irradiation temperatures, where defect annihilation would be further enhanced.[58] As a matter of fact, neutron irradiation embrittlement in a component can be reversed by annealing the damaged part at a sufficiently high temperature (above the irradiation temperature) to annihilate the defect clusters.

The amount of neutron embrittlement resulting from 288°C irradiation is found to depend strongly on the steel alloy content. For example, Hawthorne et al.[60,61] demonstrated that neutron embrittlement resulting from 288°C exposure could be eliminated completely by careful reduction in residual element content (Fig. 11.23). Although the presence of phosphorus, sulfur, and vanadium in solid solution were identified as being objectionable, copper was singled out as the most harmful element. The combined effect of irradiation temperature and copper content on the increment in yield strength resulting from a given neutron fluence is shown in Fig. 11.24. Note that low copper levels ($<0.03\%$) or high

---

*The rate of neutron irradiation or neutron flux is defined as the number of neutrons crossing a one square centimeter area in one second. The neutron fluence is defined by the product of neutron flux and time. Hence, neutron fluence is given by the number of neutrons/cm². Studies have shown that these neutrons possess a spectrum of widely varying energies, with some capable of much greater damage than others. Since we do not know the amount of damage generated by a neutron with a particular energy level, it has become the interim accepted practice to define neutron fluence as the count of neutrons possessing more than a certain minimum energy.[55] This energy level is taken usually to be 1.6×10^{-13}J (1 MeV). This counting procedure, therefore, assumes that no damage results from neutrons with energies less than this threshold level (since they are not included in the neutron count) and that all neutron energies greater than this value produce the same damage. Certainly, the current neutron counting procedure is not very discriminating but does provide some means for quantitative analysis of damage.

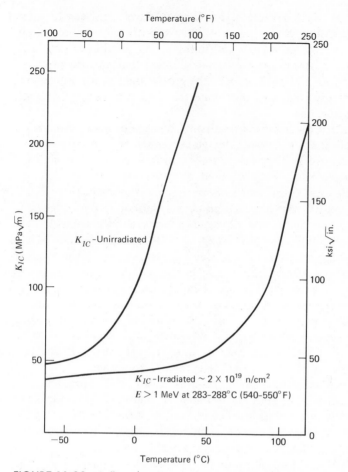

Temperature (°F)

K_{IC} (MPa√m)

ksi √in.

K_{IC} -Unirradiated

K_{IC} -Irradiated ~ 2 × 10^{19} n/cm^2
E > 1 MeV at 283–288°C (540–550°F)

Temperature (°C)

FIGURE 11.20 Influence of neutron irradiation on fracture toughness as a function of test temperature in A533-B Class I steel.[55] (Reprinted by permission of the American Society for Testing and Materials from copyright work.)

(288°C) irradiation temperatures cause the smallest increases in yield strength, while low-temperature (232°C) irradiation produces the greatest damage, regardless of the residual element level.

The nature of neutron irradiation is further complicated by size effects other than those associated with stress-state considerations. For one thing, since reactor vessels are fabricated from very thick plate (on the order of 20 cm), through thickness variations in microstructure and associated mechanical properties are expected, with minimum properties anticipated in the midthickness region. Superimposed on this gradient of K_{IC} (unirradiated) values is a separate gradient of irradiation damage that decreases continuously from the inner core

TABLE 11.3 Low Alloy Steels for Use in Nuclear Reactor Pressure Vessels[55]

Alloy	Carbon	Manganese	Phosphorus (max)	Sulfur (max)	Silicon	Molyb- denum	Nickel	Chromium	Vanadium	σ_{ts} MPa (ksi)	σ_{ys} MPa(ksi)	% Elong.
A302-C	0.25	1.1–1.55	0.035	0.040	0.13–0.32	0.41–0.64	0.37–0.73	—	—	550–690 (80–100)	345 (50)	20
A533-B Class 1	"	"	"	"	"	"	"	"	"	"	"	"
A533-B Class 2	"	"	"	"	"	"	"	"	"	620–795 (90–115)	450 (65)	16
A508 Class 2	0.27 max	0.50–0.96	0.025	0.025	0.15–0.35	0.55–0.70	0.50–0.90	0.25–0.45	0.05	620–795 (90–115)	450 (65)	16
A508 Class 4	0.23 max	0.20–0.40	0.020	0.020	0.30 max	0.40–0.60	2.75–3.90	1.50–2.00	0.03	725 (105)	585 (85)	18
A542 Class 1	0.15 max	0.27–0.63	0.035	0.035	0.50	0.90–1.10	—	1.88–2.62	—	725–860 (105–125)	585 (85)	14
A542 Class 2	"	"	"	"	"	"	"	"	"	795–930 (115–135)	690 (100)	13
A543 Class 1	0.23	0.40	0.020	0.020	0.18–0.37	0.41–0.64	2.93–4.07	1.44–2.06	0.03	725–860 (105–125)	585 (85)	14
A543 Class 2	"	"	"	"	"	"	"	"	"	795–930 (115–135)	690 (100)	14

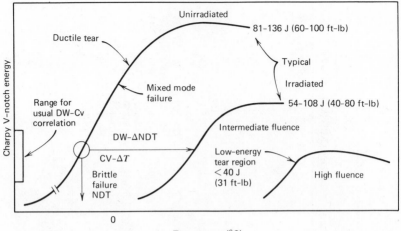

FIGURE 11.21 Diagram showing transition temperature shift resulting from neutron irradiation of typical reactor-vessel steels.[55] (Reprinted by permission of the American Society for Testing and Materials from copyright work.)

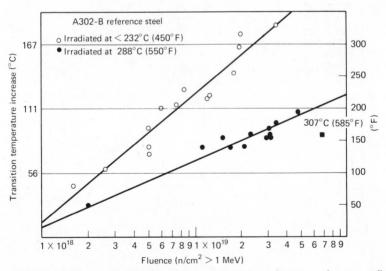

FIGURE 11.22 Transition temperature increase as function of neutron fluence for different irradiation temperatures in A302-B steel. Least damage associated with highest exposure temperature.[58] (Reprinted with permission of the American Institute of Mining, Metallurgical, and Petroleum Engineers.)

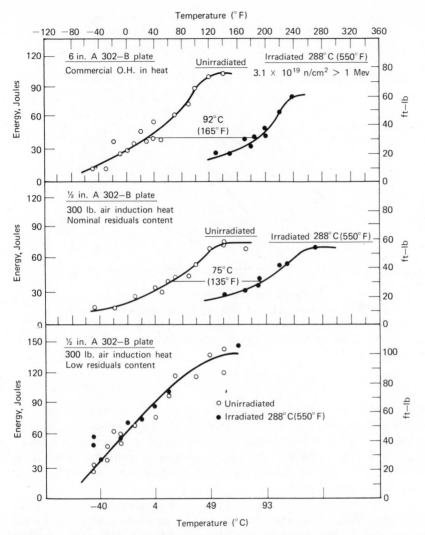

FIGURE 11.23 Decrease in neutron irradiation-induced transition temperature shift as function of residual element content in A302-B steel.[61] (Reprinted by permission of the American Society for Testing and Materials from copyright work.)

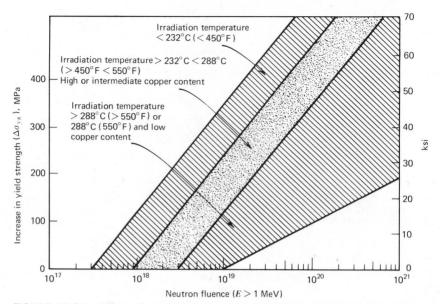

FIGURE 11.24 Effect of copper content and irradiation temperature on neutron-induced increase in room-temperature yield strength of A302-B and A533-B steels.[55] (Reprinted by permission of the American Society for Testing and Materials from copyright work.)

surface in association with an attenuation in neutron fluence through the thickness. For example, Loss et al.[62] reported a 20-fold decrease in neutron fluence from 6×10^{19} to 0.3×10^{19} neutrons/cm^2 across a 20-cm thick pressure vessel wall. From Fig. 11.24 it is seen that the degree of neutron embrittlement decreases markedly with this attenuation in neutron fluence, especially for steels containing high copper content.

Although the previous discussion focused on the embrittlement of reactor vessel low alloy steels, it should also be noted that neutron exposure in the temperature range $0.30 < T_h < 0.55$ also causes irradiation damage in stainless steels and other alloys used to contain the nuclear fuel. It is beli ed that hydrogen and helium—produced by nuclear transmutations—seg. egate to vacancy clusters and stabilize internal voids.[55] (Neutron irradiation produces both interstitials and vacancies, but there is preferential recombination of interstitials, leaving an excess of vacancies in the lattice.) Since the excess vacancy and gaseous element concentration increases with increasing neutron fluence, it is found that the relative volume change also increases (Fig. 11.25). The particular temperature regime $0.3 < T_h < 0.55$, where void-induced swelling is most prevalent, arises from the fact that the kinetics of vacancy condensation are too slow at temperatures below $T_h = 0.3$; above about $T_h = 0.55$, the super-

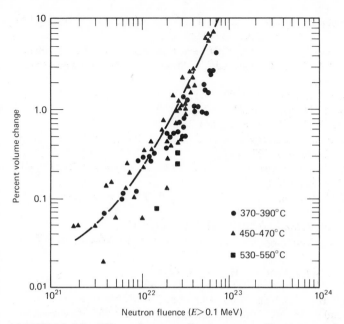

FIGURE 11.25 Effect of neutron fluence on swelling in annealed type 304 stainless steel in the 370 to 550°C temperature range.[55] (Reprinted by permission of the American Society for Testing and Materials from copyright work.)

saturation of vacancies is inadequate to sustain the voids.[55] Similarly, it has been found that swelling can be suppressed by addition of refractory elements to the alloy which attenuate vacancy mobility. In addition to swelling of the fuel cladding alloy, it is known that the uranium fuel itself undergoes volumetric expansion resulting from the precipitation of krypton and xenon gas bubbles, which form as a result of the fission process.[56] Both swelling of the fuel and cladding are highly undesirable and lead to reductions in the operating life of the fuel and metal cladding.

REFERENCES

1. R. W. Staehle, Proceedings of Conference, *Fundamental Aspects of Stress Corrosion Cracking*, R. W. Staehle, A. J. Forty, and D. van Rooyen, eds. NACE, Houston, Texas, 1969.
2. H. H. Uhlig, *Fracture*, Vol. 3, H. Liebowitz, ed., Academic Press, 1971, p. 643.
3. H. H. Johnson and P. C. Paris, *Eng. Fract. Mech.*, **1**, 1967, p. 3.
4. B. F. Brown, *Mater. Res. Stand.*, **6**, 1966, p. 129.
5. B. F. Brown and C. D. Beachem, *Corr. Sci.*, **5**, 1965, p. 745.
6. B. F. Brown, *Met. Rev.*, **13**, 1968, p. 171.

7. R. P. Wei, S. R. Novak and D. P. Williams, *Mater. Res. Stand.*, **12**, (9) 1972, p. 25.
8. S. R. Novak and S. T. Rolfe, *J. Mater.*, **4**, 1969, p. 701.
9. S. M. Wiederhorn, *Environment-Sensitive Mechanical Behavior*, Vol. 35, Metallurgical Society Conf., A. R. C. Westwood and N. S. Stoloff, eds., Gordon and Breach, New York, 1966, p. 293.
10. S. M. Wiederhorn, *Int. J. Fract. Mech.*, **42**, 1968, p. 171.
11. S. J. Hudak, Jr., M. S. Thesis, Lehigh University (1972).
12. J. D. Landes and R. P. Wei, *Int. J. Fract.*, **9**(3), 1973, p. 277.
13. W. D. Benjamin and E. A. Steigerwald, *Trans.*, ASM, **60**, 1967, p. 547.
14. M. O. Speidel, *Met. Trans.*, **6A**, 1975, p. 631.
15. M. O. Speidel, *The Theory of Stress Corrosion Cracking in Alloys*, J. C. Scully, ed., NATO, Brussels, Belgium, 1971, p. 289.
16. M. H. Peterson, B. F. Brown, R. L. Newbegin, and R. E. Groover, *Corrosion*, **23**, 1967, p. 142.
17. *Damage Tolerant Design Handbook*, MCIC-HB-01, Sept. 1973.
18. G. E. Linnert, *Welding Metallurgy*, Vol. 2, American Welding Society, New York, 1967.
19. W. Hofmann and W. Rauls, *Weld. J.*, **44**, 1965, p. 225s.
20. J. B. Steinman, H. C. VanNess and G. S. Ansell, *ibid.*, p. 2215.
21. G. G. Hancock and H. H. Johnson, *Trans. Met. Soc.*, AIME, **236**, 1966, p. 513.
22. D. O. Hayward and B. M. W. Trapnell, *Chemisorption*, 2nd Ed., Butterworths, Washington, D.C., 1964.
23. F. J. Norton, *J. Appl. Phys.*, **11**, 1940, p. 262.
24. G. L. Hanna, A. R. Troiano, and E. A. Steigerwald, *Trans. Quart.*, ASM, **57**, 1964, p. 658.
25. C. A. Zapffe, *JISI*, **154**, 1946, p. 123.
26. A. S. Tetelman and W. D. Robertson, *Trans.*, AIME, **224**, 1962, p. 775.
27. A. R. Troiano, *Trans.*, ASM, **52**, 1960, p. 54.
28. J. G. Morlet, H. H. Johnson and A. R. Troiano, *JISI*, **189**, 1958, p. 37.
29. A. R. C. Westwood, C. M. Preece, and M. H. Kamdar, *Trans. Quart.*, ASM, **60**, 1967, p. 723.
30. A. Kelly, W. R. Tyson and A. H. Cottrell, *Phil. Mag.*, **15**, 1967, p. 567.
31. N. J. Petch and P. Stables, *Nature*, **169**, 1952, p. 842.
32. *Effect of Hydrogen on Behavior of Materials*, AIME, Sept. 7–11, 1975, Moran, Wyoming.
33. D. P. Williams, *Int. J. Fract.*, **9**(1), 1973, p. 63.
34. W. Beck, J. O'M. Bockris, J. McBreen, and L. Nanis, *Proc. Roy. Soc.*, **A290**, 1966, p. 221.
35. D. P. Williams and H. G. Nelson, *Met. Trans.*, **1**, 1970, p. 63.
36. W. F. Brown, Jr., *Trans.*, ASM, **42**, 1950, p. 452.
37. E. J. Ripling, *Trans.*, ASM, **42**, 1950, p. 439.
38. M. A. Grossman, *Trans.*, AIME, **167**, 1946, p. 39.
39. B. S. Lement, B. L. Averbach, and M. Cohen, *Trans.*, ASM, **46**, 1954, p. 851.
40. J. M. Capus and G. Mayer, *Metallurgia*, **62**, 1960, p. 133.
41. C. H. Shih, B. L. Averbach, and M. Cohen, *Trans.*, ASM, **48**, 1956, p. 86.
42. J. H. Hollomon, *Trans.*, ASM, **36**, 1946, p. 473.
43. B. C. Woodfine, *JISI*, **173**, 1953, p. 229.

44. J. R. Low, Jr., *Fracture of Engineering Materials*, ASM, Metals Park, Ohio, 1964, p. 127.
45. C. J. McMahon, Jr., ASTM *STP 407*, 1968, p. 127.
46. D. Kalderon, *Proc. Inst. Mech. Eng.*, **186**, 1972, p. 341.
47. K. Balajiva, R. M. Cook, and D. K. Worn, *Nature*, **178**, 1956, p. 433.
48. W. Steven and K. Balajiva, *JISI*, **193**, 1959, p. 141.
49. J. M. Capus, ASTM *STP 407*, 1968, p. 3.
50. J. R. Low, Jr., D. F. Stein, A. M. Turkalo, and R. P. LaForce, *Trans. Met. Soc.*, AIME, **242**, 1968, p. 14.
51. H. L. Marcus and P. W. Palmberg, *Trans. Met. Soc.*, AIME, **245**, 1969, p. 1665.
52. H. L. Marcus, L. H. Hackett, Jr., and P. W. Palmberg, ASTM *STP 499*, 1972, p. 90.
53. D. F. Stein, A. Joshi, and R. P. LaForce, *Trans.*, ASM, **62**, 1969, p. 776.
54. J. R. Rellick and C. J. McMahon, Jr., *Met. Trans.*, **5**, 1974, p. 2439.
55. S. H. Bush, *J. Test. Eval.*, **2**(6), 1974, p. 435.
56. A. Tetelman and A. J. McEvily, *Fracture of Structural Materials*, John Wiley, New York, 1967.
57. D. McClean, *Mechanical Properties of Metals*, John Wiley, New York, 1962.
58. L. E. Steele, *Nucl. Mater.*, **16**, 1970, p. 270.
59. L. E. Steele, ASTM *STP 484*, 1971, p. 164.
60. U. Potapovs and J. R. Hawthorne, *Nucl. Appl.* **6**(1), 1969, p. 27.
61. J. R. Hawthorne, ASTM *STP 484*, 1971, p. 96.
62. F. J. Loss, J. R. Hawthorne, C. Z. Serpan, Jr., and P. P. Puzak, *NRC Report 7209*, Mar. 1, 1971.

PROBLEMS

11.1 An investigation was made of the rate of crack growth in a 7079-T651 aluminum plate exposed to an aggressive environment under a static stress σ. A large test sample was used with a single-edge notch placed in the T-L orientation. As indicated in the accompanying table, the rate of crack growth under sustained loading was found to vary with the magnitude of the applied stress and the existing crack length. The material exhibits Region I and II EAC but not Region III.

Cracking Rate (m/sec)	Applied Stress (MPa)	Crack Length (mm)
10^{-9}	35	5
32×10^{-9}	35	10
1×10^{-6}	70	5
1×10^{-6}	70	7.5

If the K_{IC} for the materials is 20 MPa$\sqrt{m}$, how long would it take to break a sample containing an edge crack 5 mm long under a load of 50 MPa? *Hint:* First establish the crack growth rate relationships.

11.2 To avoid slowly cooling through the 400 to 600°C temper embrittlement range, one engineer recommended that a thick section component be water quenched from 850°C to room temperature. Are there any problems with this procedure?

11.3 For the 18 Ni (300)-maraging steel listed in Table 11.2, calculate the stress level to cause failure in a center-notched sample containing a crack 5 mm long. What stress level limit would there have to be to insure that EAC did not occur in a 3 $\frac{1}{2}$% NaCl solution?

11.4 How much faster than the room temperature value would a crack grow in a high-strength steel submerged in water if the temperature were raised 100°C?

11.5 Assuming for the moment that impurity elements do segregate in advance of newly formed and growing carbides according to the hypothesis of Rellick and McMahon, discuss how the temper embrittlement reaction in alloy steels might be suppressed by changes in alloy chemistry and prior heat treatment.

CHAPTER
TWELVE
CYCLIC STRESS
AND STRAIN
FATIGUE

Daydreamers have two options for supplementary entertainment: doodling or paper clip bending. The doodler is limited by the amount of paper available, while the paper clip bender's amusement is tragically short-lived—the clip breaks after only a few reversals! This simple example describes a most insidious fracture mechanism—failure does not occur when the component is loaded initially; instead, failure occurs after a certain number of similar load fluctuations have been experienced. The author of a book about metal fatigue[1] began his treatise by describing a photograph of his car's steel rear axle, which had failed: "the final fracture occurring at 6:00 AM just after setting out on holiday." Somewhat less expensive damage, but saddening nonetheless, was the failure of my son's vehicle (Fig. 12.1). From an examination of the fracture surfaces, it was concluded that this failure originated at several sites and traveled across the section, with occasional arrest periods prior to final separation. Another fatigue failure generated in my household is shown in Fig. 12.2. The failure of this zinc die-cast door stop nearly destroyed the lovely crystal chandelier in the front hall of my home. Regardless of the material—steel paper clips and car axles, plastic tricycles, zinc doorstops—fatigue failures will occur when the component experiences cyclic stresses or strains that produce permanent damage. Since the majority of engineering failures involve cyclic loading of one kind or another, it is appropriate to devote considerable attention to this subject in this chapter and in Chapter Thirteen. The reader is also referred to Chapter Fourteen for descriptions of case histories of fatigue failures.

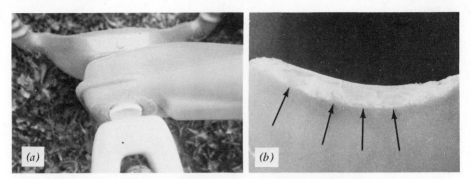

FIGURE 12.1 Fatigue fracture of plastic tricycle. (a) General location of failure; (b) several origins are identified by arrows. Note characteristic fatigue ring-like markings emanating from each origin, which represent periods of growth during life of component. (Courtesy Jason and Michelle Hertzberg.)

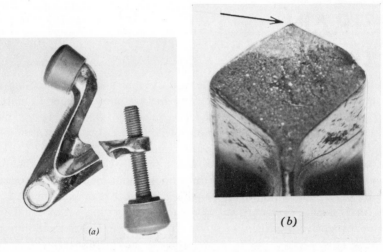

FIGURE 12.2 Zinc die-cast door stop fatigue fracture. Fatigue crack grew from corner until reaching a critical size and causing failure. Arrow in (b) indicates crack origin.

12.1 MACROFRACTOGRAPHY OF FATIGUE FAILURES

A macroscopic examination of many service failures generated by cyclic loading reveals distinct fracture surface markings. For one thing, the fracture surface is generally flat, indicating the absence of an appreciable amount of gross plastic deformation during service life. In many cases, particularly failures occurring over a long period of time, the fracture surface contains lines referred to in the literature as "clam shell markings," arrest lines, and/or "beach markings" (Fig. 12.3). These markings have been attributed to different periods of crack extension, such as during one flight or one sequence of maneuvers of an aircraft or the operation of a machine during a factory work shift. It is to be emphasized that these bands reflect *periods* of growth and are not representative of *individual* load excursions. Unique markings associated with the latter are discussed in Chapter Thirteen. It is believed that these alternate crack growth and dormant periods cause regions on the fracture surface to be oxidized and/or corroded by differing amounts, resulting in the formation of a fracture surface containing concentric rings of nonuniform color. Similar bands resulting from variable amplitude block loading have been found on fracture surfaces (see Fig. 13.9). Since these "beach markings" often are curved, with the center of curvature being at the origin, they serve as a useful guide to direct the investigator to the fracture initiation site.

As shown in Fig. 12.4, the fracture surface may exhibit any one of several

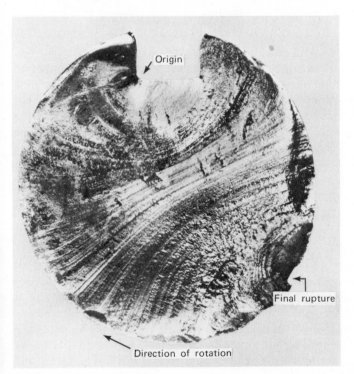

FIGURE 12.3 Fatigue fracture surface of rotating steel shaft. Center of curvature of earlier "beach markings" locate crack origin at corner of spline.[2] (By permission from D. J. Wulpi, *How Components Fail*, copyright American Society for Metals, 1966.)

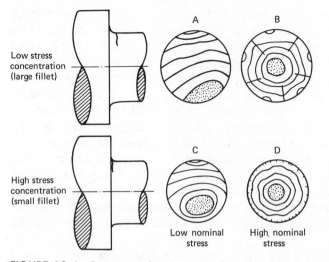

FIGURE 12.4 Diagrams showing typical fatigue fracture surface appearance for varying conditions of stress concentration and stress level.[2] (By permission from D. J. Wulpi, *How Components Fail*, copyright American Society for Metals, 1966.)

patterns, depending on such factors as the applied stress and the number of potential crack nucleation sites. For example, we see that as the severity of a design imposed stress concentration and/or the applied stress increases, the number of nucleation sites increase. Either of these conditions should be avoided if at all possible. In fact, most service failures exhibit only one nucleation site, which eventually causes total failure. The size of this fatigue crack at the point of final failure is related directly to the applied stress level and the fracture toughness of the material. Hence, from Eq. 8-22 one may conclude that a very low cyclic stress was responsible for the failure shown in Fig. 12.3.

12.2 CYCLIC STRESS CONTROLLED FATIGUE

Many engineering components must withstand numerous load or stress reversals during their service lives. Examples of this type of loading include alternating stresses associated with a rotating shaft, pressurizing and depressurizing cycles in an aircraft fuselage at take-off and landing, and load fluctuations affecting the wings during flight. Depending on a number of factors, these load excursions may be introduced either between fixed strain or fixed stress limits; hence, the fatigue process in a given situation may be governed by a strain or stress controlled condition. Discussion in this section is restricted to stress controlled fatigue; strain controlled fatigue is considered in Section 12.3.

One of the earliest investigations of stress controlled cyclic loading effects on fatigue life was conducted by Wöhler,[3] who studied railroad wheel axles that were plagued by an annoying series of failures. Several important facts emerged from this investigation, as may be seen in the plot of stress versus the number of cycles to failure (a so-called S-N diagram) given in Fig. 12.5. First, the cyclic life of an axle increased with decreasing stress level and below a certain stress level, it seemed to possess infinite life—fatigue failure did not occur (at least not before 10^6 cycles). Second, the fatigue life was reduced drastically by the presence of a notch. These observations have led many current investigators to view fatigue as a three-stage process involving initiation, propagation, and final failure stages (Fig. 12.6). When design defects or metallurgical flaws are preexistent, the initiation stage is shortened drastically or completely eliminated, resulting in a reduction in potential cyclic life.

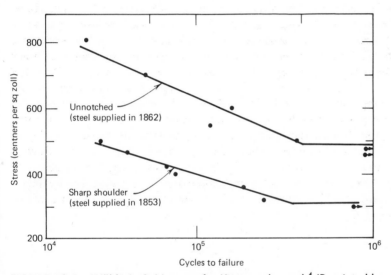

FIGURE 12.5   Wöhler's S-N curves for Krupp axle steel.[4] (Reprinted by permission of the American Society for Testing and Materials from copyright work.)

Over the years, laboratory tests have been conducted in bending (rotating or reversed flexure), torsion, pulsating tension, or tension-compression axial loading. Such tests have been conducted under conditions of constant load or moment (to be discussed in this section), constant deflection or strain (Section 12.3), or a constant stress intensity factor (Chapter Thirteen). Examples of different loading conditions are shown in Fig. 12.7. In rotating bending with a single load applied at the end of the cantilevered test bar (Fig. 12.7a), the bending moment increases with increasing distance from the applied load point

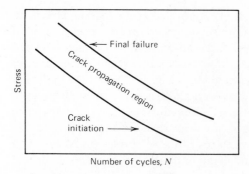

FIGURE 12.6 Fatigue life depends on relative extent of initiation and propagation stages.

and precipitates failure at the base of the fillet at the end of the gage section. In effect, this represents a notched fatigue test, since the results will depend strongly on fillet geometry. The rotating beam-loaded case (Fig. 12.7b) produces a constant moment in the gage section of the test bar that can be used to generate either unnotched or notched test data. Notched test data are obtained by the addition of a circumferential notch in the gage section. Both specimen types generally represent zero mean load conditions or a load ratio of $R = -1$($R \equiv$ minimum load/maximum load). These test specimen configurations and modes of loading may be suitable for evaluating the fatigue characteristics of a component subjected to simple rotating loads. However, it is often more realistic to use the axially loaded specimen (Fig. 12.7c) to simulate service conditions that involve direct loading when mean stress is an important variable. Such is the case for aircraft wing loads where fluctuating stresses are superimposed on both a tensile (lower wing skin) and compressive (upper wing skin) mean stress.

Standard definitions regarding key load or stress variables are shown in Fig. 12.8 and defined by

$$\Delta\sigma = \sigma_{max} - \sigma_{min} \tag{12-1a}$$

$$\sigma_a = \frac{\sigma_{max} - \sigma_{min}}{2} \tag{12-1b}$$

$$\sigma_m = \frac{\sigma_{max} + \sigma_{min}}{2} \tag{12-1c}$$

$$R = \frac{\sigma_{min}}{\sigma_{max}} \tag{12-1d}$$

Most often, $S-N$ diagrams similar to that shown in Fig. 12.5 are plotted, with the stress amplitude given as half the total stress range. Another example of constant load amplitude fatigue data for 7075-T6 aluminum alloy notched specimens is shown in Fig. 12.9. Note the considerable amount of scatter in

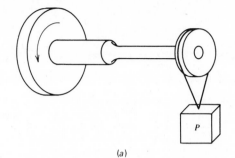

(a)

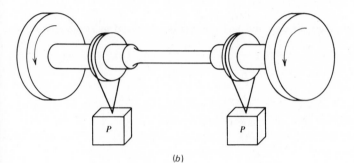

(b)

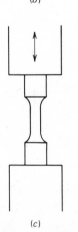

(c)

FIGURE 12.7   Various loading configurations used in fatigue testing. (a) Single-point loading, where bending moment increases toward the fixed end; (b) beam loading with constant moment applied in gage section of sample; (c) pulsating tension or tension-compression axial loading.

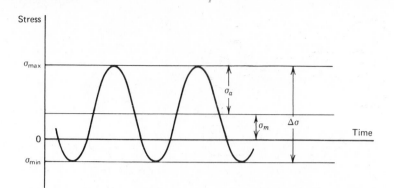

FIGURE 12.8 Nomenclature to describe test parameters involved in cyclic stress testing.

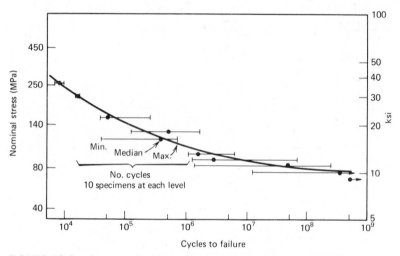

FIGURE 12.9 Constant load amplitude fatigue data for 7075-T6 aluminum alloy notched specimens (0.25 mm root radius).[5] (Reprinted from Hardrath et al., NASA TN D-210.)

fatigue life found among the 10 specimens tested at each stress level. The smaller scatter at high stress levels is believed to result from a much shorter initiation period prior to crack propagation. The existence of scatter in fatigue test results is common and deserving of considerable attention, since engineering design decisions must be based on recognition of the statistical character of the fatigue process. The origins of test scatter are manifold. They include variations in testing environment, preparation of the specimen surface, alignment of the test machine, and a number of metallurgical variables. With regard to the testing apparatus, the least amount of scatter is produced by rotating bending machines, since misalignment is less critical than in axial loading machines. Standard statistical procedures are used to obtain mean values $\bar{x}$ and associated standard deviations s for a given data population where

$$\bar{x} = \sum \frac{x}{n} \tag{12-2}$$

$$s = \left[\frac{\sum (x - \bar{x})^2}{n - 1} \right]^{1/2} \tag{12-3}$$

where

$$x = \text{test value}$$

$$n = \text{number of test values}$$

When considering alternating stress-cyclic life data, the test value x usually represents the range of cyclic lives experienced at a given stress level. With these statistical parameters, the confidence limits of the probability of survival may be determined. That is, the fatigue life anticipated with a desired confidence level of $\gamma\%$ that at least $P\%$ of the samples will survive may be given by[1]

$$\text{anticipated life } (\gamma, P) = \bar{x} - qs \tag{12-4}$$

where

$$q = f \text{ (confidence level, } \gamma\%; \text{ probability of survival, } P\%; \text{ and}$$

$$\text{number of samples used to determine } \bar{x} \text{ and } s)$$

The selection of a particular confidence limit depends on the importance of the component to the overall integrity of the structure. The more important the component, the higher the required confidence limit would have to be and the lower the operating stress. It may be seen from Table 12.1 that q increases with increasing confidence level and probability of survival, and with a decrease in the number of test values in the data population. Therefore, to have a high degree of confidence of component survival, the design stress must be reduced, especially if few data are available for the analysis. With the aid of Eq. 12-4 and Table 12.1, it is possible to develop families of curves representing the probability of component survival (or failure) for use in engineering design decision making (Fig. 12.10).

TABLE 12.1 Values of q for $S-N$ Data Assuming Normal Distribution[a]

p	75	90	95	99	99.9	75	90	95	99	99.9
n		$\gamma = 0.50$					$\gamma = 0.75$			
4	0.739	1.419	1.830	2.601	3.464	1.256	2.134	2.680	3.726	4.910
6	0.712	1.360	1.750	2.483	3.304	1.087	1.860	2.336	3.243	4.273
8	0.701	1.337	1.719	2.436	3.239	1.010	1.740	2.190	3.042	4.008
10	0.694	1.324	1.702	2.411	3.205	0.964	1.671	2.103	2.927	3.858
12	0.691	1.316	1.691	2.395	3.183	0.933	1.624	2.048	2.851	3.760
15	0.688	1.308	1.680	2.379	3.163	0.899	1.577	1.991	2.776	3.661
18	0.685	1.303	1.674	2.370	3.150	0.876	1.544	1.951	2.723	3.595
20	0.684	1.301	1.671	2.366	3.143	0.865	1.528	1.933	2.697	3.561
25	0.682	1.297	1.666	2.357	3.132	0.842	1.496	1.895	2.647	3.497
		$\gamma = 0.90$					$\gamma = 0.95$			
4	1.972	3.187	3.957	5.437	7.128	2.619	4.163	5.145	7.042	9.215
6	1.540	2.494	3.091	4.242	5.556	1.895	3.006	3.707	5.062	6.612
8	1.360	2.219	2.755	3.783	4.955	1.617	2.582	3.188	4.353	5.686
10	1.257	2.065	2.568	3.532	4.629	1.465	2.355	2.911	3.981	5.203
12	1.188	1.966	2.448	3.371	4.420	1.366	2.210	2.736	3.747	4.900
15	1.119	1.866	2.329	3.212	4.215	1.268	2.068	2.566	3.520	4.607
18	1.071	1.800	2.249	3.106	4.078	1.200	1.974	2.453	3.370	4.415
20	1.046	1.765	2.208	3.052	4.009	1.167	1.926	2.396	3.295	4.319
25	0.999	1.702	2.132	2.952	3.882	1.103	1.838	2.292	3.158	4.143

[a]Data obtained from ASTM *STP 91*, 1963, p. 67.

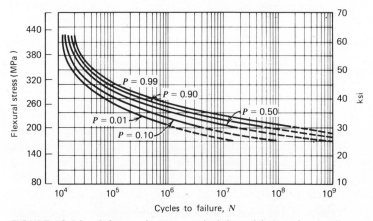

FIGURE 12.10 Schema showing probability of failure for small, unnotched specimens of 7075-T6 aluminum alloy.[6] (Reprinted from *1953 Transactions of the ASME* with permission of the American Society of Mechanical Engineers.)

EXAMPLE 1   From Fig. 12.10, at a stress level of 215 MPa, 10 specimens failed after 19,200, 17,700, 17,600, 17,100, 16,400, 16,300, 16,100, 16,000, 15,900, and 15,400 cycles, respectively. For this stress level, compute the cyclic life that provides 95% confidence that 99.9% of the components would survive.

From Eqs. 12-2 and 12-3, the mean value and standard deviation are found to be

$$\bar{x} = 16,770 \text{ cycles}$$

$$s = 1,135 \text{ cycles}$$

For a 95% confidence level that 99.9% of the samples would survive cycling at 215 MPa, the safe life is determined from Eq. 12-4, where $q = 5.203$ (Table 12.1).

$$\text{safe life} = \bar{x} - qs$$

$$= 16,770 - 5.203 \, (1,135)$$

$$= 10,865 \text{ cycles}$$

12.2.1   Effect of Mean Stress on Fatigue Life

As mentioned in the previous section, mean stress can represent an important test variable in the evaluation of a material's fatigue response. It then becomes necessary to portray fatigue life data as a function of two stress variables from the ones defined in Eq. 12-1. Sometimes this is done by plotting S–N data for a given material at different σ_m values, as shown in Fig. 12.11a. Here we see a

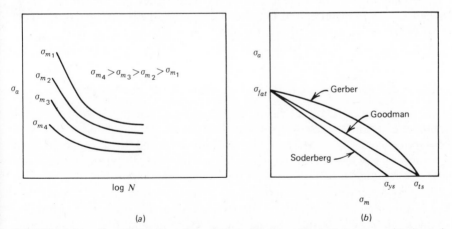

(a)   (b)

FIGURE 12.11   Representative plots of data showing effect of stress amplitude and mean stress on fatigue life. (a) Typical S–N diagrams with differing σ_m levels; (b) Gerber, Goodman, and Soderberg diagrams showing combined effect of alternating and mean stress on fatigue endurance.

trend of decreasing cyclic life with increasing σ_m level for a given σ_a level. Alternately, empirical relationships have been developed to account for the effect of mean stress on fatigue life.

$$\text{Goodman relation:} \quad \sigma_a = \sigma_{fat}\left(1 - \frac{\sigma_m}{\sigma_{ts}}\right) \tag{12-5}$$

$$\text{Gerber relation:} \quad \sigma_a = \sigma_{fat}\left[1 - \left(\frac{\sigma_m}{\sigma_{ts}}\right)^2\right] \tag{12-6}$$

$$\text{Soderberg relation:} \quad \sigma_a = \sigma_{fat}\left(1 - \frac{\sigma_m}{\sigma_{ys}}\right) \tag{12-7}$$

where

σ_a = fatigue strength in terms of stress amplitude, where $\sigma_m \neq 0$

σ_m = mean stress

σ_{fat} = fatigue strength in terms of stress amplitude, where $\sigma_m = 0$

σ_{ts} = tensile strength

σ_{ys} = yield strength

These relationships are shown in Fig. 12.11b and illustrate the relative importance of σ_a and σ_m on fatigue endurance. Experience has shown that most data lie between the Gerber and Goodman diagrams; the latter, then, represents a more conservative design criteria for mean stress effects.

12.2.2 Stress Fluctuation and Cumulative Damage

Much of the fatigue data discussed thus far were generated from constant amplitude, constant frequency tests, but these results are not realistic in actual field service conditions. Many structures are subjected to a range of load fluctuations, mean levels, and frequencies. The task, then, is to predict, based on constant amplitude test data, the life of a component subjected to a variable load history. A number of cumulative damage theories, proposed during the past few decades, describe the relative importance of stress interactions and the amount of damage—plastic deformation, crack initiation, and propagation—introduced to a component. For example, if the same amount of damage is done to a component at any stress level as a result of a given fraction of the number of cycles required to cause failure, we see from Fig. 12.12 that $n_1/N_1 + n_2/N_2 +$

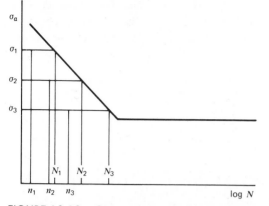

FIGURE 12.12 Component cyclic life determined from $\sum n_i / N_i = 1$ if damage at σ_i is linear function of n_i and damage is not a function of block sequencing.

$n_3 / N_3 = 1$. This may be described in more general form by

$$\sum_{i=1}^{k} \frac{n_i}{N_i} = 1 \qquad (12\text{-}8)$$

where

k = number of stress levels in the block loading spectrum

i = ith stress level

n_i = number of cycles applied at σ_i

N_i = fatigue life at σ_i

Equation 12-8 is the work of Palmgren[7] and Miner[8] and is often referred to as the Palmgren–Miner cumulative damage law. Note that Eq. 12-8 shows no dependence on the order in which the block loads are applied and, as such, represents interaction-free behavior as well. In reality, the Palmgren–Miner law is unrealistic, since the amount of damage accumulated does depend on block sequencing and varies nonlinearly with n_i. For example, if a high load block is followed by a low load block, experimental data in *unnotched* specimens generally indicate $\sum n / N < 1$. (The reverse is true for the case of *notched* samples as will be described in Chapter Thirteen; the opposite trend reflects different effects of load interactions on the initiation and propagation stages in the fatigue process). Since crack propagation begins sooner at the higher stress levels, it is argued that the initial cycles at the second block of lower stress excursions would do more damage than normally anticipated, since the initiation process would have been truncated by the high load block. The deleterious effect of overstressing in unnotched testing is shown in Fig. 12.13. Alternately, when σ_1/σ_2 is less than unity, $\Sigma n/N > 1$ for some alloys. Such understressing is seen to "coax" the fatigue limit of certain steels that strain age to somewhat higher levels.

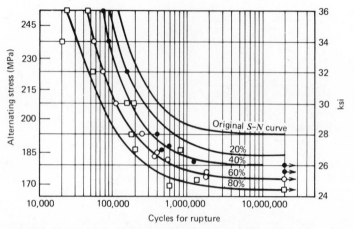

FIGURE 12.13   S–N diagrams showing decreased cyclic life after initial cycling at ±250 MPa for 20, 40, 60 and 80% of anticipated life at that stress for SAE 1020 steel.[9] (Reprinted by permission of the American Society for Testing and Materials from copyright work.)

12.2.3   Stress Concentration and Specimen Size Effects

Fatigue failures almost always initiate at the surface of a component. What are some of the factors contributing to this behavior? First, many stress concentrations, such as surface scratches, dents, machining marks, and fillets, are unique to the surface as is corrosion attack, which roughens the surface. In addition, cyclic slip causes the formation of surface discontinuities, such as intrusions and extrusions, that are precursors of actual fatigue crack formation. (The processes involved in intrusion and extrusion formation are discussed in Section 12.4.) The data shown in Fig. 12.14 clearly show the serious loss in fatigue limit associated with a deterioration in surface quality. Recall that a similar response was recognized by Wöhler more than 100 years ago (Fig. 12.5).

In order to quantitatively evaluate the severity of a particular stress concentration, many investigators adopted the stress concentration factor K_t as the comparative key parameter.[4] [From Chapter Eight, it is not surprising to find the stress intensity factor also being used in this fashion (see Chapter Thirteen).] Assuming elastic response, the fatigue strength at N cycles in a notched

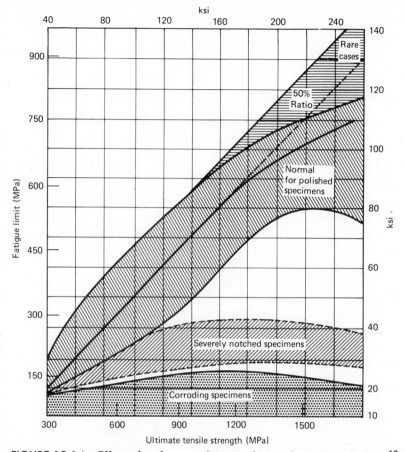

FIGURE 12.14 Effect of surface condition on fatigue limit in steel alloys.[10] (Reprinted with permission from John Wiley and Sons, Inc.)

component would be expected to decrease by a factor equal to K_t. For example, if a material exhibits a smooth bar fatigue limit of 210 MPa, the same material would have a fatigue limit of 70 MPa if a theoretical stress concentration factor of 3 were present. In reality, the reduction in fatigue strength at N cycles is less than that predicted by the magnitude of K_t. Rather, the fatigue strength is reduced by a factor K_f, which represents the effective stress concentration factor as affected by plastic flow. The relative notch sensitivity q for a given material and notch root detail may be given by

$$q = \frac{K_f - 1}{K_t - 1} \tag{12-9}$$

CYCLIC STRESS AND STRAIN FATIGUE / **429**

where

$$q = \text{notch sensitivity factor, wherein } 0 \leqslant q \leqslant 1$$

$$K_t = \text{theoretical stress concentration factor}$$

$$K_f = \text{effective stress concentration factor defined by the ratio}$$

$$\frac{\text{unnotched fatigue strength at } N \text{ cycles}}{\text{notched fatigue strength at } N \text{ cycles}}$$

From Fig. 12.15, it is seen that the relative notch sensitivity increases with increasing tensile strength and severity of notch root. As might be expected, the relative notch sensitivity decreases with increasing plastic deformation capacity.

Another factor that controls fatigue strength at N cycles is the size of the test bar. Although no size effect is observed in *axial* loading of *smooth* bars, a strong size effect is noted in smooth and notched samples subjected to *bending* and in *axially* loaded *notched* bars. In all cases, the section size effect is related to a stress gradient existing in the sample, which in turn controls the volume of material subjected to the highest stress levels. For the case of bending, the smaller the cross section of the test bar, the higher the stress gradient and the smaller the volume of material experiencing maximum stress. Comparing this situation to that of axially loaded smooth specimens where no stress gradient exists and the entire cross section is stressed equally, one finds bending fatigue

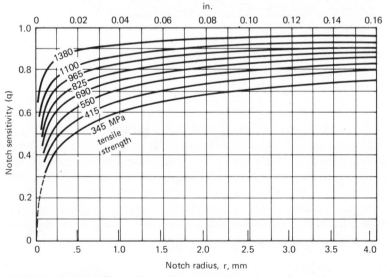

FIGURE 12.15 Effect of tensile strength and notch acuity on relative notch sensitivity.[11] (Reprinted with permission from McGraw-Hill Book Co.)

strengths to be higher than values obtained from axially loaded samples. From a statistical viewpoint, the larger the volume of material experiencing maximum stress, the greater the probability of finding a weak area that would lead to more rapid failure. As a final note, it is helpful to consider a reappraisal by Findley[12] of the specimen size effect in fatigue testing. If one assumes that fatigue crack initiation will occur when cyclic slip develops over some critical region requiring a minimum driving force σ', then slip can occur only in the outer fibers, where the applied stress is greater than σ'. Since the stress at the outer fiber in bending is

$$\sigma = \frac{Mr}{I} \qquad (12\text{-}10)$$

where

σ = flexural stress

M = bending moment

I = moment of inertia

r = radius of circular rod or distance from neutral

axis to outer fiber

the flexural stress decreases to σ' when one moves a distance Δr from the outer surface, so that

$$\sigma' = \frac{M}{I}(r - \Delta r) \qquad (12\text{-}11)$$

and

$$\sigma = \frac{\sigma'}{1 - \dfrac{\Delta r}{r}} \qquad (12\text{-}12)$$

From Eq. 12-12, we see that no size effect is predicted when uniaxial tension is applied to an unnotched specimen. Here $\sigma = \sigma'$. However, a size effect is anticipated in specimens possessing a large stress gradient (that is, in small specimens subjected to either bending or torsion, and in notched, axially loaded samples). Finally, it is apparent from Eq. 12-12 that the size effect disappears for large samples since $r \gg \Delta r$, whereupon σ' approaches σ.

12.2.4 Material Behavior

This section presents an overview of the effect of mechanical properties on material fatigue response. Since detailed discussions of the effect of microstructure and thermomechanical treatment on fatigue behavior in various alloy systems would be beyond the scope of this book, the reader is referred to numerous articles in the open literature. Books by Forrest,[1] Sines and Waisman,[11] and Forsyth[13] should provide an excellent starting point for such an investigation.

Generally, materials tend to exhibit $S-N$ plots of two basic shapes. Either the plot shows a well-defined fatigue limit (Fig. 12.5) below which the material would appear to be immune from cyclic damage or a continually decreasing curve (Fig. 12.10) with no apparent lower stress limit below which the material could be considered completely "safe." (Note the strong resemblance to environment assisted cracking behavior discussed in the previous chapter.) In materials that possess a "knee" in the $S-N$ curve, the *fatigue limit* is readily determined as the stress associated with the horizontal portion of the $S-N$ curve. It has been found that many steel alloys exhibit this type of behavior. Furthermore, the fatigue limit of these alloys often is approximately one-half the tensile strength of the material. Consequently, it would appear to be good design practice to use a material with as high a tensile strength as possible to maximize fatigue resistance. Unfortunately, this can get you in a lot of trouble, since (as shown in Chapters Ten and Eleven) fracture toughness decreases and environmental sensitivity increases with increasing tensile strength. Since tensile strength and hardness are related, it is possible to estimate the fatigue limit in a number of steels simply by determining the hardness level—a very inexpensive test procedure, indeed. We see from Fig. 12.16 that a good correlation exists up to a hardness level of about $40R_c$. Above $40R_c$, test scatter becomes considerable and the fatigue limit, hardness relationship becomes suspect. As we saw from Eq. 12-4, when the standard deviation associated with the test data is large, the operating stress or allowable service life would have to be reduced to insure the desired probability of uneventful component service.

The fatigue behavior of nonferrous alloys usually follow the second type of $S-N$ plot, and no clear fatigue limit is defined. Consequently, the "fatigue limit" for any such alloy would have to be defined at some specific cyclic life—usually 10^7 cycles. Examining various aluminum alloys reveals that "fatigue limit"/tensile strength ratios are lower than those found in steel alloys and range from about 0.3 to 0.35. Many studies have been conducted and theories proposed to account for the relatively poor fatigue response shown by this important group of engineering materials. It is presently felt[15] that extremely fine and atomically ordered precipitates, contained within Al-Cu alloys, are penetrated by dislocations moving back and forth along active slip planes. This action produces an initial strain hardening response followed by local softening, which serves to concentrate additional deformation in narrow bands and leads to crack initiation. Localized softening is believed to occur by a disordering process resulting from repeated precipitate cutting by the oscillating dislocations. To offset this, it has been suggested that additional platelike particles that are impenetrable by dislocations be added to the microstructure to arrest the mechanically induced disordering process. In this manner, the fine, ordered particles, penetrable by dislocations, would act as precipitation hardening agents while relatively larger,

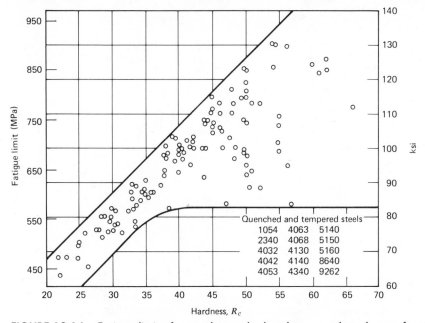

FIGURE 12.16 Fatigue limit of several quenched and tempered steels as a function of hardness level. Considerable uncertainty exists in determining fatigue limits at hardness levels in excess of 40 R_c (about 1170 MPa).[14] (By permission from *Metals Handbook*, Vol. 1, copyright American Society for Metals, 1961.)

platelike particles that are not cut by dislocations, would enhance fatigue behavior.

Since the fatigue limit associated with long cyclic life is strongly dependent on tensile strength, it follows that fatigue behavior should be sensitive to alloy chemistry and thermomechanical treatment. Since large inclusions do not significantly alter tensile strength but do serve as potential crack nucleation sites, their presence in the microstructure is undesirable. By eliminating them through more careful melting practice and stricter alloy chemistry, one finds a reduction in early life failures and a concomitant reduction in the amount of scatter in test results. In this instance, a reduction in inclusion content has a favorable effect on both fatigue behavior and fracture toughness.

Although inclusions and some other internal metallurgical flaws have a deleterious effect on unnotched fatigue response, they have an interesting influence on notched fatigue behavior. That is, the notch sensitivity associated with an external notch is lower in a material that already contains a large population of internal flaws. An example of this is seen from a comparison of relative notch sensitivity between flake-graphite grey cast iron and nodular cast iron. From Table 12.2 we see that the fatigue limit/tensile strength ratio is

| Material | Tensile Strength, (MPa) | Bending Fatigue Limit, MPa | | σ_{fat}/σ_{ts} | K_f | K_t | q |
		Plain	Notched				
Flake-graphite							
grey cast iron	315	150	95	0.48	1.57	3.1	0.27
	305	110	110	0.36	1.00	3.1	0
	360	165	150	0.46	1.09	3.1	0.04
	330	140	140	0.42	1.00	3.1	0
	345	145	145	0.42	1.00	3.1	0
	285	125	105	0.43	1.20	3.1	0.1
	420	160	145	0.38	1.09	3.1	0.04
				Avg. 0.42			Avg. 0.06
Nodular cast iron							
(as cast)	550	260	150	0.48	1.73	3.1	0.35
	590	255	145	0.43	1.76	3.1	0.36
	630	260	180	0.42	1.46	3.1	0.22
	535	260	140	0.49	1.90	3.5	0.26
	550	240	130	0.44	1.84	3.5	0.34
	560	255	145	0.46	1.76	3.5	0.30
	590	235	110	0.40	2.13	3.5	0.45
				Avg. 0.45			Avg. 0.33
Nodular cast iron							
(heat treated)	320	165	145	0.52	1.14	3.1	0.07
	425	200	115	0.47	1.70	3.1	0.33
	505	205	115	0.41	1.77	3.1	0.37
	535	220	115	0.41	1.88	3.1	0.42
	290	150	105	0.52	1.47	3.5	0.19
	315	160	125	0.50	1.28	3.5	0.11
	355	180	110	0.50	1.63	3.5	0.25
				Avg. 0.48			Avg. 0.25

lowest in the flake-graphite cast iron, which probably reflects the damaging effect of the sharp graphite flakes. Conversely, the notch sensitivity to an external circumferential V-notch is lowest in this alloy. You might say that with all the graphite flakes present to create a multitude of stress concentrations, one more notch is not that harmful.

12.2.4.1 SURFACE TREATMENT

Although changes in overall material properties do influence fatigue behavior (for example, see Fig. 12.16), greater property changes are effected by localized modification of the specimen or component surface, since most fatigue cracks originate in this region. To this end, a number of surface treatments have been

TABLE 12.2b Effect of a Circumferential V-notch on the Bending Fatigue Limit of Cast Iron[1]

Material	Tensile Strength, ksi	Bending Fatigue Limit, ksi		σ_{fat}/σ_{ts}	K_f	K_t	q
		Plain	Notched				
Flake-graphite							
grey cast iron	46	22	14	0.48	1.57	3.1	0.27
	44	16	16	0.36	1.00	3.1	0
	52	24	22	0.46	1.09	3.1	0.04
	48	20	20	0.42	1.00	3.1	0
	50	21	21	0.42	1.00	3.1	0
	41.6	18	15	0.43	1.20	3.1	0.1
	60.6	23	21	0.38	1.09	3.1	0.04
				Avg. 0.42			Avg. 0.06
Nodular cast iron							
(as cast)	80	38	22	0.48	1.73	3.1	0.35
	86	37	21	0.43	1.76	3.1	0.36
	91.4	38	26	0.42	1.46	3.1	0.22
	77.8	38	20	0.49	1.90	3.5	0.26
	79.6	35	19	0.44	1.84	3.5	0.34
	81	37	21	0.46	1.76	3.5	0.30
	85.4	34	16	0.40	2.13	3.5	0.45
				Avg. 0.45			Avg. 0.33
Nodular cast iron							
(heat treated)	46.2	24	21	0.52	1.14	3.1	0.07
	61.4	29	17	0.47	1.70	3.1	0.33
	73	30	17	0.41	1.77	3.1	0.37
	77.6	32	17	0.41	1.88	3.1	0.42
	42	22	15	0.52	1.47	3.5	0.19
	46	23	18	0.50	1.28	3.5	0.11
	51.6	26	16	0.50	1.63	3.5	0.25
				Avg. 0.48			Avg. 0.25

developed; they may be classified in three broad categories: mechanical treatments, including shot peening, cold rolling, grinding, and polishing; thermal treatments, such as flame and induction hardening; and surface coatings, such as case hardening, nitriding, and plating. One of the most widely used mechanical treatment involves the use of shot peening. In this process, small, hard particles (shot) about 0.08 to 0.8 mm diameter are blasted onto the surface that is to be treated. This action results in a number of changes to the condition of the material at and near the surface.[16] First, and most importantly, a thin layer of compressive residual stress is developed that penetrates to a depth of about one-quarter to one-half the shot diameter (Fig. 12.17). Since the peening process involves localized plastic deformation, it is believed that the surrounding elastic

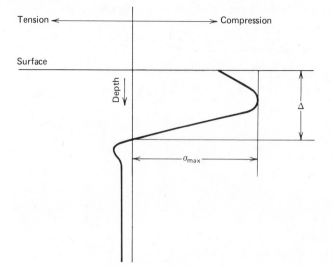

Tension ← → Compression

Surface

Depth

Δ

σ_{max}

FIGURE 12.17 Diagram showing residual stress distribution after shot peening process. Compressive residual stress extends from surface to a depth Δ.

material forces the permanently strained peened region back toward its original dimensions, thereby inducing a residual compressive stress. Depending on the type of shot, shot diameter, pressure and velocity of shot stream, and duration of the peening process, the maximum compressive stress can reach about one-half the material yield strength. Consequently, the peening process benefits higher strength alloys more than the weaker ones. Since the peened region has a localized compressive mean stress, it acts to reduce the most damaging tensile portion of the applied alternating stress range (Fig. 12.8), resulting in a substantial improvement in fatigue life.

It should be emphasized that shot peening is effective only in specimens or components that contain a stress concentration or stress gradient; the peening process is of minimal utility in unnotched components. Also, shot peening is of limited use when high applied stresses are anticipated (that is, in the low cycle fatigue regime), since large stress excursions, particularly those in the plastic range, cause rapid "fading" of the residual stress pattern. On the other hand, shot peening is very useful in the high cycle portion of fatigue life associated with lower stress levels.

Another beneficial effect of shot peening, though of secondary importance, is the work hardening contribution in the peened material that results from plastic deformation. Particularly in cases involving low-strength alloys with high strain hardening capacity, the material strain hardens, thereby contributing to a higher fatigue strength associated with the higher tensile strength. Finally, the shot peening process alters the surface by producing small "dimples" which, by themselves, would have a deleterious effect on fatigue life by acting as countless local stress concentrations. Fortunately, the negative aspect of this surface roughening is more than counterbalanced by the concurrent favorable residual compressive stress field. To be sure, the fatigue properties of a component may be improved still further if the part is polished after a shot peening treatment.

Surface rolling also produces a favorable residual stress that can penetrate deeper than that produced by shot peening and which does not roughen the component surface. Surface rolling finds extensive use in components possessing surfaces of rotation, such as in the practice of rolling machine threads.

Flame and induction hardening heat treatments in certain steel alloys are intended to make the component surface harder and more wear-resistant. This is done by heating the surface layers into the austenite phase region and then quenching rapidly to form hard, untempered martensite. Since the tensile strength and hardness of this layer is markedly increased, the fatigue strength likewise is improved [though at the expense of reliability when the hardness exceeds 40 R_c or about 1170 MPa (Fig. 12.16)]. In addition, since the austenite to martensite phase transformation involves a volume expansion that is resisted by the untransformed core, a favorable compressive residual stress is developed in this layer, which contributes an additional increment to the improved fatigue response of steel alloys heat treated in this manner.

Like flame and induction hardening, case hardening by either carburizing or nitriding is intended primarily to improve wear-resistance in steels but simultaneously improves fatigue strength. Components to be carburized are treated in a high-temperature carbonaceous atmosphere to form a carbide-rich layer some 0.8 to 2.5 mm deep (Fig. 12.18), while nitrided samples are placed in a high-temperature ammonia atmosphere, where nitrogen reacts with nitride-forming elements within the steel alloy to form a nitrided layer about 0.5 mm deep. In both instances, the improvement in fatigue strength results from the intrinsic strength increase within the carburized or nitrided case and also from the favorable residual compressive stress pattern that accompanies the process. The latter factor can be compared to similar residual stress patterns arising from the mechanical and thermal treatments described above, but it is in sharp contrast to the unfavorable residual tensile stresses resulting from nickel and chromium plating procedures. In these two cases, fatigue resistance is definitely impaired. Such problems are not found with cadmium, zinc, lead, and tin platings, but one must be wary of any electrolytic procedure, since the component may become charged with hydrogen gas and be susceptible to hydrogen embrittlement-induced premature failure (see Chapter Eleven).

The improvement in fatigue resistance afforded by case hardening is considerable enough to transfer the fatigue initiation site from the component surface to the case-core boundary region, where (1) the residual stress shifts abruptly to a tensile value, and (2) the intrinsic strength of the core is considerably less than that associated with the case material. As one might expect, case hardening

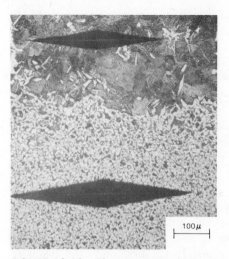

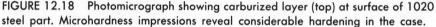

100μ

FIGURE 12.18   Photomicrograph showing carburized layer (top) at surface of 1020 steel part. Microhardness impressions reveal considerable hardening in the case.

imparts a significant improvement in fatigue resistance to components experiencing a stress gradient, such as those in plain bending or in any notched sample. Here, the applied stress is much lower in the area of the weak link in a case hardened part—the case-core boundary. By contrast, less improvement is anticipated when an axially loaded unnotched part is case hardened, since failure can occur anywhere within the uniformly loaded cross section and will do so at the case-core boundary.

Although case hardening considerably improves fatigue resistance, inadvertent decarburizing in steel alloys during a heat treatment can seriously degrade hardness, strength, and fatigue resistance (Fig. 12.19). Logically, decarburizing results in a loss of intrinsic alloy strength, since carbon is such a potent strengthening agent in most iron based alloys. In addition, the propensity for a volumetric contraction in the low carbon surface region, which is restrained by the higher carbon interior regions, may produce an unfavorable residual tensile stress pattern.

From the above, considerable improvement in fatigue properties may be achieved by introducing a favorable residual compressive stress field and avoiding any possibility for decarburization. In fact, Harris[17] showed that when decarburization was avoided and machine threads were rolled rather than cut, the fatigue endurance limit of threaded steel bolts increased by over 400% (Table 12.3).

A number of other conditions may degrade fatigue behavior. These include: inadequate quenching, which produces local soft spots that have poorer fatigue

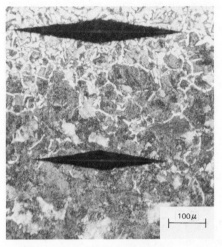

FIGURE 12.19 Photomicrograph showing decarburized layer (top) at surface of 1080 steel part. Microhardness impressions reveal softening in the decarburized zone.

100μ

TABLE 12.3 Fatigue Strength in Threaded[a] Bolts[17]

| | Fatigue Strength[b] | |
Manufacturing Procedure	MPa	(ksi)
Thread rolling of unground stock + additional heat treatment	55–125	(8–18)
Machine cut threads	195–220	(28–32)
Thread rolling of ground stock with no subsequent heat treatment	275–305	(40–44)

[a]K_t 3.5–4.0
[b]Tensile strength of material = 760–895 MPa (110–130 ksi)

resistance; excessive heating during grinding, resulting in reversion of the steel to austenite, which forms a brittle martensite upon quenching; and splatter from welding, which creates local hot spots and causes local metallurgical changes that adversely affect fatigue response.

12.3 CYCLIC STRAIN CONTROLLED FATIGUE

Localized plastic strains can be generated by loading a component that contains a notch. Regardless of the external mode of loading (cyclic stress or strain controlled), the plasticity near the notch root experiences a strain controlled condition dictated by the much larger surrounding mass of essentially elastic material. Consequently, strain controlled fatigue test results are useful in evaluating component life where notches are present. Other examples of strain controlled cyclic loading include thermal cycling, where component expansions and contractions are dictated by the operating temperature range, and reversed bending between fixed displacements, such as in the reciprocating motion shown in Fig. 12.20.

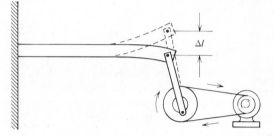

FIGURE 12.20   Reciprocating action produces fixed beam displacements. Compare this case to the stress controlled condition shown in Fig. 12-7a.

By monitoring strain and stress during a cyclic loading experiment, the response of the material can be clearly identified. For example, for a material exhibiting Type I stress-strain behavior (Chapter One) involving only elastic deformation under the applied loads, a hysteresis curve like that shown in Fig. 12.21 is produced. Note that the material's stress-strain response is retraced completely; that is, the elastic strains are completely reversible. For Type II behavior involving elastic-homogeneous plastic flow, the complete load excursion (positive and negative) produces a curve similar to Fig. 12.22 that reflects both elastic and plastic deformation. The area contained within the hysteresis loop represents a measure of plastic deformation work done on the material. Some of this work is stored in the material as cold work and/or associated with configurational changes (entropic changes), such as in polymer chain realignment, and the remainder is emitted as heat. From Fig. 12.22, the elastic strain

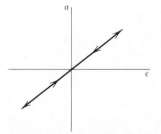

FIGURE 12.21   Hysteresis loop for cyclic loading in ideally elastic material.

range in the hysteresis loop is given by

$$\Delta \epsilon_e = XT + QY = \frac{\Delta \sigma}{E} \qquad (12\text{-}13)$$

The plastic strain range is equal to TQ or equal to the total strain range minus the elastic strain range. Hence

$$\Delta \epsilon_p = \Delta \epsilon_T - \frac{\Delta \sigma}{E} \qquad (12\text{-}14)$$

Note that as the amount of plastic strain diminishes to zero, the hysteresis loop in Fig. 12.22 shrinks to that shown in Fig. 12.21. Consequently, the elastic strain approaches the total strain. It is important to recognize that *fatigue damage will occur only when cyclic plastic strains are generated*. This basic rule should not be construed as a "security blanket" whenever nominal applied stresses are below the material yield strength, since stress concentrations readily elevate local stresses and associated strains into the plastic range.

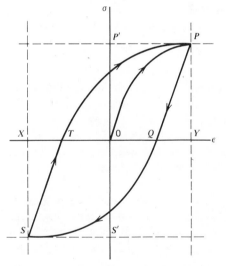

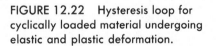

FIGURE 12.22 Hysteresis loop for cyclically loaded material undergoing elastic and plastic deformation.

12.3.1 Cycle Dependent Material Response

Cycle dependent material responses under stress and strain control are shown in Figs. 12.23 and 12.24, respectively, which reflect changes in the shape of the hysteresis loop. It is seen that, in both cases, the material response changes with

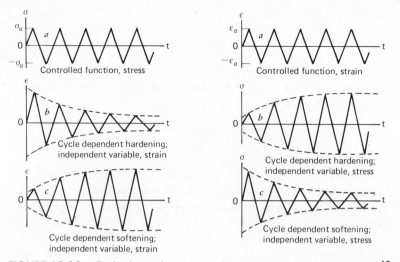

FIGURE 12.23 Cycle dependent material response under stress control.[18] (Reprinted with permission of the University of Wisconsin Press and the Regents of the University of Wisconsin System.)

FIGURE 12.24 Cycle dependent material response under strain control.[18] (Reprinted with permission of the University of Wisconsin Press and the Regents of the University of Wisconsin System.)

continued cycling until cyclic stability is reached.* That is, the material becomes either more or less resistant to the applied stresses or strains. Therefore, the material is said to cyclically strain harden or strain soften. Referring again to Fig. 12.22 for the case of stress control, where the fatigue test is conducted in a stress range between P' and S', the width of the hysteresis loop TQ (the plastic strain range) contracts when cyclic hardening occurs and expands during cyclic softening. Cyclic softening under stress control is a particularly severe condition because the constant stress range produces a continually increasing strain range response, leading to early fracture (Fig. 12.23). Under cyclic strain conditions within limits of strains X and Y, the hysteresis loop expands above P and below S for cyclic hardening and shrinks below P and above S for cyclic softening. An example of cyclic strain hardening and softening under strain controlled test conditions is shown in Fig. 12.25.

After cycling a material for a relatively short duration (often less than 100

---

*Although most of our discussions will focus on symmetrical loading about zero, it is important to appreciate what happens to a sample when a nonzero mean stress is superimposed during a cyclic strain experiment. The specimen is found to accumulate strains as a result of each cycle. This accumulation has been termed "cyclic-strain-induced creep" and will contribute to either an extension or contraction of the sample, depending on the sense of the applied mean stress.

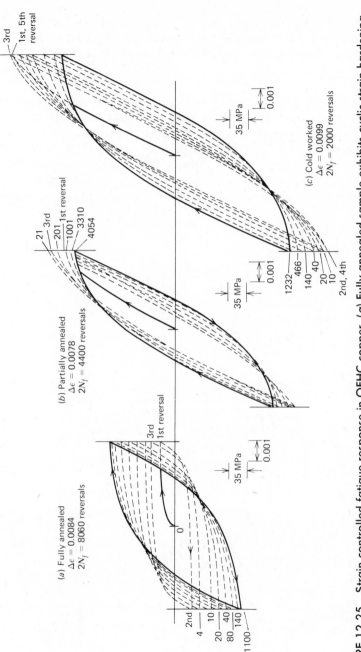

FIGURE 12.25 Strain controlled fatigue response in OFHC copper. (a) Fully annealed sample exhibits cyclic strain hardening; (b) partially annealed sample exhibits cyclic stability; (c) severely cold-worked sample exhibits cyclic strain softening.[19] (Reprinted by permission of the American Society for Testing and Materials from copyright work.)

cycles), the hysteresis loops generally stabilize and the material achieves an equilibrium condition for the imposed strain limits. The cyclically stabilized stress-strain response of the material may then be quite different from the initial monotonic response. Consequently, cyclically stabilized stress-strain curves are important characterizations of a material's cyclic response. These curves may be obtained in several ways. For example, a series of companion samples may be cycled within various strain limits until the respective hysteresis loops become stabilized. The cyclic stress-strain curve is then determined by fitting a curve through the tips of the various superimposed hysteresis loops, as shown in Fig. 12.26.[20] This procedure involves many samples and is expensive and time-consuming. A faster method for obtaining cyclic stress-strain curves is by multiple step testing, wherein the same sample is subjected to a series of

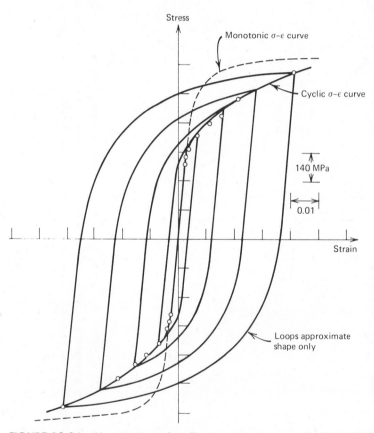

FIGURE 12.26 Monotonic and cyclic stress-strain curves for SAE 4340 steel. Data points represent tips of stable hysteresis loops from companion specimens.[22] (Reprinted by permission of the American Society for Testing and Materials from copyright work.)

alternating strains in blocks of increasing magnitude. In this manner, one specimen yields several hysteresis loops, which may be used to construct the cyclic stress-strain curve.[21] An even quicker technique involving only one sample has been found to provide excellent results and is used extensively in current cyclic strain testing experiments. As seen in Fig. 12.27, the specimen is subjected to a series of blocks of gradually increasing and then decreasing strain excursions.[21] It has been found that after a relatively few such blocks (the greater the number of cycles within each block, the fewer the number of blocks needed for cyclic stabilization), the material reaches a stabilized condition. At this point, the investigator simply draws a line through the tips of each hysteresis loop, from the smallest strain range to the largest. As such, each loop contained within the hysteresis envelope represents the cyclically stabilized condition for the material at that particular strain range. By initiating the test with the maximum strain amplitude in the block, the monotonic stress-strain curve is automatically determined for subsequent comparison with the cyclically stabilized curve. In this manner, both the monotonic and cyclic stress-strain curves can be determined from the same sample. Obviously, this method results in savings in test time and money. It should be noted that if a specimen subjected to either multiple or incremental step testing were to be pulled to fracture after cyclic stabilization, the resulting stress-strain curve would be virtually coincident with the one generated by the locus of hysteresis loop tips.

By comparing monotonic and cyclically stabilized stress-strain curves, Landgraf et al.[21] demonstrated that certain engineering alloys will cyclically strain harden and others will soften (Fig. 12.28). From the Holloman relation-

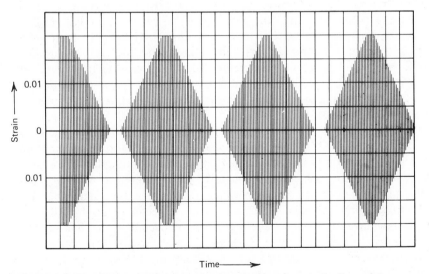

FIGURE 12.27   Incremental step test program showing strain time plot.

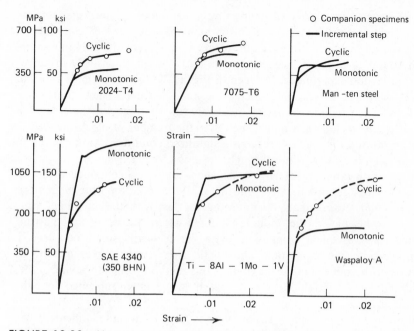

FIGURE 12.28 Monotonic and cyclic stress-strain curves for several engineering alloys.[21] (Reprinted by permission of the American Society for Testing and Materials from copyright work.)

ship given by $\sigma = K\epsilon^n$, it is possible to mathematically describe the material's stress-strain response in either the monotonic or cyclically stabilized state. Consequently, one may define the strain hardening coefficient for both monotonic (n) and cyclic (n') conditions as well as the monotonic yield strength (σ_0) and cyclic (σ_0') counterpart. (Typical values for a number of engineering alloys are given in Table 12.4.) Although large changes occur in these properties as a result of cyclic hardening or softening, it is worth noting that, for most metals, n' ranges from 0.1 to 0.2.

Can one determine in advance which alloys will cyclically harden and which will soften? Manson et al.[20,24] observed that the propensity for cyclic hardening or softening depends on the ratio of monotonic ultimate strength to 0.2% offset yield strength. When $\sigma_{ult}/\sigma_0 > 1.4$, the material will harden, but when $\sigma_{ult}/\sigma_0 < 1.2$, softening will occur. For ratios between 1.2 and 1.4, forecasting becomes difficult, though a large change in properties is not expected. Also, if $n > 0.20$ the material is likely to strain harden, and softening will occur if $n < 0.10$. Therefore, *initially hard and strong materials will generally cyclically strain soften, and initially soft materials will harden.* Manson's rule makes use of monotonic properties to determine *whether* cyclic hardening or softening will occur. However, the *magni-*

tude of the cyclically induced change can be determined only by comparison of the monotonic and cyclic stress-strain curves (see Table 12.4 and Figs. 12.26 and 12.28).

But why do these materials cyclically harden or soften? The answer to this question appears to be related to the nature and stability of the dislocation substructure of the material. For an initially soft material, the dislocation density is low. As a result of plastic strain cycling, the dislocation density increases rapidly, contributing to significant strain hardening. At some point, the newly generated dislocations assume a stable configuration for that material and for the magnitude of cyclic strain imposed during the test. When a material is hard initially, subsequent strain cycling causes a rearrangement of dislocations into a new configuration that offers less resistance to deformation—that is, the material strain softens.

As we saw in Chapters Two and Three, dislocation mobility that strongly affects dislocation substructure stability depends on the material's stacking fault energy (SFE). Recall that when SFE is high, dislocation mobility is great because of enhanced cross-slip; conversely, cross-slip is restricted in low SFE materials. As a result, some materials cyclically harden or soften more completely than others. For example, in a relatively high SFE material like copper, initially hard samples cyclically strain soften, and initially soft samples cyclically harden; thus, the cyclically stabilized condition is the same *regardless* of the initial state of the material (Fig. 12.29*a*). In this case, the mechanical properties of the material in the stabilized state are *independent* of prior strain history. This is not true for a low stacking fault energy material, where restricted cross-slip will prevent the development of a common dislocation state from an initially hard and soft condition, respectively. In addition, the low SFE material will harden or soften more slowly than the high SFE alloy. We see from Fig. 12.29*b* that the material will cyclically soften and harden, but a final stabilized state is never completely achieved and is not equivalent for the two different starting conditions. For such materials, the "final" cyclically stabilized state is *dependent* on prior strain history.

One might then expect to find dislocation substructures in cyclically loaded samples similar to those found as a result of unidirectional loading. In fact, Feltner and Laird[26] observed that "those factors which give rise to certain kinds of dislocation structures in unidirectional deformation affect the cyclic structures in the same way." For example, we see from Fig. 12.30 that at high cyclic strains a cell structure is developed in high SFE alloys. If cyclic straining causes coarsening of a preexistent cell structure, then softening will occur. If the cell structure gets finer, then cyclic straining results in a hardening process. In low SFE alloys, dislocation planar arrays and stacking faults are present. These findings are similar to that discussed in Chapters Two and Three for monotonic loading. A parallel condition is found in monotonic and cyclically induced dislocation structures produced at low strains.

Material	Condition	σ_0/σ_0' (MPa)	n/n'	ϵ_f/ϵ_f'	σ_f/σ_f' (MPa)	b	c
Steel							
SAE 1015	Normalized, 80 BHN	225/240	0.26/0.22	1.14/0.95	725/825	−0.11	−0.64
SAE 950X	As received, 150 BHN	345/335	0.16/0.134	1.06/0.35	750/625	−0.075	−0.54
VAN-80	As received, 225 BHN	565/560	0.13/0.134	1.15/0.21	1220/1055	−0.08	−0.53
SAE 1045	Q+T (650°C), 225 BHN	635/415	0.13/0.18	1.04/1.0	1225/1225	−0.095	−0.66
SAE 1045	Q+T (370°C), 410 BHN	1365/825	0.076/0.146	0.72/0.60	1860/1860	−0.073	−0.70
SAE 1045	Q+T (180°C), 595 BHN	1860/1725	0.071/0.13	0.52/0.07	2725/2725	−0.081	−0.60
AISI 4340	Q+T (425°C), 409 BHN	1370/825	—/0.15	0.48/0.48	1560/2000	−0.091	−0.60
AISI 304 ELC	BHN 160	255/715	—/0.36	1.37/1.02	1570/2415	−0.15	−0.77
AISI 304 ELC	Cold drawn, BHN 327	745/875	—/0.17	1.16/0.89	1695/2275	−0.12	−0.69
AISI 305[b]	0% C.W.	250/405	—/0.05	—	—	—	—
AISI 305[b]	50% C.W.	850/710	—/0.11	—	—	—	—
AM 350	Annealed	440/1350	—/0.13	0.74/0.33	2055/2800	−0.14	−0.84
AM 350	Cold drawn 30%, BHN 496	1860/1620	—/0.21	0.23/0.098	2180/2690	−0.102	−0.42
18 Ni maraging	ST(790°C)/1 hr)+ 480°C (4 hr), BHN 480	1965/1480	$\dfrac{0.015\text{–}0.030}{0.008}$	0.81/0.60	2240/2240	−0.07	−0.75

Aluminum							
2014-T6	BHN 155	460/415	—/0.16	0.29/0.42	600/850	−0.106	−0.65
2024-T4	—	305/440	0.20/0.08	0.43/0.21	635/1015	−0.11	−0.52
5456	H31, 95 BHN	235/360	—/0.16	0.42/0.46	525/725	−0.11	−0.67
7075-T6	—	470/525	0.113/0.146	0.41/0.19	745/1315	−0.126	−0.52
Copper							
OFHC[c]	Annealed	20/140	0.40/0.16	—	—	—	—
70/30 brass[b]	Annealed	140/240	—/0.08	—	—	—	—
70/30 brass[b]	82% C.W.	570/475	—/0.11	—	—	—	—
Nickel							
Waspalloy[c]	—	545/705	0.11/0.17	—	—	—	—
MP35N[b]	0% C.W.	350/625	—/0.06	—	—	—	—
MP35N[b]	20% C.W.	910/745	—/0.10	—	—	—	—
MP35N[b]	40% C.W.	1180/1850	—/0.14	—	—	—	—

[a] L. E. Tucker, R. W. Landgraf and W. R. Brose, SAE Report 740279, Automotive Engineering Congress, Feb. 1974.
[b] Ref. 23.
[c] Ref. 21.

TABLE 12.4b Monotonic and Cyclic Properties of Selected Engineering Alloys[a]

Material	Condition	σ_0/σ_0' (ksi)	n/n'	ϵ_f/ϵ_f'	σ_f/σ_f' (ksi)	b	c
Steel							
SAE 1015	Normalized, 80 BHN	33/35	0.26/0.22	1.14/0.95	105/120	−0.11	−0.64
SAE 950X	As received, 150 BHN	50/49	0.16/0.134	1.06/0.35	109/91	−0.075	−0.54
VAN-80	As received, 225 BHN	82/81	0.13/0.134	1.15/0.21	177/153	−0.08	−0.53
SAE 1045	Q+T (1200°F), 225 BHN	92/60	0.13/0.18	1.04/1.0	178/178	−0.095	−0.66
SAE 1045	Q+T (700°F), 410 BHN	198/120	0.076/0.146	0.72/0.60	270/270	−0.073	−0.70
SAE 1045	Q+T (360°F), 595 BHN	270/250	0.071/0.13	0.52/0.07	395/395	−0.081	−0.60
AISI 4340	Q+T (800°F), 409 BHN	199/120	—/0.15	0.48/0.48	226/290	−0.091	−0.60
AISI 304 ELC	BHN 160	37/104	—/0.36	1.37/1.02	228/350	−0.15	−0.77
AISI 304 ELC	Cold drawn, BHN 327	108/127	—/0.17	1.16/0.89	246/330	−0.12	−0.69
AISI 305[b]	0% C.W.	36/59	—/0.05	—	—	—	—
AISI 305[b]	50% C.W.	123/103	—/0.11	—	—	—	—
AM 350	Annealed	64/196	—/0.13	0.74/0.33	298/406	−0.14	−0.84
AM 350	Cold drawn 30%, BHN 496	270/235	—/0.21	0.23/0.098	316/390	−0.102	−0.42
18 Ni maraging	ST (1450°F/1 hr) + 900°F (4 hr), BHN 480	285/215	$\frac{0.015\text{–}0.030}{0.008}$	0.81/0.60	325/325	−0.07	−0.75

Aluminum							
2014-T6	BHN 155	67/60	—/0.16	0.29/0.42	87/123	−0.106	−0.65
2024-T4	—	44/64	0.20/0.08	0.43/0.21	92/147	−0.11	−0.52
5456	H31, 95 BHN	34/52	—/0.16	0.42/0.46	76/105	−0.11	−0.67
7075-T6	—	68/76	0.113/0.146	0.41/0.19	108/191	−0.126	−0.52
Copper							
OFHC[c]	Annealed	3/20	0.40/0.16	—	—	—	—
70/30 brass[b]	Annealed	20/35	—/0.08	—	—	—	—
70/30 brass[b]	82% C.W.	83/69	—/0.11	—	—	—	—
Nickel							
Waspalloy[c]	—	79/102	0.11/0.17	—	—	—	—
MP35N[b]	0% C.W.	51/91	—/0.06	—	—	—	—
MP35N[b]	20% C.W.	132/108	—/0.10	—	—	—	—
MP35N[b]	40% C.W.	171/123	—/0.14	—	—	—	—

[a]L. E. Tucker, R. W. Landgraf and W. R. Brose, SAE Report 740279, Automotive Engineering Congress, Feb. 1974.

[b]Reference 23

[c]Reference 21

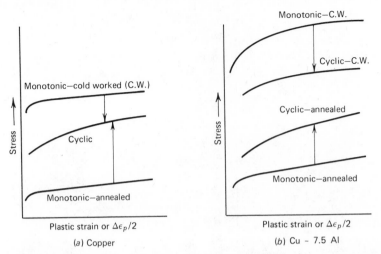

FIGURE 12.29   Cyclic response of (*a*) high stacking fault energy copper and (*b*) low stacking fault energy Cu-7.5% Al alloy. Cyclically stabilized state in high SFE alloy is path independent.[25] (Reprinted from C. E. Feltner and C. Laird, *Acta Metallurgica, 15,* 1967, with permission of Pergamon Press.)

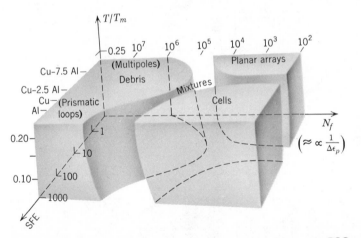

FIGURE 12.30   Schema showing dislocation substructures in FCC metals as a function of stacking fault energy, strain amplitude, and temperature.[26] (Reprinted with permission from the American Institute of Mining, Metallurgical and Petroleum Engineers.)

Some cyclic strain tests have been conducted on amorphous and crystalline polymers.[27-29] To date, all results indicate that these materials cyclically strain soften, but no evidence has been found for cyclic hardening response. Consequently, the final mechanical state of the polymers tested appears to be path-dependent.

12.3.2　Strain-Life* Curves

Having identified the response of a solid to cyclic strains, it is now appropriate to consider how cyclically stabilized material properties affect the life of a specimen or engineering component subjected to strain controlled loading. To accomplish this, it is convenient to begin our analysis by considering the elastic and plastic strain components separately. The elastic component is often described in terms of a relationship between the true stress amplitude and number of load reversals

$$\frac{\Delta\epsilon_e E}{2} = \sigma_a = \sigma_f'(2N_f)^b \tag{12-15}$$

where

$\dfrac{\Delta\epsilon_e}{2} =$ elastic strain amplitude

$E =$ modulus of elasticity

$\sigma_a =$ stress amplitude

$\sigma_f' =$ fatigue strength coefficient, defined by the stress

intercept at one load reversal $(2N_f = 1)$

$N_f =$ cycles to failure

$2N_f =$ number of load reversals to failure

$b =$ fatigue strength exponent

This relationship, which represents an empirical fit of data above the fatigue limit (see Fig. 12.5), is similar in form to that proposed in 1910 by Basquin.[30] A sampling of test results is shown in Fig. 12.31a and fitted to Eq. 12-15. Additional test results from a variety of steel alloys (see Fig. 12.32) demonstrate increased fatigue life with decreasing fatigue strength exponent. Based on an

---

*"Life" is generally defined as the time necessary to develop a crack of 3 to 6 mm in length. Consequently, "life" when used in this context means time for crack initiation plus early growth.

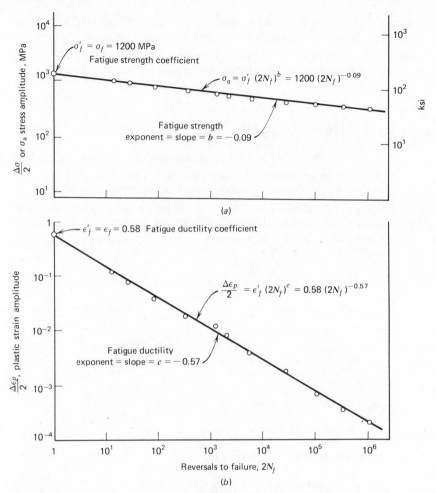

FIGURE 12.31 Fatigue properties of SAE 4340 steel. (*a*) Fatigue strength properties; (*b*) fatigue ductility properties.[19,20] (Reprinted by permission of the American Society for Testing and Materials from copyright work.)

energy argument, Morrow[19] determined that

$$b = \frac{-n'}{1 + 5n'} \tag{12-16}$$

Consequently, fatigue life under essentially elastic stress (or strain) cycling is enhanced by a low value of n'. One would also want the material to have a high fatigue strength coefficient σ_f'. Many observations have shown that σ_f' is approximately equal to the monotonic fracture strength σ_f (Fig. 12.33) so that a

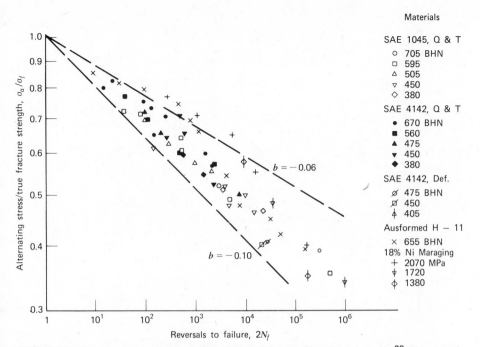

Materials

SAE 1045, Q & T
○ 705 BHN
□ 595
△ 505
▽ 450
◇ 380

SAE 4142, Q & T
● 670 BHN
■ 560
▲ 475
▼ 450
◆ 380

SAE 4142, Def.
⌀ 475 BHN
⌀ 450
⚲ 405

Ausformed H − 11
× 655 BHN

18% Ni Maraging
+ 2070 MPa
⚲ 1720
⌀ 1380

$b = -0.06$

$b = -0.10$

FIGURE 12.32 Stress amplitude-fatigue life behavior of hardened steels.[22] (Reprinted by permission of the American Society for Testing and Materials from copyright work.)

very strong alloy would be preferred for predominantly elastic strains associated with high cycle fatigue (HCF).

The plastic component of strain is best described by the Manson–Coffin relationship[20,24,31,32]

$$\frac{\Delta\epsilon_p}{2} = \epsilon_f'(2N_f)^c \qquad (12\text{-}17)$$

where

$\dfrac{\Delta\epsilon_p}{2}$ = plastic strain amplitude

ϵ_f' = fatigue ductility coefficient, defined by the strain

intercept at one load reversal $(2N_f = 1)$

$2N_f$ = total strain reversals to failure

c = fatigue ductility exponent, a material property in the

range -0.5 to -0.7

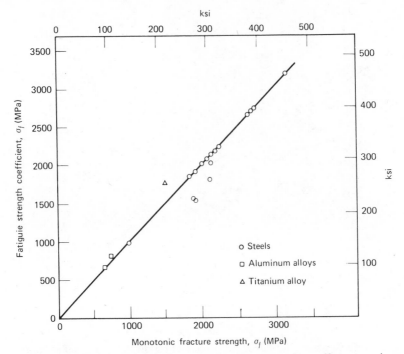

FIGURE 12.33 Correlation between fatigue strength coefficient and monotonic fracture strength.[22] (Reprinted by permission of the American Society for Testing and Materials from copyright work.)

Data from SAE 4340 steel are plotted in Fig. 12.31b and fitted to Eq. 12-17. Note the extrapolation of the curve to $2N_f = 1$, where ϵ_f' is defined. Additional test results for a number of steel alloys are given in Fig. 12.34 and show superior fatigue life for alloys with low fatigue ductility exponents c. Morrow[19] also determined that

$$c = \frac{-1}{1 + 5n'}\tag{12-18}$$

which indicates that plastic strain fatigue life should be greater in materials with high values of n'. It is obvious from Eq. 12-17 that the plastic strain capacity for a given cyclic life also depends strongly on the fatigue ductility coefficient ϵ_f'. From a number of studies, the relationship between ϵ_f' and the monotonic true fracture ductility ϵ_f is not as exact as noted for the case of σ_f and σ_f'. Estimates ranging from $\epsilon_f' = 0.35\ \epsilon_f$ to $1.0\ \epsilon_f$ have been used to analyze fatigue data, since extrapolation of the curve in Fig. 12.31b does not always coincide with the monotonic fracture strain determined from a tensile test. Additional attempts have been made to define ϵ_f' mathematically in terms of both monotonic and

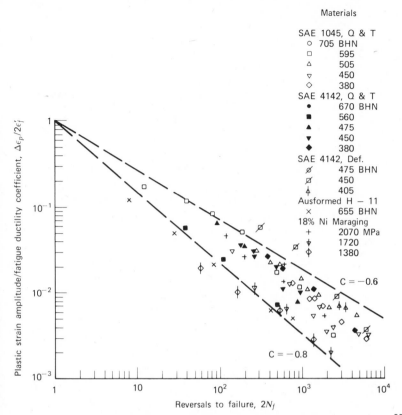

Materials

SAE 1045, Q & T
○   705 BHN
□   595
△   505
▽   450
◇   380
SAE 4142, Q & T
●   670 BHN
■   560
▲   475
▼   450
◆   380
SAE 4142, Def.
⌀   475 BHN
⌀   450
⊿   405
Ausformed H − 11
×   655 BHN
18% Ni Maraging
+   2070 MPa
⌽   1720
⟡   1380

$C = -0.6$

$C = -0.8$

Plastic strain amplitude/fatigue ductility coefficient, $\Delta\epsilon_p/2\epsilon_f'$

Reversals to failure, $2N_f$

FIGURE 12.34   Strain amplitude-fatigue life behavior of hardened steels.[22] (Reprinted by permission of the American Society for Testing and Materials from copyright work.)

cyclic properties.[22] For convenience, however, ϵ_f' is generally taken to be equal to ϵ_f for a first approximation. From the above discussion it may be concluded that plastic strain fatigue or low cycle fatigue (LCF) resistance should be greater in more ductile materials and in materials possessing high cyclic strain hardening exponents.

Perhaps you noted the contradiction arising from the preference for high n' materials in plastic strain dominated (LCF) situations but low n' materials for elastic strain (HCF) conditions. In light of this fact, design engineers would have to base their choices of materials on the strain range experienced by the component: ductile alloys for low cycle fatigue and high-strength alloys for high cycle fatigue. The best choice for overall high fatigue resistance would be a tough alloy possessing an optimum combination of strength and ductility.

Manson et al.[24] argued that the fatigue resistance of a material subjected to a given strain range could be estimated by superposition of the elastic and plastic

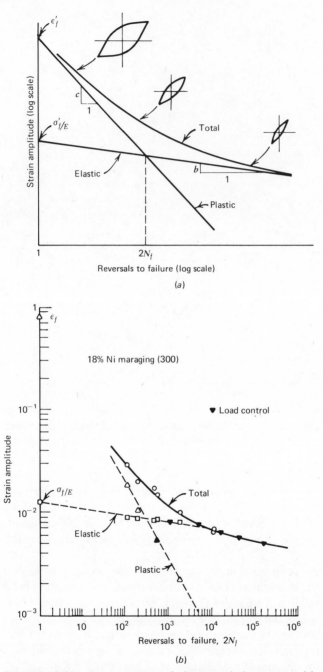

FIGURE 12.35 Superposition of plastic and elastic strain life curves to produce the total strain life fatigue relationship. (*a*) Schema; (*b*) actual data from 18% Ni maraging steel.[22] (Reprinted by permission of the American Society for Testing and Materials from copyright work.)

strain components. Therefore, by combining Eqs. 12-14, 12-15, and 12-17, the total strain amplitude may be given by

$$\frac{\Delta\epsilon_T}{2} = \frac{\Delta\epsilon_e}{2} + \frac{\Delta\epsilon_p}{2} = \frac{\sigma_f'}{E}(2N_f)^b + \epsilon_f'(2N_f)^c \qquad (12\text{-}19)$$

It would be expected, then, that the total strain life curve would approach the plastic strain life curve at large strain amplitudes and approach the elastic strain life curve at low total strain amplitudes. This is shown schematically in Fig. 12.35a, and for an actual case involving fatigue data from a high-strength steel alloy in Fig. 12.35b. Finally, when the total strain life plots for strong, tough, and ductile alloys are compared, the aforementioned trends in material selection are verified. We see from Fig. 12.36 that ductile alloys are best for high cyclic strain applications, and strong alloys are superior in the region of low strains. Note that around 10^3 cycles (2×10^3 reversals), corresponding to a total strain of about 0.01, there appears to be no preferred material. That is to say, *it makes no sense to attempt to optimize material properties for applications if strains of about 0.01 are encountered*—just about any alloy will serve the purpose.

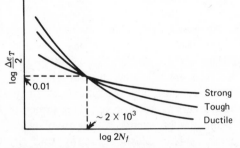

FIGURE 12.36   Schema showing optimum material for given total strain amplitude. Note lack of material property effect for cyclic life of about 10^3 (2×10^3 reversals) corresponding to a total strain range of about 0.01.[22] (Reprinted by permission of the American Society for Testing and Materials from copyright work.)

12.4   FATIGUE CRACK INITIATION

Fatigue cracks are initiated at heterogeneous nucleation sites within the material whether they be preexistent (associated with inclusions, gas pores, or local soft spots in the microstructure) or are generated during the cyclic straining process itself. As one might expect, elimination of preexistent flaws can result in a substantial improvement in fatigue life. A good illustration of this is found when steels for roller bearings are vacuum melted as opposed to air melted. The much lower inclusion level in the vacuum melted steel enables these bearings to withstand many more load excursions than the air melted ones.

The most intriguing aspect of the fatigue crack initiation process relates to the generation of the nucleation sites. Although strains under monotonic loading produce surface offsets that resemble a staircase morphology, cyclic strains produce sharp peaks (extrusions) and troughs (intrusions)[33,34] resulting from nonreversible slip (Fig. 12.37). Many investigators have found that these surface notches serve as fatigue crack nucleation sites. Consequently, when the surface is periodically polished to remove these offsets, fatigue life is improved.[35] These surface upheavals represent the free surface terminations of dense bands of highly localized slip, the number of which increase with strain range. Careful studies have demonstrated these bands to be softer than the surrounding matrix material,[36] and it is believed that they cyclically soften relative to the matrix, resulting in a concentration of plastic strain. These bands are called "persistent slip bands" because of two main results. First, when a metallographic section is prepared from a damaged specimen, the deformation bands persist after etching, indicating the presence of local damage. Second, when the surface offsets are removed by polishing and the specimen cycled again, new surface damage occurs at the same sites. Consequently, although cracking begins at the surface, it is important to recognize that the material within these persistent bands and below the surface is also damaged and will control the location of the surface nucleated cracks.[37]

Various dislocation models have been proposed to explain the fatigue crack nucleation process. All of these models are based on the same basic assumptions: slip is produced and then localized in such a way as to crystallographically notch the specimen surface. If slip offsets are to occur, the Burgers vector must not lie in the plane of the surface.[38] Correspondingly, the magnitude of the surface displacements will increase as the Burgers vector approaches a direction normal to the surface. A reasonable sequence of events would involve the following stages:[33,34] slip is produced as a result of plastic straining, and slip bands of highly localized deformation are generated; extrusions and intrusions appear at the surface which then penetrate along the length of the persistent slip band to become sharp fissures; at a certain point, the fissure breaks through into the matrix and a "crack" is developed.

More recently, Kramer[39] has suggested an alternate mechanism for fatigue crack initiation. He proposes that a surface layer of high dislocation density is formed during cycling which eventually becomes strong enough to support a dislocation pileup. The stress concentration associated with this pileup is then thought to trigger the development of a crack within the hardened layer. It is interesting to note that the improvement in fatigue life by periodic surface polishing, mentioned above, could also be rationalized by Kramer's model; the polishing action would remove the hardened layer, thereby arresting the formation of dislocation pileups believed responsible for the cracking process. Additional studies are currently in progress to further clarify this model.

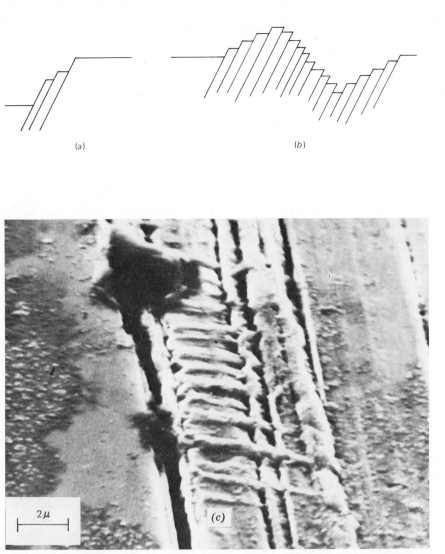

FIGURE 12.37 Plastic strain-induced surface offsets. (a) Monotonic loading giving rise to staircase morphology slip offsets; (b) cyclic loading which produces sharp peaks (extrusions) and troughs (intrusions); (c) photomicrograph showing intrusions and extrusions on prepolished surface.[34] (Reprinted with permission from W. A. Wood and Academic Press.)

461

12.5 GENERAL COMMENTARY

It has been the purpose of this chapter to identify some aspects of the fatigue process. Fatigue crack propagation will be considered in Chapter Thirteen. The reader should now recognize the importance of identifying the cyclic conditions experienced by an engineering component—whether they be cyclic strain or cyclic stress—since they will affect the proper selection of material for the intended application. The following generalizations for improved fatigue resistance are presented as a review:

1. Avoid stress concentrations wherever possible.
2. Introduce favorable residual compressive stresses by various mechanical, thermal, and metallurgical procedures.
3. Eliminate metallurgical defects, such as inclusions, pores, and "soft spots" (the latter arise from improper heat treatment).
4. For low cycle fatigue applications involving cyclic strain control, a material possessing high ductility and a high cyclic strain hardening coefficient is preferred.
5. For high cycle fatigue applications involving predominantly elastic cyclic strains, a strong alloy with a low cyclic strain hardening exponent is more desirable.
6. Where possible, introduce alloying elements that lower the stacking fault energy to homogenize and maintain the reversible nature of cyclic slip, thereby minimizing nonreversible slip processes responsible for slip band extrusion and intrusion generation.
7. Develop stable microstructures that resist cyclically induced disordering of fine precipitates.

REFERENCES

1. P. J. Forrest, *Fatigue of Metals*, Addison-Wesley, Reading, Mass., 1962.
2. D. J. Wulpi, *How Components Fail*, American Society for Metals, Metals Park, Ohio, 1966.
3. A. Wöhler, *Zeitschrift für Bauwesen*, **10**, 1860.
4. R. E. Peterson, Edgar Marbury Lecture, ASTM, 1962, p. 1.
5. H. F. Hardrath, E. C. Utley, and D. E. Guthrie, NASA *TN D-210*, 1959.
6. G. M. Sinclair and T. J. Dolan, *Trans.*, ASME, **75**, 1953, p. 867.
7. A. Palmgren, *Bertschrift des Vereines Ingenieure*, **58**, 1924, p. 339.
8. M. A. Miner, *J. Appl. Mech.*, **12**, 1954, p. A-159.
9. J. B. Kommers, *Proc.*, ASTM, **45**, 1945, p. 532.
10. D. K. Bullens, *Steel and Its Heat Treatment*, **1**, 1938, p. 37.

11. G. Sines and J. L. Waisman, *Metal Fatigue*, McGraw-Hill Book Co., New York, 1959, p. 298.

12. W. N. Findley, *J. Mech. Eng. Sci.*, **14** (6), 1972, p. 424.

13. P. J. E. Forsyth, *The Physical Basis of Metal Fatigue*, American Elsevier Publishing Co., Inc., New York, 1969.

14. *Metals Handbook*, 8th Ed., Vol. 1, "Properties and Selection of Metals, " ASM, Novelty, Ohio, 1961, p. 217.

15. C. Calabrese and C. Laird, *Mater. Sci. Eng.*, **13** (2), 1974, p. 141.

16. F. Sherrett, "The Influence of Shot-Peening and Similar Surface Treatments on the Fatigue Properties of Metals," Part I, S&T Memo 1/66, Ministry of Aviation, U.S. Gov't. Report 487487, Feb. 1966.

17. W. J. Harris, "The Influence of Decarburization on the Fatigue Behavior of Steel Bolts," S&T Memo 15/65, Ministry of Aviation, U.S. Gov't. Report 473394, August 1965.

18. B. I. Sandor, *Fundamentals of Cyclic Stress and Strain*, University of Wisconsin Press, Madison, Wisc., 1972.

19. Jo Dean Morrow, *Internal Friction, Damping and Cyclic Plasticity*, ASTM *STP 378*, 1965, p. 45.

20. R. W. Smith, M. H. Hirschberg, and S. S. Manson, NASA *TN D-1574*, NASA, April 1963.

21. R. W. Landgraf, Jo Dean Morrow, and T. Endo, *J. Mater., JMLSA*, **4** (1), 1969, p. 176.

22. R. W. Landgraf, *Achievement of High Fatigue Resistance in Metals and Alloys*, ASTM *STP-467*, 1970, p. 3.

23. J. P. Hickerson and R. W. Hertzberg, *Met. Trans.*, **3**, 1972, p. 179.

24. S. S. Manson and M. H. Hirschberg, *Fatigue: An Interdisciplinary Approach*, Syracuse University Press, Syracuse, N.Y., 1964, p. 133.

25. C. E. Feltner and C. Laird, *Acta Met.*, **15**, 1967, p. 1621.

26. C. E. Feltner and C. Laird, *Trans.*, AIME, **242**, 1968, p. 1253.

27. B. Tomkins and W. D. Biggs, *J. Mater. Sci.*, **4**, 1969, p. 532.

28. S. Rabinowitz and P. Beardmore, *J. Mater. Sci.*, **9**, 1974, p. 81.

29. P. Beardmore and S. Rabinowitz, *Polymeric Materials*, ASM, Metals Park, Ohio, 1975, p. 551.

30. O. H. Basquin, *Proc.*, ASTM, **10**, Part II, 1910, p. 625.

31. L. F. Coffin, Jr., *Trans.*, ASME, **76**, 1954, p. 931.

32. J. F. Tavernelli and L. F. Coffin, Jr., *Trans.*, ASM, **51**, 1959, p. 438.

33. W. A. Wood, *Fracture*, Technology Press of M.I.T. and John Wiley, New York, 1959, p. 412.

34. W. A. Wood, *Treatise on Materials Science and Technology*, Vol. 5, H. Herman, ed., Academic Press, New York, 1974, p. 129.

35. T. H. Alden and W. A. Backofen, *Acta Met.*, **9**, 1961, p. 352.

36. O. Helgeland, *J. Inst. Met.*, **93**, 1965, p. 570.

37. C. Roberts and A. P. Greenough, *Phil. Mag.*, **12**, 1965, p. 81.

38. W. A. Backofen, *Fracture*, Technology Press of M.I.T. and John Wiley, New York, 1959, p. 435.

39. I. R. Kramer, *Met. Trans.*, **5**, 1974, p. 1735.

PROBLEMS

12.1 Calculate the total strain range that a smooth bar cyclic strain specimen would be able to tolerate before failing after 1000 cycles (2000 load reversals). Show all your computations and assumptions.

$$E = 205 \text{ GPa}$$

$$\sigma_f \text{ (monotonic)} = 1850 \text{ MPa}$$

$$\epsilon_f \text{ (monotonic)} = 0.7$$

$$n' \text{ (cyclic)} = 0.15$$

12.2 Two investigators independently reported fatigue test results for Zeusalloy 300. Both reported their data in the form of σ–N curves for notched bars. The basic difference between the two results was that Investigator I reported inferior behavior of the material compared with the results of Investigator II (i.e., lower strength for a given fatigue life) but encountered much less scatter. Can you offer a possible explanation for this observation? Describe the macroscopic fracture surface appearance for the two sets of test bars.

12.3 Bend eight paper clips to failure. Determine the cyclic life that would give 90% confidence that 90% of the paper clips will not fail. Break 10 more clips and check your answer. Finally, take all 18 results and determine the cyclic life for 90% confidence of 90% survival. How does this answer differ from the initial finding?

12.4 Calculate the fatigue life of SAE 1015 and 4340 steel tempered at 425°C when the samples experience total strain ranges of 0.05, 0.01, and 0.001. Which alloy is best at each of these applied strain ranges?

CHAPTER
THIRTEEN
FATIGUE CRACK
PROPAGATION

As discussed in Chapter Seven, a number of engineering system breakdowns can be attributed to preexistent flaws that caused failure when a certain critical stress was applied. In addition, these defects may have grown to critical dimensions prior to failure. The latter case—that involving subcritical flaw growth—is important to guard against for a number of reasons. First, if a structure or component contained a defect large enough to cause immediate failure upon loading, the defect quite likely could be detected by a number of nondestructive test (NDT) procedures and repaired before damaging loads would be applied. Assuming such a defect was not detected, the procedure of proof testing (subjecting a structure, such as a pressure vessel, to a preservice simulation test at a stress level equal to or slightly higher than the design stress) certainly would cause the structure to fail but, at least, under controlled conditions with minimum risk to human lives and damage to other equipment in the engineering system. On the other hand, were the crack to be subcritical in size and undetected by NDT, a successful proof test would prove only that a flaw of critical dimensions did not exist at that time. *No guarantee could be given that the flaw would not grow during service to critical dimensions and would later precipitate a catastrophic failure.* This chapter is concerned with factors that control the fatigue crack propagation (FCP) process in engineering materials.

13.1 STRESS AND CRACK LENGTH CORRELATIONS WITH FCP

Crack propagation data may be obtained from a number of specimens such as many of those shown in Fig. 8.7. Starting with a mechanically sharpened crack, cyclic loads are applied and the resulting change in crack length monitored and recorded as a function of the number of load cycles. Many monitoring techniques have been employed, such as the use of a calibrated traveling microscope (Fig. 13.1), eddy current techniques, electropotential measurements, compliance measurements, and acoustic emission detectors. A typical plot of such data is shown in Fig. 13.2, where the crack length is seen to increase with increasing number of loading cycles. The fatigue crack growth rate is determined from such

FIGURE 13.1   Monitoring crack length with traveling microscope.

a curve either by graphical procedures or by computation. From these methods, the crack growth rates resulting from a given cyclic load are $(da/dn)_{a_i}$ and $(da/dn)_{a_j}$ when the crack is of length a_i and a_j, respectively.

It is important to note that the crack growth rate most often increases with increasing crack length. (This is generally the case, though not always, as will be discussed below.) It is most significant that the crack becomes longer at an increasingly more rapid rate, thereby shortening component life at an alarming rate. An important corollary of this fact is that most of the loading cycles involved in the total life of an engineering component are consumed during the early stages of crack extension when the crack is small and, perhaps, undetected. The other variable that controls the rate of crack propagation is the magnitude of the stress level. It is clearly evident from Fig. 13.2 that FCP rates increase with increasing stress level.

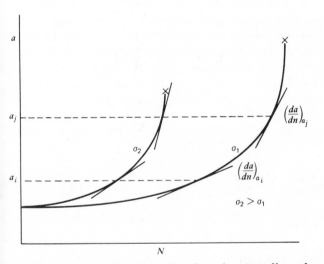

FIGURE 13.2   Crack propagation data showing effect of applied stress level. FCP rate increases with stress and crack length.

Since many researchers have probed the nature of the fatigue crack propagation process, it is not surprising to find in the literature a number of empirical and theoretical "laws," many of the form

$$\frac{da}{dn} \propto f(\sigma, a) \tag{13-1}$$

reflecting the importance of the stress level and crack length on FCP rates. Quite often, this function assumes the form of a simple power relationship wherein

$$\frac{da}{dn} \propto \sigma^m a^n \tag{13-2}$$

where

$$m \approx 2\text{-}7$$

$$n \approx 1\text{-}2$$

For example, Liu[1] theorized m and n to be 2 and 1, respectively, while Frost[2] found empirically for the materials he tested that $m \approx 3$ and $n \approx 1$. Weibull[3] accounted for the stress and crack length dependence of the crack growth rate by assuming the FCP rate to be dependent on the net section stress in the component. Paris[4] postulated that the stress intensity factor—itself a function of stress and crack length—was the overall controlling factor in the FCP process. This postulate appears reasonable, since one might expect the parameter K, which controlled static fracture (Chapter Eight) and environment assisted cracking (Chapter Eleven) to control dynamic fatigue failures as well. From Chapter Eight the stress intensity factor levels corresponding to the crack growth rates identified in Fig. 13.2 would be $Y_i \sigma_1 \sqrt{a_i}$ and $Y_j \sigma_1 \sqrt{a_j}$ at σ_1 and $Y_i \sigma_2 \sqrt{a_i}$ and $Y_j \sigma_2 \sqrt{a_j}$ at σ_2, respectively. By plotting values of da/dn and ΔK at the associated values of $a_{i,j...n}$, a strong correlation was observed (Fig. 13.3), which suggested a relationship of the form

$$\frac{da}{dn} = A \Delta K^m \tag{13-3}$$

where

$\dfrac{da}{dn} =$ fatigue crack growth rate

$\Delta K =$ stress intensity factor range ($\Delta K = K_{max} - K_{min}$)

$A, m = f$ (material variables, environment, frequency, temperature, stress ratio)

It is encouraging to note that the FCP response of many materials is correlated with the stress intensity factor range, even though FCC, BCC, and HCP metals all were included in the data base.

Thus, for an interesting period during the early 1960s, the battle of the crack growth rate relationships began. Although experimental evidence was mounting in favor of the stress intensity factor approach, it was not until two critical sets of experiments were reported and fully appreciated that the importance of the stress intensity factor in controlling fatigue crack propagation rates was fully accepted. In one paper, Swanson et al.[5] reasoned that if any of the various

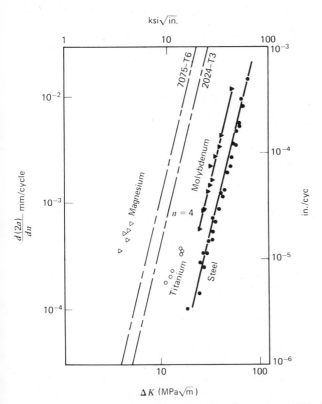

FIGURE 13.3 Fatigue crack propagation for various FCC, BCC, and HCP metals. Data verify power relationship between ΔK and da/dn.[4] (With permission of Syracuse University Press.)

relationships between stress and crack length was truly the critical parameter controlling crack growth rate behavior, then keeping that parameter constant during a fatigue test would cause the crack to grow at a constant velocity. In other words, the crack length versus number of cycles curve would appear as a straight line. To achieve this condition, Swanson decreased the cyclic load level incrementally in varying amounts with increasing crack length to maintain constant magnitudes of $\sigma^3 a$, $\sigma^2 a$, σ_{net}, and K, respectively, deemed to be the controlling parameters as discussed above. It is demonstrated clearly in Fig. 13-4 for the case of σ_{net} and K, that while each parameter when held constant did produce a constant growth rate over a limited range, maintaining a constant stress intensity factor caused the crack to grow at a constant velocity throughout the entire fatigue test life. The stress intensity factor clearly was the key parameter controlling crack propagation. The second set of critical experiments was reported by Paris and Erdogan,[6] who analyzed test data from center

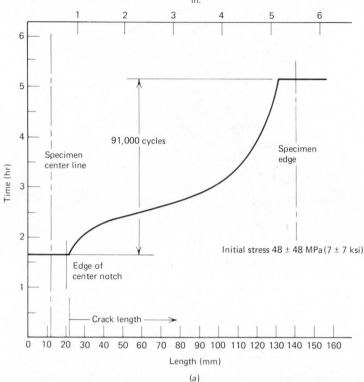

FIGURE 13.4 Crack length versus time plots revealing crack growth rate behavior in 7079-T6 aluminum alloy. (a) Linear load shedding to achieve constant net section stress; (b) load shedding to achieve constant ΔK conditions. Note constant growth rate over entire specimen life in (b)[5]. (Reprinted with permission of the American Society for Testing and Materials from copyright work.)

cracked panels of a high-strength aluminum alloy. In one instance, the loads were applied uniformly and remote from the crack plane (Fig. 13.5). From Fig. 8.7c, the stress intensity factor for this configuration is found to increase with increasing crack length. Therefore, if a constant load range were applied during the life of the test, the stress intensity level would increase continually because of the increasing crack length. Correspondingly, crack growth rates would be low at first and increase gradually as the crack extended. In the other set of experiments, center notched panels were loaded by concentrated forces acting at the crack surfaces (Fig. 13.5). For this configuration, the stress intensity factor is found to be[7]

$$K = \frac{P}{\sqrt{\pi a}} \tag{13-4}$$

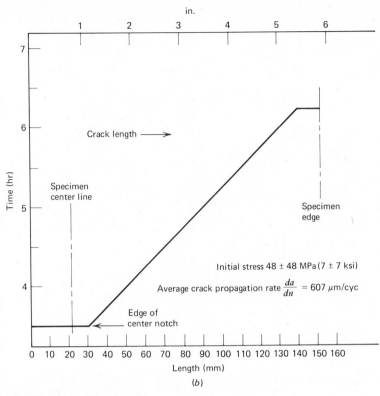

Crack length ⟶

Specimen
center line

Time (hr)

Specimen
edge

Initial stress 48 ± 48 MPa (7 ± 7 ksi)

Average crack propagation rate $\dfrac{da}{dn}$ = 607 μm/cyc

Edge of
center notch

Length (mm)

(b)

FIGURE 13.4   (*Continued*)

Most interestingly, the stress intensity factor is observed to be large when the crack length is small but decreases with increasing crack length. Consequently, one would predict that if the stress intensity factor did control FCP rates, the crack growth rate would be large initially but would *decrease* with increasing crack length. Such a reversal in FCP behavior would not be predicted if crack growth were controlled by the net section stress, for example, since the latter term would increase with both specimen load configurations. In fact, the experimental results showed the crack growth rate reversal anticipated by the sense of the crack length dependent change in stress intensity factor, dK/da. The crack-line loaded sample exhibited the highest growth rates at the outset of the test when the crack was small, with progressively slower growth rates being monitored with increasing crack length. The opposite was true for the grip loaded sample (Fig. 13.5).

Not only did these experiments verify the importance of the stress intensity factor in controlling the fatigue crack propagation process, they illustrated again

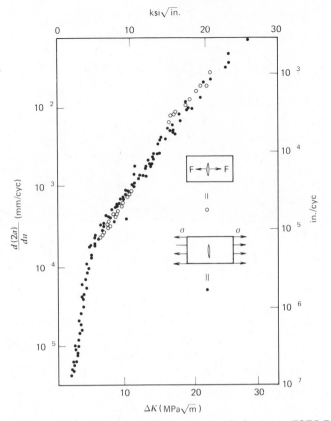

FIGURE 13.5 Fatigue crack propagation behavior in 7075-T6 for remote and crack line loading conditions.[4] (With permission of Syracuse University Press.)

the interchangeability of specimen geometry and load configuration in the determination of material properties such as crack growth rate or fracture toughness values. As a final note, over the growth range from about 2×10^{-6} to 2×10^{-3} mm/cyc, crack growth rate behavior in relatively inert environments as described by Eq. 13-3 is not strongly dependent on the mean stress intensity level. To a first approximation, crack growth rates are found to double when K_{mean} is doubled. By contrast, when the ΔK level is doubled, the crack would propagate 16 times faster (assuming $m = 4$). Consequently, mean stress effects in this growth rate range are considered to be of secondary importance. (Crack growth rate conditions where mean stress effects are more important are discussed in Section 13.4.) It is to be noted that the secondary importance of mean stress in controlling FCP response is similar to the smaller role played by mean stress in cyclic life tests as portrayed with Gerber and Goodman diagrams (Chapter Twelve).

13.1.1 Fatigue Life Calculations

When conducting a failure analysis, it is often desirable to compute component life for comparison with the actually recorded service life. Alternately, if one were designing a new part and wished to establish for it a safe operating service life, fatigue life calculations would be required. Such a computation can be performed by integrating Eq. 13-3 (where $\Delta K = Y \Delta \sigma \sqrt{a}$) with the starting and final flaw size as limits of the integration. When the geometrical correction factor Y does not change within the limits of integration [e.g., for the case of a circular flaw where $K = \frac{2}{\pi} \sigma \sqrt{\pi a}$ (Fig. 8.7i)], the cyclic life is given by

$$N_f = \frac{2}{(m-2) A Y^m \Delta \sigma^m} \left[\frac{1}{a_0^{(m-2/2)}} - \frac{1}{a_f^{(m-2/2)}} \right] \quad \text{for } m \neq 2 \quad (13\text{-}5)$$

where

$$N_f = \text{number of cycles to failure}$$

$$a_0 = \text{initial crack size}$$

$$a_f = \text{final crack size at failure}$$

$$\Delta \sigma = \text{stress range}$$

$$A, m = \text{material constants}$$

$$W = \text{specimen width}$$

Usually, however, this integration cannot be performed directly, since Y varies with the crack length. Consequently, cyclic life may be estimated by numerical integration procedures by using different values of Y held constant over a number of small crack length increments. It is seen from Eq. 13-5 that when $a_0 \ll a_f$ (the usual circumstance) the computed fatigue life is not sensitive to the final crack length a_f but, instead, is strongly dependent on estimations of the starting crack size a_0.

To illustrate the use of Eq. 13-5 in the computation of fatigue life, let us reconsider the material selection problem described in Chapter Eight (Example 2). Let us suppose that the 0.45C-Ni-Cr-Mo steel is available in both the 2070 MPa and 1520 MPa tensile strength levels, and a design stress level of one-half tensile strength is required. It is necessary to estimate the fatigue life of a component manufactured from the material in the two strength conditions. Using the design stress levels, a stress range of 1035 MPa and 760 MPa would be experienced by the 2070 MPa and 1520 MPa materials, respectively. It is immediately obvious from Eq. 13-5 that, all things being equal, the total fatigue life would be lower in the higher strength material because it would experience a higher stress range. Using a value of $m = 2.25$ as found by Barsom et al.[8] for 19 steels, the fatigue life in the stronger material would be reduced by almost a factor of two. This should be considered as a minimum estimate of the reduction in fatigue life, since there is evidence to indicate that the exponent m increases with decreasing fracture toughness.[9] Furthermore, recalling that the critical flaw size in the 2070 MPa level material was only one-fifth that found in the 1520 MPa alloy, the computed service life in the stronger alloy would be reduced further. This would be true especially if the initial crack was not much smaller than the critical flaw size. Therefore, one concludes that the stronger material is inferior in terms of potential fatigue life, critical flaw size, associated fracture toughness, and environment assisted cracking sensitivity (Fig. 11.10).

13.2 MACROSCOPIC FRACTURE MODES IN FATIGUE

As discussed in Chapter Twelve, the fatigue fracture process can be separated into three regimes: crack initiation (sometimes obviated by preexistent defects), crack propagation, and final fracture (associated with crack instability). The existence and extent of these stages depends on the applied stress conditions, specimen geometry, flaw size, and the mechanical properties of the material. Stage I, representing the initiation stage, usually extends over only a small percentage of the fracture surface but may require many loading cycles if the nucleation process is slow. Often, Stage I cracks assume an angle of about 45 degrees in the x–y plane with respect to the loading direction.[10] After a relatively short distance, the orientation of a Stage I crack shifts to permit the crack to propagate in a direction normal to the loading direction. This transition has been associated with a changeover from single to multiple slip.[11,12] The plane on which the Stage II crack propagates depends on the relative stress state; that is, the extent of plane strain or plane stress conditions. When the stress intensity factor range is low (resulting from a low applied stress and/or small crack size), a small plastic zone is developed (Eq. 8-33). When the sheet thickness is large compared to this zone size, plane strain conditions prevail and flat fracture usually results. With subsequent fatigue crack extension, the stress intensity factor and the plastic zone size increase. When the zone is large compared to specimen thickness, plane stress conditions and slant fracture are dominant. Therefore, depending on the stress level and crack length, the fractured component will possess varying amounts of flat and slant fracture. Consequently, a fatigue crack may start out at 90 degrees to the plate surface but complete its propagation at 45 degrees to the surface (Fig. 13.6). Alternately, the crack could propagate immediately at 45 degrees if the plastic zone size to plate thickness ratio were high enough, reflecting plane stress conditions. It is important to recognize that both unstable, fast-moving cracks and stable, slow-moving fatigue cracks may assume flat, slant, or mixed macromorphologies.

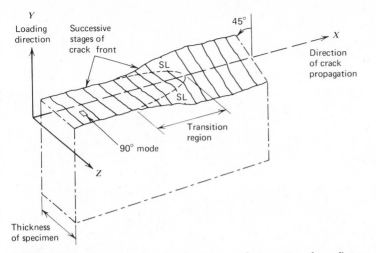

Y
Loading
direction

Successive
stages of
crack front

SL

SL

45°

X

Direction
of crack
propagation

Transition
region

90° mode

Z

Thickness
of specimen

FIGURE 13.6 Diagram showing fracture mode transition from flat to slant fracture appearance. (Adapted from J. Schijve).

As was just discussed, the plastic zone (defined by Eq. 8-33) can be used to estimate the relative amount of flat and slant fracture surface under both static and cyclic loading conditions. From Fig. 13.7, this plastic zone is developed by the application of a stress intensity factor of magnitude K_1. However, when the latter is reduced by h_k because the direction of loading is reversed, the local stress is reduced to a level corresponding to a stress intensity level of K_2. Since the elastic stress distribution associated with K_1 was truncated at σ_{ys} by local yielding, subtraction of an elastic stress distribution in going from K_1 to K_2 will cause the final crack tip stress field to drop sharply near the crack tip and even go into compression. At K_2, a smaller plastic zone is formed in which the material undergoes compressive yielding. Paris[4] showed that the size of this smaller plastic zone, which experiences alternate tensile and compressive yielding, may be estimated by substituting h_k for K and $2\sigma_{ys}$ for σ_{ys} in Eq. 8-33. As a result, the size of the reversed plastic zone may be given by

$$r_y = \frac{1}{8\pi} \left(\frac{h_k}{\sigma_{ys}} \right)^2 \tag{13-6}$$

or four times smaller than the comparable monotonic value.

Since the material within this smaller plastic zone experiences reversed cyclic straining, it might be expected that cyclic strain hardening or softening would result, depending on the starting condition of the material. This has been borne out by microhardness measurements made by Bathias and Pelloux[13] near the tip

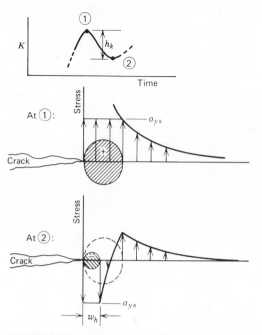

FIGURE 13.7 Monotonic and reversed plastic zone development at tip of advancing fatigue crack.[4] (With permission of Syracuse University Press.)

of a fatigue crack. Evidence of extensive cyclic softening and hardening within the reversed plastic zone in two different steels is shown in Fig. 13.8. The significance of material property changes on fatigue crack propagation in this zone is examined further in Section 13.7.

13.3 MICROSCOPIC FRACTURE MECHANISMS

A high-magnification examination of the clam shell markings found on service failure fracture surfaces (Figs. 12.1 and 12.3) and the growth bands developed in laboratory samples subjected to block loading conditions (Fig. 13.9) reveals the presence of many smaller parallel lines, referred to as fatigue striations (Fig. 13.10). Several important facts are known about these markings. First, they appear on fatigue fracture surfaces in many materials, such as BCC, HCP, and

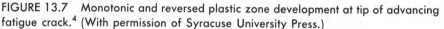

FIGURE 13.8 Cyclically induced material property changes occurring at fatigue crack tip. (a) Diagram showing monotonic and cyclic plastic zone boundaries; (b) cyclic strain softening in two maraging steels; (c) cyclic strain hardening in austenitic steel.[13] (Reprinted with the permission of the American Society of Metals. Copyright 1973.)

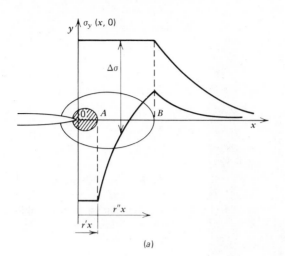

(a)

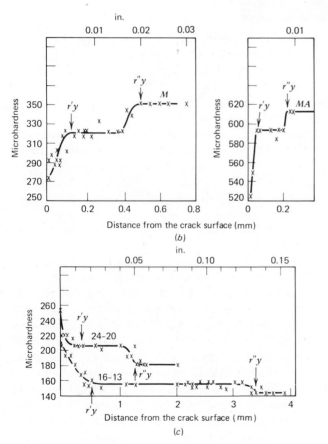

Distance from the crack surface (mm)

(b)

Distance from the crack surface (mm)

(c)

FIGURE 13.9 Photograph showing macrobands on fatigue fracture surface resulting from variable amplitude block loading.[14] (Courtesy of H. I. McHenry.)

FCC metals, and some amorphous polymers, and are oriented parallel to the advancing crack front. In a quantitative sense, Forsyth and Ryder[16] provided critical evidence that each striation represents the incremental advance of the crack front as a result of one loading cycle and that the extent of this advance varies with the stress range. This is shown clearly in Fig. 13.10c, which reveals striations of differing width that result from a random loading pattern.

It is appropriate, then, to emphasize the clear distinction between macroscopically observed clam shell markings, which represent periods of growth during which thousands of loading cycles may have occurred, and microscopic striations, which represent the extension of the crack front during one load excursion. *There can be thousands or even tens of thousands of striations within a single clam shell marking.*

Although these striations provide evidence that fatigue damage was accumulated by the component during its service life, fatigue crack propagation can occur without their formation. Usually, microvoid coalescence occurs at high ΔK levels,[17] and a cleavage-like faceted appearance dominates in many materials at very low ΔK levels[12] (Fig. 13.11). It is generally observed that the relative striation density found at intermediate ΔK values seems to vary with stress state and alloy content. Although striations are most clearly observed on flat surfaces associated with plane strain conditions, elongated dimples and evidence of abrasion are the dominant fractographic features of plane stress slant fracture surfaces. In terms of metallurgical factors, it is much easier to find striations on fatigue surfaces in aluminum alloys than in high-strength steels. In some cases, it is virtually impossible to identify clearly defined areas of striations in the latter material, thereby making fractographic examination most difficult.

Fatigue striations can assume many forms, such as the highly three-dimensional or flat ones seen in Fig. 13.10d, e. It is not absolutely clear why there are

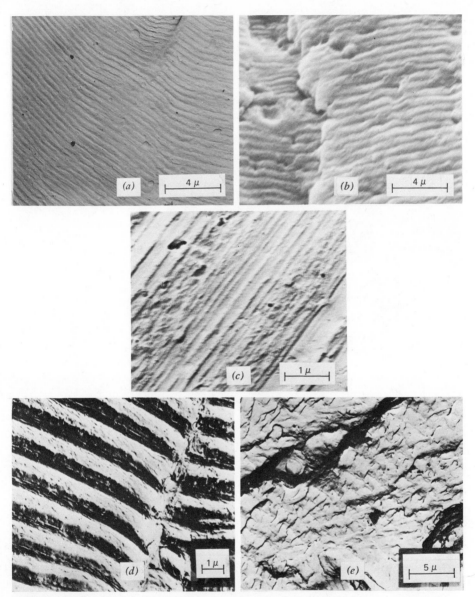

FIGURE 13.10   Electron fractographs revealing fatigue striations found on fracture surface and within macroscopic bands (Figs. 12.1, 12.3, 13.9). (a) TEM, constant load range; (b) SEM, constant load range; (c) TEM, random loading; (d) TEM, ductile striations;[15] (e) TEM, brittle striations.[15] (Reprinted with permission of the American Society for Testing and Materials from copyright work.)

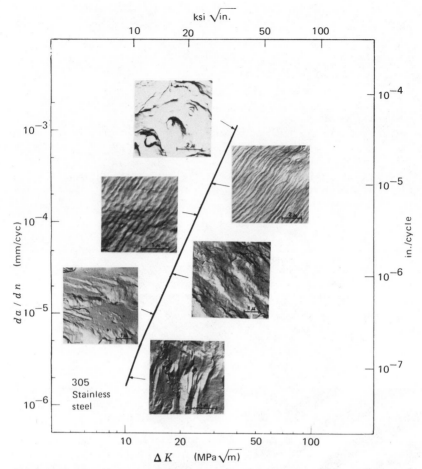

FIGURE 13.11 Change in fracture surface appearance in 305 stainless steel as a function of ΔK level.[12] (Reprinted with permission of the American Society for Testing and Materials from copyright work.)

different morphologies, but they are often associated with the test environment during crack propagation. Fatigue striations are relatively flat and assume a cleavage-like appearance when formed in an aggressive environment, but tend to appear more ductile when formed in an inert environment. Although striation morphology may be affected by the service environment, definite and pronounced changes have been observed in the appearance of fatigue striations after exposure to oxidizing or corroding atmospheres. For example, fatigue striations can be completely obliterated as a result of exposure to a high-temperature, oxidizing environment. Even at room temperature, fatigue striations become increasingly more difficult to detect with time. As a result, the amount of fractographic information to be gleaned from a fracture surface decreases with time, particularly in the case of steel alloys; the fracture surface details in aluminum alloys are maintained for a longer period because of the protective nature of the aluminum oxide film that forms quickly on the fracture surface.

The reader should recognize that even when striations are expected to form (Fig. 13.11), they are not always as clearly defined as those in Fig. 13.10. Whether due to environmental and/or metallurgical effects or related to service conditions, such as abrasion of the mating fracture surfaces, striations may appear either continuous or discontinuous, clearly or poorly defined, and either straight or curved.

From metallographic sections and electron fractographic examination, three basic interpretations of the morphology of fatigue striations have evolved. The striations are considered to be undulations on the fracture surface with (1) peak-to-peak and valley-to-valley matching of the two mating surfaces, (2) matching crevices separating flat facets, or (3) peak-to-valley matching. Based on these interpretations of striation morphology, different mechanisms have been proposed for striation formation. One mechanism involves plastic blunting processes at the crack tip,[18] which occur regardless of material microscopic slip character; another model takes account of crystallographic considerations, wherein striations are thought to form by sliding off on preferred slip planes.[15,17,19,20]

The effect of crystallography on striation formation can be supported by both direct and circumstantial evidence. Pelloux[20] demonstrated in elegant fashion with the aid of etch pit studies that striation orientation in an aluminum alloy was sensitive to changes in crystal orientation and that striations tended to form on (111) slip planes and parallel to $\langle 110 \rangle$ directions. The latter is in agreement with theoretical considerations[17,20] (Fig. 13.12) and experimental findings, which show a strong tendency for the macroscopic fracture plane to lie parallel to $\{100\}$ or $\{110\}$ planes.[21] It might then be argued that when slip planes are oriented favorably with respect to the maximum resolved shear stresses at the advancing crack tip, a clearly defined striation could be formed. Alternately, poorly defined striations or none at all might be found when slip planes were unfavorably oriented. It is quite probable that crystallographic considerations dominate striation formation at low ΔK levels where few slip systems are operative, while the plastic blunting model would provide a better picture of events at high ΔK levels.

13.3.1 Correlations with the Stress Intensity Factor

More quantitative information has been obtained from the measurement of fatigue striations than from any other fracture surface detail. Since the striation

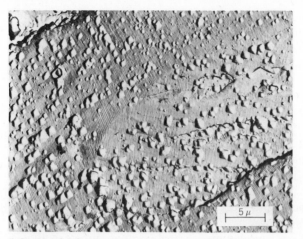

FIGURE 13.12 Fatigue striations in 2024-T3 aluminum alloy. Note concurrent change in striation and etch pit orientation.[20] (Reprinted with permission of the American Society of Metals. Copyright 1969.)

represents the position of the crack front after each loading cycle, its width can be used to measure the FCP rate at any given position on the fracture surface. It is not surprising, then, to find reasonable correlation at a given ΔK level between the macroscopically determined growth rate as measured with a traveling microscope and the microscopic growth rate as measured by the width of individual striae[22] (Fig. 13.13). Additional correlations[23] between striation spacing and ΔK have been found for a number of materials (Fig. 13.14). Here, ΔK has been normalized with respect to the elastic modulus of the respective materials examined (see Section 13.7).

The practical significance of the data correlations found in Figs. 13.13 and 13.14 cannot be overemphasized, since such data are very useful in a failure analyses.

EXAMPLE 1   After a certain period of service, a 15-cm wide panel of 2024-T3 aluminum alloy was found to contain a 5-cm long edge crack oriented normal to the stress direction. The crack was found to have nucleated from a small, preexistent flaw at the edge of the panel. The magnitude of the cyclic stress was analyzed to be less than 20% of the yield strength ($\sigma_{ys} \approx 345$ MPa) and was believed to be distributed uniformly along the plane of the crack. Since the crack had reached dangerous proportions, the panel was removed from service and examined fractographically. Average striation widths of 10^{-4} mm and 10^{-3} mm were found at distances of 1.5 and 3 cm, respectively, from the origin of the crack. Was the premature failure caused by the existence of the surface flaw or related to a much higher cyclic stress level than originally estimated?

For this configuration, the stress intensity factor is given from Fig. 8.7f

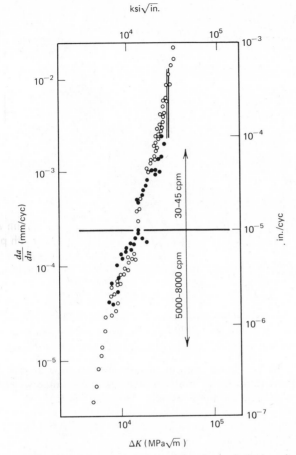

ksi$\sqrt{\text{in.}}$

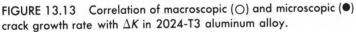

FIGURE 13.13   Correlation of macroscopic (○) and microscopic (●) crack growth rate with ΔK in 2024-T3 aluminum alloy.

as

$$K = Y\sigma\sqrt{a}$$

with $Y(a/W)$ being defined at $Y(1.5/15)$ and $Y(3/15)$ to be 2.1 and 2.43, respectively. From Fig. 13.13, the apparent stress intensity range based on the two striation measurements is found to be 12.7 and 20.9 MPa$\sqrt{\text{m}}$, corresponding to crack lengths of 1.5 and 3 cm, respectively. Therefore, two independent estimates of the actual stress range can be obtained directly from the above equation:

$$\Delta\sigma = \frac{\Delta K}{Y\sqrt{a}} = \frac{12.7}{(2.1)\sqrt{0.015}} = 49.4 \text{ MPa}$$

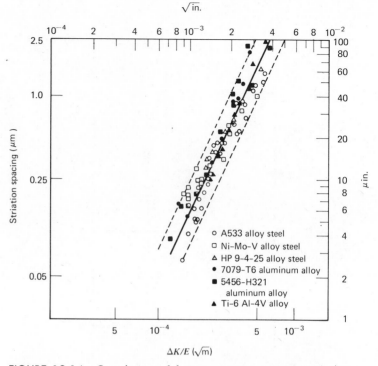

FIGURE 13.14 Correlation of fatigue striation spacing with ΔK normalized with respect to elastic modulus.[23] (Reprinted with permission of the American Society of Metals. Copyright 1969.)

and

$$\frac{20.9}{(2.43)\sqrt{0.03}} = 49.7 \text{ MPa}$$

Since both numbers are self-consistent and are in agreement with the original design estimates, the striation data appear valid. Therefore, it is concluded that premature failure was caused by early crack propagation from the small edge flaw.

Although this procedure is extremely useful, its implementation should be exercised with deliberate caution. First, it is critically important to accurately identify the crack length position where the striation spacing measurements were made. The stress level cannot be computed if the crack length is not known. In many service failure reports, striation photographs are presented without identification of the precise location of the region of the fracture surface. Without such information, the photograph serves only to identify the mechanism of failure but does not enable the examiner to perform any meaningful calculations.

Since striation formation is a highly localized event, it is dependent on both stress intensity factor and metallurgical conditions. It has been shown repeatedly in laboratory experiments that for constant stress intensity conditions, striation spacings in a local region may vary by a factor of two to four. Therefore, to arrive at a meaningful estimate of crack growth rate at a particular crack length, many measurements of striation spacing should be made. In addition, measurements should be made at different crack length positions, as done in Example 1, to serve as a comparative check on the computation.

The prevailing stress intensity factor range could also be estimated with the aid of an empirical correlation identified by Bates and Clark,[23] who showed that

$$\text{striation spacing} \approx 6\left(\frac{\Delta K}{E}\right)^2 \tag{13-7}$$

where

$$\Delta K = \text{stress intensity factor range}$$

$$E = \text{modulus of elasticity}$$

It is particularly intriguing that Eq. 13-7 can be used to estimate ΔK based on fractographic information for any metallic alloy (Fig. 13.14). Since the exponent in Eq. 13-7 is approximately two, while the exponent in the Paris-type Eq. 13-3 depends on material variables, environment, frequency, and temperature and can vary from about 2 to 7, agreement between macroscopic and microscopic crack growth rates should not be expected in the majority of instances. Consequently, striation spacing measurements should be used in conjunction with Eq. 13-7 or compared with previously determined fractographic information, rather than macroscopic data, whenever estimations of the prevailing ΔK level are desired.

13.4 CRACK GROWTH BEHAVIOR AT ΔK EXTREMES

13.4.1 High ΔK Levels

Although Eq. 13-3 does provide a simple relationship by which crack growth rates may be correlated with the stress intensity factor range, it does not account for crack growth characteristics at both low and high levels of ΔK. If enough data are obtained for a given material—say, four to five decades of crack growth rates—the da/dn versus ΔK curve assumes a sigmoidal shape, as shown in Fig. 13.15. That is, the ΔK dependence of crack growth rate increases markedly at

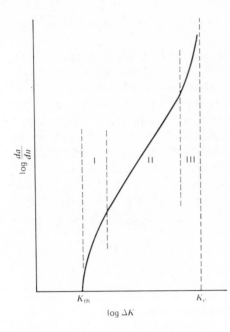

FIGURE 13.15 Diagram showing three regimes of fatigue crack growth response. Region I, crack growth rate decreases rapidly with decreasing ΔK and approaches lower limit at ΔK_{th}; Region II, midrange of crack growth rates where "power law" dependence prevails; Region III, acceleration of crack growth resulting from local fracture as K_{max} approaches K_c.

both low and high ΔK values. At the high growth rate end of the spectrum, part of this deviation sometimes may be accounted for by means of a plasticity correction, since the plastic zone becomes large at high ΔK levels. This has the effect of increasing ΔK_{eff} (Eq. 8-36) and thus tends to straighten out the curve. Another factor to be considered is that as K_{max} approaches K_c, local crack instabilities occur with increasing frequency, as evidenced by increasing amounts of microvoid coalescence and/or cleavage on the fracture surface. As might be expected, this effect is magnified with increasing mean stress. Characterizing the mean stress level by R, the ratio of minimum to maximum loads, it is seen from Fig. 13.16a that crack growth rates at high ΔK values increase with increasing mean stress, and little mean stress sensitivity is observed at lower ΔK levels. One relationship expressing crack growth rates in terms of $\Delta K, K_c$, and a measure of K_{mean} was proposed by Forman et al.,[24] in the form

$$\frac{da}{dn} = \frac{C\Delta K^n}{(1-R)K_c - \Delta K} \tag{13-8}$$

where

$$C, n = \text{material constants}$$

$$K_c = \text{fracture toughness}$$

$$R = \text{load ratio}\left(\frac{K_{min}}{K_{max}}\right)$$

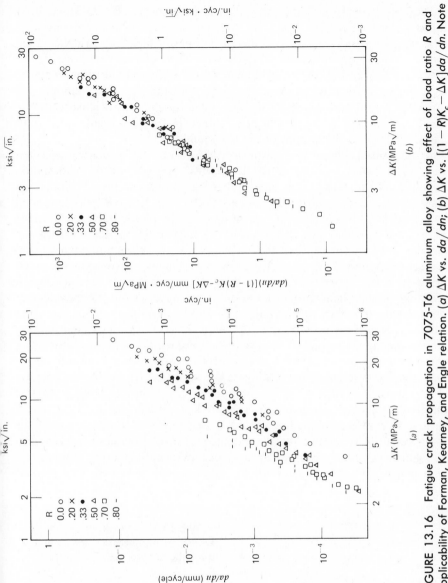

FIGURE 13.16 Fatigue crack propagation in 7075-T6 aluminum alloy showing effect of load ratio R and applicability of Forman, Kearney, and Engle relation. (a) ΔK vs. da/dn; (b) ΔK vs. $[(1-R)K_c - \Delta K]da/dn$. Note less scatter in b. (Data from Hudson.[28])

From Eq. 13-8, we see that the simple power relationship (Eq. 13-3) has been modified by the term $[(1 - R)K_c - \Delta K]$, which decreases with increasing load ratio R and decreasing fracture toughness K_c, both of which lead to higher crack growth rates at a given ΔK level. A typical plot of normalized data according to Eq. 13-8 is shown in Fig. 13.16b. Although Eq. 13-8 correctly identifies material FCP response under combinations of high ΔK and K_{mean} conditions, the relationship is difficult to apply because of difficulties associated with the determination of the K_c value, which, as shown in Chapter Eight, varies with planar and thickness dimensions of the test sample.

Other relationships describing mean stress effects on FCP response have taken account of the plastic zones at the crack tip[25] and the plastic deformation process itself. With regard to the latter, Elber[26] proposed that the crack might be partially closed during part of the loading cycle, even when $R > 0$. He argued that residual tensile displacements, resulting from the plastic damage of fatigue crack extension, would interfere along the crack surface in the wake of the advancing crack front and cause the crack to close above the minimum applied load level. This hypothesis was verified with compliance measurements taken from fatigued test panels that showed that an *effective* change in crack length (i.e., change in compliance) occurred prior to any *actual* change in crack length. In other words, the crack was partially closed for a portion of the loading cycle and did not open fully until a certain opening K level, K_{op}, was applied. As a result, the damaging portion of the cyclic load excursion would be restricted to that part of the load cycle which acted on a fully opened crack. From Fig. 13.17, the effective stress intensity factor range ΔK_{eff} would be denoted by the opening level K_{op} to K_{max}, rather than by the applied ΔK level $K_{max} - K_{min}$. In this connection, it is interesting to note that crack growth rates are relatively insensitive to compressive loading excursions where $R < 0$. In fact, a number of investigators[27,28] have shown that the fatigue response of materials subjected to $R < 0$ loading conditions can be approximated by simply ignoring the negative portion of the load excursion, since the crack would be closed. The marked

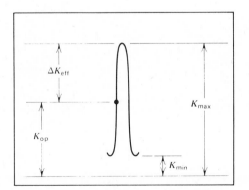

FIGURE 13.17 Crack surface interference results in crack opening K_{op} to be above zero. ΔK_{eff} defined as $K_{max} - K_{op}$.

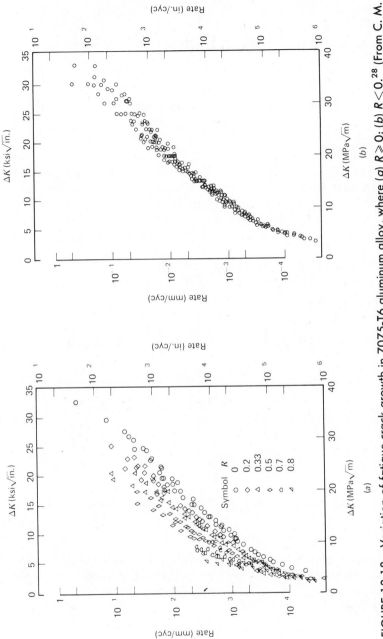

FIGURE 13.18 Variation of fatigue crack growth in 7075-T6 aluminum alloy, where (a) $R \geq 0$; (b) $R < 0$.[28] (From C. M. Hudson, NASA TN D-5390, 1969.)

change in importance of mean stress when negative load excursions are encountered is shown in Fig. 13.18. We see in Fig. 13.18*a*, where $R \geqslant 0$, that FCP rates in 7075-T6 aluminum alloy are affected by changes in mean load level. By contrast, no significant change is found in crack growth rates when $R \leqslant 0$, (Fig. 13.18*b*). (Note that in the latter figure, ΔK was defined as K_{max}; that is, $K_{min} = 0$.)

Recent studies by Lindley and Richards[29] suggest that crack closure and its effect on crack propagation may not be important in plane strain situations but, instead, restricted to plane stress conditions, where much larger amounts of crack tip plastic deformation are available for subsequent crack front interference. The applicability of the crack closure model in the latter case was verified by Hertzberg and von Euw[30] who showed that the fracture mode transition (FMT) was dependent on ΔK_{eff}. From Chapter Eight, this transition is related to a critical ratio of plastic zone size to panel thickness. Therefore, it was surprising to find the transition occurring instead at a constant crack growth rate[31,32] (Fig. 13.19).

From Table 13.1, the FMT did not occur at a specified value of K_{max} or ΔK. However, it did occur at a constant ΔK_{eff} level, which would account for the FMT being observed at a constant growth rate.

TABLE 13.1   Fracture Mode Transition in 2024-T3 Aluminum Alloy[30,31]

	K_{max}		ΔK_{app}		ΔK_{eff}^a	
R	$MPa\sqrt{m}$	$kg/mm^{3/2}$	$MPa\sqrt{m}$	$kg/mm^{3/2}$	$MPa\sqrt{m}$	$kg/mm^{3/2}$
0.1	10.4	33.5	9.4	30.2	5.1	16.3
0.2	11.2	36	8.9	28.8	5.2	16.7
0.3	12.1	39	8.5	27.3	5.2	16.9
0.4	13	42	7.8	25.2	5.1	16.6
0.5	14.3	46	7.1	23	5.0	16.1

$^a\Delta K_{eff}$ in 2024-T3 aluminum alloy was calculated from the relationship[26] $\Delta K_{eff} = \Delta K_{app}$ $(0.5 + 0.4 \, R)$.

Considerably more work is needed to verify the crack closure model for other materials and to determine its sensitivity to stress state. Consequently, the model is currently used only as a framework for understanding the physical process of fatigue crack growth.

13.4.2   Low ΔK Levels

At the other end of the crack growth rate spectrum, the simple power relationship (Eq. 13-3) is violated again for low ΔK conditions, where the FCP rate

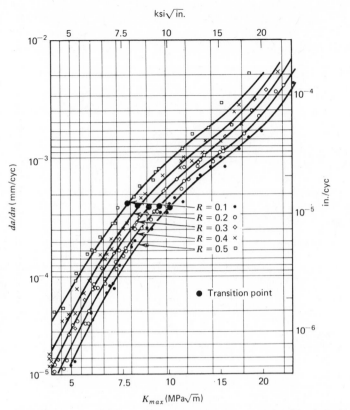

FIGURE 13.19 Crack propagation behavior in 2024-T3 Al clad sheet. Note fracture mode transition (●) at constant crack growth rate.[32] (Reprinted by permission of the American Society for Testing and Materials from copyright work.)

diminishes rapidly to a vanishingly small level (Fig. 13.15). From such data[33,34] as shown in Fig. 13.20, a limiting stress intensity factor range (the threshold level ΔK_{th}) is defined and represents a service operating limit below which fatigue damage is highly unlikely. In this sense, ΔK_{th} is much like K_{IEAC}, the threshold level for environment assisted cracking (see Chapter Eleven). Designing a component such that $\Delta K \leqslant \Delta K_{th}$ would be a highly desirable objective, but it is sometimes not very realistic in the sense that ΔK_{th} for engineering materials often represents only 5 to 10% of anticipated fracture toughness values (Table 13.2). Therefore, to operate under $\Delta K \leqslant \Delta K_{th}$ conditions would require that

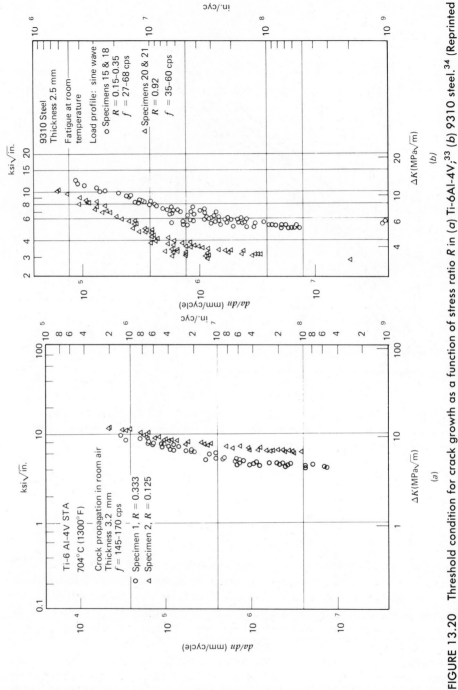

FIGURE 13.20 Threshold condition for crack growth as a function of stress ratio R in (a) Ti-6Al-4V;[33] (b) 9310 steel.[34] (Reprinted with permission of the American Society for Testing and Materials from copyright work.)

TABLE 13.2 Threshold Data in Engineering Alloys

Material	R	ΔK_{th} MPa$\sqrt{m}$	ΔK_{th} ksi$\sqrt{in.}$	Ref.
9310 Steel	0.25	~6.1	~5.5	a
	0.9	~3.3	~3	a
A533B Steel	0.1	8	7.3	b
	0.3	5.7	5.2	b
	0.5	4.8	4.4	b
	0.7	3.1	2.8	b
	0.8	3	2.75	b
A508	0.1	6.7	6.1	b
	0.5	5.6	5.1	b
	0.7	3.1	2.8	b
T-1	0.2	~5.5	~5	c
	0.4	~4.4	~4	c
	0.9	~3.3	~3	c
Ti-6 Al-4 V	0.15	~6.6	~6	d
	0.33	~4.4	~4	d
18/8 Austenitic steel	0	6.1	5.5	e
	0.33	5.9	5.4	e
	0.62	4.6	4.2	e
	0.74	4.1	3.7	e
Copper	0	2.5	2.3	e
	0.33	1.8	1.6	e
	0.56	1.5	1.4	e
	0.80	1.3	1.2	e
60/40 brass	0	3.5	3.2	e
	0.33	3.1	2.8	e
	0.51	2.6	2.4	e
	0.72	2.6	2.4	e
Nickel	0	7.9	7.2	e
	0.33	6.5	5.9	e
	0.57	5.2	4.7	e
	0.71	3.6	3.3	e

[a]P. C. Paris, *MTS Closed Loop Magazine*, **2** (5), 1970.
[b]P. C. Paris, et al., ASTM *STP 513*, 1972, p. 141.
[c]R. J. Bucci, et al., *op. cit.*, p. 177.
[d]R. J. Bucci, et al., *op. cit.*, p. 125.
[e]L. D. Pook, *op. cit.*, p. 106.

virtually all defects be eliminated from a component and/or the design stress be extremely low. This is desirable in the design of nuclear power generation equipment where safety is of prime concern; however, designing an aircraft such that $\Delta K \leqslant \Delta K_{th}$ is highly impractical. Theoretically, you could design an airplane that would not fatigue but the beefed-up structure necessary to reduce the stress level to below the ΔK_{th} level would weigh so much that the plane would not be able to take off! Since many engineering structures do fulfill their intended service life without incident, it is apparent that some components do operate under $\Delta K \leqslant \Delta K_{th}$ conditions.

As seen in Table 13.2, the effect of K_{mean} (i.e., R ratio) on crack propagation becomes important once again at very low ΔK levels. Clearly, the mean load level has a strong effect on ΔK_{th}. At progressively higher ΔK levels (though still in the low crack growth regime), the effect of R ratio diminishes (Fig. 13.20). Consequently, mean stress effects are seen to be important at both very low and very high ΔK levels but not at intermediate levels.

Before concluding this section, it is appropriate to consider the similarity between the threshold stress intensity range, defined from FCP data, and the fatigue limit, determined from stress-cyclic life plots. In fact, it is intriguing to consider a possible correlation between these two fatigue parameters. For example, if one considers an unnotched sample to have a characteristic internal defect with a penny-shaped crack morphology (Fig. 8.7i), the associated stress intensity factor would be[7]

$$K = \frac{2}{\pi} \sigma \sqrt{\pi a} \qquad (13\text{-}9)$$

where a = radius of a thin, circular disc. If we let $K = \Delta K_{th}$ and $\sigma = \sigma_{fat}$, then for a material such as T-1 steel, ΔK_{th} is found to be 5.5 (Table 13.2) and σ_{fat} is approximated to be 0.5 σ_{ts} (Section 12.2.4), or 415 MPa. Therefore

$$5.5 = \frac{2}{\pi}(415)\sqrt{\pi a}$$

$$a \approx 0.14 \text{ mm}$$

which might represent the size of a typical microstructural defect in this material.

As a final note, it is suggested that fatigue limit information be viewed with caution, since these data are defined usually at 10^7 cycles to denote the stress level below which fatigue failures should not occur. Since FCP rates have been measured as low as 5×10^{-9} mm/cyc,[35] it is quite likely that additional failures would have occurred in unnotched specimens had they been subjected to further cycling.

13.5 INFLUENCE OF LOAD INTERACTIONS

Much of the FCP data discussed thus far were gathered from specimens subjected to simple loading patterns without regard to load fluctuations. Although this may provide a reasonable simulation condition for components experiencing nonvarying load excursions, constant amplitude testing does not simulate variable load-interaction effects, which, in some cases, extend fatigue life measurably. In the most simple case, involving superposition of single-peak tensile overloads on a regular sinusoidal wave form, laboratory tests[15,36-41] have demonstrated significant FCP delay after each overload, with the amount of delay increasing with both magnitude and number of overload cycles (see Table 13.3).

TABLE 13.3 Cyclic Delay Resulting from Simple and Multiple Overloads in 2024-T3 Aluminum Alloy[38,39]

ΔK				
MPa$\sqrt{m}$	ksi$\sqrt{in.}$	% Peak Load	Number of Peak Loads	$N_d^*(10^3$ cycles)
15	13.65	53	1	6
15	13.65	82	1	16
15	13.65	109	1	59
16.5	15	50	1	4
16.5	15	50	10	5
16.5	15	50	100	9.9
16.5	15	50	450	10.5
16.5	15	50	2000	22
16.5	15	50	9000	44
23.1	21	50	1	9
23.1	21	75	1	55
23.1	21	100	1	245

The retarding effect of a peak overload is demonstrated clearly in Fig. 13.21 for a constant ΔK loading situation, where the crack growth rate associated with the invarient ΔK level is given by the constant slope b_2. Obviously, the FCP rate is depressed after the overload for a distance a^* from the point of the overload. Hertzberg and co-workers[38,41] have shown that this distance corresponds to the plastic zone dimension of the overload. Therefore, once the crack grows through the overload plastic zone, resumption of normal crack propagation is expected. Since N^* represents the total number of cycles necessary to traverse the overload plastic zone at an attenuated velocity, the actual number of cycles associated with retardation is shown in Fig. 13.21 and given by

$$N_d^* = N^* - \frac{a^*}{b_2}$$ (13-10)

where a^*/b_2 is the number of cycles necessary for a crack to traverse a distance a^* at a fixed rate b_2.

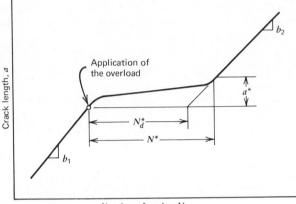

FIGURE 13.21 Crack growth rate plot illustrating effect of single-peak overload. Note delayed crack growth retardation.

Although the average crack growth rate within the affected region may be estimated crudely as a^*/N_d^*, the actual growth rate within a^* (found by differentiation of Fig. 13.21) is far more complex (Fig. 13.22). Perhaps the most significant feature of this plot is the fact that the crack growth rate decelerates after the overload, but does not reach a minimum level until the crack has progressed one-eighth to one-fourth the distance into the overload plastic zone, (i.e., a'_{min}). This characteristic of decelerating crack growth after overload cycling is called *delayed retardation*. Mills and Hertzberg[41] demonstrated that the greatest delay resulting from a second overload excursion occurred when the second overload was placed at a distance a'_{min} (corresponding to the point of minimum crack growth rate) from the initial overload. It is encouraging to note that the macroscopic crack growth rates recorded in the affected region are verified by electron fractographic observations of striation spacings.[15,38]

The importance of large overload cycles in affecting fatigue life of engineering components is illustrated by the findings of Schijve et al.[42] They found that, under aircraft flight simulation conditions involving a random load spectrum, when the highest wind-related gust loads from the laboratory loading spectrum

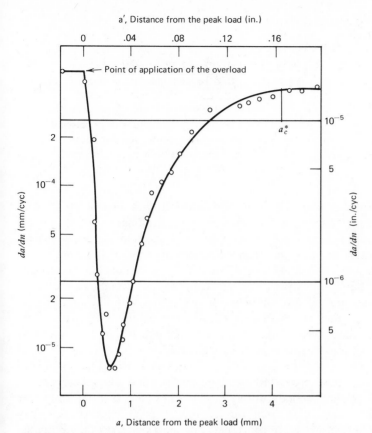

FIGURE 13.22 Crack growth rate behavior through the overload affected region revealing minimum growth rate occurring after crack had propagated part way through affected region. Resumption of normal crack growth rate occurs at distance comparable to overload plastic zone size, a_c*.[38] (Reprinted with permission of the American Society for Testing and Materials from copyright work.)

were eliminated, the specimens showed lower test sample fatigue life than did specimens that experienced some of the more severe load excursions. This fascinating load interaction phenomenon has led some investigators to conclude that an aircraft that logged some bad weather flight time could be expected to possess a longer service life than a plane having a better flight weather history.

Similar temporary FCP attenuations have been observed under block loading conditions. With high–low block loading sequences, fatigue crack propagation delay occurred immediately and increased with load block ratio $\Delta\sigma_1/\Delta\sigma_2$ (Fig. 13.23).[38,44,45] Note that high–low block loading in the presence of a notch can be beneficial, whereas the reverse is true in smooth specimen testing (Chapter

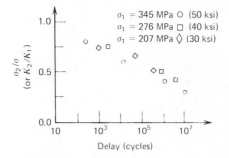

FIGURE 13.23 Delay effect in 7075-T6 aluminum alloy due to reduction in stress range from σ_1 to σ_2. (From Rice[51] with data from Hardrath[43]; reprinted with permission of the American Society for Testing and Materials from copyright work.)

Twelve). In the latter instance, the load interaction had an adverse effect, with $\Sigma n/N < 1$ (see Fig. 12.13). On the basis of these findings, it is tentatively concluded that overload cycling is damaging in initiation-dominated situations (e.g., unnotched samples), and overloads are beneficial in the FCP regime.

Immediate delay in block load tests differed from that found in single peak overload results, but the delay was found again to be temporary; the crack eventually resumed normal growth behavior associated with the applied ΔK stress level. Although they represent only a transient response, load interactions can account for a large percentage of the total fatigue life if the temporary attenuation of crack growth rate is large enough and/or developed a number of times during the service life of a component. To this end, it may be possible in some instances to impose proof test overloads before and during component service to introduce time periods during which crack growth rates would be suppressed, thereby enhancing total service life. Though a proof test may enhance fatigue crack propagation resistance, one must be alert to the potential in certain material-environment systems for environment assisted cracking during hydrostatic proof testing (Chapter Eleven). Consequently, when multiple mechanisms exist for subcritical flaw growth, hydrostatic proof tests should be conducted with great caution.

By contrast, negative peak loads and low–high block loading produce very little change in FCP rates, although a transient *increase* in crack growth rate is found with low–high block load sequencing. It has been argued that little damage is to be expected when negative loads are applied, since they tend to close the crack and make it less harmful. This point was considered in Section 13.4. Negative loads are damaging in a relative sense when they immediately follow a positive load excursion. Apparently, the beneficial effects of positive overloading are minimized by the subsequent negative load excursions. This is unfortunate since the air-to-ground cycle involved in landing an aircraft does represent a compressive overload on the lower portion of the wings. Consequently, whatever beneficial load interaction effects that might have been introduced during the prior flight pattern could be partially wiped out upon landing.

Such load interaction effects are not accounted for by current fatigue crack extension relationships such as Eq. 13-3, or by linear cumulative damage laws (see Chapter Twelve). Since linear cumulative damage laws assume that damage occurs at a rate equal to the percentage of life consumed at a given stress level, such laws predict neither accelerations or decelerations in FCP rates nor the associated decrease or increase in fatigue life. Several models discussed in the next section attempt to overcome this problem.

13.5.1 Load Interaction Models

Several models have been proposed to explain load interaction phenomena. It has been argued by some that crack growth delay is caused by residual compressive stresses acting at the crack tip.[44-46] These compressive stresses are believed to result from elastic unloading of the specimen which squeezes the overload plastic zone to produce a crack tip compressive residual stress field. Willenborg, Engle, and Wood[47] developed a model based on this physical picture. They assumed fatigue crack growth was dependent on an effective stress intensity factor range and defined ΔK_{eff} by the sum of applied and residual stress intensity factor levels:

$$\Delta K_{eff} = \Delta K_{app} + \Delta K_{res} \qquad (13\text{-}11)$$

where

$$\Delta K_{eff, app, res} = \text{effective, applied, and residual stress intensity}$$

$$\text{levels, respectively}$$

For the tensile overload condition, ΔK_{res} is negative such that $\Delta K_{eff} < \Delta K_{app}$. Other refinements and additional models have been proposed to predict delay behavior,[48,49] but no completely satisfactory model is available to date.

Residual stress arguments were used by Elber[50] to explain fatigue crack growth behavior of shallow flaws embedded within a shot peened surface layer. He found that if residual compressive stresses near the surface (Fig. 12.17) were present, shallow cracks would grow more slowly than if the residual stress field were absent. Conversely, deeper cracks grew more rapidly once the crack had penetrated into the subsurface tensile stress field. Furthermore, the apparent K_{IC} value for the shot peened D6AC steel alloy was both higher and lower than that found in samples that were not shot peened when the crack was embedded in the compressive and tensile residual stress zones, respectively.

Fatigue crack growth delay has also been attributed to crack tip blunting resulting from large load excursions.[51] Since the blunted crack would represent a less severe crack tip state, additional fatigue damage should be less severe and the crack should grow more slowly—consistent with experimental findings.

Unfortunately, this model runs afoul of the observed phenomenon of delayed retardation. Were crack tip blunting to be a viable mechanism, it would be expected that maximum delay would occur immediately after the overload cycle, rather than after the crack had propagated part way into the overload zone.

A third model proposed to explain load interaction effects is based on crack closure effects[26] (see Section 13.4 for a discussion of this subject). It is argued that permanent tensile displacements resulting from the overload cycle(s) generate crack surface interference in the wake of the advancing crack front.* Not only does this model explain attenuated crack growth rates caused by a temporary elevation in the opening stress intensity level, but it can predict the delayed retardation found in single peak overload tests. In addition, it also predicts immediate delay in the case of high–low block loading situations. From Fig. 13.24a, the tensile displacements associated with cycling at ΔK_1, which produce an opening level of K_{op}, are already behind the crack tip and act to reduce the ΔK_{eff} level (in this case to zero). Also, the transient increase in crack growth rates associated with a low–high block loading sequence may be explained with the aid of the closure model (Fig. 13.24b). Here, during the first few cycles of the high load block, the material experiences a low opening level and hence a greater ΔK_{eff}. Once these larger displacements begin to interfere in the wake of the crack front, the opening level begins to rise and eventually reaches its new equilibrium value.

Additional experimental findings lend further credence to the crack closure model. It has been shown that when the fracture surfaces in the wake of the fatigue crack are removed with a narrow grinding wheel, the crack growth rate upon subsequent load cycling is higher than before the machining operation.[39,41] Obviously, this effect may be rationalized in terms of the elimination of fracture surface material that was causing interference. It would be expected that if crack surface interference really does occur then some evidence of abrasion should be found on the fracture surface. As shown in Fig. 13.25 for both single-peak overload and high–low block loading sequences, extensive abrasion and obliteration of fracture surface detail is readily apparent. (Note the large striation or stretch band associated with the single-peak overload in Fig. 13.25a.)

As mentioned in Section 13.4, the importance of crack closure may be restricted to the plane stress regime of the fatigue crack propagation process, where greater plastic deformation is found. Indeed, the substantially greater delays found upon overloading at higher ΔK levels (Table 13.3) may well reflect a larger crack closure effect associated with plane stress conditions at the higher ΔK level. This hypothesis was substantiated by additional overload tests performed on the same alloy, at the same ΔK levels, but with different sheet

---

*Since crack surface interference generates compressive stresses, it is difficult to distinguish between crack closure and residual compressive stress contributions to FCP delay. Consequently, although the discussion in this section treats the two concepts separately, it should be recognized that they are interrelated.

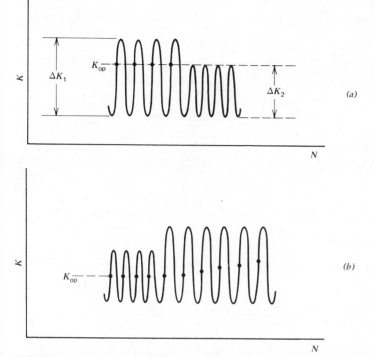

FIGURE 13.24   Block loading. (a) High–low block loading situation where $K_{max(low\ block)} = K_{op(high\ block)}$. This would result in immediate crack arrest. (b) Low–high block loading situation that would result in transient high growth rates at beginning of high block cycling. Subsequently, crack interference would raise closure stress to equilibrium level.

thicknesses. From Fig. 13.26, it is seen that much larger delays are encountered by overloading thin sheets (plane stress condition) than in the thicker samples (plane strain condition). Consistent with these findings is the observation that the extent of delay increases with decreasing yield strength (i.e., larger plastic zones) for a given set of overload test conditions.[49]

13.6   ENVIRONMENTALLY ENHANCED FCP (CORROSION FATIGUE)

Recalling from Chapter Eleven that cracks can grow in many materials as a result of sustained loading conditions in an aggressive environment, it is not surprising to find that fatigue crack propagation rates also are sensitive to

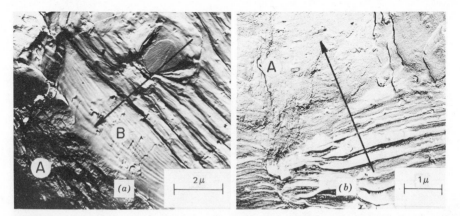

FIGURE 13.25 Abrasion in regions A resulting from overload cycling. (*a*) Single-peak overload. Note stretch zone at B. (*b*) High–low block loading sequence. Arrow indicates crack direction.[38] (Reprinted with permission of the American Society for Testing and Materials from copyright work.)

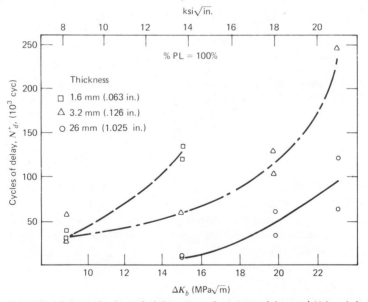

FIGURE 13.26 Cycles of delay as a function of base ΔK level for single-peak overloads. Delay increases with decreasing thickness in 2024-T3 aluminum alloy.[40] (Reprinted with permission from W. J. Mills and R. W. Hertzberg, *Engineering Fracture Mechanics*, 1975, Pergamon Press.)

environmental influences. An example of this behavior, the effect of water partial pressure on fatigue behavior in 7075-T6 aluminum alloy, is shown in Fig. 13.27.[52] For the two stress intensity levels examined, a large shift in FCP rates is noted over a small change in water content, resulting in an order of magnitude change in crack growth between very dry and wet test conditions. Tests conducted on several aluminum alloys in other environments, such as wet and dry oxygen, wet and dry argon and dry hydrogen, indicate that enhanced crack growth in aluminum alloys is due to the presence of moisture.[52–54] This is consistent with static test results reported in Section 11.1.1. This has led investigators to reexamine the relative fatigue behavior of many engineering alloys to determine the relative contribution of environmental effects. For example, earlier test results generated in uncontrolled laboratory environments (Fig. 13.3) indicated a marked superiority in fatigue performance of 2024-T3 versus 7075-T6 aluminum alloys. By conducting tests in these two alloys in both dry and wet argon atmospheres, Hartman et al.[52,53] determined that FCP differences in these alloys were minimized greatly by eliminating moisture from the test environment. Consequently, the superiority of 2024-T3 over 7075-T6 in uncontrolled test atmospheres is due mainly to a much greater environmental sensitivity in 7075-T6. This is consistent with the fact that 7075-T6 is more susceptible to environment assisted cracking than 2024-T3.

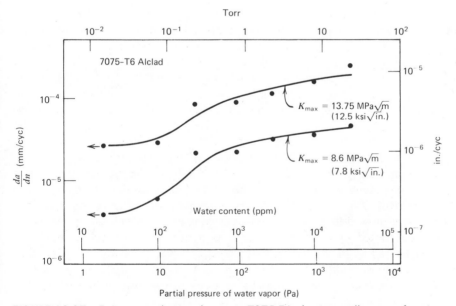

FIGURE 13.27 Fatigue crack growth rate in 7075-T6 aluminum alloy as a function of water vapor content in test atmosphere. Data from Ref. 52 and redrawn by Wei.[54] (Reprinted from R. P. Wei, *Engineering Fracture Mechanics*, 1970, Pergamon Press.)

During the past few years, many more material-environment systems have been identified as being susceptible to corrosion fatigue. It is found that many aluminum, titanium, and steel alloys are adversely affected during fatigue testing by the presence of water, and titanium and steel alloys (but not aluminum alloys) are affected by dry hydrogen. In these studies, test frequency, load ratio, load profile, and temperature have been identified as major variables affecting FCP response of a material subjected to an aggressive environment. For example, the harmful effects of a 3.5% saline solution on fatigue performance in a titanium alloy are shown in Fig. 13.28 as a function of test frequency. As might be expected, FCP rates increase when more time (i.e., lower frequencies) is allowed for environmental attack during the fatigue process. It should be pointed out that no important frequency effects are found in metals when tested in an inert atmosphere. Also note the negligible environmental effect on FCP at high crack growth rates where the mechanical process of fatigue damage is probably taking place too quickly for chemical effects to be important.

Fatigue crack growth rate sensitivity to environment and test frequency has also been found in ferrous alloys (Fig. 13.29). By describing the fatigue behavior of a 12 Ni-5 Cr-3 Mo maraging steel when fatigued in a 3% NaCl solution by the relationship

$$\frac{da}{dn} = D(t)\Delta K^2 \tag{13-12}$$

where $D(t) =$ material parameter dependent on the material-environment system, Barsom showed $D(t)$ to be a strong function of frequency (Fig. 13.30) and also found this parameter to increase with decreasing values of K_{IEAC}.[*56,57] What is most intriguing about these data is the fact that they reveal a significant environmental sensitivity, even though all tests were conducted with K_{max} being maintained *below* the K_{IEAC} level for the material. As a result, several inconsistencies are apparent which are hard to rationalize. First, why should there be any environmental effect during fatigue if the tests were conducted below K_{IEAC}? Perhaps a protective film, developed at the crack tip under sustained loading conditions acts to protect the material from the environment but is ruptured by fatigue cycling, thereby permitting the corrodent to reattack the crack tip region. Such a hypothesis is supported by observations made by Bucci,[55] who showed that environment assisted cracking does occur below K_{IEAC} when the arrested crack is subjected to a period of load cycling and then reloaded below the previously established K_{IEAC} level. This would suggest that constant loading conditions used to obtain a K_{IEAC} value for a given material represent a *metastable* condition, easily upset by the imposition of a number of load fluctuations.

---

*Recall from Chapter Eleven that the threshold level for stress corrosion cracking, K_{ISCC}, has been redefined by the more general term for environment assisted cracking, K_{IEAC}.

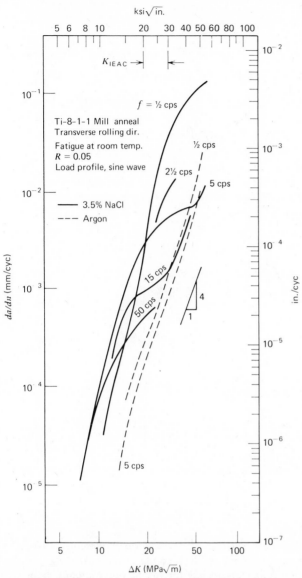

FIGURE 13.28 Effect of frequency on fatigue crack growth in Ti-8 Al-1 Mo-l V alloy in 3.5% NaCl and argon atmospheres.[55] (Courtesy of R. J. Bucci, Alcoa Research Laboratories.)

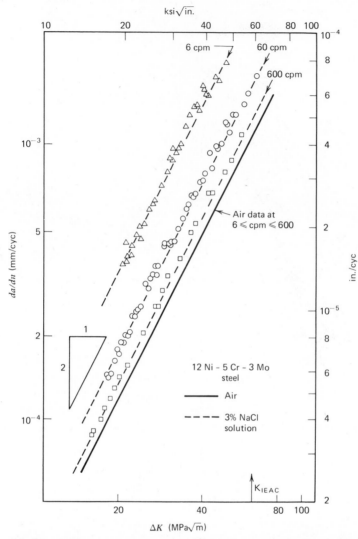

FIGURE 13.29 Corrosion fatigue (3% NaCl solution) crack growth data in 12 Ni-5 Cr-3 Mo steel as a function of test frequency. All tests conducted with $\Delta K < K_{IEAC}$.[57] (Reprinted with permission of the American Society for Testing and Materials from copyright work.)

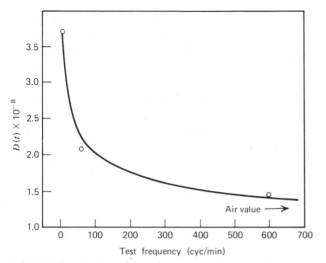

FIGURE 13.30 Correlation between time dependent function $D(t)$ and fatigue test frequency.[56] (Reprinted with permission from J. M. Barsom, *Engineering Fracture Mechanics*, 1971, Pergamon Press.)

A second difficulty arises when one attempts to extrapolate the data in Fig. 13.30 to establish $D(t)$ values at very low cyclic frequencies. It would appear that $D(t)$ would rise asymptotically to extremely high values at ever decreasing test frequencies, but boundary conditions imposed by the environment assisted cracking test itself (one cycle in 1000 hr) require that $D(t)=0$ whenever $K_{max} < K_{IEAC}$. Consequently, expansion of the data base in Fig. 13.30 in the low frequency region would be expected to reveal a frequency where $D(t)$ is maximized and below which $D(t)$ would drop precipitously to zero.

The effect of environment on crack propagation in ceramic materials is profound relative to the effect of cyclic loading. In fact, Evans and Fuller[58] found that load cycling produces no significant enhancement of environmentally induced crack growth in porcelain and glass. Restating this experimental finding in different terms, it is seen that without an aggressive environment, ceramic materials are not subject to pure cyclic damage; that is, they do not fatigue. This should come as no surprise since fatigue damage involves localized yielding, and most ceramics possess negligible plastic deformation capacity.

Corrosion fatigue is found to be sensitive to subtleties of the cyclic wave form. For the 12 Ni-5 Cr-3 Mo steel-3% NaCl solution material-environment system, Imhof and Barsom[57] showed that when square and negative sawtooth wave forms were used, the material responded as though the tests were conducted in air. On the other hand, when the load profile was that of a sine, triangle, or positive sawtooth wave form, a strong environmental influence of the salt solution was observed (Fig. 13.31). By contrast, no effect of load profile was detected when the steel was fatigued in air. On the basis of these and other supporting results, these investigators concluded that environmental attack occurred only during the tensile loading portion of the loading cycle, or when plastic deformation occurs, and was not affected by hold time. They found that when the load rise time was small, the influence of an aggressive environment was minimized and vice-versa.

Although the effect of mean stress on FCP in the intermediate growth rate regime was found generally to be of secondary importance (Section 13.4), it does become a major variable during corrosion fatigue conditions (Fig. 13.32). From these data, it would appear that high R ratio conditions enhance the corrosion

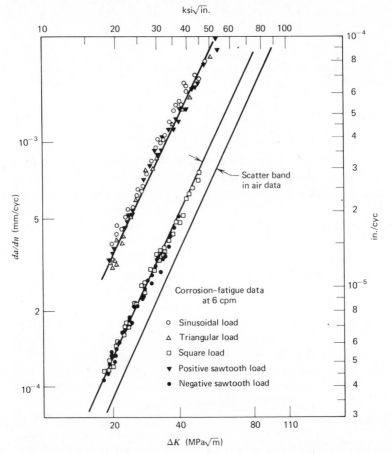

FIGURE 13.31 Effect of cyclic wave form on fatigue crack propagation in 12 Ni-5 Cr-3 Mo steel in 3% NaCl solution. Negligible environmental effect is noted in square and negative sawtooth wave form.[57] (Reprinted with permission of the American Society for Testing and Materials from copyright work.)

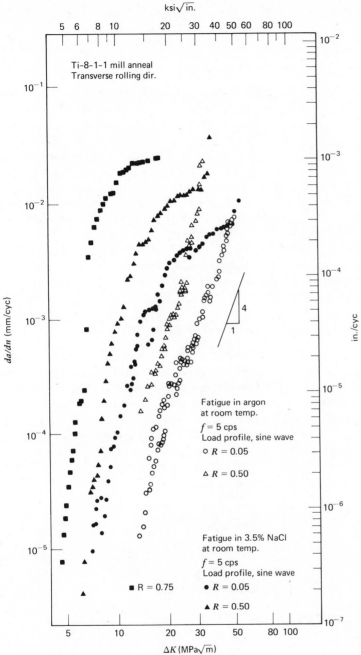

FIGURE 13.32 Effect of load ratio R on fatigue crack propagation in Ti-8 Al-1 Mo-1 V alloy. Tests conducted in 3.5% NaCl solution and in argon.[55] (Courtesy of R. J. Bucci, Alcoa Research Laboratories.)

component of crack growth, while low R ratio testing reflects more of the intrinsic fatigue response of the material. The greater importance of mean stress effects during environmentally enhanced fatigue crack propagation may be rationalized with the aid of the superposition model described in the next section.

As might be expected, the other major test variable relating to corrosion fatigue is that of test temperature. Many investigators have found FCP rates to increase with increasing temperature, as shown in Fig. 13.33 for A212B steel. For many years, a controversy has existed concerning the origin of this FCP temperature sensitivity. Is it due to a creep component or to an environmental component, both of which increase with increasing test temperature? In a series of recent experiments conducted at elevated temperatures in inert environments and in vacuum,[60-62] it was shown that neither temperature nor frequency had any effect on fatigue crack propagation rates. In fact, test results were comparable to room-temperature results. This was the case even when the inert environment was liquid sodium.[62] On the basis of these results, it is concluded that higher FCP rates at higher temperatures mainly result from material-environment interactions, rather than a creep contribution.

13.6.1   Corrosion Fatigue Superposition Model

Wei and Landes[63] and Bucci[55] developed a model that would account for effects of environment, test frequency, wave form, and load ratio on corrosion fatigue crack propagation behavior. They approximated the total crack extension rate under corrosion fatigue conditions by a simple superposition of the intrinsic fatigue crack growth rate (determined in an inert atmosphere) and the crack extension rate due to a sustained load applied in an aggressive environment (determined as the environment assisted crack growth rate). Therefore

$$\left(\frac{da}{dn}\right)_T = \left(\frac{da}{dn}\right)_{fat} + \int \frac{da}{dt} K(t)dt \tag{13-13}$$

where

$\left(\dfrac{da}{dn}\right)_T$ = total corrosion fatigue crack growth rate

$\left(\dfrac{da}{dn}\right)_{fat}$ = fatigue crack growth rate defined in an

   inert atmosphere

$\left(\dfrac{da}{dt}\right)$ = crack growth rate under sustained loading

$K(t)$ = time-dependent change in stress intensity factor

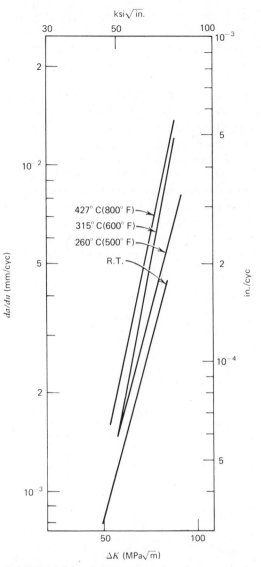

FIGURE 13.33   Schema revealing fatigue crack growth in A212B steel at various test temperatures.[59] (Reprinted with permission of the American Society for Testing and Materials from copyright work.)

Two important aspects of this model should be emphasized. First, its linear character implies that there is no interaction (or synergism) between the purely mechanical and environmental components. Second, the model also depends on the assumption that the same mechanisms control the fracture process in both environment assisted cracking and corrosion fatigue. For example, in the case of steels, both processes are believed to be controlled by hydrogen embrittlement. The application of the superposition model is demonstrated with the aid of Fig. 13.34. The FCP of the material in an inert environment $(da/dn)_{fat}$ is determined first and plotted in Fig. 13.34a. Next, the sustained loading crack growth rate component developed during one load cycle is obtained by integrating the product of the environment assisted cracking rate da/dt and the time-dependent change in stress intensity level $K(t)$ over the time period for one load cycle (Fig. 13.34b,c,d). This increment then is added to $(da/dn)_{fat}$ to obtain $(da/dn)_T$, the corrosion fatigue crack growth rate. Modest success has been achieved when this model is used to analyze some titanium and ferrous alloys; predicted values of $(da/dn)_T$ have been in reasonable agreement with experimentally determined date (Fig. 13.35).

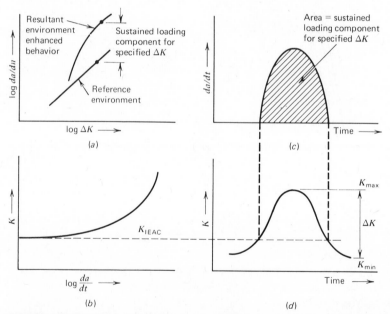

FIGURE 13.34 Schematic diagram of superposition model. (a) Fatigue behavior in both reference and aggressive environments; (b) sustained loading environment cracking behavior; (c) K versus time for one cycle of loading; (d) sustained loading crack velocity versus time. Note that the model cannot predict environment enhanced fatigue crack propagation below K_{IEAC}.[55] (Courtesy of R. J. Bucci, Alcoa Research Laboratories.)

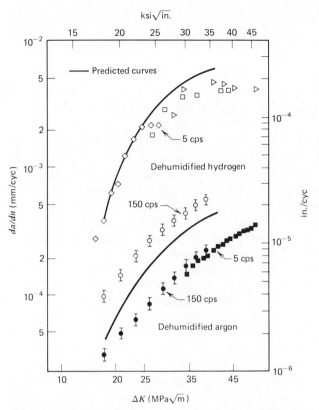

FIGURE 13.35 Application of superposition model to fatigue crack growth behavior in 18Ni (250) maraging steel tested in dehumidified argon and hydrogen.[63] (Reprinted with permission of the American Society for Testing and Materials from copyright work.)

Unfortunately, there are limitations to the use of the model, as well as problems in explaining certain data trends. Most importantly, the superposition model fails to predict environmentally enhanced fatigue crack growth behavior where $K_{max} < K_{IEAC}$ (see Fig. 13.29). From Chapter Eleven and Fig. 13.34b, this would correspond to a condition where $da/dt = 0$. Indeed, the environmental effect on fatigue crack growth is greatest at these low stress intensity levels (Fig. 13.28). Perhaps the problem of applying the superposition model when $K_{max} < K_{IEAC}$ may not be so much a problem with the model but rather with the definition of K_{IEAC}, since some sustained crack growth does occur when $K < K_{IEAC}$ if the crack is resharpened by fatigue cycling. At any rate, basic changes in material response occur above and below K_{IEAC}. Below K_{IEAC}, any hold time would not be expected to contribute any additional crack growth increment, but the reverse would be true above this level. Consequently, for the present, the superposition model should be used with caution and in the regime where $K > K_{IEAC}$.

13.7 METALLURGICAL ASPECTS OF FCP

Many studies conducted in the intermediate growth rate regime reveal that monotonic yield strength, thermomechanical treatment, and preferred orientation do not have a pronounced effect on FCP rates in aluminum and steel alloys; that is, fatigue crack propagation is relatively structure insensitive.* Such a generalization must be tempered somewhat by the fact that alloy chemistry and microstructure do influence corrosion fatigue behavior. However, the latter sensitivity may be interpreted as a metallurgical effect on environment assisted cracking susceptibility; it affects total fatigue behavior as described in Section 13.6 by the superposition model. (For example, the dominant reasons why 7075-T6 has poorer fatigue resistance than 2024-T3 may be traced to the greater environmental sensitivity of the 7075-T6 alloy.)

It is interesting to find, however, that FCP rates in engineering alloys are dependent on a structure insensitive property—the modulus of elasticity.[65] By normalizing the stress intensity factor range ΔK by E, differences in fatigue response such as those shown in Fig. 13.3 are largely eliminated (Fig. 13.36). Similarly, normalization of fatigue striation spacing data from different alloys is possible (Fig. 13.14) and, as shown in Section 13.3.1, is useful in estimating stress intensity values for any material. Since at least some materials show a ΔK^2 dependence of crack growth rate and are dependent on $1/E$, it has been suggested that FCP rates may be related to $\mathcal{G}$ or the crack opening displacement. The potential of such relationships needs to be explored more fully.

Metallurgical factors do affect fatigue crack propagation rates at high ΔK levels. This is because as ΔK becomes very large, K_{max} approaches K_{IC} or K_c where local fractures occur with increasing frequency and produce accelerated growth (recall Eq. 13-8). Since tougher materials are typically cleaner and will exhibit fewer local instabilities, their crack growth rates should be lower at high K levels. This is consistent with the well-established rule of thumb regarding low cycle fatigue (analogous to high FCP rates); low cycle fatigue resistance is enhanced by improvements in toughness and/or ductility. A number of investigators have verified this relationship and have rationalized differences in macroscopic and microscopic crack growth rates. For example, in the high ΔK regime, FCP rates in a banded steel [consisting of alternate layers of ferrite and pearlite (Fig. 13.37)] increased when tested in the arrester, divider, and short transverse

---

*Recent studies have shown the FCP response of titanium alloys to be sensitive to certain metallurgical variables. See the review by Stoloff and Duquette.[64]

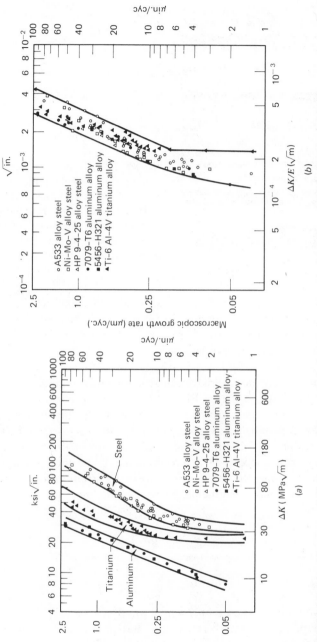

FIGURE 13.36 Macroscopic fatigue crack growth behavior in several aluminum, titanium and steel alloys. Data plotted versus (a) ΔK; (b) $\Delta K/E$.[23] (Reprinted with permission of the American Society of Metals. Copyright 1969.)

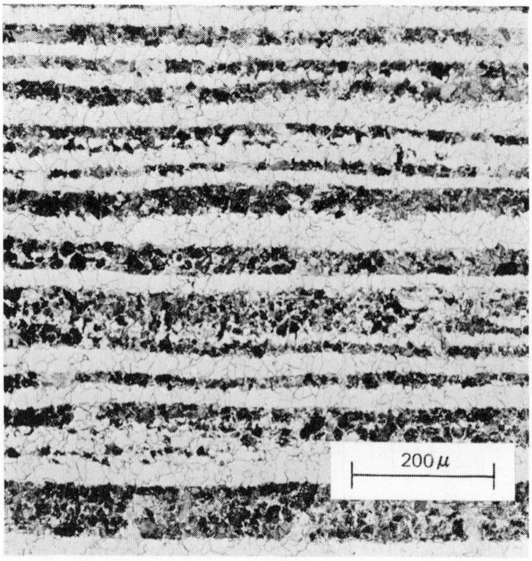

FIGURE 13.37 Microstructure in banded steel revealing alternating layers of ferrite and pearlite.[66] (From F. Heiser and R. W. Hertzberg, *Transactions of the ASME*, 1971, with permission of the American Society of Mechanical Engineers.)

directions, respectively (Fig. 13.38).[66] Although little difference in fatigue response was found at low ΔK levels among the three orientations tested, a 40-fold difference in macroscopic crack growth rate was observed in going from the least to the most damaging loading direction.

It is important to note that FCP anisotropy was also found in homogenized samples of this alloy when the layered microstructure was eliminated [but not the alignment of sulfide particles (see Fig. 4.17)]. Since the *microscopic* growth rate (i.e., fatigue striation spacings) was the same in the three crack plane orientations, it was concluded that the anisotropy in *macroscopic* FCP was related to different amounts of sulfide inclusion fracture in the three orientations. Consequently, macroscopic FCP is considered to be the summation of several fracture mechanisms, the most important being striation formation and local fracture of brittle microconstituents. From this, macroscopic growth rates may be described by

$$\left(\frac{da}{dn}\right)_{\text{macro}} = \sum Af(K)_{\substack{\text{striation} \\ \text{mechanism}}} + Bf'(K)_{\substack{\text{void} \\ \text{coalescence}}}$$

$$+ Cf''(K)_{\text{cleavage}} + Df'''(K)_{\substack{\text{corrosion} \\ \text{component}}} + \cdots \quad (13\text{-}14)$$

From Eq. 13-14, we know at least that $A \approx 6/E^2$ and $f(K) \approx \Delta K^2$.

In another attempt to correlate macroscopic and microscopic crack growth rates above the cross-over point, Bates[67] adjusted Eq. 13-7 so that

$$\left(\frac{da}{dn}\right)_{\text{macro}} \approx \frac{6}{f_s}\left(\frac{\Delta K}{E}\right)^2 \quad (13\text{-}15)$$

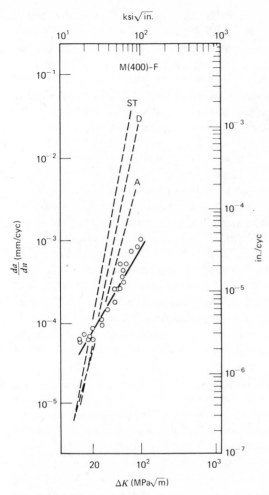

FIGURE 13.38 Fatigue crack propagation in banded steel as a function of specimen orientation. (ST, short transverse; D, divider; A, arrester geometry). Striation spacing (data points and solid line) is seen to be independent of specimen orientation.[66] (From F. Heiser and R. W. Hertzberg, *Transactions of the ASME*, 1971, with permission of the American Society of Mechanical Engineers.)

where f_s = percentage of striated area on the fracture surface. This relationship is consistent with observations by Broek[68] and Pelloux et al.,[69] who also found increasing amounts of particle rupture and associated void coalescence on the fracture surface with increasing stress intensity levels. The latter investigators noted a marked increase in particle rupture when the plastic zone dimension grew to a size comparable to the particle spacing. Although Eq. 13-15 provides a rationale for differences in macroscopic and microscopic FCP rates, it may be too impractical to use because an extensive amount of fractographic information is required.

Monotonic yield strength variations have been shown to exert little influence on fatigue crack growth behavior. As shown in Fig. 13.39, a number of steel alloys were found to possess comparable FCP rates even though their respective yield strengths differed markedly.[70] Similarly, only modest shifts in the slopes of $\log \Delta K - \log da/dn$ plots were found in studies of brass and stainless steel alloys in both cold-worked and annealed condition where four- to ten-fold differences in monotonic yield strength were reported.[21] Fairly strong crystallographic textures were developed in both the cold-worked and recrystallized conditions, but, again, little effect was noted on FCP response for both brass and steel specimens oriented so as to present maximum densities of {111}, {110}, and {100} crystallographic planes, respectively, on the anticipated crack plane. It was noted, however, that the actual fracture plane and crack direction were affected strongly by crystallographic texture in that the gross crack plane avoided a {111} orientation,[21,71] consistent with expectations of the striation formation model discussed in Section 13.3.

It has been suggested that FCP does not depend on typical tensile properties because monotonic properties are not the controlling parameter. Instead, cyclically stabilized properties may hold the key to fatigue crack propagation behavior. Starting or monotonic properties between two given alloys may differ widely, but their final or cyclically stabilized properties would not. For example, soft alloys would strain harden and hard ones would strain soften; as a result, the materials would be more similar in their final state than at the outset of testing. Consequently, if fatigue crack propagation response were dependent on cyclically stabilized properties, smaller differences in FCP behavior would be expected than that based on a comparison of monotonic values. A number of studies[72–74] have been conducted to establish correlations between cyclic strain and FCP data. For example, the slope m of the $da/dn - \Delta K$ plot is seen to decrease with increasing cyclic yield strength σ_0' and cyclic strain hardening exponent n'.[73] Although it is encouraging to find such a correlation, which may well represent a bridge between cyclic strain and fatigue crack propagation studies, more work is needed before it will be possible to predict FCP rates from cyclic strain data. For one thing, it is not clear from Fig. 13.40 whether a high or low slope is desirable for optimum fatigue performance.[75] Obviously, the intercept "A" from Eq. 13-3 is equally important in this determination. For example,

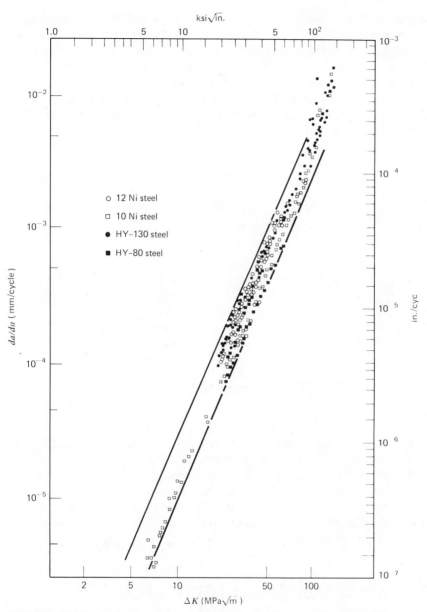

FIGURE 13.39 Fatigue crack growth behavior in several martensitic steels.[57] (Reprinted with permission of the American Society for Testing and Materials from copyright work.)

alloy A would be better than alloy C but alloy D, which has the same slope as A, would be worse than C. Furthermore, the choice between alloy B and C would depend on the anticipated crack growth rate regime for the engineering component. If many fatigue cycles were anticipated, the designer should opt for alloy B, since fatigue cracks would propagate more slowly over most of the component service life and allow for much greater fatigue life. In a low cycle fatigue situation, representative of conditions to the right of the cross-over point, alloy C would be preferred.

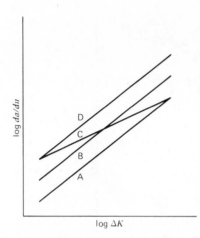

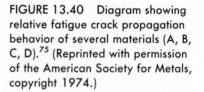

FIGURE 13.40 Diagram showing relative fatigue crack propagation behavior of several materials (A, B, C, D).[75] (Reprinted with permission of the American Society for Metals, copyright 1974.)

In summary, fatigue crack propagation response in metal alloy systems is far more sensitive to environmental and load interaction effects than to metallurgical variations. Since metallurgical factors appear to affect $S-N$ response (Chapter Twelve) but not FCP behavior, these results suggest that metallurgy influences fatigue crack initiation more than the propagation stage. The relative importance of mean stress on FCP is complex, depending on test environment and the ΔK regime of the test.

13.8 FATIGUE OF ENGINEERING PLASTICS

As we see in Fig. 13.41, fatigue failures in polymers may be induced either by large-scale hysteretic heating, resulting in actual polymer melting, or by fatigue crack initiation and propagation to final failure.[76] Over the years, the basic difference and importance of polymer fatigue failures induced by these two processes has become a source of controversy among researchers. Consequently, the remainder of this chapter is devoted to this intriguing subject.

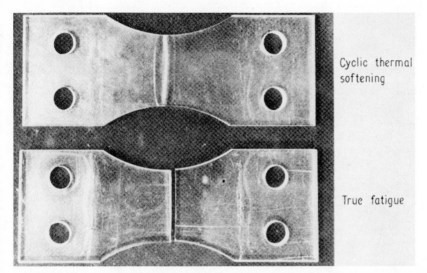

Cyclic thermal
softening

True fatigue

FIGURE 13.41   Typical fatigue and cyclic thermal softening failures in poly(methyl methacrylate).[76] (Reproduced by courtesy of The Institution of Mechanical Engineers from an article by I. Constable, J. G. Williams, and D. J. Burns from *JMES, 12*, p. 20, 1970.)

13.8.1   Thermal Failure

The major cause of thermal failure is believed to involve the accumulation of hysteretic energy generated during each loading cycle. Since this energy is dissipated largely in the form of heat, an associated temperature rise will occur for every loading cycle when isothermal conditions are not met. As shown in Fig. 13.42, the temperature rise can be great enough to cause the sample to melt, thereby preventing it from carrying any load.[77] Failure is presumed, therefore, to occur by viscous flow, although the occurrence of some bond breakage cannot be excluded.

From the work of Ferry,[79] the energy dissipated in a given cycle may be described by

$$\dot{E} = \pi f J''(f, T)\sigma^2 \qquad (13\text{-}16)$$

where

$$\dot{E} = \text{energy dissipated}$$

$$f = \text{frequency}$$

$$J'' = \text{the loss compliance}$$

$$\sigma = \text{the peak stress}$$

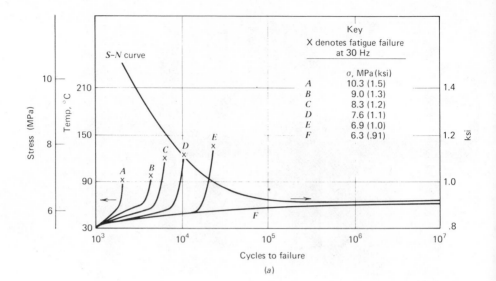

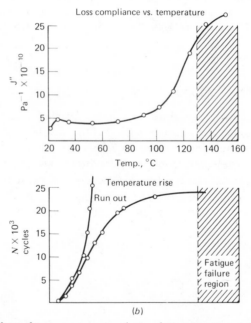

FIGURE 13.42 Effect of temperature rise during fatigue testing, (a) Temperature rise to cause thermal failure at different stress levels (no failure seen in *F* where sample temperature stabilized);[77] (b) loss compliance and temperature rise.[78] (Reprinted with permission of the Society of Plastics Engineers.)

Neglecting heat losses to the surrounding environment, Eq. 13-16 may be reduced to show the temperature rise per unit time as

$$\Delta \dot{T} = \frac{\pi f J''(f, T)\sigma^2}{\rho c_p} \tag{13-17}$$

where

$\Delta \dot{T}$ = temperature change/unit time

ρ = density

c_p = specific heat

Equation 13-17 is useful in identifying the major variables associated with hysteretic heating. For example, the temperature rise is seen to increase rapidly with increasing stress level. Figure 13.42a illustrates a typical curve of stress versus number of cycles to failure for poly(tetrafluoroethylene) (PTFE), along with the superposition of temperature rise curves corresponding to the various stress levels. Note that for all stress levels above the endurance limit (the stress level below which fatigue failure was not observed) the polymer heated to the point of melting (shown by the temperature rise curves A, B, C, D, and E). Evidently, heat was generated faster than it could be dissipated to the surrounding environment. When a stress level less than the endurance limit was applied, the temperature rise became stabilized at a maximum intermediate level below the point where thermal failure was observed (Curve F). These specimens did not fail after 10^7 cycles.

From Eq. 13-17, the rate of temperature rise depends on the magnitude of the loss compliance J'', which is itself a function of temperature. As we see in Fig. 13.42b, the loss compliance rises rapidly in the vicinity of the glass transition temperature after a relatively small change at lower temperatures. Consequently, one would expect the temperature rise in the sample to be moderate during the early stages of fatigue cycling but markedly greater near the final failure time. It is concluded, therefore, that thermal failure describes an event primarily related to the lattermost stages of cyclic life. In further support of this last statement, other tests have been reported wherein intermittent rest periods were interjected during the cyclic history of the sample. In this manner, any temperature rise resulting from adiabatic heating could be dissipated periodically. It would be expected, then, that the fatigue life of specimens allowed intermittent rest periods would be substantially greater than that of uninterrupted test samples. Indeed, several investigators[77,80,81] have shown significant improvement in fatigue life when intermittent rest periods were introduced during testing. On the basis of these test results, it is concluded that linear cumulative damage laws cannot be applied to thermal failures.

Finally, the fatigue lives of polymers subjected to isothermal test conditions are superior to those exhibited by samples examined under adiabatic test conditions, consistent with the absence of hysteretic heating in the former case.[82]

From Eq. 13-17, the fatigue life of a sample should decrease with increasing frequency, since the temperature rise per unit time is proportional to the frequency. Test results shown in Fig. 13.43a clearly support the anticipated effect of this variable. Also, thermal failures are affected by specimen configuration. As mentioned above, the temperature rise resulting from each loading cycle

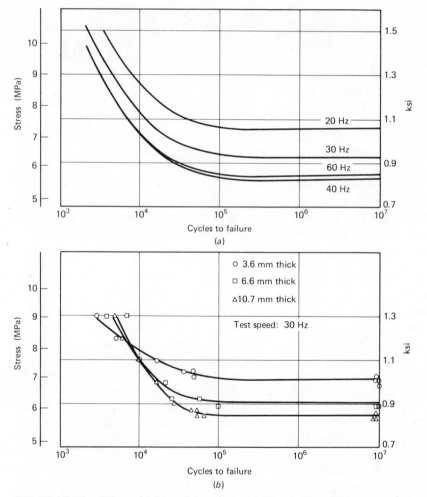

FIGURE 13.43 Effect of (a) test frequency and (b) specimen thickness on fatigue life in Halon TFE.[77] (Reprinted with permission of the Society of Plastics Engineers.)

depends on the amount of heat dissipated to the surroundings. Consequently, the fatigue life of a given sample should be dependent on the heat transfer characteristics of the sample and the specimen surface to volume ratio. For example, the endurance limit in PTFE increases with decreasing specimen thickness (Fig. 13.43b). This sensitivity to specimen shape constitutes a major drawback to unnotched specimen tests involving thermal failure, since the test results are a function of specimen geometry and, therefore, do not reflect the intrinsic response of the material being evaluated.

It is clear from the above that thermal fatigue may be suppressed by several factors, such as limiting the applied stress, decreasing test frequency, allowing for periodic rest periods, or cooling the test sample, and by increasing the sample's surface to volume ratio. It is extremely important to recognize, however, that suppressing thermal fatigue by any of the above procedures may not preclude mechanical failure caused by crack initiation and propagation. The corollary is true though; if stresses are reduced to the point where mechanical failure does not occur, this stress level certainly will be low enough such that thermal failure will not occur either.

Although I choose to treat mechanical and thermal fatigue failures as distinctly different events, there are points of common ground. For example, it would be expected that hysteretic heating would take place within the plastic zone at the tip of a crack. Since this heat source is small compared to the much larger heat sink of the surrounding material, it would be expected that any temperature rise would be limited and restricted to the proximity of the crack tip.

It would appear then that the likelihood of thermal failure would depend on the size of the heated zone in relation to the overall specimen dimensions. When this ratio is large, as in the case of unnotched test bars, thermal failures are distinctly possible. When the ratio is very small, say, in the case of a notched bar, thermal failures would not be expected.

13.8.2 Mechanical Failure—FCP

During the past few years, a number of FCP studies of engineering plastics have been conducted, and data are now available for 15 to 20 different materials.[83] With such a body of data, certain conclusions and generalities may be drawn. As in metals, the FCP rates of polymers are strongly dependent on the magnitude of the stress intensity factor range, regardless of polymer chemistry or long-range architecture (Fig. 13.44). Note the data correlation for both amorphous, crystalline, and rubber-modified amorphous polymers on the same plot of ΔK versus da/dn. In a sense, this is analogous to plots of data from metal alloys possessing various crystal structures (Fig. 13.3). Here, again, fatigue crack growth rates may be correlated with Eq. 13-3. On the basis of these initial results, it was concluded that the superior FCP behavior exhibited by poly-

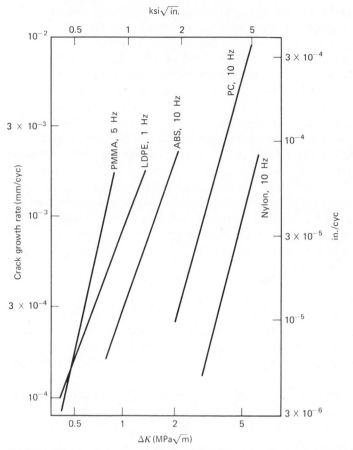

FIGURE 13.44   Relationship between crack growth rate in several polymers as a function of stress intensity factor ΔK. Poly(methyl methacrylate) (PMMA), low-density polyethylene (LDPE), acrylonitrile-butadiene-styrene (ABS), polycarbonate (PC), and Nylon 66.[84] (Reproduced with permission from *Journal of Materials Science*, **5**, 1970.)

carbonate and Nylon 66 was related to the availability of significant energy dissipation mechanisms that are linked to main chain segmental motions.[84] For the case of Nylon 66, superior FCP behavior also may have been due to crystalline regions acting in a manner that would retard crack advance, for example, by deformation and disruption of the crystallites. Indeed, Vadimsky et al.[85] have shown that greater local strains are observed in the crystalline regions of polyethylene than in the amorphous matrix regions. To explore the potentially beneficial role of crystalline regions in FCP, additional tests were conducted[86] on four other polymers: Nylon 6, poly(vinylidene fluoride) (PVF2), polyacetal

(Celcon), and poly(tetrafluorethylene) (Teflon)—all possessing significant crystalline regions. Nylon 6 was found to be almost as good as Nylon 66, and PVF2 proved to be the best polymer investigated thus far (Fig. 13.45). (Recent studies in my laboratory have shown Celcon to exhibit FCP rates even lower than those in PVF2 for the same ΔK level.) Finally, Teflon proved to be so fatigue resistant that stable cracks could not be grown at otherwise reasonable loads.

One may speculate whether polystyrene (PS) and poly(methyl methacrylate) PMMA, the worst materials shown in Fig. 13.45, possess this dubious distinction as a result of their low fracture energy, amorphous structure, and/or tendency to craze. Since PMMA and PS possess comparably low fracture toughness

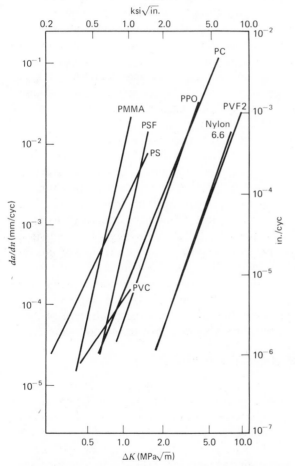

FIGURE 13.45 Comparative fatigue crack propagation behavior in crystalline and amorphous polymers.[87] (Reprinted with permission of the Society of Plastics Engineers.)

$(K_{IC} = 1.32$ and 1.76 MPa$\sqrt{m}$, respectively),[88,89] are both amorphous, and craze readily, comparable FCP rates might be expected. As seen in Fig. 13.45, this was shown to be the case. In pursuing the matter further, it would appear that fracture toughness represents the most important parameter of the three. For example, polycarbonate (PC) which is amorphous, does not craze readily at room temperature, and is fracture tough, has excellent fatigue behavior. On the other hand, polysulfone (PSF), similar to PC in that it is amorphous and does not craze readily at room temperature,[90] has much lower fracture toughness *and* fatigue resistance. Consequently, a material does not necessarily have to craze readily to possess inferior fatigue behavior. In fact, in the case of styrene based polymers, the reverse is true. Using PS as the reference point, it is seen from Fig. 13.46 that cross-linking adversely affects fatigue behavior, but mechanically mixing micron-sized rubber spheres (HIPS) or polymer blending submicron-sized butadiene spheres (ABS) in the polystyrene matrix produces a more fatigue resistant material. From Chapter Six, we may conclude that crazing would be suppressed by cross-linking, while rubber modification enhances craze development on a broader scale. Since fracture toughness in these materials varies directly with total craze volume, one may conclude that increasing fracture toughness holds the key to improvement in fatigue behavior. Whether higher toughness is achieved through enhanced localized deformation (crazing) or homogeneous deformation (viscous flow) appears to be less important than the fact that the toughness is increasing. This latter point is confirmed by the cross-plot in Fig. 13.47, which reveals that the higher the toughness of the material, the greater the ΔK level necessary to drive the crack forward at an arbitrary growth rate. As such, polymer FCP sensitivity to microstructure and fracture toughness represents a significant departure from results in metal alloy systems, where FCP response was found to be relatively insensitive to metallurgical variables and fracture toughness levels.[64,70] In this regard, a comparison between PC and PMMA fatigue behavior provides a useful example of this fact. Whereas the FCP behavior of all metal alloys may be lumped together in one band of data by normalizing ΔK with respect to the modulus of elasticity of the respective alloy, this is not possible when PC and PMMA are compared. Since both materials possess comparable moduli of elasticity, normalization of ΔK is unnecessary, leaving a 1300-fold difference in crack growth rate to be explained.

13.8.2.1 POLYMER FCP FREQUENCY SENSITIVITY

One is faced with an interesting challenge when trying to explain the effect of test frequency on polymer fatigue performance. Although hysteretic heating arguments appear sufficient to explain a *diminution* of fatigue resistance with increasing cyclic frequency in *unnotched* polymer test samples, the fatigue resistance of several polymers in the *notched* condition is *enhanced* with increasing cyclic frequency. Note the pronounced decrease in FCP rate with increasing test frequency for a given ΔK level in polystyrene (Fig. 13.48a). As summarized

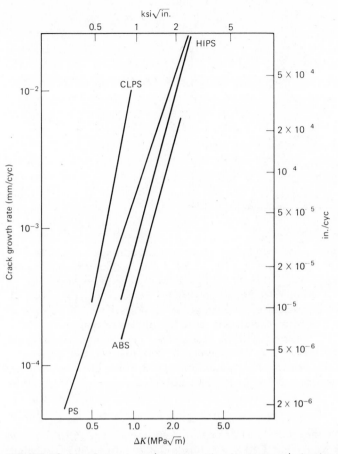

ksi√in.

FIGURE 13.46 Fatigue crack propagation in styrene derivative polymers: cross-linked (CLPS), high-impact (HIPS), regular grade (PS), and acrylonitrile-butadiene-styrene (ABS).

in Table 13.4, some engineering plastics exhibit a large frequency sensitivity factor (defined as the multiple by which the FCP rate changes per decade change in frequency) and others do not.

One intriguing correlation is apparent from these data: Every polymer examined that exhibited a strong FCP frequency sensitivity also exhibited a strong tendency to undergo heterogeneous deformation by crazing; negligible frequency sensitivity was associated with materials reported to be much more resistant to crazing (at least under the conditions experienced during the test program). In some manner, then, crazing propensity is tied to the sensitivity of crack growth rate to frequency. To examine the generality of this statement

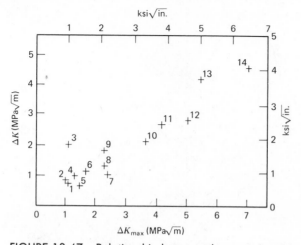

FIGURE 13.47 Relationship between the stress intensity factor range, ΔK (ΔK level corresponding to an arbitrary value of da/dn, 7×10^{-4} mm/cyc), and the maximum stress intensity factor range ΔK_{max} observed at failure for a group of polymers. The polymers are (1) cross-linked polystyrene, (2) poly(methyl methacrylate), (3) poly(vinyl chloride), (4) low-density polyethylene, (5) polystyrene, (6) polysulfone, (7) high-impact polystyrene, (8) acrylonitrile-butadiene-styrene resin, (9) chlorinated polyether, (10) poly(phenylene oxide), (11) Nylon 6, (12) polycarbonate, (13) nylon 66, and (14) poly(vinylidene fluoride).[83] (From J. A. Manson and R. W. Hertzberg, *CRC Critical Reviews in Macromolecular Science, 1* (4), pp. 433 ff., 1973, © The Chemical Rubber Co., 1973. Used by permission of The Chemical Rubber Co.)

more closely, consider the FCP data for cross-linked polystyrene (CLPS) shown in Fig. 13.48*b*. It is clear that CLPS exhibits a smaller frequency sensitivity factor (1.5) than PS (2.3), which is consistent with a more restricted crazing tendency in the cross-linked material. These data provide further confirmation that frequency sensitivity is related to the crazing tendency of the material. To examine the potential role of environment on FCP in PMMA, PVC, and polycarbonate (PC), tests have been conducted in water, laboratory air, and dry nitrogen at several test frequencies. The results indicate that for the material-environment frequency combinations examined, frequency sensitivity was not related to environmental effects. Instead, the frequency sensitivity in a given polymer could reflect a competition between strain rate and creep effects. A change in strain rate will affect the stiffness of the material as well as its yield strength, especially with highly viscoelastic materials. Correspondingly, changes in test frequency alter the time at load and the number of load applications per unit time. This would affect the amount of polymer creep.

An intriguing correlation has been found between the relative FCP frequency sensitivity in polymers and the frequency of movement of main chain segments

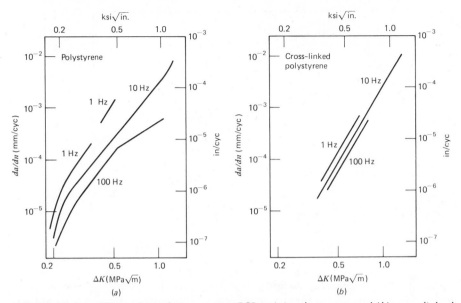

FIGURE 13.48 Effect of test frequency on FCP in (*a*) polystyrene and (*b*) cross-linked polystyrene. Note the larger frequency sensitivity in polystyrene.[87] (Reprinted with permission of the Society of Plastics Engineers.)

TABLE 13.4 Frequency Sensitivity of Polymer FCP[87]

Sensitive		Insensitive	
Material	FSF[a]	Material	FSF
PMMA	3.2	PC	~1
PS	2.3	PSF	~1
PPO	2.	Nylon 66	~1
Noryl	2.	PVF2	~1
PVC	1.8		
CLPS	1.5		

[a]FSF = Frequency sensitivity factor.

responsible for generating the β transition peak (see Chapter Six) at room temperature.[87] Data for several polymers listed in Table 13.4 are shown in Fig. 13.49, along with the fatigue test frequency range. Note the greatest frequency sensitivity in the material that revealed its β peak at a frequency comparable to the fatigue test frequency range. This resonance condition suggests the possibility that localized crack tip heating may be responsible for polymer FCP frequency sensitivity. One may then speculate whether other materials that were

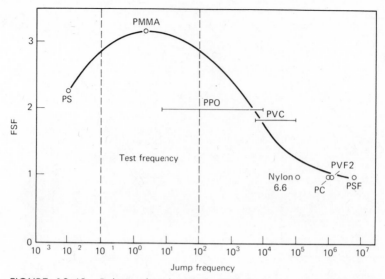

FIGURE 13.49 Relationship between FCP frequency sensitivity and the room temperature jump frequency for several polymers.[87] (Reprinted with permission of the Society of Plastics Engineers.)

not FCP frequency sensitive at room temperature might be made so at other temperatures, if the necessary segmental motion jump frequency were comparable to the mechanical test frequency. Indeed, this has been verified for PC and PSF under low temperature test conditions; conversely, the FCP response of PMMA was found to be *less* frequency sensitive at $-50°C$ than at room temperature, consistent with expectations.

13.8.3 Fracture Surface Micromorphology

At least two distinctly different sets of parallel markings have been found on the fatigue fracture surfaces of amorphous plastics such as PMMA, PS, and PC. At relatively large ΔK levels, striations are found that correspond to the incremental advance of the crack as a result of one load cycle (Fig. 13.50a). Similar markings have been reported for rubber.[91] The dependence of fatigue striation spacing on the stress intensity factor range and the excellent correlation with associated macroscopic crack growth rates in PC may be seen in Fig. 13.51.[83] The essentially exact correlation between macroscopic and microscopic growth rates reflects the fact that 100% of the fracture surface in this ΔK regime is striated; that is, only one micromechanism is operative. Contrast this with the results for metals, where several micromechanisms operate simultaneously (Eq.

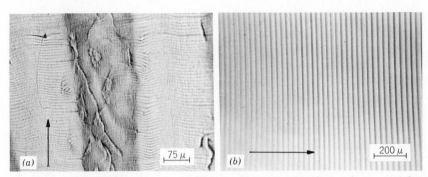

FIGURE 13.50 Fatigue fracture surface markings in amorphous plastics. (*a*) Striations associated with crack advance during one load cycle; (*b*) discontinuous growth bands equal in size to crack tip plastic zone. Arrow indicates crack direction.

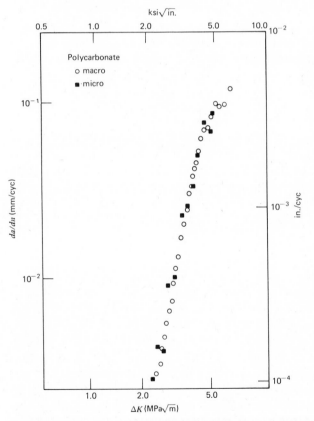

FIGURE 13.51 Dependence of fatigue striation spacings on ΔK and correlation with macroscopic growth rate in polycarbonate.[83] (From J. A. Manson and R. W. Hertzberg, *CRC Critical Reviews in Macromolecular Science*, 1 (4), pp. 433 ff., 1973, © The Chemical Rubber Co., 1973. Used by permission of The Chemical Rubber Co.)

13-14). In the latter case, the two measurements of crack growth rate do not always agree (see Fig. 13.38).

The other sets of parallel fatigue markings have been found at low ΔK levels and at high test frequencies in PS, PC, PSF, PMMA and at all stress intensity levels in PVC.[92–95] These bands (Fig. 13.50b) are too large to be caused by the incremental extension of the crack during one loading cycle. Instead, they correspond to discontinuous crack advance following several hundred loading cycles during which the crack tip remains stationary. The fatigue fracture sequence that produces these markings is shown in Fig. 13.52. The plastic zone —actually a long, thin craze—is seen to grow continuously, although it is characterized by a decreasing rate with increasing craze length. When the craze reaches a critical length, the crack advances abruptly across the entire craze and

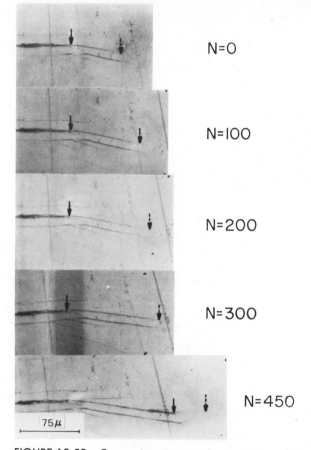

N=0

N=100

N=200

N=300

N=450

75μ

FIGURE 13.52 Composite micrograph revealing position of craze ($\downarrow$) and crack ($\downarrow$) tip after fixed cyclic increments in PVC.[93] (Reproduced with permission from *Journal of Materials Science, 8,* 1973.)

arrests. With further cycling, a new craze is developed and the process is repeated. Close examination of the fracture surface reveals that the growth bands consist of equiaxed dimples, which decrease in diameter from the beginning to the end of each band (Fig. 13.53). The variable dimple size is believed to reflect the void size distribution within the craze prior to crack instability; it also parallels the extent of crack opening displacement with increasing distance from the crack tip.[93]

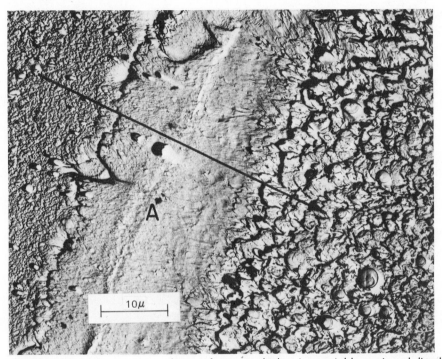

FIGURE 13.53 Transmission electron fractograph showing variable equiaxed dimple size in PVC discontinuous growth bands. Largest dimples are found near beginning of band. Region A is the arrest line between bands. Arrow indicates crack propagation direction.[93] (Reproduced with permission from *Journal of Materials Science, 8,* 1973.)

The size of these bands increase with ΔK (Fig. 13.54) and correspond to the dimension of the crack tip plastic zone[92,94,95] as computed by the Dugdale plastic strip model

$$R \approx \frac{\pi}{8} \frac{K^2}{\sigma_{ys}^2} \qquad (8\text{-}41)$$

This calculation can be reversed to compute the yield strength that controls the

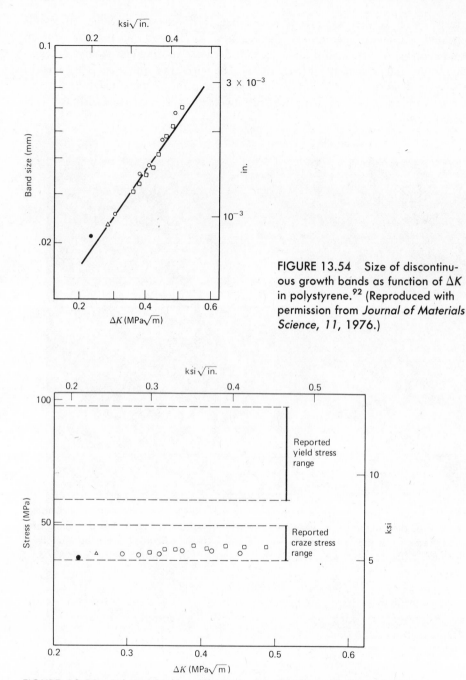

FIGURE 13.54 Size of discontinuous growth bands as function of ΔK in polystyrene.[92] (Reproduced with permission from *Journal of Materials Science, 11, 1976.*)

FIGURE 13.55 Relationship between inferred yield strength from discontinuous growth band measurements and ΔK. Also shown are yield and craze strength values for polystyrene as reported in the literature.[92] (Reproduced with permission from *Journal of Materials Science, 11, 1976.*)

crack tip deformation process. By setting R equal to the band width for a given ΔK level, an inferred yield strength was determined in PS and other materials and found equal to the craze stress for that material (Fig. 13.55).

In summary, two different sets of fatigue markings may be found on the fracture surfaces of the same plastic, each band corresponding to either one or as many as 1000 load cycles. Without accurate macroscopic growth rate data or fractographic analysis, it is difficult to distinguish between them. To do so is highly desirable, since the markings would provide valuable information concerning the number of cycles associated with the fatigue crack propagation process.

REFERENCES

1. H. W. Liu, *Appl. Mater. Res.*, **3**(4), 1964, p. 229.
2. N. E. Frost, *J. Mech. Eng. Sci.*, **1**(2), 1959, p. 151.
3. W. Weibull, *Acta Met.*, **11**(7), 1963, p. 725.
4. P. C. Paris, *Fatigue—An Interdisciplinary Approach*, Proceedings 10th Sagamore Conference, Syracuse University Press, Syracuse, N.Y., 1964, p. 107.
5. S. R. Swanson, F. Cicci, and W. Hoppe, ASTM *STP 415*, 1967, p. 312.
6. P. C. Paris and F. Erdogan, *J. Basic Eng. Trans.*, ASME, Series D, **85**(4), 1963, p. 528.
7. P. C. Paris and G. C. M. Sih, ASTM *STP 381*, 1965, p. 30.
8. J. M. Barsom, E. M. Imhof, and S. T. Rolfe, *Eng. Fract. Mech.*, **2**(4), 1971, p. 301.
9. G. Miller, *Trans. Quart.*, ASM, **61**, 1968, p. 442.
10. P. J. E. Forsyth, *Acta Met.*, **11**(7), 1963, p. 703.
11. D. O. Swenson, *J. Appl. Phys.*, **40**, 1969, p. 3467.
12. R. W. Hertzberg and W. J. Mills, ASTM *STP 600*, 1976, p. 220.
13. C. Bathias and R. M. Pelloux, *Met. Trans.*, **4**(5), 1973, p. 1265.
14. H. I. McHenry, Ph.D. Dissertation, Lehigh University, 1970.
15. J. C. McMillan and R. W. Hertzberg, ASTM *STP 436*, 1968, p. 89.
16. P. J. E. Forsyth and D. A. Ryder, *Metallurgia*, **63**, 1961, p. 117.
17. R. W. Hertzberg, ASTM *STP 415*, 1967, p. 205.
18. C. Laird and G. C. Smith, *Phil. Mag.*, **7**, 1962, p. 847.
19. G. Jacoby, *Exp. Mech.*, **5**(3), 1965, p. 65.
20. R. M. Pelloux, *Trans. Quart.*, ASM, **62**(1), 1969, p. 281.
21. J. H. Weber and R. W. Hertzberg, *Met. Trans.*, **4**, 1973, p. 595.
22. R. W. Hertzberg and P. C. Paris, *Proceedings, International Fracture Conference*, Sendai, Japan, **1**, 1965, p. 459.
23. R. C. Bates and W. G. Clark, Jr., *Trans. Quart.*, ASM, **62**(2), 1969, p. 380.
24. R. G. Forman, V. E. Kearney, and R. M. Engle, *J. Basic Eng. Trans.*, ASME, **89**, 1967, p. 459.
25. R. Roberts and F. Erdogan, *J. Basic Eng. Trans.*, ASME, **89**, 1967, p. 885.
26. W. Elber, ASTM *STP 486*, 1971, p. 230.
27. W. Illg and A. J. McEvily, Jr., NASA *TND-52*, 1959.
28. C. M. Hudson, NASA Tech. Note *D-5390*, 1969.
29. T. C. Lindley and C. E. Richards, *Mater. Sci. Eng.*, **14**, 1974, p. 281.

30. R. W. Hertzberg and E. F. J. vonEuw, *Int. J. Fract. Mech.*, **7**, 1971, p. 349.
31. D. Broek and J. Schijve, National Aeronautical and Astronautical Research Institute, Amsterdam, Holland, NLR-TR-M 2.111, 1963.
32. J. Schijve, ASTM *STP 415*, 1967, p. 415.
33. R. J. Bucci, P. C. Paris, R. W. Hertzberg, R. A. Schmidt, and A. F. Anderson, ASTM *STP 513*, 1972, p. 125.
34. R. J. Bucci, W. G. Clark, Jr., and P. C. Paris, ASTM *STP 513*, 1972, p. 177.
35. B. M. Linder, M. S. Thesis, Lehigh University, 1965.
36. J. Schijve and D. Broek, *Aircr. Eng.*, **34**, 1962, p. 314.
37. J. Schijve, D. Broek, and P. deRijk, NLR Report *M2094*, Jan. 1962.
38. E. F. J. vonEuw, R. W. Hertzberg, and R. Roberts, ASTM *STP 513*, 1972, p. 230.
39. V. W. Trebules, Jr., R. Roberts, and R. W. Hertzberg, ASTM *STP 536*, 1973, p. 115.
40. W. J. Mills and R. W. Hertzberg, *Eng. Fract. Mech.*, **7**, 1975, p. 705.
41. W. J. Mills and R. W. Hertzberg, *Eng. Fract. Mech.* **8**, 1976, p. 000.
42. J. Schijve, F. A. Jacobs, and P. J. Tromp, NLR *TR 69050 U*, June 1969.
43. H. F. Hardrath, *Fatigue—An Interdisciplinary Approach*, Syracuse University Press, Syracuse, N. Y., 1964.
44. H. F. Hardrath and A. J. McEvily, Jr., *Proceedings of the Crack Propagation Symposium*, **1**, Cranfield, England, Oct. 1961.
45. C. M. Hudson and H. F. Hardrath, NASA Technical Note *D-960*, Sept. 1961.
46. J. Schijve and D. Broek, *Aircr. Eng.*, **34**, 1962, p. 314.
47. J. Willenborg, R. M. Engle, and H. A. Wood, Tech. Mem. 71-1-FBR, 1971.
48. D. E. Wheeler, *J. Basic Eng. Trans.*, ASME, **94**, 1972, p. 181.
49. J. P. Gallagher and T. F. Hughes, AFFDL-TR-74-27, 1974.
50. W. Elber, ASTM *STP 559*, 1974, p. 45.
51. J. Rice, ASTM *STP 415*, 1967, p. 247.
52. A. Hartman, F. J. Jacobs, A. Nederveen, and R. deRijk, NLR *TN/M 2182*, 1967.
53. A. Hartman, *Int. J. Fract. Mech.*, **1**(3), 1965, p. 167.
54. R. P. Wei, *Eng. Fract. Mech.*, **1**, 1970, p. 633.
55. R. J. Bucci, Ph.D. Dissertation, Lehigh University, 1970.
56. J. M. Barsom, *Eng. Fract. Mech.*, **3**(1), 1971, p. 15.
57. E. J. Imhof and J. M. Barsom, ASTM *STP 536*, 1973, p. 182.
58. A. G. Evans and E. R. Fuller, *Met. Trans.*, **5**, 1974, p. 27.
59. H. I. McHenry and A. W. Pense, ASTM *STP 520*, 1973, p. 345.
60. H. D. Solomon and L. F. Coffin, ASTM *STP 520*, 1973, p. 112.
61. M. W. Mahoney and N. E. Paton, *Nucl. Tech.*, **23**, 1974, p. 290.
62. L. A. James and R. L. Knecht, *Met. Trans.*, **6A**, 1975, p. 109.
63. R. P. Wei and J. D. Landes, *Mater. Res. Stand.*, **9**, 1969, p. 25.
64. N. S. Stoloff and D. J. Duquette, *CRC Crit. Rev. Sol. State Sci.*, **4**, 1974, p. 615.
65. S. Pearson, *Nature*, **211**, 1966, p. 1077.
66. F. A. Heiser and R. W. Hertzberg, *J. Basic Eng. Trans.*, ASME, **93**, 1971, p. 71.
67. R. C. Bates, Westinghouse Scientific Paper *69-1D9-RDAFC-P2*, 1969.
68. D. Broek, Paper 66, *Second International Fracture Conference*, Brighton, 1969, p. 754.
69. S. M. El-Soudani and R. M. Pelloux, *Met. Trans.*, **4**, 1973, p. 519.
70. J. M. Barsom, *J. Eng. Ind.*, ASME, Series B, **93**(4), 1971, p. 1190.
71. J. H. Weber and R. W. Hertzberg, *Met. Trans.*, **2**, 1971, p. 3498.

72. B. Tomkins, *Phil. Mag.*, **18**, 1968, p. 1041.
73. J. P. Hickerson, Jr. and R. W. Hertzberg, *Met. Trans.*, **3**, 1972, p. 179.
74. S. Majumdar and J. D. Morrow, ASTM *STP 559*, 1974, p. 159.
75. R. W. Hertzberg, *Met. Trans.*, **5**, 1974, p. 306.
76. I. Constable, J. G. Williams, and D. J. Burns, *J. Mech. Eng. Sci.*, **12**, 1970, p. 20.
77. M. N. Riddell, G. P. Koo, and J. L. O'Toole, *Polym. Eng. Sci.*, **6**, 1966, p. 363.
78. G. P. Koo, M. N. Riddell, and J. L. O'Toole, *Polym. Eng. Sci.*, **7**, 1967, p. 182.
79. J. D. Ferry, *Viscoelastic Properties of Polymers*, John Wiley, New York, 1961.
80. L. J. Broutman and S. K. Gaggar, *Proceedings of the Twenty Seventh Annual Technical Conference*, 1972, Society of the Plastics Industry, Inc., Section 9-B, p. 1.
81. A. V. Stinkas and S. B. Ratner, *Plasticheskie Massey*, **12**, 1962, p. 49.
82. L. C. Cessna, J. A. Levens, and J. B. Thomson, *Polym. Eng. Sci.*, **9**, 1969, p. 339.
83. J. A. Manson and R. W. Hertzberg, *CRC Crit. Rev. Macromol. Sci.*, **1**(4), 1973, p. 433.
84. R. W. Hertzberg, H. Nordberg, and J. A. Manson, *J. Mater. Sci.*, **5**, 1970, p. 521.
85. R. G. Vadimsky, H. D. Keith, and F. J. Padden, Jr., *J. Polym. Sci.*, *A-2*, **7**, 1969, p. 1367.
86. R. W. Hertzberg, J. A. Manson, and W. C. Wu, ASTM *STP 536*, 1973, p. 391.
87. R. W. Hertzberg, J. A. Manson, and M. D. Skibo, *Polym. Eng. Sci.*, **15**(4), 1975, p. 252.
88. N. H. Watts and D. J. Burns, *Polym. Eng. Sci.*, **1**, 1967, p. 90.
89. Y. Katz, P. L. Key, and E. R. Parker, *Trans.*, ASME, Ser. D, **90**, 1968, p. 622.
90. S. Rabinowitz and P. Beardmore, *CRC Crit. Rev. Macromol. Sci.*, **1**, 1972, p. 1.
91. E. H. Andrews, *J. Appl. Phys.*, **32**(3), 1961, p. 542.
92. M. D. Skibo, R. W. Hertzberg, and J. A. Manson, *J. Mater. Sci.*, **11**, 1976, p. 479.
93. R. W. Hertzberg and J. A. Manson, *J. Mater. Sci.*, **8**, 1973, p. 1554.
94. J. P. Elinck, J. C. Bauwens, and G. Homes, *Int. J. Fract. Mech.*, **7**(3), 1971, p. 227.
95. M. D. Skibo, R. W. Hertzberg, J. A. Manson, and S. Kim, *J. Mater. Sci.*, **12**, 1977, p. 000.

PROBLEMS

13.1 During the course of a simple sinusoidal wave form fatigue test, the machine command signal was changed for one cycle at position A. The resulting data are shown below. Describe what changes were made in the machine signal and the corresponding effect on the specimen.

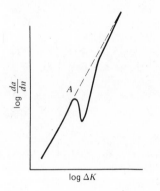

13.2 (a) A 10-cm square, 20-cm long extruded bar of 7075-T6511 is hollowed out to form a thin-walled cylinder (closed at one end), 20-cm long with an outer diameter of 9-cm. The cylinder is fitted with a 7-cm diameter piston designed to increase pressure within the cylinder to 55 MPa. On one occasion, a malfunction in the system caused an unanticipated pressure surge of unknown magnitude, and the cylinder burst. Examination of the fracture surface revealed a metallurgical defect in the form of an elliptical flaw 0.45-cm long at the inner diameter wall and 0.15-cm deep. This flaw was oriented normal to the hoop stress of the cylinder. Compute the magnitude of the pressure surge responsible for failure. (For mechanical property data see Tables 10.2 and 10.8.)

(b) Assume that another cylinder had a similarly oriented surface flaw but with a semicircular shape ($a = 0.15$-cm). How many pressure cycles could the cylinder withstand before failure? Assume normal operating conditions for this cylinder and that the material obeys a fatigue crack propagation relationship

$$\frac{da}{dn} = 5 \times 10^{-39} (\Delta K)^4$$

where da/dn and ΔK have the units of m/cyc and Pa, respectively.

13.3 A large steel plate is used in an engineering structure. A radical metallurgy graduate student intent on destroying this component decides to cut a very sharp notch in the edge of the plate (perpendicular to the applied loading direction). If he walks away from the scene of his dastardly deed at a rate of 5 km/hr, how far away will he get by the time his plan succeeds? Here are hallowed hints for the hunter:

(a) The plate is cyclically loaded uniformly from zero to 80 KN at a frequency of 25 Hz.

(b) The steel plate is 20-cm wide and 0.3-cm thick.

(c) The yield strength is 1400 MPa and the plane strain fracture toughness is 48 MPa$\sqrt{m}$.

(d) The misled metallurgist's multilating mark was measured to be 1-cm long (through thickness).

(e) A janitor noted, in subsequent eyewitness testimony, that the crack was propagating at a velocity proportional to the square of the crack tip plastic zone size. (The janitor had just completed a correspondence course entitled "Relevant Observations on the Facts of Life" and was alerted to the need for such critical observations.)

(f) Post-failure fractographic examination revealed the presence of fatigue striations 2.5×10^{-4}-mm in width where the crack was 2.5-cm long.

13.4 If the plate in the previous problem had been 0.15- or 0.6-cm thick, respectively, would the villain have been able to get farther away before his plan would have succeeded? (Assume that the load on the plate was also adjusted so as to maintain a constant stress.)

13.5 Estimate the stress intensity factor range corresponding to an observed striation spacing of 10^{-4} mm/cyc in the steel alloy shown in Fig. 13.38. Compare the results you would get when ΔK is determined from the striation data and the *macroscopic* data in Fig. 13.38. Also, compute ΔK from Eq. 13-7.

13.6 The Liberty Bell, like most other bells, contains a large percentage of tin (18–25 w/o) in a copper based alloy. From the copper-tin phase diagram (see any number of metallurgy texts or handbooks), describe the probable mechanical properties for this alloy. From this decide whether a deliberate overload would have prolonged the bell's service life.

13.7 Many years ago, a crack was discovered in "Big Ben," the bell located in the Parliament building in London. To avoid its catastrophic failure or complete replacement, it was decided to replace the clapper with a smaller one and to rotate the bell to change the point of clapper impact. Using fracture mechanics concepts, explain how this alteration has succeeded to this day in prolonging "Big Ben's" life.

CHAPTER FOURTEEN
ANALYSES OF ENGINEERING FAILURES

We have come now to the moment of truth—we must now use our knowledge and understanding of fracture mechanics and the relationship between mechanical properties and microstructure to analyze actual service failures. However, before discussing recommended procedures for failure analyses and the details of several case histories, it is best to stand back for a moment and view component failures in a broader sense. To begin, we must ask who bears responsibility for these failures? Is it the company or individual that manufactured the component or engineering system, or the company or person that operated it when it failed? Such is the basis for debate in many product liability law suits. For example, opposing lawyers might ask of manufacturer and user the following questions:

1. Were engineering factors such as stress, potential flaw size, material, and environment considered in the design of the part?
2. Was the part underdesigned?
3. Was a proper material selection made for the manufacture of the part?
4. Was the part manufactured properly?
5. What limits were placed upon the use of the part and what, if any, service life was guaranteed?
6. Were these limits conservative or unconservative?
7. Were these limits respected during the operation of the part?

A product liability case often becomes entangled in a number of ambiguities arising from incomplete or unsatisfactory answers to these questions. As such, it is important for the practicing engineer called in to analyze a failure and, perhaps, testify in court, to identify the major variables pertaining to the design and service life of the component. Recognizing that an individual from one field may be reluctant to challenge the conclusions drawn by an expert in another discipline, it becomes difficult to reconcile the two points of view without an overview of the facts involved. In many cases, these differences contrast the importance of the continuum versus the microstructural approach to the under-standing of the component response (or failure). The most valuable expert witness is one who can appreciate and evaluate the input from different disciplines and educate the court as to their respective significance in the case under study. On the basis of such expert testimony, the courts are able to draw conclusions and render judgments. A delightful statement, made by the auditor for the 1919 Boston molasses tank law suit, relates to this decision-making process:

> "Weeks and months were devoted to evidence of stress and strain, of the strength of materials, of the force of high explosives, of the bursting power of gas and of similar technical problems.... I have listened to a demonstra-tion that piece "A" could have been carried into the playground only by the force of a high explosive. I have thereafter heard an equally forcible demonstration that the same result could be and in this case was produced by the pressure caused by the weight of the molasses alone. I have heard that the presence of Neumann bands in the steel herein considered along the line of fracture proved an explosion. I have heard that Neumann bands proved nothing. I have listened to men upon the faith of whose judgment any capitalist might well rely in the expenditure of millions in structural steel, swear that the secondary stresses in a structure of this kind were negligible and I have heard from equally authoritative sources that these same secondary stresses were undoubtedly the cause of the accident. Amid this swirl of polemical scientific waters it is not strange that the auditor has at times felt that the only rock to which he could safely cling was the obvious fact that at least one-half the scientists must be wrong. By degrees, however, what seem to be the material points in the case have emerged...."[1]

In the following sections, attention is given to fracture surface examination techniques, estimation procedures for determination of stress intensity factors for complex crack geometries, identification of data needed for successful failure analyses, and, finally, to discussions of numerous service failures.

14.1 MACROSCOPIC FRACTURE SURFACE EXAMINATION

The functions of a macroscopic fracture surface examination are to locate the crack origin, determine its size and shape, characterize the texture of the fracture surface, and note any gross markings suggestive of a particular fracture mechanism. To begin, one should attempt to identify whether there are one or more crack origins, since this may provide an indication of the magnitude of stress in the critical region. In general, the number of crack nuclei increases with increasing applied stress and magnitude of an existing stress concentration factor (Fig. 12.4). Even when one crack grows to critical dimensions, secondary cracks can develop before final failure because of load adjustments that may accommodate the presence of the primary defect.

Whether there are one or more fracture nucleation sites, it is of utmost importance to locate them and identify precisely the reason for their existence. When the fracture mechanism(s) responsible for growth of the initial defect to critical proportions is known, the engineer can recommend "fixes" or changes in component design.

The task at this point is to find the origin. This was not difficult in the case of the tricycle and doorstop failures mentioned in Chapter Twelve, but one can well imagine the difficulty of sifting through the wreckage of a molasses tank, ship, or bridge failure (see Chapter Seven) to find their respective fracture origins. For these situations, there could be literally thousands of linear meters of fracture surface to examine. Where does one begin? Once begun, how do you know the direction in which to proceed to locate the origin? Fortunately, the crack often leaves a series of fracture markings in its wake that may indicate the relative direction of crack motion. For example, the curved lines (called "chevron" markings) which seem to converge near midthickness of the fracture surface shown in Fig. 14.1 have been shown to point back toward the crack origin. Another example of "chevron" markings was shown in Fig. 8.15c. It is

FIGURE 14.1   Chevron markings curve in from the two surfaces and point back to the crack origin. (Courtesy of Roger Slutter, Lehigh University.)

believed that within the material localized separations ahead of the crack grow back to meet the advancing crack front and form these curved tear lines. All one needs to do is follow the "chevron" arrow. When a crack initiates in a large component, it sometimes branches out in several directions as it runs through the structure. Although the "chevron" markings along each branch will point in different directions relative to the component geometry, it is important to recognize that the different sets of "chevron" markings all point in the same *relative* direction—back toward the origin (Fig. 14.2). "Chevron" markings grow out radially from an internally located origin, as shown in Fig. 14.3 for the case of a 6-cm diameter steel ($\sigma_{ys} = 550$ MPa) reinforcing bar. These markings may sometimes be obscured by other fracture markings, such as by secondary fractures in anisotropic materials. A crack "divider" orientation Charpy specimen of banded steel (Fig. 13.37) reveals many fracture surface delaminations caused by σ_z stresses acting parallel to the crack front, but they cloud the expected "chevron" pattern (Fig. 14.4).

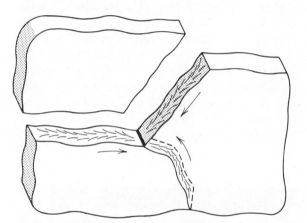

FIGURE 14.2   Multiple chevron patterns emanating from crack origin. Each pattern points back to crack nucleation site.

As one follows the path of the crack, shear lips are often found that represent the regions on the fracture surface which correspond to plane stress conditions. As such, it is tempting to relate the size of the plastic zone to the amount of shear lip found on the fracture surface.[2] Since $r_y \approx (1/2\pi)(K^2/\sigma_{ys}^2)$ at the surface of the plate and the shear lips form on ± 45 degree bands to the sheet thickness, it is seen from Fig. 14.5 that the depth D of the shear lips can be approximated by the plastic zone radius. Hence

$$D \approx r_y \approx \frac{1}{2\pi} \left(\frac{K}{\sigma_{ys}} \right)^2 \tag{14-1}$$

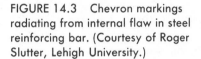

FIGURE 14.3 Chevron markings radiating from internal flaw in steel reinforcing bar. (Courtesy of Roger Slutter, Lehigh University.)

FIGURE 14.4 Crack "divider" fracture surface. Delaminations obscure anticipated "chevron" pattern. (Reprinted with permission from Metals Society, *JISI, 209*, 1971, p. 975.)

Combining Eqs. 8-22 and 14-1, we find

$$\text{shear lip} \approx \frac{1}{2\pi} \frac{Y^2\sigma^2 a}{\sigma_{ys}^2} \tag{14-2}$$

Knowing the geometrical correction factor Y for the component and the crack length where the shear lip was measured, Eq. 14-2 can be used to estimate the prevailing stress level. This approach to determining the stress level is highly empirical and appears to work satisfactorily only for certain materials such as high-strength aluminum and certain alloy steels but not for lower strength steels. When the correlation does not hold, it usually results in too low an estimate of K

from Eq. 14-1; that is, the shear lip is smaller than that would be expected from the actual plastic zone. Because Eq. 14-2 is a highly empirical determinant of stress level, computed values must be considered tentative until corroborated by additional findings. The use of shear lip measurements to determine the stress level is discussed in case history 1 in Section 14.5.

It is possible that a crack may initiate by one mechanism and propagate by one or more different ones. For example, a crack may begin at a metallurgical defect, propagate for a certain distance by a fatigue process, and then continue growing by a combination of fatigue crack propagation and environment assisted cracking when the stress intensity factor exceeds K_{IEAC}. Such mechanism changes may be identified by changes in texture of the fracture surface. For example, the fracture surface shown in Fig. 14.6a reveals the different textures associated with fatigue and stress corrosion subcritical flaw growth. In the broken wing strut (Fig. 14.6b), we find the regions of FCP (shiny areas) interrupted by two separate localized crack instabilities (dull areas), which probably were caused by two high load excursions during the random loading life history of the strut. Another example of a plane strain "pop-in" is found on the fracture surface of a fracture toughness test sample that exhibited Type II (Fig. 8.16) load-deflection response (Fig. 14.6c).

"Pop-in" can also result from the presence of local residual stresses. Assuming that a crack is embedded within a localized stress concentration region, the application of a moderate load could develop a stress intensity level (magnified by the local stress concentration) equal to K_{IC} or K_c (depending on the prevailing stress state), which would cause the crack to run unstably through the

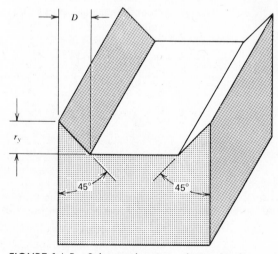

FIGURE 14.5 Schema showing relationship between shear lip depth and estimated plane stress plastic zone size.

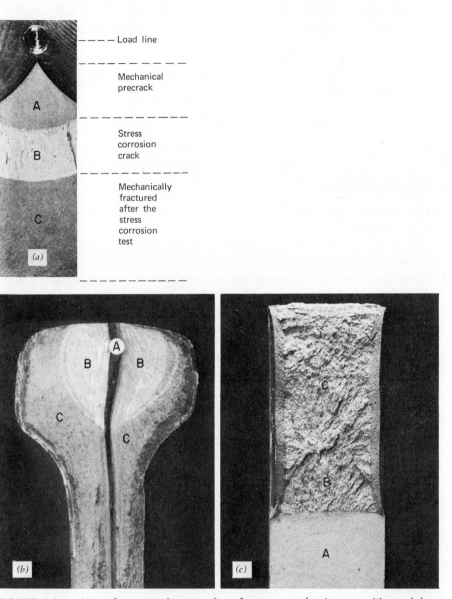

- - - - Load line

- - - - - - - -

Mechanical
precrack

- - - - - - - -

Stress
corrosion
crack

- - - - - - - -

Mechanically
fractured
after the
stress
corrosion
test

(a)

- - - - - - - -

FIGURE 14.6 Macrofractographs revealing fracture mechanism transitions. (*a*) Transition from fatigue A to stress corrosion cracking B to fast fracture C;[26] (*b*) wing strut with metallurgical delamination A, fatigue B, and static fracture C; (*c*) fracture toughness sample revealing fatigue precracking zone A, pop-in instability B and fast fracture C. (Fig. 14.6*a* reprinted with permission of Markus O. Speidel, Brown Bovari Co.)

component. However, the crack would soon run out of the region of high stress concentration and arrest (and produce a fracture surface marking), since the moderate load without the stress concentration does not possess the necessary driving force to sustain crack growth.

Other macroscopic arrest lines, such as the fatigue crack propagation "beach mark" and load block band, were discussed in Chapter Twelve (see Fig. 12.3) and Chapter Thirteen (see Fig. 13.9), respectively. The reader should review these illustrations.

14.2 METALLOGRAPHIC AND FRACTOGRAPHIC EXAMINATION

Having located the crack origin, a typical failure analysis would proceed normally with two main interim objectives: (1) identification of the micro-mechanism(s) of subcritical flaw growth and (2), estimation of the stress intensity conditions prevailing at the crack tip when failure occurred. As was discussed in Chapter Seven, metallographic techniques have often been used to determine the crack path relative to the component microstructure. In addition to identifying the microscopic fracture path (transgranular versus intergranular), metallographic sections are useful in establishing the metallurgical condition of the material. Grain size and shape offer important clues to the thermomechanical history of the component (see Chapter Four). For example, a coarse-grained structure is indicative of a very high-temperature annealing process, while an elongated grain structure indicates not only the application of a deformation process in the history of the material, such as rolling, forging, and drawing, but also the deformation direction. Such information allows one to anticipate the presence of anisotropic mechanical properties that must be identified in relation to the predominant stress direction. Examples of mechanical property anisotropy were given in Tables 10.2 and 10.8.

Determination of the microstructural constituents enables the examiner to determine whether the component had been heat treated properly. Identification of a possible grain boundary phase, for example, can explain the occurrence of an intercrystalline fracture. Finally, with the aid of an inclusion count, the relative cleanliness of the metallurgical structure can be determined. Although it is not possible to express the fracture toughness of a material in terms of some measure of inclusion content [recently, some progress has been achieved in this regard (see Chapter Ten)], it is known that fracture toughness decreases with increasing inclusion content. Hence, a trained metallographer may ascertain from metallographic examination whether the material in question is representative of good or bad stock.

A fractographic study with either a scanning or transmission electron microscope reveals the microscopic character of the fracture surface. Since this topic has been discussed at considerable length in earlier chapters, the reader is referred specifically to Chapters Seven and Thirteen and Appendix A before proceeding further.

14.3 STRESS INTENSITY FACTOR ESTIMATIONS

It is critical in any failure analysis where fracture mechanics concepts may be applied to estimate the crack tip stress intensity factor. A number of typical crack configurations commonly found in real engineering components were given in Fig. 8.7. In addition, reference was made to three sources of stress intensity factor solutions for literally hundreds of different crack configurations and loading conditions.[3-5] And, yet, situations arise where a specific K calibration is not known. Consequently, the failure analysis cannot proceed until some approximation of K is obtained. Sometimes this represents a forbidding task, but at other times estimates can be made with considerable confidence. Let us now consider a few complex crack configurations and estimates of the associated stress intensity factors.

CASE 1: Crack emanating from a round hole[3,6]

This configuration (Fig. 14.7a) is commonly found in engineering practice (especially in aircraft components, which contain many rivet holes) since cracks often emanate from regions of high stress concentration. For the case of a round hole, K_t equals 3 at position A and is minus 1 at B. For a shallow crack ($L \ll R$), the crack tip is embedded within this local stress concentration and may be considered to be a shallow surface flaw. The stress intensity factor is then estimated to be

$$K \approx 1.12(3\sigma)\sqrt{\pi L} \ (L \ll R) \tag{14-3}$$

As one might expect, K drops quickly when L increases because the crack runs out of the high stress concentration region. Eq. 14-3, therefore, represents an upper bound solution for this crack configuration. A lower bound solution may be estimated for conditions where $L > R$. Here we may estimate the crack length to be $L + 2R$. Hence

$$K \approx \sigma \sqrt{\pi \left(\frac{L+2R}{2} \right)} \ (L > R) \tag{14-4}$$

Both upper and lower bound solutions are given in Table 14.1 for several L/R ratios. Also shown are correction factors $F(L/R)$ for this crack configuration based on the solution by Bowie[6] where

$$K = F(L/R)\sigma\sqrt{\pi L} \tag{14-5}$$

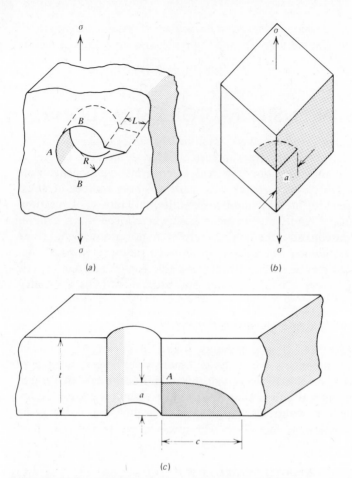

FIGURE 14.7 Complex crack configurations. (a) Crack emanating from hole; (b) circular corner crack; (c) elliptical corner crack emanating from through-thickness hole.

Note the excellent agreement between Eqs. 14-3 and 14-4 at both $L/R \approx 0$ and $L/R > 1$ extremes. Paris and Sih[3] also tabulated the correction factors for the case involving cracks emanating from both sides of the hole.

The only difficulty in using the engineering approximations for K (i.e., Eqs. 14-3, 14-4) arises in the region $0 < L/R < 0.5$, where the stress concentration at the hole decays rapidly. Broek[7,8] considered this problem and concluded that the residual strength of cracked sheets and the crack propagation rate of cracks emanating from holes could be accounted for in reasonable fashion by considering the hole as being part of the crack (i.e., by using Eq. 14-4).

TABLE 14.1 Stress Intensity Correction Factors for a Single Crack Emanating from a Hole

L/R	Eq. 14-3	Eq. 14-4	$F(L/R)^6$
0.00	3.39	—	3.39
0.10	—	—	2.73
0.20	—	—	2.30
0.30	—	1.96	2.04
0.40	—	1.73	1.86
0.50	—	1.58	1.73
0.60	—	1.47	1.64
0.80	—	1.32	1.47
1.0	—	1.22	1.37
1.5	—	1.08	1.18
2.0	—	1.00	1.06
3.0	—	0.91	0.94
5.0	—	0.84	0.81
10.0	—	0.77	0.75
∞	—	0.707	0.707

CASE 2: Semicircular corner crack[9]

This configuration (Fig. 14.7b) involves geometries shown in Figs. 8.7b and 8.7i. Since the crack is circular and lies along two free surfaces, the prevailing stress intensity level may be approximated by

$$K \approx (1.12)^2 \frac{2}{\pi} \sigma \sqrt{\pi a} \qquad (14\text{-}6)$$

where $(1.12)^2$ represents two surface flaw corrections and $2/\pi$ represents the correction for a penny-shaped embedded crack.

CASE 3: An elliptical corner crack growing from one corner of a through thickness hole.[9]

The solution to this crack configuration (Fig. 14.7c) incorporates many of the factors discussed in the previous two examples as well as some configurations shown in Fig. 8.7. The maximum stress intensity condition in this instance is located at A, since this part of the crack experiences the maximum stress concentration caused by the hole and because A is located at $\beta = 90°$ (see Fig. 8.7j). An approximate solution may be given by

$$K_A \approx 1.12(3\sigma)\sqrt{\pi a/Q} \cdot \sqrt{\left(\frac{2t}{\pi a}\tan\frac{\pi a}{2t}\right)} \qquad (14\text{-}7)$$

where

$$K_A = \text{maximum stress intensity condition along}$$
$$\text{elliptical surface at } A$$
$$a = \text{depth of elliptical flaw}$$
$$c = \text{half width of elliptical flaw}$$
$$t = \text{plate thickness}$$
$$1.12 = \text{surface flaw correction at } A$$
$$3\sigma = \text{stress concentration effect at } A$$
$$Q = \text{elliptical flaw correction} = f(a/2c)$$

$$\sqrt{\frac{2t}{\pi a} \tan \frac{\pi a}{2t}} = \text{finite panel width correction accounting for}$$
$$\text{relatively large } a/t \text{ ratio (see also Eq. 14-12)}$$

14.4   COMPONENT FAILURE ANALYSIS DATA

Having been introduced to the fundamentals of fracture mechanics analysis, stress intensity analysis of cracks, macroscopic and microscopic features of the fracture surface, and the pertinent mechanical property data necessary to adequately characterize the performance of a given material, the reader should be able to synthesize this information and solve a given service failure problem. The following checklist will assist the investigator in this task. The checklist, which takes into account the component geometry, stress state, flaw characterization, fractographic observations, metallurgical information (including component manufacture and other service information), summarizes the raw data desirable for a complete failure analysis of a fractured component. Experience has shown, however, that in the majority of instances, certain facts are never determined, and educated guesses or estimates (such as the K calibration estimations described in Section 14.3) must be introduced to complete the analysis. The objective of a checklist is to minimize the amount of guesswork while maximizing the opportunity for firm quantitative analysis.

SUGGESTED CHECKLIST OF DATA DESIRABLE
FOR COMPLETE FAILURE ANALYSIS

I. Description of Component Size, Shape, and Use_____

_____

   A. Specify areas of design stress concentrations_____

      _____

      1. Magnitude of stress concentration at failure site_____

         _____

II. Stress State for Component

   A. Type of stresses

      1. Magnitude of design stress levels

         a. Mean stress_____

         b. Stress range_____

      2. Type of stress (e.g., Mode I, II, III, or combinations)_____

         _____

      3. Presence of stress gradient_____

   B. State of stress: plane strain vs. plane stress

      1. Fracture surface appearance: percent shear lip_____

         _____

      2. Estimation from calculated plastic zone size to thickness ratio__

         _____

   C. Nature of load variations

      1. Hours of component operation_____

      2. Load cycle frequency_____

      3. Type of loading pattern

         a. Random loading_____

b. Existence of overloads resulting from abnormal service life events _____

III. Details of Critical Flaw

    A. Date of previous inspection _____

        1. Findings of previous inspection _____

    B. Nature of critical flaw leading to fracture (make use of clearly labeled sketches and/or macrophotographs with accurate magnifications)

        1. Location of critical flaw by macroscopic examination _____

        2. Critical flaw size, shape, and orientation at instability _____

        3. Surface or imbedded flaw _____

        4. Direction of crack propagation as determined by

           Chevron markings _____   Pop-in _____

           Beach markings _____

           Direction _____

    C. Manufacturing flaws related to crack initiation

        Scratches _____   Misfits _____

        Undercuts _____   Others _____

        Weld defects _____

    D. Metallurgical flaws related to crack initiation

        Inclusions _____   Voids _____

        Second-phase particles _____   Weak interfaces _____

        Entrapped slag _____   Others _____

    E. Fractographic observations

        1. Qualitative observations

           Dimpled rupture _____   Fatigue striations _____

           Cleavage _____   Corrosion _____

           Intercrystalline _____   Fretting _____

2. Quantitative observations

    a.   Striation spacings at known crack length positions_____

        _____

    b.   Striation spacing evidence of uniform or random loading___

        _____

    c.   Stretch zone width at onset of unstable crack extension____

        _____

IV. Component Material Specifications

    A.   Alloy designation_____

    B.   Mechanical Properties

	σ_{ys}	σ_{ts}	% elong.	% R.A.	K_{IC}	K_{IEAC}	Fatigue Characterization
Specified							
Actual							

    C.   Alloy Chemistry

Elements

	A	B	C	D	E	F	G
Specified							
Actual							

    D.   Melting practice:

        Air melted_____

        Vacuum melted_____

        Other_____

    E.   Ingot breakdown

        Hot rolled_____

        Cold rolled_____

        Cross rolled_____

F. Thermomechanical treatment

    1. Annealing or solution treatment condition_____

    2. Tempering or aging treatment_____

    3. Intermediate mechanical working_____

G. Component manufacture

    1. Forged_____ cast_____ machined_____

       spun_____ extruded_____ other_____

    2. Joint detail

       Welded_____   Bolted_____

       Brazed_____   Other_____

       Adhesive bonded_____

H. Surface treatment

      Shot peened_____   Flame or induction

      Cold rolled_____   hardened_____

      Carburized_____   Plated_____

      Nitrided_____   Pickled_____

                              Other_____

I. Component microstructure

    1. Presence of mechanical fibering and/or banding from chemical segregation_____

    2. Grain size and shape_____

       a. Elongated with respect to stress axis_____

       b. Grain run-out in forgings_____

    3. Inclusion count and classification_____

14.5 CASE HISTORIES

In this section six actual case histories are discussed; all involve the application of fracture mechanics principles to failure analysis. Although not all component failures require a fracture mechanics analysis, the latter approach is emphasized here to demonstrate the applicability of the fundamental concepts introduced in Chapters Seven through Thirteen. The interested reader is referred to compilations of nearly 100 additional failure analyses, which have been reported elsewhere.[10,11]

CASE 1: Analysis of Crack Development During Structural Fatigue Test[2]

This failure analysis reported by Paris in Ref. 2 represents an excellent, well-documented example of the use of several different and independent fracture mechanics procedures in the solution of a fracture problem. A program load fatigue test was conducted on a 1.78-cm thick plate of D6AC steel that had been tempered to a yield strength of 1500 MPa. Fracture of the plate occurred when fatigue cracks that had developed on both sides of a drilled hole grew into a semicircular configuration, as shown in Fig. 14.8. Note the growth rings within

FIGURE 14.8 Two corner cracks emanating from through-thickness hole, revealing fatigue growth bands and shear lips.[2] (From R. J. Gran, F. D. Orazio, Jr., P. C. Paris, G. I. Irwin, and R. W. Hertzberg, AFFDL-TR-70-149, March 1971.)

the two corner cracks produced by fatigue block loading conditions. These may be compared with similar markings shown in Fig. 13.9. The stress at failure was reported to be 830 MPa, and the maximum and minimum stresses in each of the load blocks were also known.

The stress intensity factor at fracture was determined by three separate methods. First, the stress intensity factor solution for the given crack configuration was estimated in two ways. The actual hole-crack combination was approximated by a semicircular surface flaw with a radius of 0.86 cm and by a through thickness flaw with a total length of 1.73 cm. These estimates reflect lower and upper bound solutions, respectively, since the former solution does not account for the hole passing through the entire plate thickness, and the latter solution indicates more fatigue crack growth than was actually observed. The lower bound of the stress intensity factor may be given by[3]

$$K_L = \left[1 + 0.12\left(1 - \frac{a}{c}\right)\right]\sigma\sqrt{\frac{\pi a}{Q}} \cdot \sqrt{\frac{2t}{\pi a}\tan\frac{\pi a}{2t}} \qquad (14\text{-}8)$$

where

K_L = lower bound stress intensity solution

a = crack depth—0.86 cm or 0.0086 m

c = half flaw width—0.86 cm

σ = applied stress—830 MPa

Q = elliptical flaw correction—2.5

t = plate thickness—1.78 cm

$$K_L = \left[1 + 0.12\left(1 - \frac{0.86}{0.86}\right)\right][830]\sqrt{\frac{\pi(0.0086)}{2.5}} \sqrt{\frac{2(1.78)}{\pi(0.86)}\tan\frac{\pi(0.86)}{2(1.78)}}$$

$$K_L = 96.6 \text{ MPa}\sqrt{m}$$

The upper bound solution is given by

$$K_U = \sigma\sqrt{\pi a}$$

$$= 830\sqrt{\pi(0.0086)}$$

$$= 136 \text{ MPa}\sqrt{m}$$

From these results, the actual K level at fracture may be bracketed by

$$96.6 < K_c < 136 \text{ MPa}\sqrt{m}$$

with the correct value being more closely given by the lower bound solution because of a smaller error in this estimation. Consequently, K_c (or K_{IC})≈ 110 MPa$\sqrt{m}$.

The stress intensity factor was then estimated by measurement of the shear lip depth (about 0.8 mm) along the surface of the hole (Fig. 14.8). From Eq. 14-1

$$\text{shear lip depth} \approx \frac{1}{2\pi} \frac{K^2}{\sigma_{ys}^2}$$

$$8 \times 10^{-4} \approx \frac{1}{2\pi} \left(\frac{K}{1500} \right)^2$$

$$K \approx 106 \text{ MPa}\sqrt{m}$$

which agrees extremely well with the previous estimate of 110 MPa$\sqrt{m}$.

Two additional estimates of the critical stress intensity factor were obtained by using measurements of the fatigue growth bands. It was known that the last band was produced by 15 load fluctuations between stress levels of 137 and 895 MPa. Measuring this growth band to be 0.32 mm, the average crack growth rate was found to be

$$\frac{da}{dn} \approx \frac{\Delta a}{\Delta n} \approx 3.2 \times 10^{-4}/15 \approx 2.1 \times 10^{-5} \text{ m/cyc}$$

From the fatigue crack growth rate data of Carmen and Katlin[12] the corresponding ΔK level was determined to be about 77 MPa$\sqrt{m}$. The maximum K level was then given by

$$K_{max} = \Delta K \left(\frac{\sigma_{max}}{\Delta \sigma} \right) \tag{14-9}$$

$$K_{max} = 77 \left(\frac{895}{758} \right)$$

$$K_{max} = 91 \text{ MPa}\sqrt{m}$$

A similar calculation was made for the next to last band where

$$\Delta n = 2$$

$$\Delta a = 0.16 \text{ mm}$$

$$\sigma_{min} = 138 \text{ MPa}$$

$$\sigma_{max} = 992 \text{ MPa}$$

$$\frac{da}{dn} \approx 1.6 \times 10^{-4}/2 \approx 8 \times 10^{-5} \text{ m/cyc}$$

From Carmen's results, the ΔK level corresponding to this crack growth rate was found to be 82.5 MPa$\sqrt{m}$. Again using Eq. 14-9

$$K_{max} = 82.5\left(\frac{992}{854}\right) = 95.8 \text{ MPa}\sqrt{m}$$

In both instances, estimates of K_c from fatigue growth bands were in excellent agreement with values based on estimates of the prevailing stress intensity factor and shear lip measurements. Finally, the average critical stress intensity factor (101 MPa$\sqrt{m}$) is almost identical with the known K_{IC} level for this material (see Table 10.8). To summarize, the analysis of this laboratory failure clearly demonstrates a number of different and *independent* approaches based on fracture mechanics concepts that one can employ in solving a service failure. Ideally, one should use a number of these procedures to provide cross-checks for each computation.

CASE 2: Analysis of Aileron Power Control Cylinder Service Failure[2]

Several failures of an aileron hydraulic power control unit were experienced by a certain fighter aircraft. These units consisted of four parallel chambers, pressurized by two separate pumps. Failures occurred by cracking either through the inner or outer chamber walls. In either case, the resulting loss of pressure contributed to an aircraft malfunction. Test results indicated the normal mean pressure in these chambers to be about 10.3 MPa, with fluctuations between 5.2 and 15.5 MPa caused by aerodynamic loading fluctuations. Furthermore, during an in-flight aileron maneuver, the pressure was found to rise sometimes to 20.7 MPa, with transient pulses as high as 31 MPa resulting from hydraulic surge conditions associated with rapid commands for aileron repositioning. In one particular case, an elliptical surface flaw grew from the inner bore of one cylinder toward the bore of the adjacent cylinder. A series of concentric markings suggested the initial fracture mode to be fatigue. At this point, the crack had grown to be 0.64 cm deep and 1.42 cm long. Subsequently, the crack appeared to propagate by a different mechanism (macroscopic observation) until it became a through-thickness flaw 2.7 cm long, at which time unstable fracture occurred. It was considered likely that the latter stage of subcritical flaw growth was controlled by an environment assisted cracking process that would account for the change in fracture surface appearance, similar to that shown in Fig. 14.6a. The component was made from 2014-T6 aluminum alloy and was manufactured in such a way that the hoop stress within each chamber acted perpendicular to the short transverse direction of the original forging. From the *Damage Tolerant Handbook*,[13] the yield strength and fracture toughness of the material in this direction are given as 385 MPa and 19.8 MPa$\sqrt{m}$, respectively.

Additional data concerning the geometry of the power control unit are given below.

> chamber wall thickness $(t) = 0.84$ cm
> elliptical crack depth $(a) = 0.64$ cm
> elliptical crack length $(2c) = 1.42$ cm
> $a/2c = 0.445$
> elliptical flaw correction factor $(Q) \cong 2.2$
> bore diameter $(D) = 5.56$ cm
> through thickness crack length $(2a_1) \cong 2.7$ cm

To use the plane strain fracture toughness value in subsequent fracture calculations, it is necessary to verify that t and $a \geqslant 2.5(K_{IC}/\sigma_{ys})^2$. This condition is met for this case history and supported by the observation that the fracture surface was completely flat. The stress necessary to fracture the unit may be computed by the formula for a through-thickness flaw where

$$K_{IC} = \sigma \sqrt{\pi a}$$

Setting $K_{IC} = 19.8$ MPa$\sqrt{\text{m}}$ and $a = 1.35$ cm

$$19.8 = \sigma \sqrt{\pi (1.35 \times 10^{-2})}$$

$$\sigma = 96.1 \text{ MPa}$$

The chambers have a large diameter to thickness ratio so that pressurization could be analyzed in terms of a thin-walled cylinder formulation. Since both cylinders are pressurized, the hoop stress between cylinder bores is estimated to be

$$\sigma_{\text{hoop}} = \frac{2PD}{2t}$$

where $P = $ internal fluid pressure.

Using the component dimensions and the calculated stress level at fracture (i.e., 96.1 MPa), the pressure level at fracture P is calculated to be

$$96.1 = \frac{2P(5.56 \times 10^{-2})}{2(8.4 \times 10^{-3})}$$

$$P = 14.5 \text{ MPa}$$

Since the normal mean pressure in the cylinder bores is about 10.3 MPa and reaches a maximum of about 15.5 MPa, unstable fracture could have occurred during either normal pressurization or during pressure buildups associated with an aileron repositioning maneuver.

As mentioned above, the change in fracture mechanism when the elliptical crack reached a depth and length of 0.64 cm and 1.42 cm, respectively, could have been due to the onset of static environment assisted cracking at a stress intensity where the cracking rate became independent of the K level (i.e., Stage II behavior). For such an elliptical flaw

$$K^2 = \left[1 + 0.12\left(1 - \frac{a}{c}\right)\right]^2 \sigma^2 \frac{\pi a}{Q} \left[\frac{2t}{\pi a} \tan \frac{\pi a}{2t}\right] \tag{14-8}$$

with the result that

$$K^2 = \left[1 + 0.12\left(1 - \frac{0.64}{0.71}\right)\right]^2 \sigma^2 \frac{\pi(6.4 \times 10^{-3})}{2.2} \left[\frac{2(0.84)}{\pi(0.64)}\right]$$

$$\times \left[\tan \frac{\pi(0.64)}{2(0.84)}\right]$$

$$K = 0.134\sigma$$

Assuming that the major stresses associated with static environment assisted cracking were those associated with the mean pressure level of about 10.3 MPa, the associated hoop stress is calculated to be

$$\sigma_{\text{hoop}} = \frac{2(10.3)(5.56 \times 10^{-2})}{2(8.4 \times 10^{-3})}$$

$$= 68.2 \text{ MPa}$$

Using this stress level, the stress intensity level for the onset of static environment assisted cracking is estimated to be

$$K = 0.134\sigma$$

$$= 0.134(68.2)$$

$$= 9.2 \text{ MPa}\sqrt{m}$$

Unfortunately, no environment assisted cracking (EAC) data for this material-environment system are available to check whether the number computed above is reasonable. It is known, however, that (EAC) rates in this alloy become appreciable in a salt water environment when the stress intensity level approaches 11 MPa$\sqrt{m}$. Further material evaluations would be needed to determine whether hydraulic fluid has a similar effect on the cracking response of this alloy at stress intensity levels of about 11 MPa$\sqrt{m}$.

CASE 3: Failure of Arizona Generator Rotor Forging[14–16]

This case history does not describe a true service failure, since the rotor failed during a routine balancing test *before* it was placed in service and at an operating speed *less* than that for design operation. The forged rotor, manufactured more than 20 years ago, did not possess benefits derived from current vacuum degassing melting practices as described in Chapter Ten; consequently, a large amount of hydrogen gas was trapped in the ingot as it solidified. With time, the hydrogen precipitated from the solid to form hydrogen flakes, evidenced by disc-shaped internal flaws such as the one shown in Fig. 14.9. Investigators[14,15] concluded that these 2.5 to 3.8-cm diameter circular defects existed before the balancing test and were responsible for its failure, although no specific hydrogen flake could be identified as the critical nucleation site.

The forging material contained 0.3 C, 2.5 Ni, 0.5 Mo, and 0.1 V, exhibited room temperature tensile yield and ultimate strengths of 570 and 690 MPa, respectively, and a Charpy V-notch impact energy at the fracture temperature (27°C) of 5.4 to 16.3 J. The rotor contained a central hole along its entire bore.

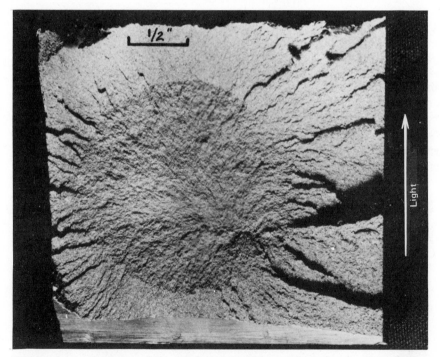

FIGURE 14.9 Hydrogen flake (dark circle) which contributed to fracture of Arizona turbine rotor.[14] (Reprinted with permission from Academic Press.)

This was done to remove the central section of the original ingot, which normally contains a relatively high percentage of inclusions and low melting point microconstituents, and to permit a more thorough examination of the rotor for evidence of any defects.[14] By introducing the bore hole, the centrifugal tangential stresses at the innermost part of the rotor are doubled according to Eqs. 14-10 and 14-11, even when the inner bore diameter is very small:

$$\sigma_{max_{\text{(solid cylinder)}}} = \frac{3+\nu}{8}\rho\omega^2 R_2^{\ 2} \tag{14-10}$$

$$\sigma_{max_{\text{(hollow cylinder)}}} = \frac{3+\nu}{4}\rho\omega^2\left(R_2^{\ 2} + \frac{1-\nu}{3+\nu}R_1^{\ 2}\right) \tag{14-11}$$

where

$$\nu = \text{Poisson's ratio}$$

$$\rho = \text{mass density}$$

$$\omega = \text{rotational speed}$$

$$R_1 = \text{inner radius}$$

$$R_2 = \text{outer radius}$$

Although one would normally try to keep stresses as low as possible, the higher stress levels associated with introduction of the bore hole are justified for the reasons cited above. Using these equations, Yukawa et al.[14] determined the maximum bore tangential stress to be 350 MPa at the fracture speed (3400 rpm).

From the above description of the Arizona rotor failure, the most reasonable stress intensity factor calibration would appear to be that associated with an internal circular flaw.[16] Assuming the worst condition, where the flaw is oriented normal to the maximum bore tangential stress, we have from Fig. 8.7

$$K_{IC} = \frac{2}{\pi}\sigma\sqrt{\pi a}$$

Using the K_{IC}-CVN relationships proposed by Barsom and Rolfe[17] and Sailors and Corten[18] (see Chapter Nine) for the transition temperature regime where

$$\frac{K_{IC}^{\ 2}}{E} = 2(CVN)^{3/2*} \quad \text{(Barsom–Rolfe)} \tag{9-7}$$

$$\frac{K_{IC}^{\ 2}}{E} = 8(CVN)^* \quad \text{(Sailors–Corten)} \tag{9-8}$$

*English units

estimates of the K_{IC} value for the rotor material were obtained and are summarized in Table 14.2.

TABLE 14.2 K_{IC}–CVN Correlations

| | Estimated K_{IC} | |
| | Barsom–Rolfe[17] | Sailors–Corten[18] |
CVN, J (ft-lb)	MPa$\sqrt{m}$ (ksi$\sqrt{in.}$)	MPa$\sqrt{m}$ (ksi$\sqrt{in.}$)
5.4–16.3	24–55	34–59
(4–12)	(22–50)	(31–54)

These values must be considered as first-order approximations in view of normal test scatter in Charpy energy measurements and the empirical nature of both Eqs. 9-7 and 9-8, but they do provide a starting point from which critical flaw sizes may be computed and compared with experimentally observed hydrogen flake sizes. (Obviously it would have been more desirable to have actual fracture toughness values to use in these computations.) For example, using the K_{IC} values derived from the Sailors–Corten relationship in Eq. 9-8, the critical flaw size range is calculated to be

$$34\text{--}59 = \frac{2}{\pi}(350)(\sqrt{\pi a})$$

$$a = 0.74 \text{ to } 2.2 \text{ cm}$$

or a hydrogen flake diameter range of about 1.5 to 4.3 cm, in excellent agreement with the observed size of these preexistent flaws. The reader should take comfort in the knowledge that hydrogen flakes have been eliminated from current large forgings by vacuum degassing techniques, and overall toughness levels of newer steels have been increased measurably.

CASE 4:   Failure of Pittsburgh Station Generator Rotor Forging[14, 16]

The Pittsburgh rotor was similar in design and material selection to the Arizona rotor described in the previous case history except that it did not contain a bore hole. Consequently, the stresses were computed from Eq. 14-10 to be roughly half those found in the Arizona rotor. On the other hand, the lack of the bore hole increased the likelihood of finding potentially damaging microconstituents along the rotor center line. As we shall see, the latter potential condition was realized and did contribute to the fracture. The Pittsburgh rotor failed on March 18, 1956 during an overspeed check. (Overspeed checks were conducted routinely after a shutdown period and before the rotor was returned to service.) The rotor was designed for 3600 rpm service and failed when being checked at

3920 rpm. It is important to note that on 10 previous occasions during its two-year life the rotor satisfactorily endured similar overspeed checks above 3920 rpm. Surely, failure during the eleventh check must have come as a rude shock to the plant engineers. One may conclude therefore, that some subcritical flaw growth must have taken place during the two-year service life to cause the rotor to fail during the eleventh overspeed test but not during any of the other 10 tests, even though these tests were conducted at higher stress levels. Macrofractographic examination revealed the probable initiation site to be an array of nonmetallic inclusions in the shape of an ellipse 5 by 12.5 cm and located nearly on the rotor center line (Fig. 14.10).[14] The maximum bore tangential stress at burst speed was found to be 165 MPa and the temperature at burst equal to 29°C. The tensile properties of the rotor material were given as 510 MPa and 690 MPa for the yield and tensile strength, respectively, with the room-temperature Charpy impact energy equal to 9.5 J.

If we take the critical flaw to be equivalent to a 5 by 12.5 cm elliptical crack—assuming that all the inclusions had linked up prior to catastrophic failure (possibly as a result of subcritical flaw growth)—the stress intensity factor at fracture could be given by

$$K = \sigma \sqrt{\pi a / Q}$$

FIGURE 14.10 Cluster of inclusions contributing to fracture of Pittsburgh turbine rotor.[14] (Reprinted with permission from Academic Press.)

The elliptical flaw shape factor Q for the condition where $a/2c = 2.5/12.5 = 0.2$ and $\sigma/\sigma_{ys} = 165/510 = 0.32$ is found from Fig. 8.5h to be 1.28. The fracture toughness of the material is then calculated to be

$$K_{IC} = 165\sqrt{\frac{\pi(2.5 \times 10^{-2})}{1.28}}$$

$$K_{IC} \approx 41 \text{ MPa}\sqrt{\text{m}}$$

This result compares very favorably with K_{IC} estimates based on the Barsom–Rolfe[17] and Sailors–Corten[18] K_{IC}–CVN correlations (Eqs. 9-7 and 9-8), where values of 37 MPa$\sqrt{\text{m}}$ and 45 MPa$\sqrt{\text{m}}$ may be computed, respectively.

Although the estimated K_{IC} value derived from the crack configuration and stress information was remarkably close to the values determined from the empirical K_{IC}–CVN correlations, it must be kept in mind that the latter values represent only a crude approximation of K_{IC}: Such derived values can vary widely because of the considerable scatter associated with Charpy energy measurements. Nevertheless, the basic merits of using the fracture mechanics approach to analyze this failure have been clearly demonstrated.

CASE 5: Stress Corrosion Cracking Failure of the Point Pleasant Bridge[19]

The failure of the Point Pleasant, West Virginia, bridge in December 1967 occurred without warning, resulting in the loss of 46 lives. Several studies were conducted immediately afterwards to determine the cause(s) of failure, since the collapse caused considerable anxiety about the safety of an almost identical bridge built around the same time and possessing a similar design and structural steel. Failure was attributed to brittle fracture of an eyebar (Fig. 14.11) that was about 17 m long, 5.1 cm thick, and 30.5 cm wide in the shank section. The ends of the bar were 70 cm in diameter and contained 29.2-cm diameter holes. It was determined that a crack had traversed one of the ligaments (the one on the top in Fig. 14.11) of the eye (along the transverse center line) with little apparent energy absorption (the fracture surface was very flat with little shear lip formation). The ligament on the opposite side of the hole suffered extensive plastic deformation before it failed, probably as a result of a bending overload. After removing the rust from the fracture surface, investigators[19] found two discolored regions covered with an adherent oxide layer. These regions were contiguous and in the shape of two elliptical surface flaws (Fig. 14.12). The size of the large flaw was

$$a = 0.3 \text{ cm}$$

$$2c = 0.71 \text{ cm}$$

$$a/2c = 0.43$$

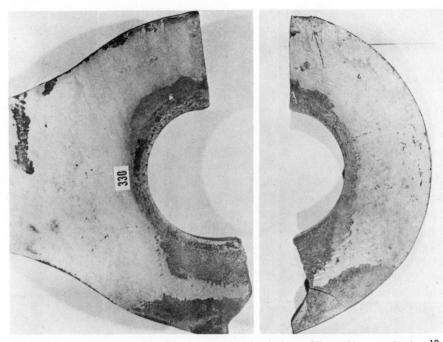

FIGURE 14.11   Fractured eyebar responsible for failure of Point Pleasant Bridge.[19] (Reprinted from *Journal of Testing and Evaluation* with permission from American Society for Testing and Materials.)

The smaller flaw had the dimensions

$$a \approx 0.1 \text{ cm}$$

$$2c = 0.51 \text{ cm}$$

$$a/2c \approx 0.2$$

Portions of the hole surface were heavily corroded, and some secondary cracks were parallel to the main fracture surface but initiated only in those regions where corrosion damage was extensive. These findings suggested the strong possibility that stress corrosion and/or corrosion fatigue mechanism(s) were involved in the fracture process. The hypothesis was further substantiated by metallographic sections which showed that the secondary cracks contained corrosion products and propagated in an irregular pattern from corrosion pits at the hole surface. Furthermore, some of these secondary cracks were opened in the laboratory, examined in the SEM and electron microprobe, and found to contain high concentrations of sulfur near the crack origin.[20] The presence of sulfur on the fracture surface was believed to be from H_2S in the air near the

FIGURE 14.12 Fracture surface of broken eyebar from Point Pleasant Bridge showing two elliptical surface flaws.[19] (Reprinted from *Journal of Testing and Evaluation* with permission from American Society for Testing and Materials.)

bridge rather than associated with manganese sulfide inclusions (commonly found in this material). The sensitivity of the bridge steel to H_2S stress corrosion cracking was verified by several tests performed on notched specimens. Fatigue crack propagation data were also obtained and used to examine the possibility that the two surface flaws had propagated instead by corrosion fatigue. Taking the maximum alternating stress on the bridge to be ± 100 MPa, Bennett and Mindlin[19] estimated that it would require over half a million load cycles to propagate a crack from a depth of 0.05 cm to one 0.25 cm deep. Since this was considered to be an unrealistically large number, it was concluded that the actual fracture mechanism was stress corrosion cracking.

Attention was then given to an evaluation of the steel's fracture toughness capacity. Using both Charpy V-notch and fracture toughness test procedures, the SAE 1060 steel (0.61 C, 0.65 Mn, 0.03 S), which had been austenitized, water quenched, and tempered for two hours at 640°C, was shown to be brittle. For example, the material was found to exhibit an average plane strain fracture toughness level of 51 MPa$\sqrt{m}$ at 0°C, the temperature of fracture. This low value is consistent with the fact that the material displayed a strong stress corrosion cracking tendency—something usually found only in more brittle engineering alloys (see Chapter Eleven). Based on a measured yield strength of 550 MPa, these results were found to reflect valid plane strain test conditions for the specimen dimensions chosen.

Estimating the stress intensity level by

$$K = 1.1\sigma\sqrt{\pi a/Q}$$

Bennett and Mindlin computed the stress level at fracture by considering only the larger surface flaw:

$$K = 1.1\sigma\sqrt{\pi a/Q}$$

$$= 1.1\sigma\sqrt{\frac{\pi(3 \times 10^{-3})}{1.92}}$$

$$= 7.7 \times 10^{-2}\sigma$$

or

$$\sigma = 13K$$

Using the range of experimentally determined K_{IC} values, (47.3 to 56.1 MPa$\sqrt{m}$), the stress level at fracture was found to be

$$\sigma = 615\text{--}730 \text{ MPa}$$

This represents an upper bound range of the fracture stress, since allowance was not made for the presence of the smaller contiguous elliptical flaw. If one assumes the crack to be elliptical with a maximum depth of 0.3 cm but with $2c = 1.6$ cm, then $a/2c \approx 0.19$ and $Q = 1.05$. This assumption should lead to a slight underestimate of the stress level.

$$K = 1.1\sigma\sqrt{\frac{\pi(3 \times 10^{-3})}{1.05}}$$

$$\sigma = 9.6K$$

Again using the K_{IC} range of 47.3 to 56.1 MPa$\sqrt{m}$, a lower bound stress range is found to be

$$\sigma = 455\text{--}540 \text{ MPa}$$

It is concluded that the actual stress range for failure was

$$455\text{--}540 < \sigma_{\text{actual}} \ll 615\text{--}730 \text{ MPa}$$

It is seen that the failure stress is approximately equal to the material yield strength. Since the shank section of the eyebar was recommended for a design stress of 345 MPa, Bennett and Mindlin concluded that stresses on the order of the yield strength could exist at the considerable stress concentration associated with this region.

On the basis of this detailed examination, it was concluded that the critical flaw was developed within a region of high stress concentration and progressed by a stress corrosion cracking mechanism to a depth of only 0.3 cm before fracture occurred. Consequently, the hostile environment, the inability to adequately paint the eyebar and thus protect it from atmospheric attack, the low fracture toughness of the material and the high design stress all were seen to contribute to failure of the bridge. It should come as no surprise that the combination of low toughness and high stress would result in a small critical flaw size (see Eq. 8-22).

CASE 6:   Weld Cold Crack-Induced Failure of Kings Bridge, Melbourne, Australia[21]

On a cold winter morning in July 1962, while a loaded truck with a total weight of 445 kN was crossing the bridge, a section of this 700-m long elevated four-lane freeway fractured, causing a portion of the bridge to drop 46 cm. Examination of the four main support girders that broke revealed that all had suffered some cracking *prior* to installation (Fig. 14.13). Indeed, subsequent welding tests established that a combination of poor detail design of the girder flange cover plate, poor weldability of the steel, poor welding procedure, and failure to properly dry low-hydrogen electrodes before use contributed to the formation of weld cold cracks located at the toe of transverse welds at the ends of the cover plates. In three of these girders, 10-cm long through-thickness cracks had developed before erection but none were ever discovered during inspection. In addition, it was determined that girder W14-2 was almost completely broken before the span failed. (The crack in this girder extended across the bottom flange and 1.12 m up the web.)

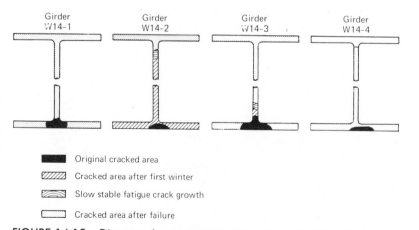

Girder W14-1   Girder W14-2   Girder W14-3   Girder W14-4

▬ Original cracked area

▨ Cracked area after first winter

▧ Slow stable fatigue crack growth

▭ Cracked area after failure

FIGURE 14.13   Diagram showing extent of cracking of girders from Kings Bridge, Australia.[21] (Courtesy of Dr. Ronald Madison.)

The collapse of the span was traced to failure of girder W14-3, which contained a T-shaped crack extending 12.5 cm across the bottom flange and 10 cm up the web (Fig. 14.13). Madison[21] postulated that the stress intensity condition at instability could be approximated by the superposition of two major components. One major K component was attributed to uniform bending loads acting along the flange and perpendicular to the 12.5-cm long flange crack. Accordingly

$$K = \sigma\sqrt{\pi a} \cdot \sqrt{\sec \pi a / W} \qquad (14\text{-}12)$$

where

$$\sqrt{\sec \pi a / W} = \text{finite width correction}$$

$$\sigma = \text{bending stress, 83 MPa}$$

$$a = 6.25 \text{ cm}$$

$$W = 41 \text{ cm}$$

$$K = 83\sqrt{\pi(6.25 \times 10^{-2})\sec \pi(6.25/41)}$$

$$= 39 \text{ MPa}\sqrt{m}$$

The secant form of the finite width correction (Eq. 14-12) is found to be more accurate than the corresponding tangent function given in Eq. 14-7, particularly at large a/W or $a/2t$ ratios.[22] Note that the panel width W and the thickness t would be the appropriate dimension in this relationship for through thickness and surface flaw configurations, respectively.

The second major K component was related to load transfer from the web, which produced wedge force loads extending 10 cm along both sides of the flange crack. These loads reflect residual stresses generated by the flange to web welds. For this configuration the K calibration is[3,4]

$$K = \frac{\sigma\sqrt{a}}{\sqrt{\pi}}\left\{ \sin^{-1}\frac{c}{a} - \left(1 - \frac{c^2}{a^2}\right)^{1/2} + 1 \right\} \qquad (14\text{-}13)$$

where

$$2a = \text{crack length, 12.5 cm}$$

$$\sigma = \text{wedge force, 262 MPa}$$

$$2c = \text{length of wedge force, 10 cm}$$

$$K = \frac{262\sqrt{6.25 \times 10^{-2}}}{\sqrt{\pi}}\left\{ \sin^{-1}\frac{5}{6.25} - \left[1 - \left(\frac{5}{6.25}\right)^2\right]^{1/2} + 1 \right\}$$

$$K = 49 \text{ MPa}\sqrt{m}$$

Therefore, $K_T = 39 + 49 = 88$ MPa$\sqrt{m}$. Note the significant contribution of the residual stresses. This value was found to be in reasonably good agreement with the dynamic fracture toughness of samples prepared from the bridge steel.

14.5.1 Additional Comments

Before concluding this section, it is important to comment further on the general problem of fatigue and fracture in welded bridges. Since these structures usually are very complicated because many cover plates, stiffeners, attachments, and splices are added to the basic beam, it is important to recognize the potential danger associated with a particular weld detail. Fisher[23] has conducted an extensive study of this problem and has proposed several categories of relative attachment detail severity as shown schematically in Fig. 14.14. Categories E and A represent the potentially most damaging and least damaging weld details, respectively. One important function of these diagrams is to direct the attention of the field engineer to the most critical details in the bridge design so that he does not waste time examining those areas experiencing a lower stress concentration. Also, the differences in stress concentration associated with categories A, B, C, D, and E have been used by the American Association of State Highway and Transportation Officials to arrive at allowable stress ranges for each detail.[23] For example, a category E detail is allowed only one-third the stress range of a category A region when a cyclic life of up to 2×10^6 cycles is anticipated and only about one-fifth that value when more than 2×10^6 cycles are desired.

As a final note, the engineer should remember the subtle but important difference between load and displacement controlled conditions governing the behavior of a given structure. For example, for the case of the two fixed-ended beams shown in Fig. 14.15, the maximum flexural stress for load-induced and displacement-induced conditions are

$$\text{load-induced:} \quad \sigma_{max} = \frac{PLc}{8I} \tag{14-14}$$

where

$$\sigma_{max} = \text{maximum flexural stress}$$

$$P = \text{point load}$$

$$L = \text{beam length}$$

$$c = \text{distance to outermost fiber}$$

$$I = \text{moment of inertia}$$

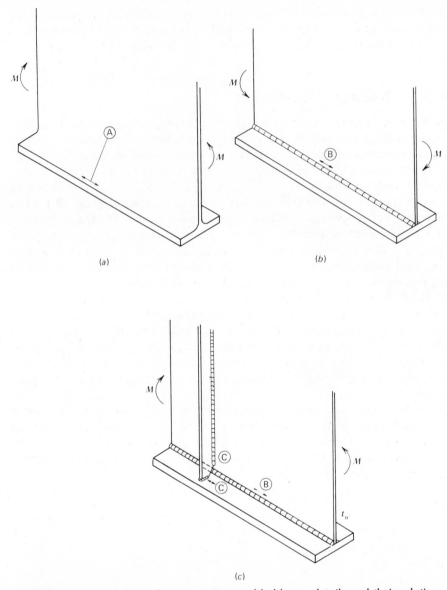

FIGURE 14.14 Drawings showing various welded beam details and their relative stress concentration severity (increasing from category A to E).[23] (Reprinted from *Research Results Digest, 59*, with permission of the National Cooperative Highway Research Program.)

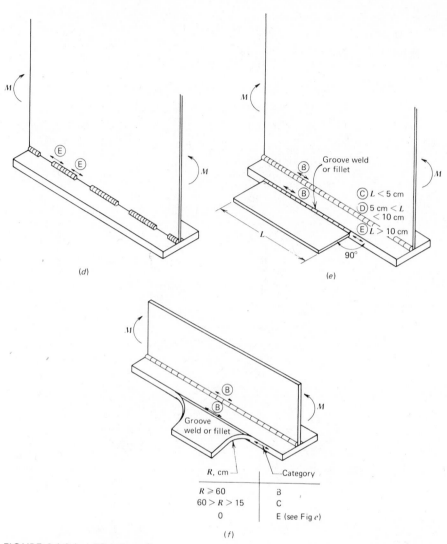

(d)

(e)

Groove weld or fillet

Ⓑ

Ⓑ

Ⓒ $L < 5$ cm

Ⓓ 5 cm $< L < 10$ cm

Ⓔ $L > 10$ cm

90°

L

Ⓔ

Ⓔ

Ⓑ

Ⓑ

Groove weld or fillet

R, cm	Category
$R \geqslant 60$	B
$60 > R > 15$	C
0	E (see Fig e)

(f)

FIGURE 14.14   *(Continued)*

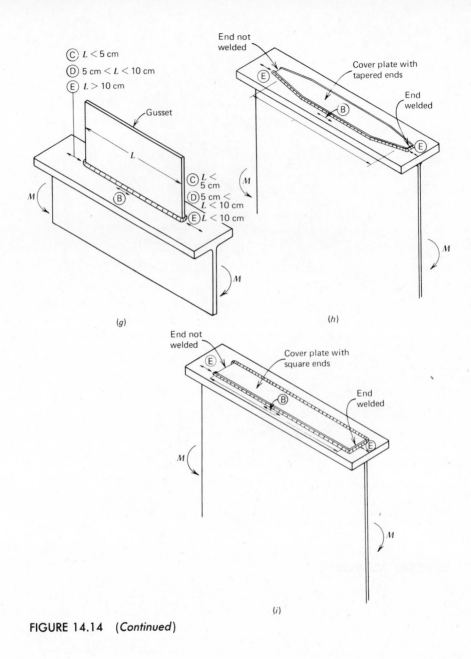

FIGURE 14.14 (*Continued*)

578

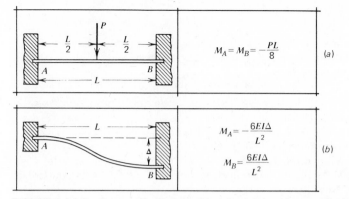

$$M_A = M_B = -\frac{PL}{8}$$ (a)

$$M_A = -\frac{6EI\Delta}{L^2}$$

$$M_B = \frac{6EI\Delta}{L^2}$$ (b)

FIGURE 14.15   Beams fixed at both ends. (a) Load controlled; (b) displacement controlled.

and

displacement-induced: $$\sigma_{max} = \frac{6Ec\Delta}{L^2}$$ (14-15)

where

$$E = \text{modulus of elasticity}$$

$$\Delta = \text{displacement}$$

The passage of a fully loaded truck across a bridge might represent a situation which could be analyzed by Eq. 14-14, but if one of the bridge foundations had settled by an amount Δ then Eq. 14-15 would prove more correct. Identification of the importance of load or displacement in controlling the bridge response is critical, since the design changes one would make to improve either load or displacement resistance of a given beam are mutually incompatible. For instance, we see from Eq. 14-14 that for a given load P, the load-bearing capacity of a beam is enhanced by *decreasing* its unsupported length and/or *increasing* its moment of inertia or rigidity. By sharp contrast, we see from Eq. 14-15 that for a given displacement Δ, the flexural stress may be reduced by *increasing* the beam length and/or *decreasing* its rigidity. For example, adding a cover plate to a flange experiencing load control would enhance its fatigue life but would prove to be counterproductive if the beam were displacement controlled. A discussion of such a dilemma with regard to the Lehigh Canal Bridge is given by Fisher et al.[24]

REFERENCES

1. *Engineering News-Record*, **94**(5), Jan. 29, 1925, p. 188.
2. R. J. Gran, F. D. Orazio, Jr., P. C. Paris, G. R. Irwin, and R. W. Hertzberg, AFFDL-TR-70-149, Mar. 1971.
3. P. C. Paris and G. C. M. Sih, ASTM *STP 381*, 1965, p. 30.
4. H. Tada, P. C. Paris, and G. R. Irwin, *The Stress Analysis of Cracks Handbook*, Del Research Corporation, Hellertown, Pa., 1973.
5. G. C. M. Sih, "Handbook of Stress Intensity Factors," Lehigh University, 1973.
6. O. L. Bowie, *J. Math. Phys.*, **35**, 1956, p. 60.
7. D. Broek and H. Vlieger, *Int. J. Fract. Mech.*, **8**, 1972, p. 353.
8. D. Broek, National Aerospace Laboratory Report NLR *TR 72134 U*, Amsterdam, The Netherlands, Nov. 1972.
9. *Fracture Mechanics Course Notes*, Del Research Corporation, Hellertown, Pa., 1973.
10. *Metals Handbook*, Vol. 10, ASM, Metals Park, Ohio, 1975.
11. *Source Book in Failure Analysis*, American Society for Metals, Metals Park, Ohio, Oct. 1974.
12. C. M. Carmen and J. M. Katlin, ASME Paper No. 66-Met-3, *J. Basic Eng.*, 1966.
13. J. E. Cambell, W. E. Berry, C. E. Feddersen, *Damage Tolerant Design Handbook*, MCIC-HB-01, Sept. 1973.
14. S. Yukawa, D. P. Timio, and A. Rubio, *Fracture*, Vol. 5, H. Liebowitz, ed., Academic Press, New York, 1969, p. 65.
15. C. Schabtach, E. L. Fogelman, A. W. Rankin, and D. H. Winnie, *Trans.*, ASME, **78**, 1956, p. 1567.
16. R. J. Bucci and P. C. Paris, Del Research Corporation, Hellertown, Pa., Oct. 23, 1973.
17. J. M. Barsom and S. T. Rolfe, *Eng. Fract. Mech.*, **2**(4) 1971, p. 341.
18. R. H. Sailors and H. T. Corten, ASTM *STP 514*, Part II, 1972, p. 164.
19. J. A. Bennett and H. Mindlin, *J. Test. Eval.*, **1**(2), 1973, p. 152.
20. D. B. Ballard and H. Yakowitz, *Scanning Electron Microscope 1970*, IITRI, Chicago, Ill., April 1970, p. 321.
21. R. Madison, Ph.D. Dissertation, Lehigh University, 1969.
22. G. R. Irwin, H. Liebowitz, and P. C. Paris, *Eng. Fract. Mech.*, **1**, 1968, p. 235.
23. J. W. Fisher, *NCHRP Research Results Digest*, **59**, Mar. 1974.
24. J. W. Fisher, B. T. Yen, and N. V. Marchica, Fritz Engineering Laboratory Report No. 386.1, Lehigh University, Bethlehem, Pa., Nov. 1974.
25. F. A. Heiser and R. W. Hertzberg, *JISI*, **209**, 1971, p. 975.
26. M. O. Speidel, *The Theory of Stress Corrosion Cracking in Alloys*, J. C. Scully, ed., NATO, Brussels, 1971.

APPENDIX A
REPLICATION TECHNIQUES AND IMAGE INTERPRETATION

Care should be exercised in cleaning fracture surfaces prior to their replication. Foreign dirt particles, grease and oil, and loosely clinging rust should be removed cautiously with an inert solvent such as acetone. Chemically active alkaline or acid solutions should not be used, since they will etch the fracture surface and obliterate important fracture markings. Similarly, debris should not be removed from the fracture surface with an abrasive instrument since this, too, will mar the fracture surface. A cellulose acetate tape, softened with acetone, when pressed onto the fracture surface and then removed strips away loosely clinging dirt. The sample can also be placed in acetone and gently vibrated in an ultrasonic cleaner. After the cleaning process, the specimen fracture surface should be preserved before replication in a dry environment or by the application of an acetone soluble lacquer spray.

Several replication procedures have been developed in the metallurgical laboratory. In the one-step process, a carbon film is vacuum deposited directly onto the fracture surface and subsequently floated free by placing the sample in an acid bath to dissolve the sample. Although this technique produces a high resolution replica, it is not employed often since the specimen is destroyed in the replication process. The most commonly used technique is a nondestructive, two-stage process that produces a replica with reasonably good resolution. A presoftened strip of cellulose acetate is pressed onto the fracture surface and allowed to dry. (To produce replicas from the fracture surfaces of polymeric materials that would dissolve in acetone, the author has found a water slurry of polyacrylic acid to be an effective replicating media.) When it is stripped from the specimen, the tape carries an impression of the fracture surface topography. Since this tape is opaque to the electron beam, further steps in the replication procedure are necessary. A layer of heavy metal is deposited at an acute angle on the side of the tape bearing the fracture impression. This is done to improve the eventual contrast of the replica. Finally, a thin layer of carbon is vapor deposited onto the tape. The plastic tape, heavy metal, carbon composite is then placed in a bath of acetone where the plastic is dissolved. (The polyacrylic acid replica would be dissolved in a water bath.) In the final step, the heavy metal, carbon replica is removed from the acetone bath and placed on mesh screens for viewing in the electron microscope. Since the viewing screens are only 3 mm in diameter, selection of the critical region(s) for examination is most important. This factor emphasizes the need for a carefully conducted macroscopic examination that should direct the examiner to the primary fracture site. By adhering to the recommended procedures for the preparation of replicas, little difficulty should be encountered. For a more extensive discussion of replica preparation techniques see Refs. 13 and 14 in Chapter Seven.

An important part of electron fractography is the interpretation of electron images in terms of actual fracture mechanisms. In addition to the references just cited, the reader is referred to the work of Beachem.[1,2] These results are summarized briefly. The fluorescent screen of the microscope (and the photographic film) will react to those electrons that are able to penetrate the replica, that is, the more electrons that penetrate the replica, the brighter the image and vice versa. We see from Fig. A.1 that the vertical walls of the raised and depressed regions at A and B, respectively, show up as black lines, since the electrons cannot penetrate the replica segment with thickness T_2. On the other hand, "shadows" are produced in the absence of the heavy metal (more resistant to electron penetration), which had been deposited at an angle to the fracture surface. These regions appear as white areas *behind* a raised particle and *within* a depressed region associated with a replica thickness T_3. The *location* of the shadow helps to determine whether the region being examined is above or below the general fracture plane. At this point, it is crucial to recognize the image reversal between one- and two-stage replicas. If the substrate in Fig. A.1 were the actual fracture surface, then region A would be correctly identified as lying above the fracture plane while region B would lie below this surface. If the substrate were the plastic, however, the previous conclusion would be completely incorrect. That is, an elevated region on the replica really represents a depressed region on the actual fracture surface. Therefore, when two-stage replicas are prepared, the images produced should be interpreted in reverse fashion—*everything that looks up is really down and vice versa.*

Finally, care should be exercised when attempting to determine the meaning of black bands on the photographs. For the example shown in Fig. A.1, these bands represent simple vertical elevations associated with regions A and B. As

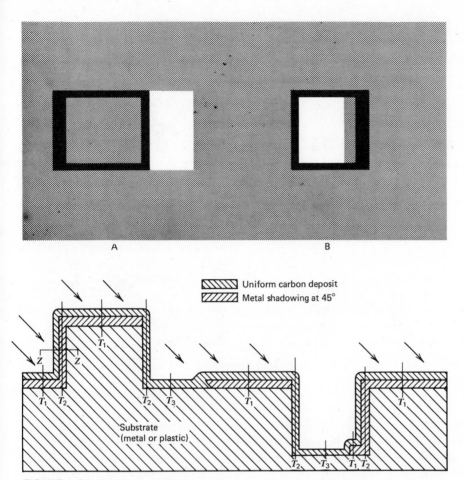

FIGURE A.1 Schema revealing stages in preparation of a fracture surface replica. Replica appears opaque to electron beam when effective thickness is T_2. Brightest image occurs in "shadow" where replica has thickness T_3.[1] (Courtesy of Cedric D. Beachem, Naval Research Laboratory.)

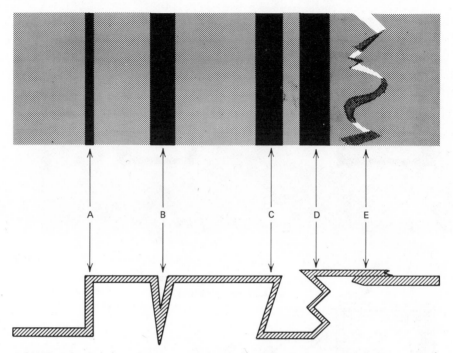

FIGURE A.2 Schema showing different interpretations of opaque black bands.[1] (Courtesy of Cedric D. Beachem, Naval Research Laboratory.)

seen in Fig. A.2, other more complex explanations are possible. In any case, these opaque bands reflect an elevation on the fracture surface that should reveal fracture mechanism details if reoriented with respect to the electron beam (Fig. A.3).

REFERENCES

1. C. D. Beachem, NRL Report *6360*, U.S. Naval Research Laboratory, Washington, D.C., Jan. 21, 1966.
2. C. D. Beachem, *Fracture*, Vol. I, H. Liebowitz, ed., Academic Press, New York, 1968, p. 243.

FIGURE A.3   Electron fractographs revealing microvoid coalescence and a large elevation. (*a*) Replica viewed at zero tilt, note large opaque band; (*b*) same area as *a* but replica tilted 24 degrees about horizontal axis. Note fracture details on fracture wall. (Courtesy of R. Korastinsky, Lehigh University.)

AUTHOR INDEX

Numbers indicate pages on which author is cited. Numbers in parentheses represent the number of author citations on that page.

Findley, W. N., 463
Fisher, J. C., 69(2), 98
Fisher, J. W., 580(2)
Flom, D. G., 69
Flory, P. J., 225
Fogelman, E. L., 580
Forman, A. J., 69
Forman, R. G., 537
Forrest, P. J., 462
Forsyth, P. J. E., 463, 537(2)
Forty, A. J., 411
Frank, F. C., 225
Frazier, R. H., 373
Freiman, S. W., 374
Frenkel, J., 69
Friedel, J., 98, 127
Frost, H. J., 175
Frost, N. E., 537
Fuller, E. R., 538

Gaggar, S. K., 539
Gallagher, J. P., 538
Garofalo, F., 174, 175
Garstone, J., 98
Geil, P. H., 224(4)
Gensamer, M., 127
Gent, A., 224
Gerberich, W. W., 374
Gertsch, W. J., 225
Gilman, J. J., 69(3), 98(2), 253
Golden, J. H., 225
Goldhoff, R. M., 175
Goodenow, R., 373
Goodier, J. N., 295
Goodman, S. R., 127(3)
Gordon, J. E., 373
Gran, R. J., 580
Grant, N. J., 174(2)
Green, D., 127
Greenough, A. P., 463
Griffith, A. A., 295
Groover, R. E., 412
Gross, J., 322
Grossman, M. A., 412
Groves, G. W., 98, 99
Guard, R. W., 69
Gurnee, E. F., 224
Guthrie, D. E., 462
Guy, A. G., 69

Hackett, L. H., Jr., 413

Hagel, W. C., 175
Hahn, G. T., 98, 295, 373, 374(2)
Hall, E. O., 98, 127
Hammant, B. L., 225
Hancock, G. G., 412
Hanna, G. L., 412
Hardrath, H. F., 462, 538(3)
Hargreaves, M. E., 98
Harrigan, M. J., 373
Harris, W. J., 463
Hartbower, C. E., 323
Hartman, A., 538(2)
Hasselman, D. P. H., 374
Hawthorne, J. R., 413(3)
Hayward, D. O., 412
Hazell, E. A., 225
Heijboer, J., 224
Heiser, F. A., 127, 373, 538, 580
Helgeland, O., 463
Hellwege, K. H., 14
Herman, H., 463
Herring, C., 175
Hertzberg, R. W., 127(2), 176, 225, 373(2),
    463, 493, 537(5), 538(8), 539(10), 580(2)
Hickerson, J. P., Jr., 463, 539
Hieronymus, W. S., 373
Hill, D. C., 253
Hirsch, P. B., 69(3), 98
Hirschberg, M. H., 463(2)
Hirth, J. P., 98(2)
Hirthe, W. M., 175
Hodge, J. M., 373
Hofmann, W., 412
Holick, A. S., 224
Hollomon, J. H., 37, 412
Homes, G., 539
Honeycombe, R. W. K., 98
Hooke, R., 37
Hoppe, W., 537
Howie, A., 69
Hoy, C. J., 323
Hren, J. J., 98
Hsiao, C. C., 224(3)
Hu, H., 127(5)
Hubert, J. F., 175
Hudak, S. J., Jr., 412
Hudson, C. M., 537, 538
Hughes, T. F., 538
Hull, D., 69, 127, 224
Hunsicker, H. Y., 374
Huntington, H. B., 14

Quist, W. E., 373

Rabinowitz, S., 224, 463(2), 539
Radon, J. C., 375
Raffo, P. L., 322
Rahm, L. F., 224
Raj, R., 175
Rankin, A. W., 580
Ratner, S. B., 539
Rauls, W., 412
Read, B. E., 224
Read, W. T., Jr., 69
Reed-Hill, R. E., 127
Rellick, J. R., 413
Rice, J. R., 295(2), 296, 538
Richards, C. E., 537
Riddell, M. N., 539(2)
Ripling, E. J., 87, 412
Roberts, C., 463
Roberts, R., 537, 538(2)
Roberts, W. T., 127
Robertson, R. E., 224
Robertson, T. S., 323
Robertson, W. D., 412
Robinson, S. L., 175
Rodriguez, F., 224
Rolfe, S. T., 37, 323(2), 412, 537, 580
Roscoe, R., 98
Rosen, B. W., 69, 224
Rosen, S. L., 224
Rosenfield, A. R., 295, 374(3)
Robinson, P. M., 98
Rubio, A., 580
Ryder, D. A., 537

Sailors, R. H., 323, 580
Sandor, B. I., 463
Sargent, C. M., 127
Sauer, J. A., 224(3)
Schabtach, C., 580
Schijve, J., 538(6)
Schilling, P. E., 373
Schmid, E., 98
Schmidt, R. A., 375, 493, 538
Scott, H. G., 98
Scully, J. C., 412, 580
Seeger, A., 98
Senz, R. R., 373
Serpan, C. Z., Jr., 413
Sherby, O. D., 174(5), 175(4)
Sherrett, F., 463

Shih, C. H., 412
Shoemaker, A. K., 37
Sih, G. C., 295(2), 537, 580(2)
Simnad, M. T., 175
Sims, C. T., 175
Sinclair, G. M., 37, 462
Sines, G., 463
Skibo, M. D., 539(3)
Smallman, R. E., 127
Smith, C. S., 127
Smith, G. C., 537
Smith, H. L., 374
Smith, J. H., 374
Smith, R. R., 225
Smith, R. W., 463
Solomon, H. D., 538
Sommer, A. W., 374
Southern, J. H., 225
Speidel, M. O., 412(2), 580
Spuhler, E. H., 373
Srawley, J. E., 295, 323
Stables, P., 412
Staehl, R. W., 411(2)
Staley, J. T., 374
Steele, L. E., 413(2)
Steigerwald, E. A., 412(2)
Stein, D. F., 98, 413(2)
Steinman, J. B., 412
Sternstein, S. S., 225(3)
Steven, W., 413
Stinkas, A. V., 539
Stokes, R. J., 373
Stoloff, N. S., 127, 412, 538
Sutton, W. H., 69
Swanson, S. R., 537
Swenson, D. O., 537

Tada, H., 295, 580
Tavernelli, J. F., 463
Taylor, G. I., 98(2)
Tegart, W. J. McG., 69, 98(2), 253
Tetelman, A. S., 323, 374, 412, 413
Thomas, D. A., 127
Thomas, G., 373, 374
Thompson, D. S., 374
Thompson, E. R., 176
Thomson, J. B., 539
Thomson, R., 69
Thornton, P. R., 98
Timio, D. P., 580
Timoshenko, S. P., 37

SUBJECT INDEX

Standard linear solid, 202-205
Steady state creep, *see* Creep rate
Stereographic projection, 86
Stereoregularity, 186
Strain, engineering, 3
   true, 4
Strain controlled-cyclic life response, 439-459
   cyclic strain hardening and softening, 441-447
   hysteresis behavior, 440-447, 452, 453
Strain hardening, crystalline polymers, 33
   metals, 18
   theories of, 91-96
Strain hardening coefficient, cyclic, 446-451, 454-459
   monotonic, 18, 25, 81
Stress, engineering, 3
   true, 4
Stress concentration factor, 236-240, 269
Stress corrosion cracking, 378, 379, 570-573
   see also Environment assisted cracking
Stress intensity factor, 265-270, 551-554
   effective, 275, 488-490, 499-501
Stress-strain curves, elastic, 6
   elastic-homogeneous plastic, 16-18
   elastic-homogeneous plastic-heterogeneous plastic, 31, 32
   elastic-heterogeneous plastic, 30, 31
   generalized single-crystal, 91-95
   serrated, 30, 31
   stress concentration effect of, 242, 243
   temperature and strain rate dependence of, 35-37, 242, 243
Superplasticity, 152-155

Temper embrittlement, 401-404
Temperature-compensated time parameter, 138
Tensile strength, metal alloys, 20
   polymers, 20, 21
   temperature and strain rate dependence of, 35-37
   ultimate, 19
Tertiary creep, 133
Theoretical crystal strength, 41, 231, 233, 234
300°C embrittlement, 400, 401
Thermal fatigue, 521-525
Tilt boundary, 60

Time-temperature superposition, 198-201
Toughness tensile specimen, 27, 30
   see also Fracture toughness
Transformation induced plasticity (TRIP), 353, 354
Transient creep, 131-133, 136, 137
Transition temperature, 300-312, 358, 361
   energy criteria, 301, 302
   fracture appearance criteria, 302
   lateral contraction criteria, 302
   size effect of, 306-311
Twin, annealing, 116
   composition plane, 105
   deformation, 101
   shape, 106
   shear strain, 106
   strain ellipsoid, 109-111
   strain energy of, 111
   thickness, 107
Twinning, atomic shuffles in, 113
   BCC crystals, 114
   FCC crystals in, 116
   HCP crystals in, 108

Vacancy diffusion, dislocation climb, 49, 50
Vacuum arc remelting, 336, 337
Viscoelasticity, creep modulus, 197
   linear, 196
   mechanical analogs, 202-205
   relaxation modulus, 197
   time-temperature superposition, 198-201
Viscosity, 203
Voigt model, 202-205

Wavy glide, 79
Weld cracking, 393, 394, 573-579

Yield point, carbon and nitrogen effect of, 91
   ceramic and ionic crystals in, 90
   iron based alloys, 31
Yield strength, critical resolved, 83
   cyclic, 446
   grain size effect of, 97
   metal alloys, 20
   0.2% offset, 19
   polymers, 20, 21
   temperature and strain rate dependence of, 35-37

MATERIALS INDEX

Acetal, applications, 188
  characteristics, 188
  fatigue crack propagation, 527
  tensile strength of, 20
Acrylonitrile-butadiene-styrene (ABS), fatigue
    crack propagation, 526, 528, 530
  tensile strength, 20
  toughness, 223
Aluminum, activation energy for creep, 141,
    142
  creep data, 146, 169, 170
  creep rate of, 144
  elastic anisotropy, 14
  elastic compliances, 13
  elastic constants, 13
  elastic modulus, 8
  rolling texture of, 124
  stacking fault energy of, 78, 81
  strain hardening coefficient, 18, 81
  theoretical and experimental yield strength,
    42
  twin elements, 105
Aluminum alloys, alloy element effect on, 347
  Charpy impact energy, 300, 311, 313

creep data, 170
environment assisted cracking, 386-390, 393,
    394
failure analyses, 562-564
fatigue crack propagation, 469-472, 482-484,
    487, 489-491, 495, 498, 502, 503, 515
fracture toughness, 285, 287, 333, 348, 356,
    357, 369
iron and silicon effect on, 344, 346-349
mechanical fibering of, 118
strain controlled-cyclic life response, 446,
    449, 456
stress controlled-cyclic life response, 422,
    424, 425
tensile and yield strength, 20
Aluminum oxide, elastic modulus, 8
  fracture toughness, 370
  slip system, 75
  whisker strength, 44, 233
Amorphous polymers, crazing, 212-216
  cyclic strain response, 453
  fatigue crack propagation, 525-537
  shear yielding, 214-216
  structure, 194

Anthracene, yield behavior, 83
Antimony, temper embrittlement effect on, 402-404
Arsenic, temper embrittlement effect on, 403

Beryllium, fracture mode transition, 281
  interplanar angles in, 110
  rolling texture of, 124
  slip plane, 73
  strain ellipsoid of, 110
  theoretical and experimental yield strength, 42
  twin elements, 105
Boron carbide (B_4C), whisker strength, 44
Brass, deformation markings in, 103
  fatigue crack propagation, 493
  rolling texture of, 124, 126
  stacking fault energy of, 78
  strain controlled-cyclic life response, 449
  strain hardening coefficient, 18

Cadmium, atom movements in twin of, 114
  elastic compliances, 14
  elastic constants, 14
  elastic modulus, 8
  rolling texture of, 124
  slip plane, 73
  theoretical and experimental yield strength, 42
  twin elements, 105
Calcium fluoride (CaF_2), slip system, 75
Carbides, role in fracture of steel, 341, 343
Carbon, steel toughness, effect of, 340-345, 351, 352
  whisker strength, 44
Cast iron fatigue, 433, 434
Cesium chloride, slip system, 75
Chromium, elastic modulus, 8
Cobalt, elastic compliances, 14
  elastic constants, 14
  high temperature strength, 172, 173
  rolling texture of, 124
  slip plane, 73
Concrete, fracture toughness, 370
Copper, creep data, 146
  cyclic strain hardening, softening, 443
  elastic anisotropy, 14
  elastic compliances, 13
  elastic constants, 13
  elastic modulus, 8
  fatigue crack propagation, 493

neutron irradiation embrittlement effect on, 409, 410
  rolling texture of, 124
  stacking fault energy of, 78, 81
  strain controlled-cyclic life response, 449, 452
  strain hardening coefficient, 18, 81
  theoretical and experimental yield strength, 42
  twin elements, 105
  whisker strength, 44
Crystalline polymers, cold drawing of, 33
  cyclic strain response, 453
  deformation mechanisms, 211, 212
  density, 187
  fatigue crack propagation, 525-527
  fibers, 221
  lamellae, 191-193
  microfibrils, 211, 212
  spherulitic structure of, 191-193
  tensile response, 32
  tie molecules, 193, 212

Diamond, elastic modulus, 8
  slip system, 75

Ethane, conformation potential energy, 183, 184

Germanium, deformation mechanism map, 157
  slip system, 75
  stress-strain curves of, 92
Glass, elastic modulus, 8
  fracture toughness, 370
Gold, creep data, 146
  elastic anisotropy, 14
  elastic compliances, 13
  elastic constants, 13
  elastic modulus, 8
  stacking fault energy of, 78
  twin elements, 105

Hydrogen, embrittlement, 392, 393-397

Iron, creep data, 147, 165
  deformation twins in, 107
  dislocation cell structure in, 95
  elastic anisotropy, 14
  elastic compliances, 13
  elastic constants, 13
  elastic modulus, 8
  Lüder strain of, 31

rolling texture of, 124
steady state creep rate of, 134
strain hardening coefficient, 18
theoretical and experimental yield strength, 42
twin elements, 105
whisker strength, 44, 233
yield point of, 31

Lead, creep data, 146
superplastic behavior, 153, 154
Lithium fluoride, dislocation etch pits in, 54
elastic compliances, 13
elastic constants, 13
slip system, 75
stress-strain curves of, 92

Magnesium, deformation twins in, 107
elastic compliances, 14
elastic constants, 14
elastic modulus, 8
rolling texture of, 124
slip plane, 73
theoretical and experimental yield strength, 42
twin elements, 105
Magnesium alloys, fatigue crack propagation, 469
tensile and yield strengths, 20
Magnesium oxide, elastic anisotropy, 14
elastic compliances, 13
elastic constants, 13
slip system, 75
Microfibrils, 211, 212
Molybdenum, creep data, 147
elastic compliances, 13
elastic constants, 13
fatigue crack propagation, 469
rolling texture of, 124
theoretical and experimental yield strength, 42

Nickel, creep data, 146
deformation mechanism map, 158, 159, 162
elastic compliances, 13
elastic constants, 13
elastic modulus, 8
rolling texture of, 124
stacking fault energy of, 78
theoretical and experimental yield strength, 42

twin elements, 105
whisker strength, 44
Nickel alloys, creep data, 170-174
deformation mechanism map, 160-163
fatigue crack propagation, 493
high temperature strength, 172-174
strain controlled-cyclic life response, 446-449
Niobium, elastic modulus, 8
rolling texture of, 124
theoretical and experimental yield strength, 42
Nitrogen, dislocation locking by, 362, 363
Nylon 66, applications, 188
characteristics, 188
elastic modulus, 8
fatigue crack propagation, 526, 527, 530-532
tensile strength of, 21

Oxygen, embrittlement of titanium alloys, 336

Phosphorus, effect on fracture toughness in steel, 336
temper embrittlement, effect on, 403
Plasticizers, 195
antiplasticizer, 222
toughness effect on, 222
Polycarbonate, elastic modulus, 8
fatigue crack propagation, 526-528, 530-534
fracture toughness, 370
properties, effect of MW, 218
tensile strength of, 21
Polychlorotrifluoroethylene, spherulites, 193
Polyethylene, applications, 188
characteristics, 188
elastic modulus, 8
extended chain conformation, 183
fatigue crack propagation, 526, 530
repeat unit, 185
single crystal, 192
tensile strength of, 21
Polyethylene terephthalate, amorphous structure, 194
Polymer additions, blowing agents, 195
crosslinking agents, 196
fillers, 195
pigments and dyestuffs, 195
plasticizers, 195
stabilizers, 195
Poly(methyl methacrylate), applications, 189
characteristics, 189
elastic modulus, 8

fatigue crack propagation, 526-528, 530-532
fracture toughness, 370
glass transition temperature, 199
thermal fatigue, 521
time-temperature superposition, 201
Polyphenylene oxide, crazes, 213
fatigue crack propagation, 527, 530-532
Polypropylene, applications, 188
characteristics, 188
elastic modulus, 8
repeat unit, 185
stereoregularity, 186
Polystyrene, dynamic response, 209
elastic modulus, 8
fatigue crack propagation, 527-532, 536
fracture toughness, 370
glass transition temperature, 199
high impact toughness of, 223
repeat unit, 185
strength-orientation dependence, 220
tensile strength of, 21
Polysulfone, fatigue crack propagation, 527,
530-532
yield strength of, 21
Poly(tetrafluorethylene) (Teflon), applications,
189
characteristics, 189
tensile strength of, 20
thermal fatigue, 522-524
Poly(vinyl chloride), applications, 189
characteristics, 189
fatigue crack propagation, 527, 530-532, 534,
535
Poly(vinyl fluoride), repeat unit, 185
Poly(vinylidene fluoride) fatigue crack propagation, 527, 530-532
tensile strength of, 21

Quartz, elastic modulus, 8

Rare earth, additions to steel, 339-342

Silica, whisker strength, 233, 234
Silicon, Frank-Read source in, 67
stress-strain curves of, 92
whisker strength, 233
Silicon carbide, elastic modulus, 8
fracture toughness, 370
whisker strength, 44
Silicon-iron, crack tip plastic zone, 276
thin film, 65

Silicon nitride, fracture toughness, 370
Silver, creep data, 146
deformation mechanism map, 157
elastic modulus, 8
rolling texture of, 124, 125
stacking fault energy of, 78
theoretical and experimental yield strength,
42
Sodium chloride, elastic compliances, 13
elastic constants, 13
slip system, 75
Spider silk, 221
Spinel (MgAl$_2$O$_4$), elastic anisotropy, 14
elastic compliances, 13
elastic constants, 13
slip system, 75
Stainless steel, creep data, 170
dislocation pile ups in, 61
fatigue crack propagation, 480, 493
fracture toughness, 285
Larson-Miller parameter constants, 167
neutron irradiation damage, 410, 411
rolling texture of, 124
stacking fault energy of, 78, 81
stacking faults in, 77
strain controlled-cyclic life response, 448
strain hardening coefficient, 18, 81
tensile and yield strengths of, 20
thin film of, 55, 77
Steels, alloying elements effect on, 350, 351
Charpy impact energy, 300, 331, 332
cleavage, 251
creep data, 170
environment assisted cracking, 382, 390, 392,
395-398
failure analyses, 559-562, 565-575
fatigue crack propagation, 469, 484, 492,
493, 504, 506-508, 511, 513, 515-517, 519
fracture mode transition, 282
fracture toughness, 285-287, 316-320, 339,
345, 353, 370
high temperature strength, 172, 173
Larson-Miller parameter constants, 167
MnS inclusions in, 120, 338-341
neutron irradiation embrittlement, 404-411
strain controlled-cyclic life response, 444,
446, 448, 454-459, 475-477
stress controlled-cyclic life response, 419,
428-430, 431-439
stress-rupture life of, 134
temper embrittlement, 401-404

tensile and yield strengths of, 20
300°C embrittlement, 400-401
transition temperature data, 301, 307-309, 314-320
whisker strength, 233
Sulfur, effect on fracture toughness in steel, 336-342

Tantalum, creep data, 147
  elastic modulus, 8
  rolling texture of, 124
Thorium oxide (ThO$_2$), slip system, 75
Tin, temper embrittlement effect on, 403
Titanium, Charpy impact energy, 300
  elastic compliances, 14
  elastic constants, 14
  elastic modulus, 8
  rolling texture of, 124
  slip plane, 73
  theoretical and experimental yield strength, 42
  twin elements, 105
Titanium alloys, environment assisted cracking, 381, 391, 398
  fatigue crack propagation, 469, 484, 492, 493, 505, 509, 515
  fracture toughness, 285, 328, 336, 355, 370
  Larson-Miller parameter constants, 167
  strain controlled-cyclic life response, 446, 456
  tensile and yield strength, 20
Titanium carbide, elastic anisotropy, 14
  elastic compliances, 13
  elastic constants, 13
Titanium oxide (TiO$_2$), activation energy for

creep, 140, 143
slip system, 75
steady state creep rate, 140, 143
Tungsten, creep data, 147
  elastic compliances, 13
  elastic constants, 13
  elastic isotropy, 14
  elastic modulus, 8
  rolling texture of, 124
  swaged wire microstructure of, 120
Tungsten carbide, elastic modulus, 8
  fracture toughness, 370

Uranium oxide (UO$_2$), slip system, 75

Vanadium, elastic modulus, 8
Vinyl polymers, 185

Whiskers, strength, 44, 233

Zinc, deformation markings in, 103
  elastic compliances, 14
  elastic constants, 14
  fatigue fracture, 416
  interplanar angles in, 110
  rolling texture of, 124
  slip plane, 73
  strain ellipsoid of, 111
  twin elements, 105
Zinc sulfide, elastic compliances, 13
  elastic constants, 13
Zirconium, rolling texture of, 124
  slip plane, 73
  twin elements, 105
  twinning modes, in, 112, 113